D0207951

# HYDROGEOLOGY

Also by R. J. M. DeWiest

GEOHYDROLOGY

# HYDROGEOLOGY

Stanley N. Davis
*Professor of Geology*
*Stanford University*

Roger J. M. DeWiest
*Professor of Hydraulics and Hydrology*
*Princeton University*

JOHN WILEY & SONS, Inc.
New York · London · Sydney

Library of Congress Catalog Card Number: 66-14133
Printed in the United States of America

SECOND PRINTING, MAY, 1967

*This Book Is Dedicated to*

Charles V. Theis

Hydrogeologist and Founder
of Modern Well Hydraulics

# PREFACE

This book is the outgrowth of our teaching experiences at Stanford University, Princeton University, the University of Chile, as well as at various summer institutes for college teachers. We have found that the contents can be covered in about thirty lectures. The text has been written so that, with the exception of the two introductory chapters, the main topics of water chemistry (Chapters 3 to 5), ground-water hydraulics (Chapters 6 and 7), exploration for ground water (Chapter 8), and ground water in various geologic environments (Chapters 9 to 12) can be read in any desired sequence. The repetition of subject matter necessary to achieve this flexibility is minimal. Although not included in our text, problem sets and laboratory exercises should be assigned in conjunction with the reading. Geologic mapping, aquifer tests, well inventories, and other field-orientated work are also needed to complete the student's background.

*Hydrogeology* is a companion volume to *Geohydrology* by Roger DeWiest. In *Hydrogeology* the emphasis has been placed on the geologic aspects of ground water, while in *Geohydrology*, as the name suggests, the emphasis is on the hydrologic or fluid-flow aspects of ground water. We have confined the inevitable overlap of the two books to the Introduction and to the chapters dealing with fluid flow. In *Hydrogeology* we have made extensive revisions so that many sections within the area of overlap are entirely different from each other.

In writing this textbook on hydrogeology we have faced the task of extracting only small fragments of knowledge from the rapidly accumulating and sometimes bewildering array of publications. The obvious solution to this problem is to present as many unifying principles as possible and to assume that further study as well as actual experience will round out the student's understanding of hydrogeology. To some extent, we have used this approach with regard to chemistry, classical mechanics, and geologic concepts. Unfortunately, the bare framework of basic principles is too abstract to inspire or even properly instruct students. We have, therefore, attempted to bridge the gap between theory and practice by using many of the utilitarian aspects of hydrogeology to illustrate the relevance of the basic principles. Attention is focused on the exploration for and the utilization of ground water. Although the scope is somewhat restricted for this reason, we hope that the inclusion of basic ideas will also make the book useful to those interested in such diverse problems as the compaction of engineered fill, water law, and mine drainage.

Our professional experience has given us a strong preference for the use of metric units. Nevertheless, all except the most provincial hydrogeologists in English-speaking countries must learn to work equally well with English and metric units, so we have deliberately intermixed them in this textbook. We hope that this mixture will not prove too distracting to those who have been trained in countries which long ago shed their medieval systems of measurement.

We gratefully acknowledge the formal as well as the informal assistance given to us in the writing of this book. Besides the important legacy from past workers and former teachers we have had the good fortune to receive a constant inflow of ideas, data, and encouragement from countless colleagues, correspondents, and students. Professor Konrad B. Krauskopf of Stanford University read Chapters 3 and 4; Professor William E. Bonini of Princeton University read Chapter 8; and Professor Francis R. Hall of the University of New Hampshire read the entire manuscript. To these individuals, specific thanks are due for numerous improvements in the final manuscript.

STANLEY N. DAVIS
ROGER DEWIEST

# CONTENTS

# *chapter 1*

# INTRODUCTION

## 1.1 *Scope of Topic*

Hydrogeology can be defined as the study of ground water with particular emphasis given to its chemistry, mode of migration, and relation to the geologic environment. This definition is in conformity with common European usage of the term [5,9,11,12,19,20,33]. Many hydrogeologists, nevertheless, concern themselves with a wide variety of activity that may range from quasilegal counseling on water rights to the specification of perforation geometry for water-well casing. Some confusion has, consequently, arisen concerning the true scope of the science. This confusion is caused by the failure to distinguish properly between the science itself and the application of that science to problems of law, public health, engineering, agriculture, and drilling technology.

The term hydrogeology was first used in 1802 by the noted French naturalist Lamarck [27]. He attached a meaning to the term that was almost identical to Powell's [38] "hydric geology" which was defined as a study of the "phenomena of degradation (erosion) and deposition by aqueous agencies." Neither Lamarck's nor Powell's terms were employed by their contemporaries so the use of the terms should not be considered to have been preempted. Lucas in 1879 was probably first to use the term hydrogeology for the geologic study of underground water [28]. The widest currency was given to this meaning of the term by Mead [30] in his classic book on hydrology, first published in 1919. He defined "hydrogeology" as the study of the laws of the occurrence and movement of subterranean waters. Engineers and geologists alike agree with him that "this must presuppose or include a sufficient study of general geology to give a comprehensive knowledge of the geological limitations which must be expected in hydrographic conditions and of the modifications due to geological changes." Mead stressed the special character of the study of "ground water as a geological agent, the understanding of which would contribute to attain a comprehension of the birth and growth of rivers and drainage systems." Meinzer [31] later subdivided the science of hydrology, which according to his definition dealt specifically with water completing the hydrologic cycle from the time it is precipitated on the land until it is discharged into the sea or returned to the atmosphere, into surface hydrology and subterranean hydrology or "geohydrology."

Obviously, some controversy has arisen to the proper meaning of the words hydrogeology and geohydrology. Mead and Meinzer, authors of classical textbooks on hydrology, have used both terms to describe the same subdivision of hydrology. Moreover, many American geologists have followed the literal meaning of the word hydrogeology and have extended its usage to cover all studies, both of surface and subsurface water, which include a substantial amount of geologic orientation. The subject matter of this book, nevertheless, is confined to subsurface water and the title of the book has been chosen in conformity with Mead's definition. The knowledge of hydrogeology is a basic requirement in the training of ground-water geologists and ground-water hydraulicians.

The best scientific research and professional work is often accomplished through combined efforts by geologists, hydraulicians, agronomists, chemists, and physicists specialized in the earth sciences. Cooperation has become imperative with the increasing complexity of the ground-water problems that remain to be solved and has been fostered in recent times by a better dissemination of scientific literature [37], avoiding thereby unnecessary duplication of research. Thus there is a common interest in the fundamentals of ground-water flow held by the hydrogeologist concerned with the evaluation of the safe yield of a ground-water basin and by the engineer in charge of drainage and irrigation projects. Principles of dispersion and diffusion in porous media applied by petroleum engineers in the study of the migration of gas and oil are also used in the analysis of salt-water intrusion in coastal aquifers. The domain of fluid flow through porous media is not confined to rock and earth materials: mechanical engineers are interested in the heat-exchange process associated with the movement of a gas through porous media; chemical engineers study the mass transfer of a gas from a mixture to a liquid solvent flowing in opposite sense through packed towers. Consequently, a vast body of technical literature dealing with the physics of the flow through porous media has become available to the hydrogeologist [3,39].

## 1.2 *Hydrogeology and Human Affairs*

Many aspects of hydrogeology are as yet of special interest only to scientists; nevertheless it is a branch of the earth sciences which has in large measure been born out of practical considerations. Indeed, many of the most important advances in hydrogeology today have been stimulated by studies designed to solve problems of great economic importance. This trend will probably continue as the demand for water will undoubtedly increase with growing population and industrialization.

An individual in an industrialized urban area may use from one million to five million gallons of water during his lifetime. If his share of industrial, agricultural, and recreational usage is counted, the total amount may well exceed ten million gallons. In contrast, an individual living under primitive conditions might easily survive a lifetime with fifty thousand gallons and would be able to maintain personal cleanliness with an additional fifty thousand gallons. Why, then, has the modern individual become so wasteful of water, using as much as perhaps a thousand times the amount required by his ancestors? The answer is simply because water is convenient, commonplace, and, especially, because it is the cheapest of all commodities. At a cost of a few cents per ton, water can be used to transport most industrial and domestic wastes. Vast quantities of water are used for nothing more than heat-exchange purposes. Unlike most other commodities, water can be stored economically in amounts of billions of tons and retained in storage for many months or even years. Cities in the western United States sell processed water that is used to keep lawns green through hot, dry summers. In some communities more than half the water that is sold is used for lawns and flower gardens. It is not surprising, therefore, that the high water consumption in the United States tends to increase at an even faster rate than does the population (Figure 1.1). The exceptionally rapid increase in ground water used for irrigation is largely due to a post-World War II expansion of irrigated lands in California, Texas, Arizona, and Louisiana combined with the utilization of larger amounts of ground water for supplemental irrigation in the eastern part of the United States.

About four-fifths of water used for all purposes (exclusive of hydroelectric power and navigation) comes from streams and lakes. Even so, the economic importance of subsurface water can be hardly emphasized too strongly. It is more desirable than surface water for at least seven reasons: (1) it is commonly free of pathogenic organisms and needs no purification for domestic or industrial uses; (2) the temperature is nearly constant which is a great advantage if the water is used for heat exchange; (3) turbidity and color are generally absent; (4) chemical composition is commonly constant; (5) ground-water storage is generally greater than surface-water storage, so ground-water supplies are not seriously affected by short droughts; (6) radiochemical and biological contamination of

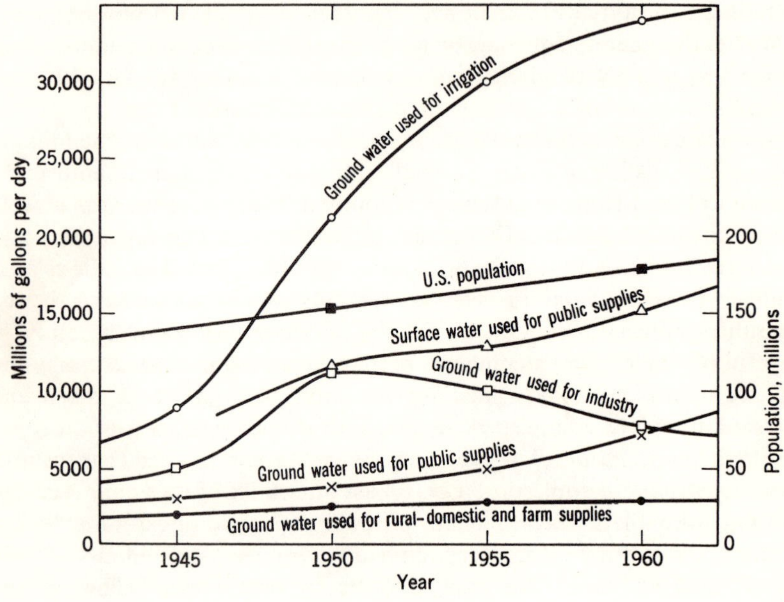


Figure 1.1

most ground water is difficult; (7) ground water, which has been stored by nature through many years of recharge, is available in many areas which do not have dependable surface-water supplies.

Three common disadvantages discourage ground-water development in some areas: (1) most important is the fact that many regions are underlain by rocks with insufficient porosity or permeability to yield much water to wells; (2) usually, but not always, ground water has a greater dissolved solids content than surface water in the same region; (3) the cost of developing wells is commonly greater than the cost of developing small streams. This is particularly true in regions of moderate to high precipitation.

## **1.3** *History of Hydrogeology* [25,32,43]

### GROUND-WATER UTILIZATION

In the dry regions of Asia, the universal scarcity of water, the locally dense population, and the dominance of agriculture resulted in an early development of the art of constructing wells and infiltration galleries [6]. Accounts of well water and well construction occur frequently in ancient literature and are especially well known from the Biblical record of Genesis.

Well construction in the Near East was accomplished by man and animal power and was aided by hoists and primitive hand tools. Despite great difficulties, a number of large-diameter wells, some large enough to accommodate paths for donkeys, attest both to the industry of the people and to the scarcity of water. These wells rarely exceeded a depth of 50 meters. There is little evidence of technological advances in well drilling during historic time in this region despite the fact that Egyptians had perfected core drilling in rock as early as 3000 B.C. [8]. This drilling was confined to stone quarry operations. Ancient Chinese, prolific in many inventions, were also responsible for the development of a churn drill for water wells which, in principle, is almost identical to modern machines [7]. The early machinery was, for the most part, made of wood and powered by human hands. Through a slow drilling rate sustained for years, and even decades, these ancient people were able to achieve wells of amazing depths. Bowman [6] reports a depth of 1200 meters and Tolman [43] a depth of 1500 meters. The deepest holes, however, were drilled to obtain brine and gas rather than potable water. The same methods, only slightly modified during the past 1500 years, are still used today in rural areas of Laos, Cambodia, Thailand, Burma, and China.

The greatest achievement in ground-water utilization by ancient peoples was in the construction of long infiltration galleries, or kanats, which collected water from alluvial fan deposits and soft sedimentary rock. These structures, commonly several kilometers long, collected water for both agricultural and municipal purposes. Kanats were probably used first more than 2500 years ago in Iran; however, the technique of construction spread rapidly eastward to Afghanistan and westward to Egypt. One extensive kanat system built about the year 500 B.C. in Egypt is said to have irrigated 4700 square kilometers of fertile land west of the Nile [43]. Many kanats are still in use today in Iran and Afghanistan, the best known of which are in Iran on the alluvial fans of the Elburz Mountains.

Owing to a lack of early cultural contact with China, modern percussion methods of well drilling were developed more or less independently in

Western Europe. The impetus for this development came largely from the discovery of flowing wells, first in Flanders about A.D. 1100, then a few decades later in eastern England and in northern Italy. One of the first wells was dug in A.D. 1126 by Carthusian monks from a convent near the village of Lillers. In Gonnehem, Flanders [4], near Bethune, four wells were drilled and were cased $11\frac{1}{3}$ feet above ground level so that they were able to deliver water at sufficient height to drive a water mill [21]. The wells were several hundred feet deep and tapped water under pressure from a formation consisting of fractured chalk that had its outcrop area in the higher plateaus of the Province of Artois. These and other similar wells in the region of Artois became so famous that flowing wells were eventually called artesian wells after the name of the region [7,35].

The widespread search for artesian water stimulated a rapid development of drilling techniques. Popular interest was so great in France that for a number of years the Royal and Central Society of Agriculture of France distributed annual medals and prizes to workers in the field, to authors, inventors, well drillers, and to those who introduced these wells in new areas [35]. Although drilling methods were more rapid and efficient in Europe than in China during the late eighteenth century, the depths of the wells rarely exceeded 300 meters. It was not until the end of the nineteenth century that the depths of water wells drilled by modern machinery exceeded the depths of the more primitive Chinese wells.

The methods of drilling for water have improved rapidly during the past 100 years, owing in part to knowledge borrowed from oil and gas drilling. The most significant single advance in drilling techniques has been the development of hydraulic rotary methods. Early rotary drilling was done with the aid of an outer casing; however, around 1890 thick mud was found to be sufficient for holding up the walls of the hole and the outer casing was no longer used. With this new efficiency and with the successful drilling by rotary methods of the Spindle Top oil field in Texas in 1901, rotary drilling has steadily continued to gain in popularity [6,8].

The perfection of the deep-well turbine pump in the years between 1910 and 1930 added a further stimulus to the well-drilling industry. Before this time, deep wells were fitted with low-capacity piston pumps of poor efficiency. The new turbine pumps made feasible irrigation by wells in many areas hitherto underdeveloped for agriculture. The large production of these wells has placed a greater demand on the well-drilling industry for bigger and more permanent wells.

Although technology is still being borrowed from the oil industry, many innovations such as reverse rotary drilling, gravel-envelope wells, and water-well cameras have come directly from the water-well industry itself.

THE ORIGIN OF SUBSURFACE WATER—EARLY THEORIES

When the importance of ground water in the Oriental civilizations is considered it is strange to note that there are so few records extant which concern theories of its origin. It remained for the inquisitive Greeks to speculate about the source of ground water. Even though considerable thought was given by the Greeks to the origin of ground water, their contributions were surprisingly sterile, particularly in light of the amazing progress they made in philosophy and mathematics. One probable reason for the lack of progress in ground-water theory was the insistence of Plato and others that philosophy and science be more or less separated from their important contact with experimentation, field observation, and practical applications. Thus, a wide gap was created between practice and theory. The significance of Greek thought is in the scientific dogma that it created during a period of almost 2000 years. The authority of Greek writings in the earth sciences reached a peak in the scholasticism of Albertus Magnus (A.D. 1206–1280) and Thomas Aquinas (A.D. 1225–1274) during the Middle Ages (roughly A.D. 1250–1450), but still persisted with considerable strength until a scant 200 years ago.

The Greeks were impressed with the large size of rivers in comparison with the observed runoff from heavy rains. They were also impressed by caves, sinks, and large springs characteristic of the limestone terrain which covers much of the Balkan Peninsula. The most common explanation for the origin of rivers was that they were fed from large springs, which in turn were fed by underground rivers or lakes nourished directly or indirectly by the ocean. Two problems confronted the early natural philosopher. (1) How did the ocean lose its salt? (2) How did ocean water rise from the level of the sea to springs high in the mountains?

Thales (640–546 B.C.), an Ionian philosopher of the School of Miletus, has been called the first true scientist. He taught that water was driven into the rocks by wind and that it was forced to the surface by rock pressure where it emerged as springs. Plato (427–347 B.C.), the great Athenian philosopher, conceived of one large underground cavern which is the source of all river water. Water was returned to the cavern from the ocean by various subsurface passages. The driving mechanism for this circulation was not fully explained. Krynine [26] has pointed out that the foregoing interpretation of Plato's ideas which influenced medieval science was probably not a correct interpretation of his more serious thoughts on the subject. Plato's "Critias" contains a description of the hydrologic cycle which is quite accurate. Although Aristotle (384–322 B.C.) was a student of Plato, he considerably modified Plato's concepts of the origin of ground water. Aristotle taught that ground water occurred in an intricate spongelike system of underground openings and that water was discharged from these openings into springs. Water vapor which emanated from the interior of the earth contributes the greatest part of the spring water. Aristotle did, however, recognize that some cavern water originated from rainwater which had infiltrated into the ground and had entered the cavern in liquid form rather than as vapor [1,2].

The Romans generally followed Greek teachings in the sciences. Marcus Vitruvius, however, who lived about 15 B.C., made several original contributions to engineering and science. He is, perhaps, best known for his contributions to architecture, particularly in the acoustics of buildings. He was also one of the first persons to have a correct grasp of the hydrologic cycle. He taught that water from melting snow seeped into the ground in mountainous areas and appeared again at lower elevations as springs. In contrast, the famous Stoic philosopher, Lucius Annaeus Seneca (4 B.C.–A.D. 65), held essentially the same theory as Aristotle but denied the reality of infiltrating rainwater. The conclusions of Seneca were taken for more than 1500 years as positive proof that rainfall was an insufficient source for spring water [1,2].

Bernard Palissy (1509–1589) [10], a French natural philosopher, was perhaps the first to have thoroughly modern views concerning the hydrologic cycle as reflected in the dialogue between "Theory" and "Practice" of his chapter "Des eaux et fontaines." Nevertheless, many of the ideas of the Greeks and Romans prevailed until the end of the seventeenth century. Two of the most influential scientists of their time, Johannes Kepler (1571–1630), German astronomer, and Athanasius Kircher (1602–1680), German mathematician, elaborated greatly on the earlier ideas of Seneca and Aristotle [1]. Kepler taught that the earth was similar to a large animal, that sea water was digested, and that fresh water from springs was the end product of the earth's metabolism. The ideas of Kircher were exposed in his *Mundus Subterraneus* which was first printed in 1664 and which soon became the standard reference work on geology for scholars of the seventeenth century. The work was ambitious in scope and unsurpassed in a display of spectacular imagination. Springs were thought to issue from large caverns in mountains. The springs were fed by subterranean channels connected to the sea. Whirlpools, particularly the somewhat mystical Maelström off the coast of Norway, were thought to mark the positions of openings to the caverns in the sea bottom.

Between the dawn of scientific thinking and the end of the Renaissance, about A.D. 1600, little advancement was made in hydrogeology. Five main facts were missed by all but a few early philosophers and scientists. (1) The earth does not contain a network of large interior caverns. (2) Although suction of the wind, capillary attraction, the forces of the waves, and other natural mechanisms exist to raise water against gravity, these mechanisms are insufficient to lift vast quantities of water in the earth's interior, (3) sea water does not loose all its salt by infiltrating through soil, (4) rainfall is sufficient to account for all water discharged by rivers and springs. (5) rainfall infiltrates into the ground in large quantities.

### THE FOUNDERS OF HYDROGEOLOGY

The true source of river water was proved by two French scientists, Pierre Perrault (1608–1680), and Edmé Mariotte (1620–1684) [1,2,32]. Perrault [25] measured rainfall in the Seine River basin for the years 1668, 1669, and 1670 and found the average to be 520 mm per year. He then estimated runoff from the basin and concluded that it was only one sixth of the total volume of rain, thus proving that rainfall was more than sufficient to account for all stream water. Studies of evaporation and capillary rise were also made by Perrault. He proved that capillary rise could never form a free body of water above the water table and that the height of capillary rise in sand was less than one meter. Mariotte measured the amount of infiltration of rainwater into a cellar at the Paris Observatory. He noted that this infiltration, as well as spring flow at other places, varied with the rainfall. He concluded, therefore, that springs were fed by rainwater which infiltrated into the ground. Mariotte's important contributions were published in Paris in 1690, after his death, and also as collected works in Leiden in 1717 [29]. The collected works contain Mariotte's essay "Du mouvement des eaux" (pp. 326–353), dealing with the properties of fluids, the origin of flowing wells, winds, storms and hurricanes, and other topics. Using the float method, Mariotte estimated the flow of the Seine River at the Pont Royal in Paris at 200,000 cubic feet per minute, or $1.05 \times 10$ cubic feet per year, less than one sixth of the total annual precipitation on the basin that provides the runoff to the Seine upstream to Paris. "It was therefore evident, if one third of the precipitation evaporated from the ground," as Mariotte assumed, "and if one third remained in the earth, that there would be enough water left to sustain the flow of wells and rivers in the basin." Thus Mariotte verified Perrault's conclusions concerning the source of water for runoff. Several years later Edmund Halley (1656–1742), the famous British astronomer, published studies of evaporation from the Mediterranean Sea and concluded that this evaporation was able to account for all the water flowing into this sea by rivers, thus adding important data in support of the two French scientists [32].

Artesian wells have excited speculation since the days of the early Greeks, but correct explanations were not widely published until the first part of the eighteenth century. The first explanation which was mechanically correct was made by the brilliant Iranian philosopher and scientist

Sheikh Abu Raihan al-Bīrūnī (973–1048) [13]. The best documented explanation came much later and was by Antonio Vallisnieri, president of the University of Padua, Italy, who published a paper in 1715 on the artesian water in northern Italy. He illustrated his paper with some of the earliest geologic cross sections which were drawn for him by Johann Scheuchzer [1].

Although anticipated somewhat by the work of Hagen and Poiseuille, Henri Darcy (1803–1858) [14] was the first person to state clearly the mathematical law which governs the flow of ground water. Darcy was a well-known French engineer whose main achievement was to develop a water supply for the city of Dijon, France. The development of his formula was the result of experimentation with filter sands and was presented in 1856 in an appendix of a report on the municipal water supply of Dijon. His report, however, resembled a scientific monograph on hydraulic engineering more than it did a present-day engineering report.

## MODERN HYDROGEOLOGY

Developments during the past century have been along three more or less separate lines: (1) elaboration of the relation between geology and ground-water occurrences, (2) development of mathematical equations to describe the movement of water through rocks and unconsolidated sediments, and (3) the study of the chemistry of ground water, or hydrogeochemistry.

The development of the relation between geology and ground-water occurrence is difficult to associate with individual names. In general, many geologists have contributed to specific problems. For example, the occurrence of ground water in areas of perennially frozen ground has been studied by a large number of Russian geologists. Many Dutch geologists have contributed to the understanding of ground water in coastal sand dunes. Japanese geologists and geophysicists have made numerous contributions to the understanding of hot springs. The English have made several contributions, one of which was an early application of geology by William Smith in 1827 to increase the water supply of Scarborough, England [40]. After a study of the local geology, he recommended that ground-water storage be increased by partially damming a spring. A. Daubrée [15] of France wrote one of the earliest general treatises on the geological aspects of ground water. For a specific example of modern contributions of an individual geologist we could cite the work of H. T. Stearns in the Hawaiian Islands. This work gives an excellent description of the relation between volcanic rocks and the occurrence of ground water. The work of W. M. Davis and J. H. Bretz on the formation of limestone caverns is another good example. Still another example is the work of

DuToit on the consolidated rocks of the Union of South Africa. Despite the great number of geologists that could be cited, one man, O. Ĉ. Meinzer, stands out as most important. Although he contributed to methods of making ground-water inventories, theory of artesian flow, and stressed the importance of phreatophytes, his main contribution was that of organizing the science of hydrogeology. He analyzed, defined, and welded together the various facets of the new branch of earth science, accomplishing most of this work between 1920 and 1940 when he was a member of the United States Geological Survey.

Advances in ground-water hydraulics can be more easily identified with individual people because specific formulas are commonly published rather than the generalized concepts which are so important in classical geology. Jules Dupuit [17], of France, was the first scientist to develop a formula for the flow of water into a well. This work was published only seven years after Darcy's monograph, yet successfully utilized Darcy's law. In 1870 Adolph Thiem [42], of Germany, modified Dupuit's formula so that he could actually compute the hydraulic characteristics of an aquifer by pumping a well and observing the effects in other wells in the vicinity. For many years Thiem continued to perfect his method and to apply it to various field situations. Modern methods of higher mathematics were first applied extensively to ground-water flow by Philip Forchheimer [18], of Austria, in 1886. He introduced the concept of equipotential surfaces and their relation to streamlines. He was also the first to apply Laplace's equation and the method of images. In the United States, C. S. Slichter, [41], published similar work thirteen years later. Slichter developed his ideas independently of Forchheimer [22]. Ground-water discharge into streams was studied by Edmond Maillet of France who expanded some of J. Boussinesq's earlier work (1905) into a general theory for base flow.

A great advance was made in hydrogeology in 1935 when C. V. Theis introduced an equation for nonsteady state flow to a well. A formula had been developed seven years earlier in Germany by Herman Weber [22]; nevertheless, the formula of Theis has proved to be of much greater utility. Theis' equation was based on an analogy with heat flow, but a few years later C. E. Jacob derived the same expression through hydraulic considerations alone. In the past years Jacob and many other workers have improved the usefulness of Theis' basic equation by modifying it for a large number of boundary conditions.

One of the more important contributions of recent times to the mathematical description of ground-water movement has been made by Morris Muskat in a number of technical papers and in a comprehensive book [34] originally published in 1937. His book still remains a valuable reference work for teaching and research. Equally valuable was the work of

M. King Hubbert [24], who has made numerous contributions to the understanding of fluid flow and the application of hydrodynamics to the exploration for petroleum. Hubbert derived Darcy's law from the Navier-Stokes equations and introduced the concept of force potential in his derivation, a concept more general and useful than that of the so-called velocity potential, which in a strict sense applies only to an ideal liquid having constant physicochemical properties, a condition closely approximated by water but not by other fluids of interest such as natural gas.

The subject of ground water has attracted the attention of many Russian researchers [23,37]; among them the name and contribution of Zhukovsky, pioneer of the air-foil theory, and even more, the name of Pavlovsky [36], is outstanding. Unfortunately, except for Polubarinova's book [37], little of this work has become available to the Western scientist in the form of unabridged translation into English. In the field of hydrogeology, Selin-Bekchurin [37] has been very prominent.

Chemical analyses of water have been routine for more than a century; however, the successful correlation of water chemistry with the hydrologic and geologic environments, or hydrogeochemistry, is a more recent development. Early studies of water chemistry were made by B. M. Lersch of Germany in 1864 and T. S. Hunt of Canada in 1865. Hunt made some of the earliest attempts to make geochemical interpretations. Modern hydrogeochemical studies in North America started with the work of F. W. Clarke whose most important contributions were published between 1910 and 1925 and included a large number of chemical analyses of water with geochemical interpretations. Another early geochemist who made detailed studies of specific areas in the United States was Herman Stabler. His regional studies of water chemistry in the western United States have been excelled only during the past decade. Modern trends in hydrogeochemistry include exhaustive studies of chemical ratios, mostly by Russian and French workers; the use of trace elements (to prospect for mineral deposits) in many countries; and detailed studies of isotopes of the common elements by workers in Japan, the United States, Russia, and many other countries. Hydrogeochemistry today is a subject of research and teaching, especially in French and German Universities [16].

## REFERENCES


1. Adams, F. D., 1938, *The birth and development of the geological sciences:* New York, Dover, Chapter 12, pp. 426–460.
2. Baker, M. N., and R. E. Horton, 1936, Historical development of ideas regarding the origin of springs and ground water: *Am. Geophys. Union Trans.*, v. 17, pp. 395–400.

3. Bird, R. B., and others, 1960, *Transport phenomena:* New York, John Wiley and Sons, 780 pp.
4. Blanchard, R., 1906, *La Flandre, étude géographique de la plaine flamande en France, Belgique et Hollande:* Lille, L. Danel, p. 8.
5. Bogomolow, G. W., 1958, *Grundlagen der Hydrogeologie* (translated from Russian): Berlin, Veb d Deutscher Verlag der Wissenschaften, 178 pp.
6. Bowman, I., 1911, Well-drilling methods: *U.S. Geol. Survey Water-Supply Paper* 257, pp. 23–30.
7. Brantly, J. E., 1961, Percussion-drilling system, in *History of petroleum engineering*, D. V. Carter (editor): Am. Petroleum Institute, pp. 133–269.
8. —— 1961, Hydraulic rotary-drilling system, in *History of petroleum engineering*, D. V. Carter (editor): Am. Petroleum Institute, pp. 271–452.
9. Cădere, R., 1963, Problem apelor subterane in R.P.R.: Inst. Studii Cercetări Hidrotehnice (Bucharest), *Studii Hidrogeologie* v. 1, pp. 9–22.
10. Cap, P. A., 1961, *Les oeuvres complètes de Barnard Palissy-Des eaux et fontaines:* Paris, Albert Blanchard, pp. 436–483.
11. Castany, G., 1963, *Traité pratique des eaux souterraines:* Paris, Dunod, 657 pp.
12. Challinor, J., 1964, *A dictionary of geology:* Cardiff, University of Wales Press, 289 pp.
13. Dampier-Whetham, W. C. D., 1931, *A history of science and its relations with philosophy and religion:* New York, The Macmillan Company, 514 pp.
14. Darcy, H., 1856, *Les fontaines publiques de la ville de Dijon:* Paris, V. Dalmont, 674 pp.
15. Daubrée, A., 1887, *Les eaux souterraines, aux époques anciennes et à l'epoque actuelle:* Paris, Dunod, 3 vols.
16. De Wiest, R. J. M., 1964, Educational facilities in ground-water hydrology and geology: *Ground Water*, v. 2, pp. 18–24.
17. Dupuit, J., 1863, *Etudes théoriques et pratiques sur le mouvement des eaux dans les canaux découverts et à travers les terrains perméables*, 2nd ed.: Paris, Dunod, 304 pp.
18. Forchheimer, P., 1886, Uber die Ergebigkeit von Brunnen Anlagen und Sickerschlitzen: *Zeitschrift des Architekten- und Ingenieur Vereins zu Hannover*, v. 32, pp. 539–564.
19. Fourmarier, P., 1958, *Hydrogéologie* (2nd ed.): Paris, Masson, 294 pp.
20. Giessler, A., 1957, *Das unterirdische Wasser:* Berlin, Veb Deutscher Verlag der Wissenschaften, 187 pp.
21. Hagen, G., 1853, *Handbuch der Wasserbaukunst*, v. 1: Koenigsberg, Bornträger, 87 pp.
22. Hall, H. P., 1954, A historical review of investigations of seepage toward wells: *Jour. Boston Soc. Civil Engr.*, July 1954, pp. 251–311.
23. Harr, M. E., 1962, *Ground water and seepage:* New York, McGraw-Hill Book Co., 315 pp.
24. Hubbert, M. K., 1940, The theory of ground-water motion: *Jour. Geol.*, v. 48, pp. 785–944.
25. Jones, P. B., and others, 1963, The development of the science of hydrology: *Texas Water Commission Circ.* 63–03, 35 pp.
26. Krynine, P. D., 1960, On the antiquity of "sedimentation" and hydrology (with some moral conclusions): *Geol. Soc. Am. Bull.*, v. 71, pp. 1721–1726.
27. Lamarck, J. B., 1964, *Hydrogeology* (translated by A. V. Carozzi): Urbana, University of Illinois Press, 152 pp.

28. Lucas, J., 1880, The hydrogeology of the lower greensands of Surrey and Hampshire: *Inst. Civil Engineers Minutes of Proc.* (London), v. 61, pp. 200–227.
29. Mariotte, E., 1717, *Oeuvres de Mr. Mariotte*, 2 vols, edited by P. Van der Aa, Leiden, 701 pp.
30. Mead, D. W., 1919, *Hydrology:* New York, McGraw-Hill Book Co., 626 pp.
31. Meinzer, O. E., 1939, Discussion of Question No. 2 of the International Commission on Subterranean Water: Definitions of the different kinds of subterranean water: *Am. Geophys. Union Trans.*, v. 4, pp. 674–677.
32. Meinzer, O. E. (editor), 1942, *Hydrology:* New York, McGraw-Hill Book Co., 712 pp.
33. Murawski, H., 1957, *Geologisches Wörterbuch:* Stuttgart, Ferdinand Enke Verlag, 203 pp.
34. Muskat, M., 1937, *The flow of homogeneous fluids through porous media:* New York, McGraw-Hill Book Co., 763 pp.
35. Norton, W. H., 1897, Artesian wells of Iowa: *Iowa Geol. Survey*, v. 6, pp. 122–134.
36. Pavlovsky, N. N., 1956, *Collected works*, 2 vols: Leningrad, Akad. Nauk, U.S.S.R.
37. Polubarinova-Kochina, P. Ya., 1962, *The theory of ground-water movement* (English translation): Princeton, New Jersey, Princeton University Press, Translators' remarks, p. ix.
38. Powell, J. W., 1885, Report of the director: *U.S. Geol. Survey Fifth Annual Report*, pp. xvii–xxxvi.
39. Scheidegger, A. E., 1960, *The physics of flow through porous media:* New York, The Macmillan Co. 313 pp.
40. Sheppard, T., 1917, William Smith, his maps and memoirs: *Proc. Yorkshire Geol. Soc.*, v. 19, new series, pp. 75–253.
41. Slichter, C. S., 1899, Theoretical investigation of the motion of ground waters: *U.S. Geol. Survey Nineteenth Annual Report*, Part 2, pp. 295–384.
42. Thiem, A., 1906, *Hydrologische Methoden:* Leipzig, Gebhardt, 56 pp.
43. Tolman, C. F., 1937, *Ground water:* New York, McGraw-Hill Book Co., 593 pp.

# *chapter* 2

# THE HYDROLOGIC CYCLE

## **2.1** *Introduction*

Long ago Solomon observed that "All streams flow into the sea, yet the sea is not full, though the streams are still flowing." (Ecclesiastes 1:7) The explanation of this enigma is so well known today, even to the school child, that we often forget that the role played by evaporation and precipitation is far from obvious and that it was not fully understood until modern times. The origin and movement of ground water was even less obvious to the ancients. Today we can visualize the ever-changing migration of atmospheric, surface, and ground water as a complex interdependent system called collectively the hydrologic cycle (Figure 2.1). Although the hydrogeologist is concerned chiefly with ground water, all aspects of the hydrologic cycle must be understood, at least in a general way, before an accurate picture of the subsurface portion of the cycle can be achieved.

The oceans are the immense reservoirs from which all water originates and to which all water returns. This statement is somewhat simplistic because not all water particles are in the process of completing the entire hydrologic cycle at all times. There are built in loops, for example, when water evaporates from land and returns to land as precipitation only to evaporate again, and so on. But in its most elaborate cycle, water evaporates from the ocean, forms clouds which move inland, and condense to fall to the earth as precipitation. From the earth, through rivers and underground, water runs off to the ocean. There is a slow addition of

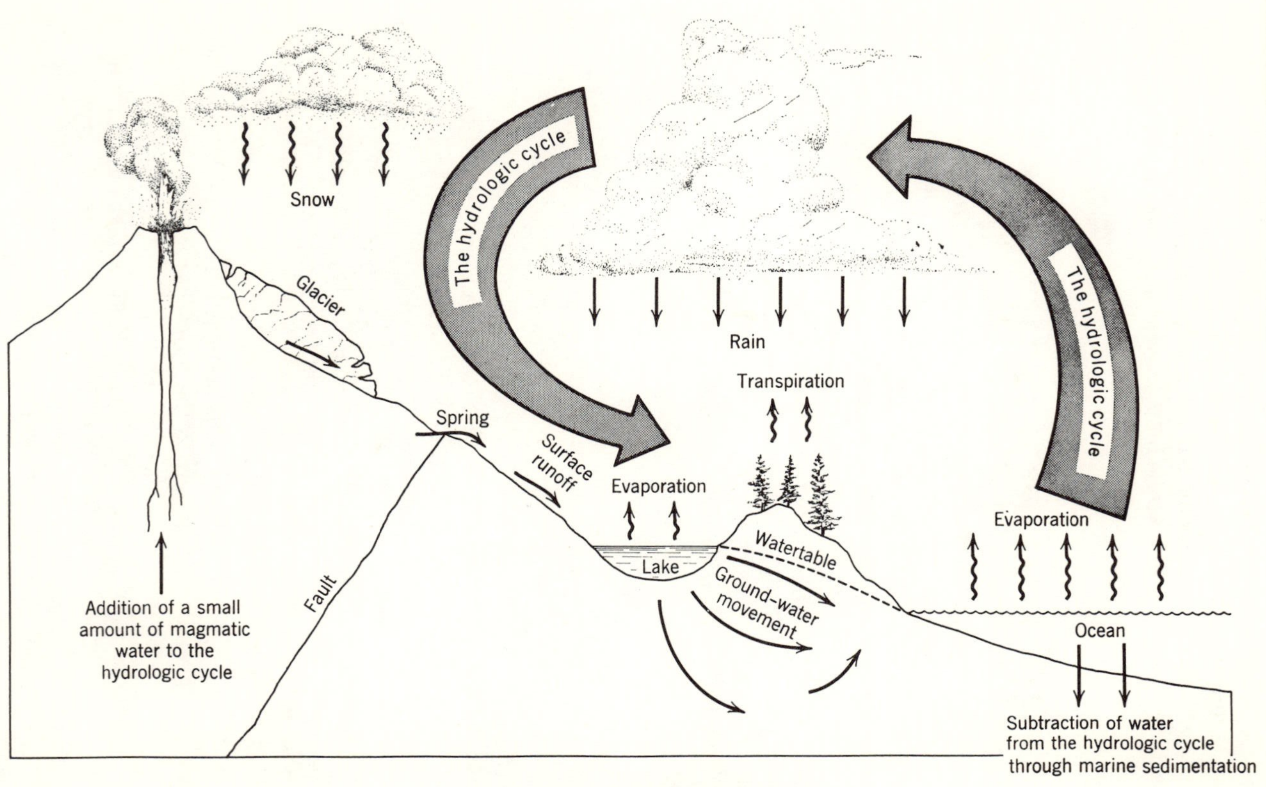


Figure 2.1 The hydrologic cycle.

water from magmatic and metamorphic sources, but there is also a constant subtraction of water that is incorporated in the structure of minerals within sedimentary deposits. Geologic evidence strongly suggests that the volume of water in the oceans has remained reasonably constant during the past 500,000,000 years, so the total amount within the hydrologic cycle must have also remained nearly constant.

## 2.2 *Precipitation*

Changes in air pressure and temperature associated with moving masses of air cause the air to be saturated with water vapor. The water vapor will then condense around minute nuclei of solid material which are in suspension in the air. Small particles of organic material such as spores and pollen, volcanic dust, salt from dried sea spray, and fine-grained minerals particularly of the clay-mineral group compose most of the nuclei. The presence of sea-born salt probably accounts for most of the sodium and chloride found in precipitation within 50 miles of the coast. The dust and smoke particles produced by factories, traffic, and household activity constitute an important source of nuclei in urban areas. These solids along with various urban gases cause a large increase in total dissolved solids of rainwater in the vicinity of cities.

When moisture is first condensed it is in very small droplets which are less than 0.04 millimeters in diameter. These droplets are kept in the air by virtue of the fact that their rate of fall is almost negligible. Clouds can be considered as colloidal suspensions of these small particles of moisture. The stability of the suspension is a function of several factors, most important of which are the size of the particles, temperature of the air, relative motion of various parts of the clouds, and the type of condensation nuclei available. One of the reasons for heavy precipitation in temperate regions appears to be the coexistence of ice and water particles in the same cloud. Vapor pressure next to the ice particles is much less than next to the water particles, hence there is a rapid transfer of vapor from the water particles to the ice particles and precipitation results.

The total amount of water entering the atmosphere by evaporation is very large. The depth of water lost annually from the ocean surface varies from about 0.5 meter in polar regions to 1.5 meters in tropical regions. The average evaporation rate for all the oceans is roughly 1.0 meter. Despite the large volume of water constantly entering the atmosphere, only a small amount of it stays in the atmosphere for more than a few weeks. Most of it is precipitated again on the ocean surfaces.

The oceanic moisture together with some continentally derived moisture will be precipitated in response to three principal meteorological controls.

The greatest amount of water is precipitated along the margins of large masses of warm, moist air which move into regions of cold air. This is usually associated with large cyclonic storms moving over long distances. A second type of control is common during warm weather. Air which is warmed at the surface of the earth tends to rise in large masses into the overlying cold air. This upward convection may be induced purely by the instability of a warm, light mass under a cool., dense mass of air, or it may be induced by rising currents of air over a mountain barrier. A third type of control is an orographic control. Precipitation is caused by adiabatic cooling of air masses as they are moved across high mountain barriers.

## 2.3 *Evapotranspiration*

Of the water that is precipitated on the earth, a large amount is returned to the atmosphere as vapor, through the combined action of evaporation, transpiration, and sublimation. These are in essence three variations of a single process due to the energy of the solar engine that keeps the hydrologic cycle running. Evaporation or vaporization is the process by which molecules of water at the surface of water or moist soil acquire enough energy through sun radiation to escape the liquid and to pass into the gaseous state. Sublimination differs from this phenomenon only in that water molecules are converted from the solid state (snow or ice) directly to vapor, without passing through the liquid state. Transpiration is the process by which plants lose water to the atmosphere. In many regions the overall amount of evaporation cannot be measured separately from transpiration, hence the two effects are considered together as "evapotranspiration." Some water vapor is also breathed into the air from the lungs of animals, but this is a minor quantity in comparison to the total effects of evapotranspiration.

The amount of solar energy received at the earth's surface may average more than 700 calories a square centimeter a day in hot, dry desert areas to less than 100 calories a square centimeter a day in cloudy polar regions. The percentage of this energy which is available for evaporation and transpiration varies with the type of natural surface. Fresh snow will reflect about 90 per cent of the energy, old snow about 50 per cent, and a free-water surface only 5 to 15 per cent. Light-colored rocky surfaces may reflect 40 to 50 per cent of the energy, but dense dark-colored vegetation will only reflect 10 to 25 per cent of the energy.

The amount of energy required to evaporate one cubic centimeter of water is 597 calories. Therefore, if all the available energy were used directly in evaporating the water, maximum annual evaporation rates of almost 400 centimeters (157 inches) should be expected for water surfaces, and maximum transpiration rates of 350 to 400 centimeters (138 to 157 inches) should be expected for dense vegetation. Actual rates are considerably lower due to energy lost by long-wave radiation and direct-energy transfer to the environment without involving evaporation. Only a few measurements of evaporation have averaged more than 320 centimeters (126 inches) per year, and only a few measurements of transpiration have averaged more than 220 centimeters (86 inches) per year. Although there is a rough correlation between solar energy received in a local area and the potential evapotranspiration [24], other factors, however, such as local turbulence and water-vapor content of the air make exact correlations difficult.

Unless the water table is within a few feet of the surface, ground water is not discharged by direct evaporation. If the soil is thoroughly saturated, however, the rate of evaporation may approach that of a free-water body.

The use of water by plants is generally much more important as a means of ground-water discharge than is direct-soil evaporation. Although direct evaporation will abstract water with about 60 times the tensional force that plants can exert, evaporation, aided by soil cracks and capillary transfer, is only effective in the upper 3 feet of sandy soils and the upper 10 feet of clayey soil. In contrast, plant roots commonly extend to depths of more than 30 feet [11], and roots at an extreme depth of 90 feet were observed in excavations for the Suez Canal [20].

### PHREATOPHYTES

Most desert plants are adjusted to an extreme economy of water. These plants, which are known as xerophytes, have a shallow but widespread root system. Another type of desert plant is localized along streams and areas with relatively shallow watertables. These plants, which are known as phreatophytes, have deeply penetrating roots that habitually reach the watertable. Phreatophytes are also found in moist environments, but the ecological classification is not distinct.

Certain phreatophytes have a low tolerance for salt and thus are valuable guides to potable water in arid and semiarid regions [13]. The ash, alder, willow, cottonwood, and aspen have been useful in this regard. These trees generally grow where the watertable is less than 30 feet deep.

Phreatophytes have assumed great importance in arid regions because of the large quantity of water which they use. Most phreatophytes have little or no economic value, so that the water is wasted. An important exception is alfalfa, *Medicago sativa*, which is a widespread phreatophyte of great economic value. The total water wasted per year in western United States alone amounts to about $25 \times 10^6$ acre feet, or about $30 \times 10^9$ cubic meters [20]. The water wasted in the arid and semiarid regions of all the world is many times this quantity.

The amount of water used by phreatophytes varies with plant species, density of growth, climatic factors, and hydrologic factors [20]. Of the climatic factors, sunlight, temperature, and humidity are most important. Transpiration will almost stop in some plants if there is no sunlight or if the temperature drops below about 40°F. Other plants will transpire slightly during the night but stop transpiring when the mean weekly temperature drops below 65°F. Transpiration will tend to be less if the relative humidity is high.

Of the hydrologic factors, depth to water and water quality are the most important. For many plants, water consumption is inversely related to the depth to the watertable. Thus the water use of some plants might be doubled if the watertable rose from a depth of 140 centimeters to a depth of 70 centimeters. The effect of water quality varies greatly from plant to plant. Most plants will use less water as the dissolved-solids concentration rises. A few plants such as salt grass, however, have an optimum growth with water having dissolved solids concentration of several thousand parts per million.

Transpiration can be most accurately determined by growing plants in large tanks in which the water intake is measured. This method has a disadvantage because large trees are virtually impossible to grow in tanks of reasonable size. If the saturated material near the watertable is of a uniform composition, a study of water-level fluctuations in shallow wells can be useful. The nightly rises in water levels shown in Figure 2.2 represent recovery of water during periods of no transpiration. Inasmuch as the water fluctuations are small in comparison with the total energy moving the water and the thickness of saturated material, the maximum velocity of water rise in the water-level record, or hydrograph, is a close approximation to the velocity of water moving continuously into the area of water withdrawal. Thus, the water use is roughly equal to this maximum velocity multiplied by the total area of open space within the saturated material.

The record of water levels in Figure 2.2 indicates a velocity of water-level recovery, $\frac{\Delta s_1}{\Delta t_1}$, of 1.5 mm/hr. The effect of water moving upward to replace transpired water, $V_{tp}$, can be calculated by correcting $\frac{\Delta s_1}{\Delta t_1}$ for the water-level change, $\frac{\Delta s_2}{\Delta t_2}$, which is caused by long-term changes in ground-water storage. This is 0.2 mm/hr in Figure 2.2. Thus,

$$V_{tp} = \frac{\Delta s_1}{\Delta t_1} \pm \frac{\Delta s_2}{\Delta t_2} \tag{2.1}$$

$$V_{tp} = (1.5 + 0.2) = 1.7 \text{ mm/hr}$$

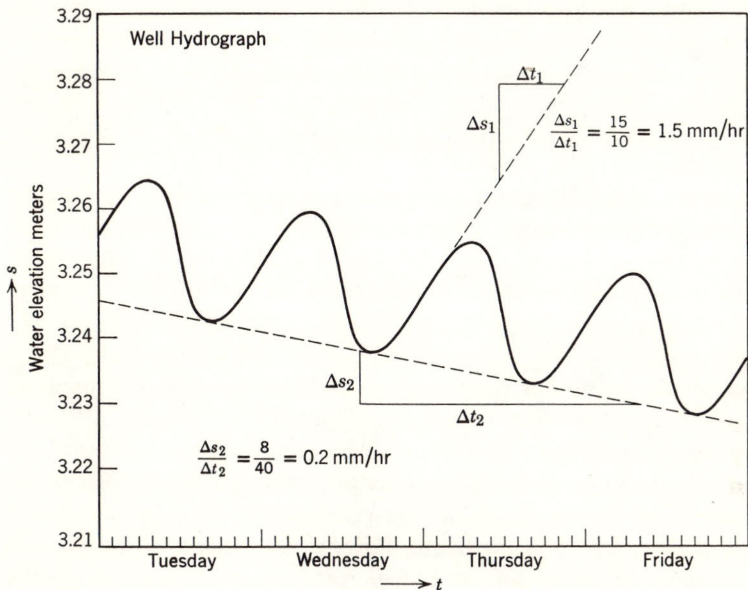


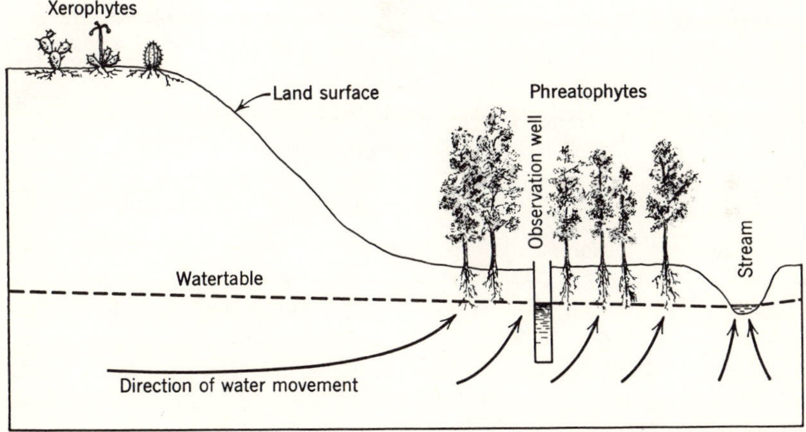


Figure 2.2 Daily water-level fluctuations in an observation well caused by phreatophytes.

The quantity of water moving per unit area, $q$, is obtained by multiplying $V_{tp}$ by 0.15, which is the assumed pore area of a unit cross section, $N_e$, in the zone near the watertable in Figure 2.2. Therefore,

$$q = N_e\left(\frac{\Delta s_1}{\Delta t_1} \pm \frac{\Delta s_2}{\Delta t_2}\right)$$

$$q = (0.15)(1.7) = 0.255 \text{ mm/hr}$$

If the vegetation of Figure 2.2 is homogeneous over an area of $4 \times 10^4$ m$^2$, the water use would be,

$$Q = qA = (4 \times 10^4)(0.255 \times 10^{-3}) = 10.2 \text{ m}^3\text{/hr}$$

or, $$Q = 2690 \text{ gal/hr}$$

If the vegetation is not homogeneous, corrections must be made for variations of density of growth and plant species.

The water used by dense growths of plants ranges from less than 1 foot per year in subarctic environments to more than $7\frac{1}{2}$ feet per year in hot, dry regions. The highest quantities measured have been used by saltceder, *Tamarix gallica*, which is native to Europe and Asia. In an extensive test in Arizona this plant consumed 270 centimeters (110 inches) of water in one year from a watertable at a depth of 123 centimeters (48 inches). Representative transpiration values for different phreatophytes are given in Table 2.1

*Table 2.1 Water Use by Dense Growths of Phreatophytes (From Robinson* [20])

| Common Name | Scientific Name | Climate | Depth to Water | | Annual Water Consumption | |
|---|---|---|---|---|---|---|
| | | | cm | in. | cm | in. |
| Saltcedar | *Tamarix gallica* | Hot dry | 123 | 48 | 270 | 110 |
| | | Hot dry | 213 | 84 | 224 | 88 |
| Greasewood | *Sacrobatus vermiculatus* | Cool dry | 50 | 20 | 66 | 26 |
| Willow | *Salix* | Hot dry | 61 | 24 | 134 | 53 |
| Cottonwood | *Populus* | Hot dry | 220 | 86 | 238 | 93 |
| Alfalfa | *Medicago savita* | Cool dry | 91 | 36 | 80 | 31 |
| | | Hot dry | 138 | 54 | 113 | 45 |
| Alder | *Alnus* | Hot dry | ... | .. | 162 | 64 |

Various possibilities exist for the salvage of ground water which is currently being wasted by useless vegetation. For very shallow water it is usually possible to lower the watertable by drainage wells and kill most grasses and small shrubs that are phreatophytes. The drainage water can then be diverted for some practical use. Certain plants can also be removed by weed killers, burning, or mechanical clearing. If this is successful, the area can be planted to useful phreatophytes such as alfalfa. The growth rate of some of the noxious plants is so great, however, that removal is quite difficult. The saltcedar, for example, can produce more than 500,000 seeds per plant per year. Many of these seeds will germinate rapidly and produce a jungle-like growth in less than five years. Saltcedar spread at an average rate of 3 square kilometers per year along the Pecos River, between 1912 and 1952.

## 2.4 *Runoff*

The term runoff is usually considered synonymous with streamflow and is the sum of surface runoff and ground-water flow that reaches the streams. Surface runoff equals precipitation minus surface retention and infiltration, the passage or movement of water through the surface of the soil; infiltration is to be distinguished from ground-water flow. For the study of precipitation-runoff relationships which are of vital importance in many hydrologic projects, a clear picture of the runoff cycle is of great help.

Surface runoff is a function of precipitation intensity, permeability of the ground surface, duration of precipitation, type of vegetation, area of drainage basin, distribution of precipitation, stream-channel geometry, depth to watertable, and the slope of the land surface. Notwithstanding this complexity, hydrologists can make meaningful analyses of runoff records and in many areas make reasonably accurate predictions of surface runoff to be expected from any given storm. Surface runoff is commonly represented in the form of a hydrograph, which is a time record of stream-surface elevation or stream discharge at a given cross section of the stream. In the case of a stream with a discharge lasting a long period after precipitation, the discharge also includes ground-water runoff.

The runoff portion of the hydrologic cycle includes the distribution of water and the path followed by water after it precipitates on the land until it reaches stream channels or returns directly to the atmosphere through evapotranspiration. The relative magnitude of the various components into which the total amount of precipitation of a given storm may be broken down depends on the physical features and conditions, natural and man-made, of the land as well as on the characteristics of the storm.

At the beginning of a storm, a large amount of precipitation is caught by trees and vegetation as interception; water thus stored on vegetation is usually well exposed to wind and offers large areas of evaporation, so that storms of light intensity and short duration may entirely be depleted by interception, and by the small amount of water that would infiltrate through the soil surface and fill puddles and surface depressions. For water to infiltrate, the soil surface must be in the proper condition. When the available interception and depression storage is completely exhausted and when the storm is such that the rainfall intensity at the soil surface exceeds the infiltration capacity of the soil, overland flow begins (Figure 2.3). The soil surface is then covered with a thin sheet of water, called surface detention. Once, the overland flow reaches a stream channel, it is called surface runoff.

Part of the water that infiltrates into the soil will continue to flow laterally as interflow at shallow depths owing to the presence of relatively impervious horizons just below the soil surface and will reach the stream channel in this capacity. Another part will percolate to the ground-water table and eventually will reach the stream channel to provide the base flow of the stream, and still a third part will remain above the watertable in the zone of unsaturated flow.

The contribution to streamflow from a storm of moderate intensity and essentially constant in time may be visualized by means of Figure 2.4. A portion of the rain falls directly on the stream channel and is indicated as channel precipitation. Initially, almost all of the rain is collected on the earth as surface retention, the sum of interception, depression storage, and evaporation. As time goes on, the storage on the foilage and depressions becomes more and more saturated and more and more water infiltrates into the soil. Finally overland flow occurs and becomes surface runoff.

## HYDROGRAPH COMPOSITION

When the drainage basin of a perennial stream is hit by a storm in the dry season, and when the river discharge decreases in time as indicated in Figure 2.5, the hydrograph of the stream will be disturbed from its smoothly leveling off curve and may assume various shapes according to the relative magnitude of rainfall intensity, rate of infiltration, volume of infiltrated water, soil-moisture deficiency, rainfall duration, and other characteristics of the storm and the basin. The most relevant parameters and their influence on the four components of runoff (surface runoff, interflow, ground-water flow, and channel precipitation) are briefly compared here. Four different hydrograph pictures result from the following possibilities.

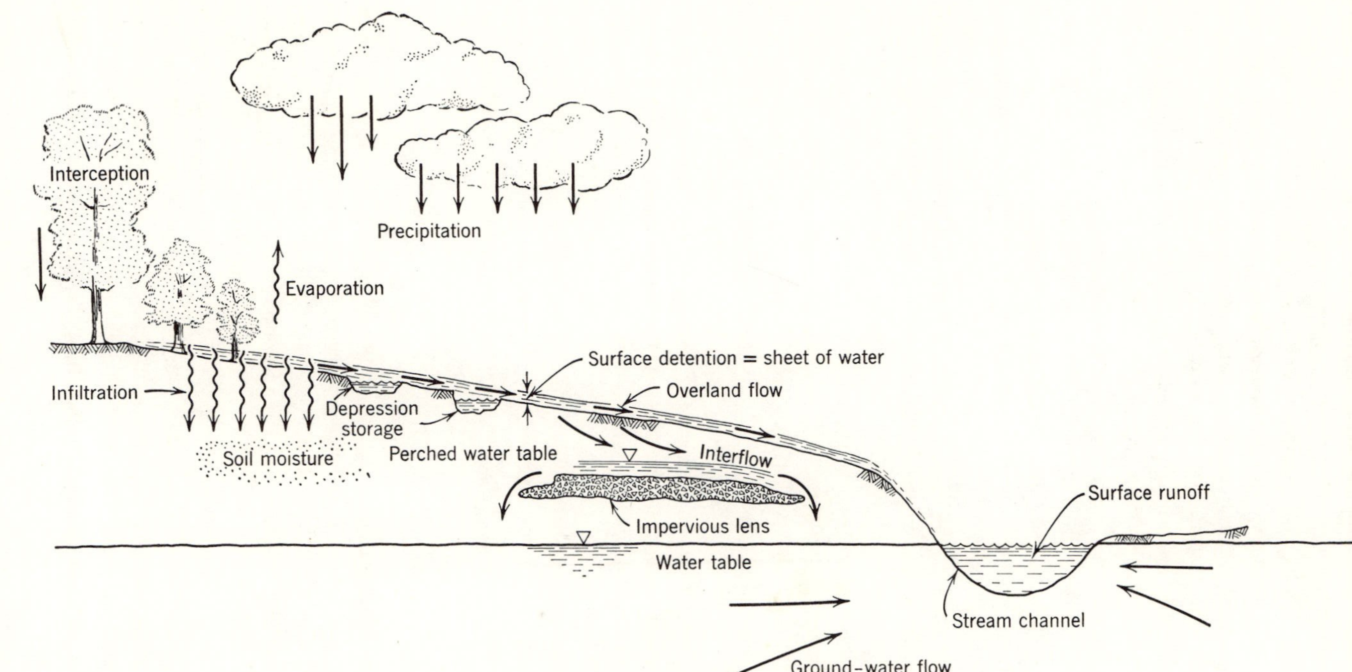


Figure 2.3 Simple picture of a runoff cycle.

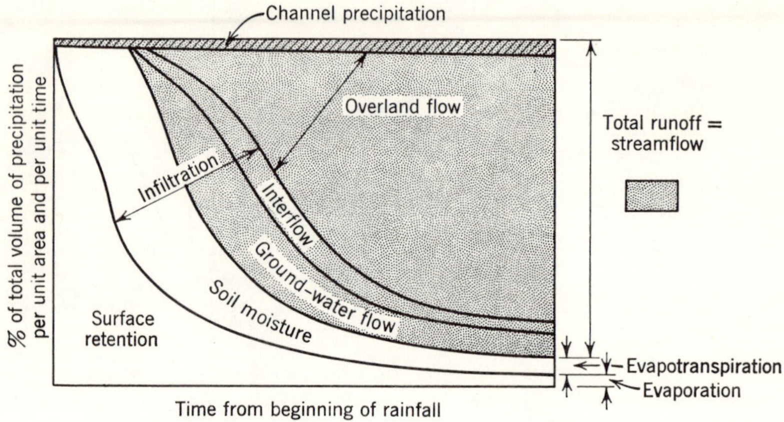


Figure 2.4 Decomposition of a storm of uniform intensity.

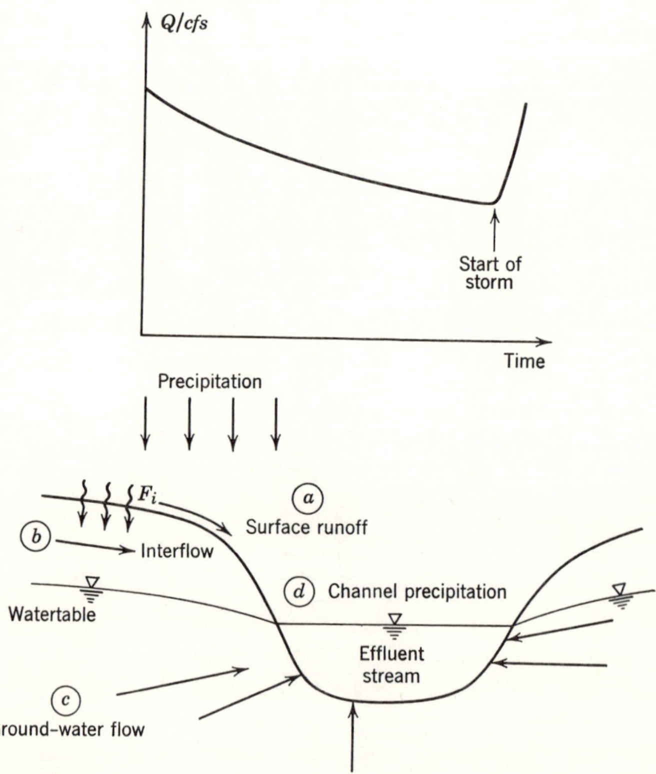


Figure 2.5 Hydrograph of effluent stream before storm. Runoff contribution starts after the storm.

NUMBER 1

Rainfall intensity $i <$ rate of infiltration $f_i$
Volume of infiltrated water $F_i <$ soil-moisture deficiency

In order to have contributions from interflow and ground-water flow due to the storm, $F_i$ must be larger than the soil-moisture deficiency. Soil-moisture deficiency is the volume of water required to bring the moisture content of the soil up to a point where any additional water will cause water to move downward through the soil.

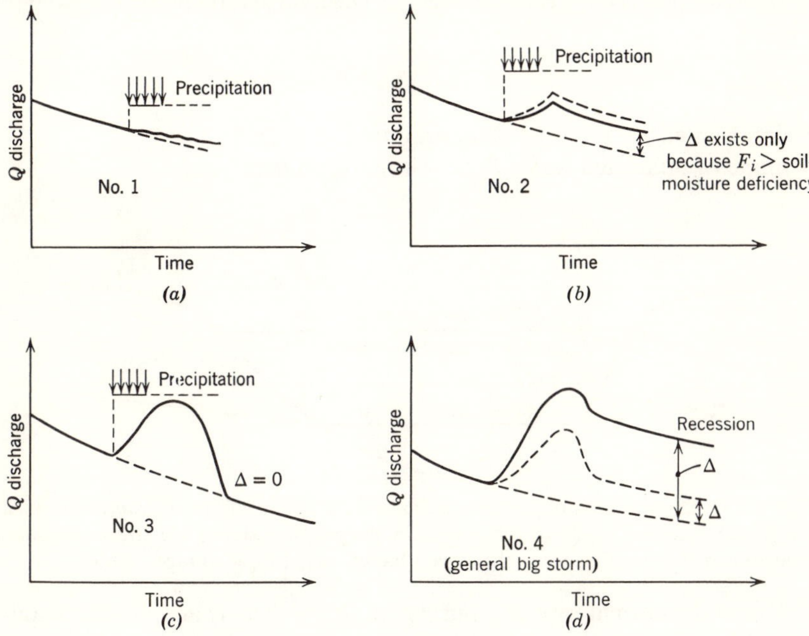


Figure 2.6 *a*, *b*, *c*, *d* Hydrograph composition.

There can only be surface runoff when $i > f_i$, and therefore the only addition to streamflow is due to channel precipitation, which leads to a slight increase of discharge $Q$ with time over the prolonged curve (Figure 2.6*a*).

NUMBER 2

Rainfall intensity $i <$ rate of infiltration $f_i$
Volume of infiltrated water $F_i >$ soil-moisture deficiency

After the moisture content of the soil reaches field capacity, interflow and ground-water flow accretion due to the storm occur. Added to channel precipitation, these flow components give the hydrograph picture of Figure 2.6*b*.

NUMBER 3

Rainfall intensity $i >$ rate of infiltration $f_i$
Volume of infiltrated water $F_i <$ soil-moisture deficiency

In this instance there are contributions from surface runoff and channel precipitation but no additional (i.e., owing to the storm) ground-water flow on top of the existing base flow sustained by the ground-water basin of the river (the river is called effluent). The hydrograph picture is given in Figure 2.6*c*.

NUMBER 4

Rainfall intensity $i >$ rate of infiltration $f_i$
Volume of infiltrated water $F_i >$ soil-moisture deficiency

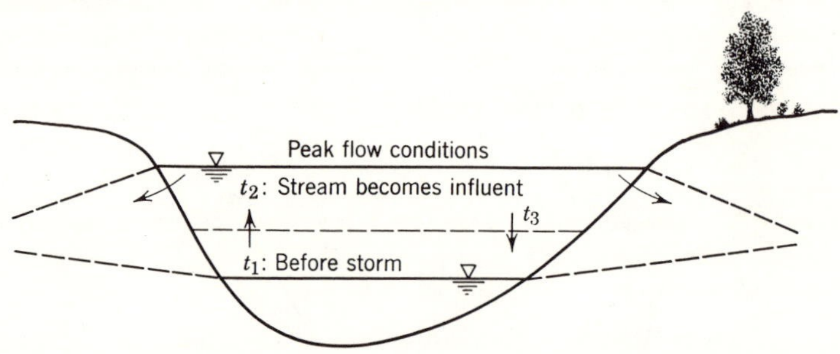


Figure 2.7 Cross section of a stream channel showing the effects of a storm in which $i > f_i$ and $F_i >$ soil moisture deficiency. In the example shown the stream becomes influent and storage of water in the stream banks, called bank storage, results.

This is the normal case of a big storm. Now there is additional streamflow due to channel precipitation, surface runoff, interflow, and ground-water flow, although the ground-water flow contribution may be negative, when the river becomes influent and the ground-water flow is reversed in sense (Figures 2.7 and 2.8). The hydrograph is given in Figure 2.6*d*.

SEPARATION OF HYDROGRAPH COMPONENTS

An enlarged picture of Figure 2.6*d* is given in Figure 2.9 and shows the contributions of the various components

*a* surface runoff
*b* interflow
*c* ground-water flow
*d* channel precipitation

Channel precipitation ends, of course, with the rainfall. The curve indicating the ground-water component may assume a variety of positions between that of Figure 2.8 and that of Figure 2.9. In these extreme cases

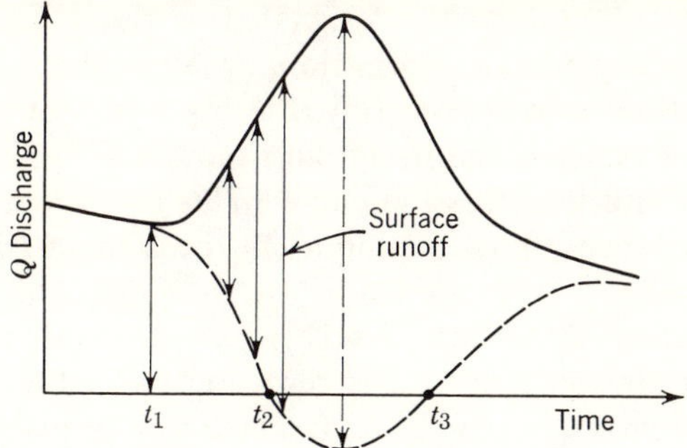


Figure 2.8 Hydrograph for influent stream conditions.

the surface runoff contribution differs significantly in magnitude for the same total hydrograph.

In practice, the problem is not to compose a hydrograph, because the hydrograph is given by measurements in a gaging station, but to separate its components. For simplicity, channel precipitation and interflow are included in surface runoff to form a single item, designated as direct or storm runoff. The problem is to separate this item from the ground-water flow component, also called base flow, given a hydrograph of which the

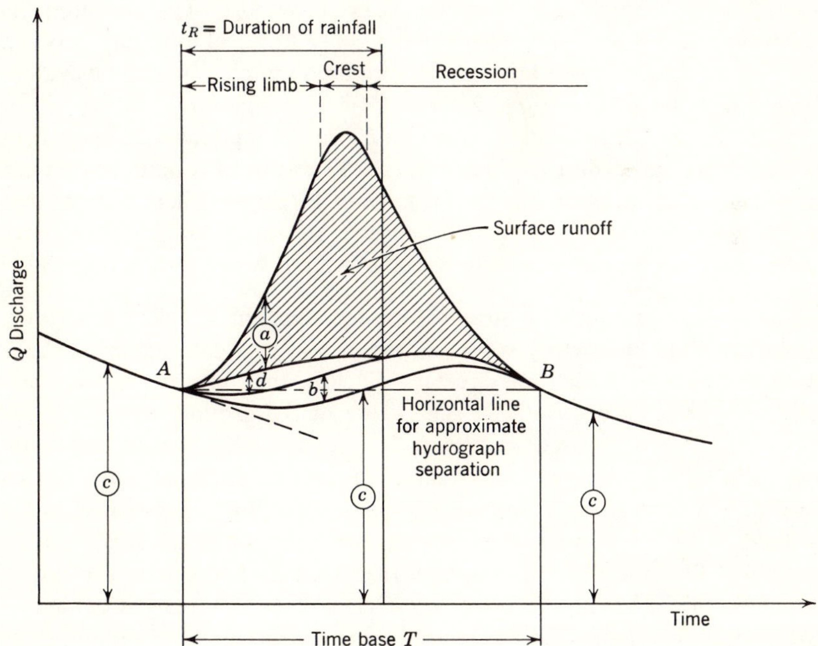


Figure 2.9 Hydrograph parts and flow contributions.

parts (Figure 2.9) are commonly designated as rising limb, or concentration curve, crest segment, and recession, or falling limb. The time base of the hydrograph may be obtained by drawing a horizontal line (Figure 2.9) through point $A$ where the rising limb starts and by finding its point of intersection $B$ with the recession curve. This horizontal line may also be considered, as a first approximation, to be the boundary between direct runoff and base flow.

A more sophisticated method of separation, however, is based on an analysis of the recession curve. The recession curve for a given basin, as suggested by Barnes, [1], may be represented by an equation which does not change in form for different storms, but only varies in the value of the recession constant $K_r$ of

$$q_1 = q_0 K_r \tag{2.3}$$

or

$$q_t = q_0 K_{rt} \tag{2.4}$$

in which $t$ is the time between the occurrence of discharge $q_0$ and $q_t$. Variations in $K_r$ are due to differences in areal-rainfall distribution and in relative magnitude of the components of flow (direct runoff versus ground-water flow), each component contributing to the final shape of the overall recession curve. Langbein [10] devised a method to filter out the ground-water component from the total recession. His technique is illustrated in Figure 2.10, where mean daily flows of one day, say $q_n$, are plotted against those of the following day, say $q_{n+1}$ ($n$ assumes values from 0 to $t - 1$ in Equations 2,3 and 2.4). For high flows, the data plot as a straight line, from which a constant $K_r$ for the total flow recession may be determined according to Equation 2.3. The value of $K_t$ may be checked by plotting log $q_t$ against $t$. Indeed, from Equation 2.4 it follows that

$$\log q_t = \log q_0 + t \log K_r \tag{2.5}$$

which is the equation of a straight line for constant $K_r$. For low flows, when the flood has clearly subsided, the $q_n$ and $q_{n+1}$ data again plot as a straight line from which the constant $K_r$, this time for the ground-water recession, may be determined. The value of the ground-water constant $K_r = 0.92$ has been used in Figure 2.11 to construct the ground-water recession curve, prepared from data at the end of the flood when the recession is known to represent only ground-water flow. Successive earlier ordinates of the ground-water recession curve are computed starting from the point $E$ of the flood hydrograph of Figure 2.11, taken as the end of direct runoff. The time of peak of the ground-water recession curve and the shape of the rising limb are selected in an arbitrary way, and although this method is classified as analytical, it is only semianalytical.

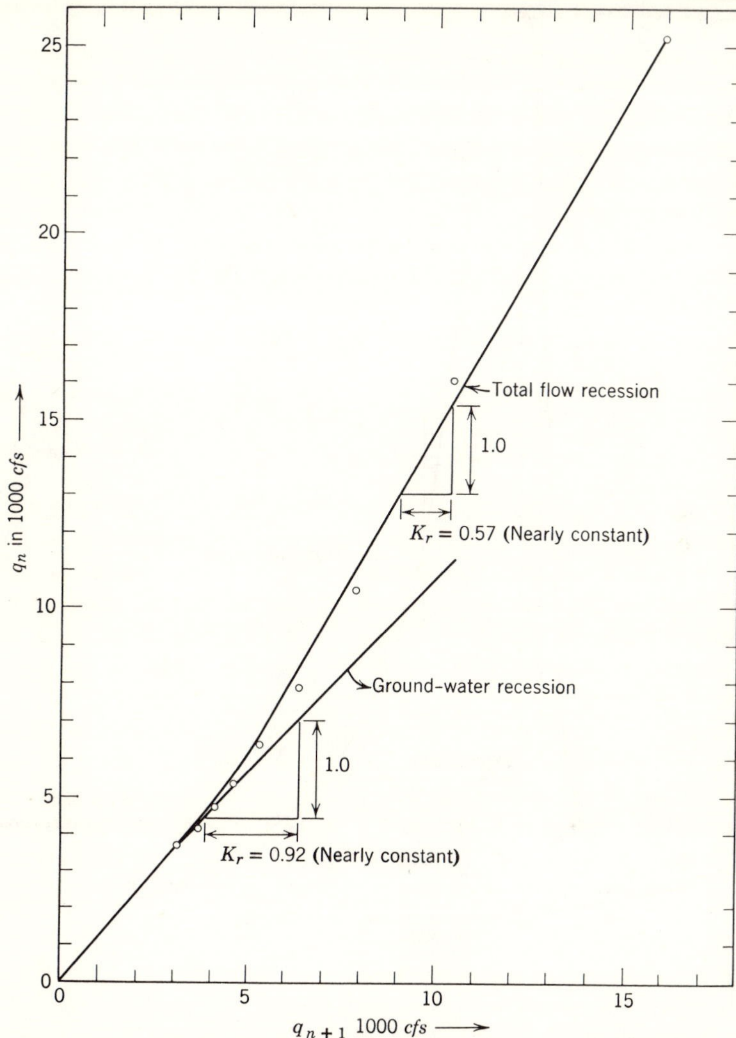


Figure 2.10 Ground-water and total recession curves, Potomac River at Paw Paw, West Virginia. Flood of April 17, 1943.

As a variance to the simple method of separating direct runoff and base flow by a horizontal line through the starting point of the rising limb, and which leads to a very long time base for the direct runoff hydrograph, the following method is often used. The dry recession curve (Figure 2.11) is extended from *A* to *B*, at the time of peak flow. From this time, a value of *N* (days) is laid off to give the point *C* on the hydrograph. *B* and *C* are

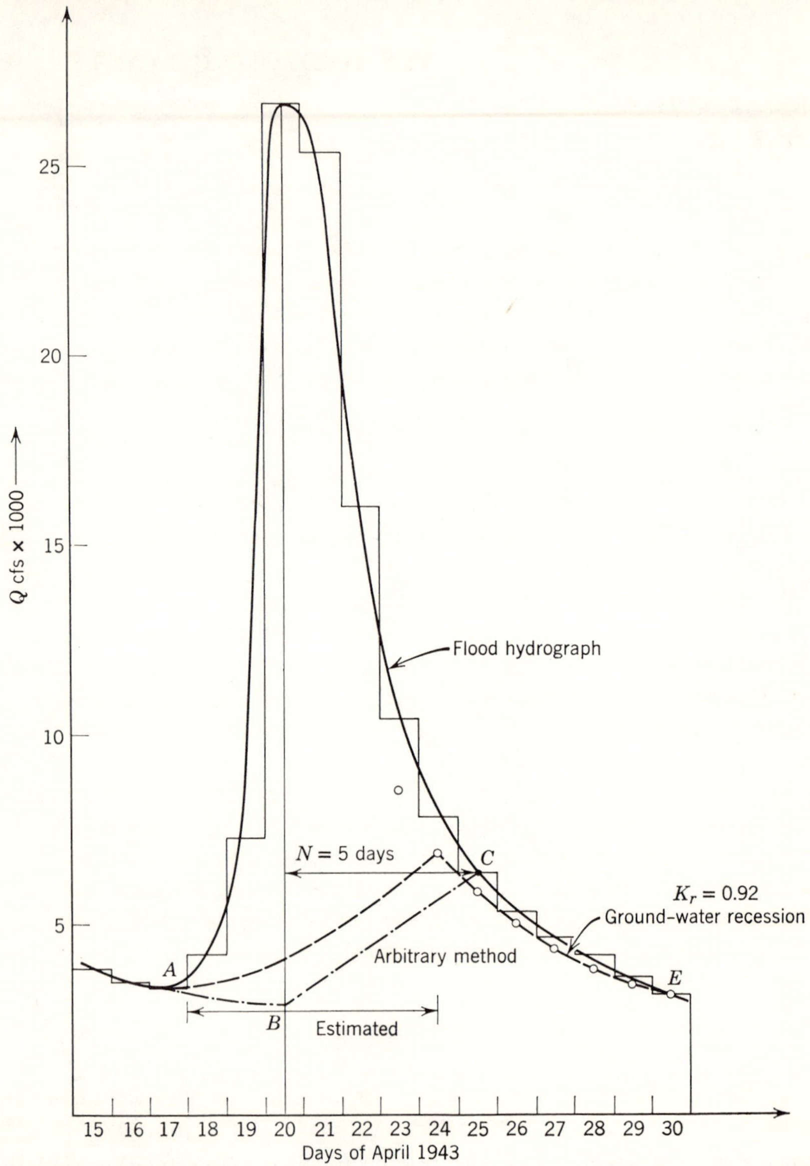


Figure 2.11 Hydrograph separation, Potomac River, West Virginia. Flood of April 17, 1943. Runoff from drainage area of 3,109 sq. mi.

1. Arbitrary method

*SFD*: $70.1 \times 10^2$

*AF*: $70.1 \times 1.98 \times 10^3 = 139 \times 10^3$

Runoff: $\dfrac{139 \text{ A ft} \times 12 \times 10^3}{3{,}109 \times 640 \text{ Acre}} = 0.84$ in.

2. Ground-water recession

*SFD*: $65.6 \times 10^3$

Runoff: 0.78 in.

then joined by a straight line. This method is completely arbitrary, and for $N$ a formula of the following form may be used

$$N = (A_d)^{0.2} \tag{2.6}$$

in which $N$ is the time in days and $A_d$ is the drainage area in square miles.

From the foregoing considerations, it is obvious that hydrograph separation is a rather arbitrary procedure, mainly because of the multitude of shapes which are acceptable for the ground-water hydrograph as a result of the possible combinations of effluent and influent conditions of the stream.

## 2.5 *Infiltration*

Horton [6] introduced the concept of infiltration in the hydrologic cycle and defined infiltration capacity $f_p$ as the maximum rate at which a given soil can absorb precipitation in a given condition. He suggested that infiltration capacity would decrease exponentially in time from a maximum initial value to a constant rate. The actual rate of infiltration $f_i$ is always smaller than $f_p$ except when the rainfall intensity $i$ equals or exceeds $f_p$; it also decreases exponentially with time, as the soil becomes saturated and as its clay particles swell.

The infiltration approach to runoff is based on the use of infiltration indices, such as the Φ-index which includes surface retention, although for storms of long duration surface retention tends to a minimum and Φ tends to the average infiltration rate $f_{i,\text{ave}}$. Therefore, the $W$-index has been defined as the average rate of infiltration during the time the rainfall intensity exceeds the infiltration capacity (Figure 2.12),

$$W = \frac{F_i}{T} = \frac{1}{T}(P - Q_s - S_e) \tag{2.7}$$

in which $F_i$ is the total amount of infiltration, $T$ the time during which rainfall intensity exceeds infiltration capacity, $P$ the precipitation, $Q_s$ the observed surface runoff from the storm, and $S_e$ the total surface retention.

Infiltration data may be used with some success to estimate maximum flood flows which consist almost entirely of surface runoff. Such flows generally occur when the initial moisture condition of the soil is quite uniform and therefore when the infiltration capacity curve is quite stable. For a simple storm distribution of continuous rainfall with intensity at all times exceeding the infiltration capacity as shown in Figure 2.13, the surface runoff may be determined from the rainfall and infiltration data as

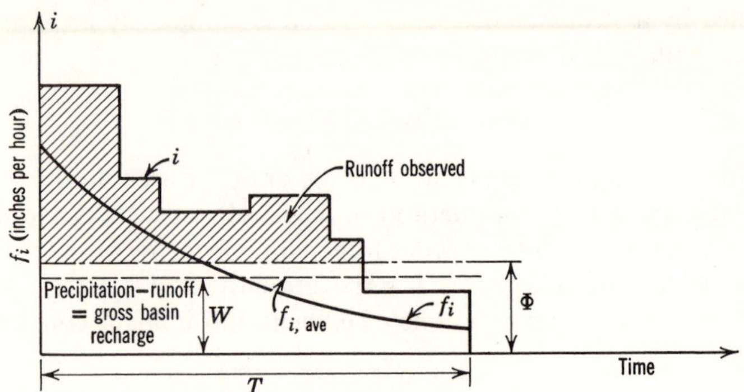


Figure 2.12 Infiltration indices Φ and *W*.

follows. The Φ-index is estimated and an infiltration curve is superimposed on the rainfall plot. The surface retention is estimated and added to the infiltration, so that the surface runoff is represented by the area between the precipitation curve and the infiltration plus surface-retention curve.

The ability to transmit water, or the hydraulic conductivity, of the soil is a highly variable quantity. If the soil is composed of well-sorted sand or gravel, the conductivity will be high and will vary only slightly with time. Most nonindurated materials at the earth's surface, however, develop a stratified structure which is called a soil profile (Figure 2.14). The soil profile is developed in response to chemical weathering, organic activity, and time. In a well-developed profile the A horizon is considerably more permeable than the B horizon. The B horizon is usually composed of masses of soil particles bound with clay and colloids. The B horizon is broken by soil fractures which form blocky or prismatic aggregates.

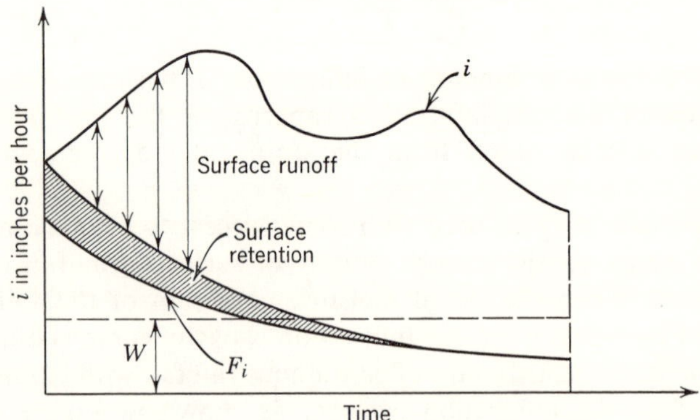


Figure 2.13 Estimation of surface runoff from precipitation and *W*-index.

When the soil is dry the fractures are more or less open, but as the soil becomes saturated by infiltrating water the clays and colloids swell so that the fractures close. Thus the soil structure has a high hydraulic conductivity at the start of infiltration and continually diminishing conductivity thereafter.

In the initial part of infiltration the attraction of the water by capillary forces of the soil is very important. The effect of capillary forces in medium- to coarse-grained soils is only minor after the infiltration front has penetrated more than two or three feet. The capillary forces are greatest with low values of initial moisture within fine-grained soils.

The effect of trapped air is opposite to that of soil structure. Initially, the advancing front of infiltrating water will be irregular and air will be expelled at various points. The energy required to force the air out of the soil will slow the rate of infiltration. As the saturated front advances, pockets of dry soil will be left and will form barriers to water movement. Continued movement of water, nevertheless, will dissolve some of the air. The effects of trapped air will, therefore, be opposite to that of soil structure. There will be an initial resistance to infiltration which will be reduced with continued passage of water. This effect can be observed during the first part of infiltration (Figure 2.15).

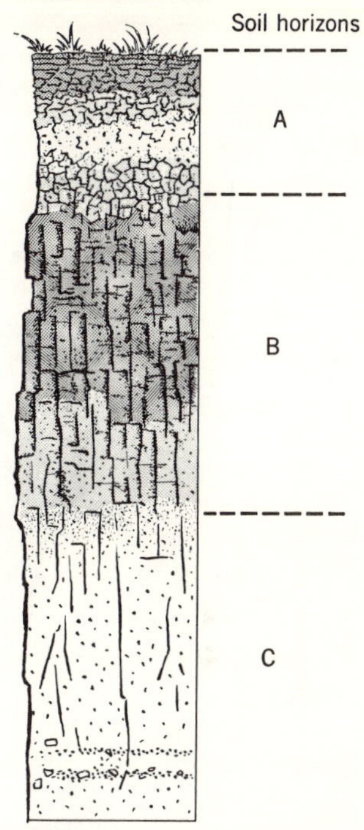


Figure 2.14 Soil profile developed on sandy alluvium. Vertical structure in the clay-rich B horizon is particularly important in controlling the infiltration of water. Initial water movement is moderately rapid until clays become wet and expand, thereby closing most of the open cracks in the soil.

The condition of the soil is also of great importance. A bare soil surface will be exposed to the direct impact of rain drops. The rain will compact the soil and also wash small particles into open cracks and holes. This has the effect of reducing infiltration as a rain continues. On the other hand, a dense cover of vegetation will protect the soil surface so compaction and rearrangement of soil particles by rain drops will be only slight. The roots of plants will also hold the soil open and increase the normal infiltration rates.

Water temperature will affect water viscosity which in turn affects infiltration rates. This effect can best be seen in artificially induced

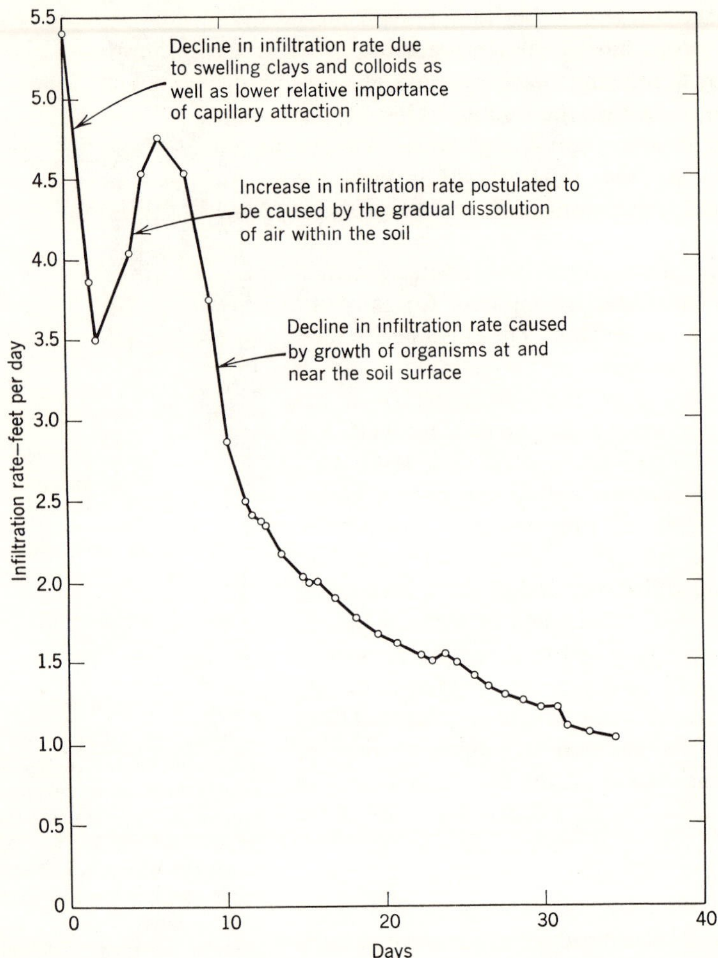


Figure 2.15 Long-term infiltration rates in a sample of soil that was flooded for more than a month. The secondary increase in infiltration rates has been observed in many, but not all, tests of various soils. The increase is tentatively explained as the result of the dissolution of air initially trapped in the soil. (Diagram redrawn from *University of California Sanitary Engineering Research Laboratory Technical Bulletin* 12 [30].)

infiltration of long duration. Other factors being constant, the infiltration will vary inversely as the viscosity or directly as the temperature. Infiltration from rain is controlled mostly by other factors that obscure the effects of temperature.

The amount of water that reaches the regional ground-water body is equal to the total infiltration minus the amount of water absorbed by the soil. Thus the moisture content of the soil before infiltration is an important factor affecting the ground-water recharge. Figure 2.16 shows the moisture conditions in a region having an annual precipitation of about 20 inches. Two curves are shown. One is the curve of monthly precipitation; the other is a curve of potential evaporation and transpiration or potential evapotranspiration. Surface runoff is assumed to be zero.

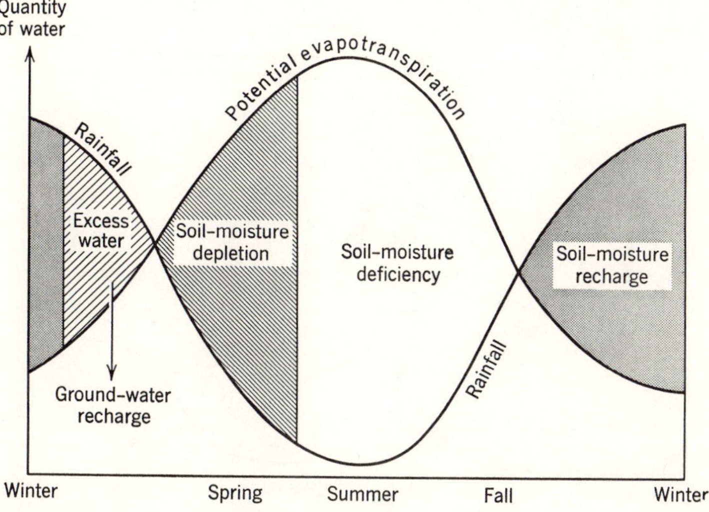


Figure 2.16 The effect of potential evapotranspiration on soil moisture in a region having little or no precipitation during the summer. (Diagram modified from Thornthwaite and Mather [28].)

During the spring the precipitation is balanced with evapotranspiration. As the precipitation drops and the temperature rises during the summer, evapotranspiration will remove moisture which is stored in the soil. During the middle part of summer the soil moisture is depleted, and water is no longer lost to the atmosphere. If water were available through irrigation, nevertheless, the amount shown in Figure 2.16 as a deficit would be used. During autumn the first rains are lost directly through evapotranspiration. As the rains increase and the temperature drops there will be a surplus of water which will be used for recharging the soil moisture. In late winter the soil moisture increases to a value beyond the field capacity and water will drain downward to the ground-water table. Figure 2.17 summarizes this sequence in a series of vertical soil-moisture profiles.

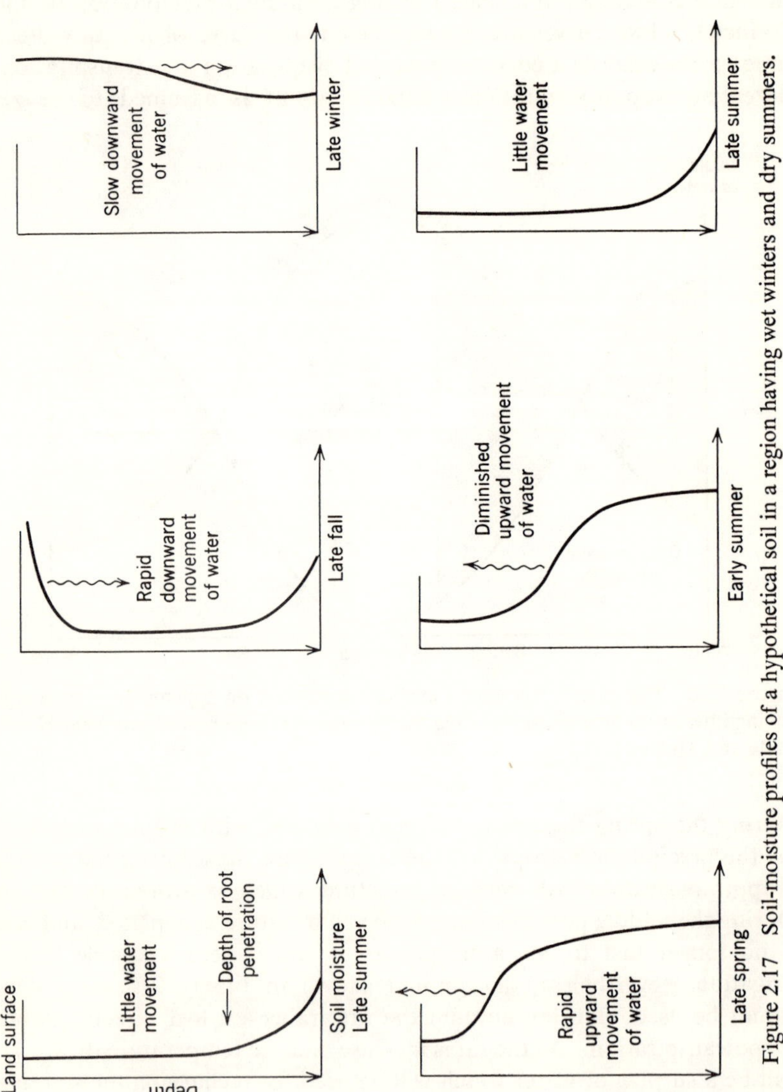


Figure 2.17 Soil-moisture profiles of a hypothetical soil in a region having wet winters and dry summers.

## 2.6 *Subsurface Movement of Water*

### ZONE OF SOIL MOISTURE

The subsurface part of the hydrologic cycle is the chief concern of the hydrogeologist. In the subsurface, all gradations exist between freely flowing water and water firmly fixed in the crystal structure of minerals.

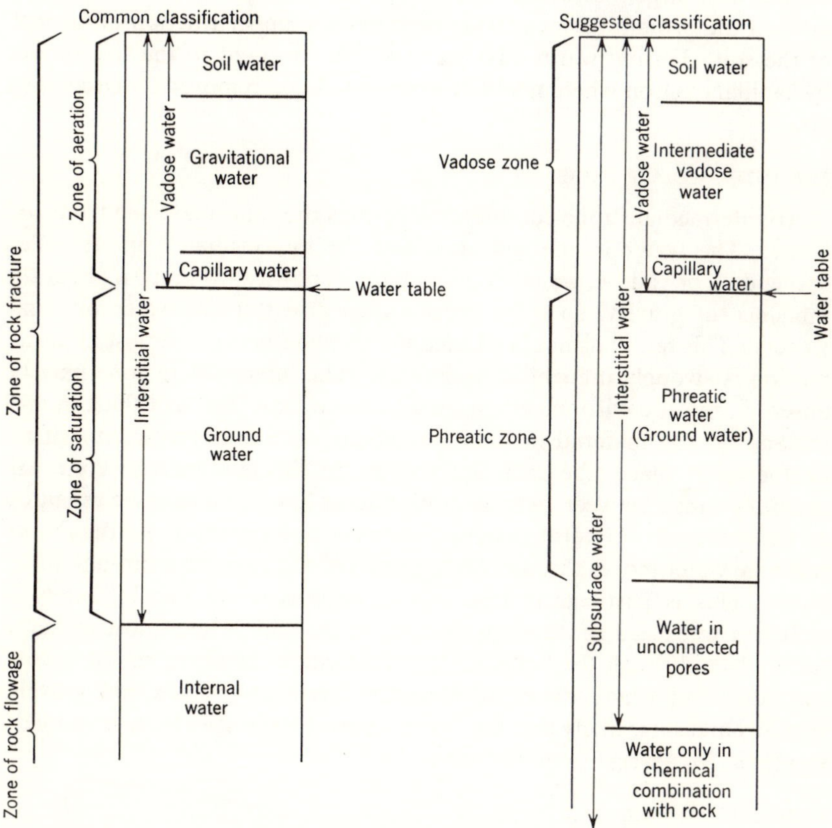


Figure 2.18 Classification of subsurface water.

Most classical discussions of subsurface water tend to emphasize several more or less distinct categories of water. Figure 2.18 shows both the classical subdivisions and suggested subdivisions that avoid some of the difficulties discussed in the following pages. There are no sharp boundaries, however, between the various types. For example, soil water is only distinguished from water in deeper unsaturated zones by the fact that it is subject to large fluctuations of quantity and quality in response to transpiration and evaporation. In a normal forest environment various trees may have some roots which extend to depths of more than 30 feet, so the moisture fluctuations induced by plants gradually diminish until a depth of more than 30 feet is reached. The moisture in the upper few inches of the soil zone also changes owing to large fluctuations of temperature and vapor pressure induced by variations of air and ground temperatures. Nights with large losses of heat through radiation cause condensation of dew on the ground surface. This may wet the upper fraction of an inch of the soil. During warm days water will be brought to the soil surface by capillary action where it will be evaporated and removed by circulating air.

### THE INTERMEDIATE ZONE

An intermediate zone commonly separates the saturated zone from soil water. The water in the soil zone and the intermediate zone is called suspended, or vadose, water. This water is in downward motion under the influence of gravity, so it has been called gravitational water by some writers. This term has not been used in this book because almost all water motion is strongly influenced by gravity. The intermediate zone may be absent in moist environments or may be more than 1000 feet thick in arid regions. If the material in the intermediate zone is isotropic, and, if recharge takes place, the moisture content in the intermediate zone will generally range between near saturation to as low as the specific retention of the material. Natural material is rarely homogeneous so that more recharge water moves through certain parts of the zone than through other parts. This is particularly true in arid regions where rainfall rarely is sufficient to exceed the storage capacity of the soil. Here, most recharge takes place through the bottoms of stream channels where runoff is concentrated and where there are abundant permeable sands and gravels. In many areas it is likely that the intermediate zone has not been thoroughly saturated for several thousand years.

### CAPILLARY FRINGE

The intermediate zone terminates below with the capillary fringe. The transition to the capillary fringe is rather abrupt in coarse-grained sediments, but is very gradual in silts and clays. If recharge is active in fine-grained soils, there may be very little difference in moisture content between the intermediate zone and the capillary fringe. The surface of the capillary fringe is irregular in detail, and its position varies constantly with changes in water level and amount of recharge. The upper part of the capillary fringe contains numerous pockets of air which slow the movement of water. In the lower part of the fringe, however, the material is just as fully saturated as it is below the watertable. Furthermore, the physical forces governing the flow of fluid in the lower part of the capillary fringe are identical to the forces below the watertable.

### WATERTABLE

The zone of ground water, or phreatic water, is divided from the capillary fringe by the watertable. The watertable is a theoretical surface which is approximated by the elevation of water surfaces in wells which penetrate only a short distance into the saturated zone. If ground-water flow is horizontal, then water levels in wells will correspond very closely to the watertable. The presence of the wells, nevertheless, will distort slightly the flow pattern and hence the water levels within the wells.

The most common definitions of the watertable state that it is the surface separating the capillary fringe from the "zone of saturation," or that it is the surface defined by the water levels in wells which tap an unconfined saturated material. A more exact definition states that the watertable is the surface in unconfined material along which the hydrostatic pressure is equal to the atmospheric pressure. The meaning of this statement is best seen by reference to Figure 2.19. Manometer A terminates in the air above ground so that the atmospheric pressure causes the water level to stand at equal elevations in both arms of the tube. Manometer B terminates in unsaturated material above the capillary fringe so that the full effect of capillary suction is registered. Manometer C terminates within the capillary fringe so that the hydraulic pressure is lower than the atmospheric pressure. Manometer D terminates at the watertable so that the water levels are equalized as in A. This indicates that the hydraulic pressure is equal to the atmospheric pressure. Manometer E terminates below the watertable. Here, the hydraulic pressure is greater than the atmospheric pressure.

### ZONE OF PHREATIC WATER

The water below the watertable is generally called ground water, and the zone below the watertable is called the zone of saturation. Both terms are confusing because in its nonscientific interpretation, ground water should refer to any water below the surface of the ground, and the zone of saturation should include all saturated material. For this reason, the term "subsurface water" must be used as a general term for all water below the land surface. Furthermore the "zone of saturation" is not an exact term because the lower part of the capillary fringe is also saturated. Inasmuch as the water in the lower part of the capillary fringe migrates with about the same velocity as the water just below the watertable, there

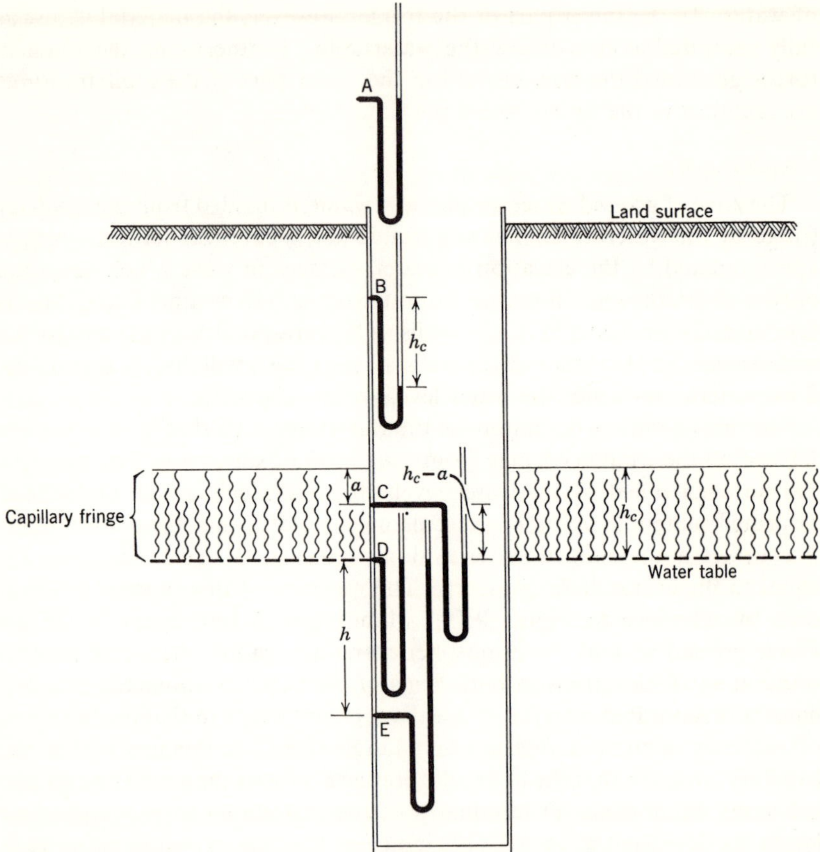


Figure 2.19 The watertable in a homogeneous medium illustrated through the use of hypothetical manometers that have an infinitesimal diameter, are filled with water, and are in contact with the porous medium outside the dry well. Manometer D registers atmospheric pressure and marks the position of the watertable.

is no justification for retaining this definition of the "zone of saturation." A more acceptable term for this zone would be the "zone of phreatic water," since phreatic water is defined as water that will enter freely into wells. Water in the capillary fringe will not drain into a well unless it is a discharging well which is terminated below the watertable.

The zone of phreatic water merges at depth into a zone of dense rock with some water in pores, although the pores are not interconnected so that water will not migrate. The depth to dense rock varies with the geologic environment. In areas of intrusive and metamorphic rocks this zone of dense rock may start at depths of less than 10,000 feet. In areas of deep sedimentary basins, the depth may be nearly 50,000 feet. At depths greater than 100,000 feet, the heat and pressure may become so great that open pores are not possible and the water may be found only in chemical combination with other material.

### AQUIFERS

Only a small fraction of most phreatic zones will yield significant amounts of water to wells. The water-bearing portions are called aquifers. A rock which neither transmits nor stores water is called an aquifuge. This is in contrast to aquifers which both transmit and store water in pore space within rocks and aquicludes which only store water but do not transmit significant amounts. The term aquitard is also used by some writers to describe natural material that stores water and also transmits enough water to be significant in the study of the regional migration of ground water but not enough water to supply individual wells.

The foregoing terms all lack precise definitions with respect to measurable physical properties. In fact, it is often pointed out that an aquifer that may yield a hundred gallons of water per day in a desert region would be classed as an aquitard or even an aquiclude if it were found in an alluvial valley filled with gravel that will yield several million gallons of water per day to large wells.

For this reason some hydrogeologists prefer defining aquifers as natural zones below the surface that yield water in sufficiently large amounts to be important economically.

Aquifers can be nonindurated sedimentary deposits, fractured zones in dense plutonic rocks, porous sandstone beds, open caverns in limestone, and many other geologic features. Although highly useful calculations can be made by assuming that aquifers are uniform in composition and horizontal and tabular, the hydrogeologist should always be aware of the almost infinite variety of shapes and water-bearing properties of aquifers. Many of the variations found in nature are discussed in detail in later chapters.

### CONFINED AND UNCONFINED WATER

Water that is in direct contact vertically with the atmosphere through open spaces in permeable material is called unconfined water. Confined water is separated from the atmosphere by impermeable material. The division between confined and unconfined water is entirely gradational. The term semiconfined is used for the intermediate conditions. In many areas the first unconfined water encountered is above the general zone of phreatic water and is a more or less isolated body of water whose position is controlled by structure or stratigraphy (Figure 2.20). This is called perched water and the upper surface is called a perched watertable.

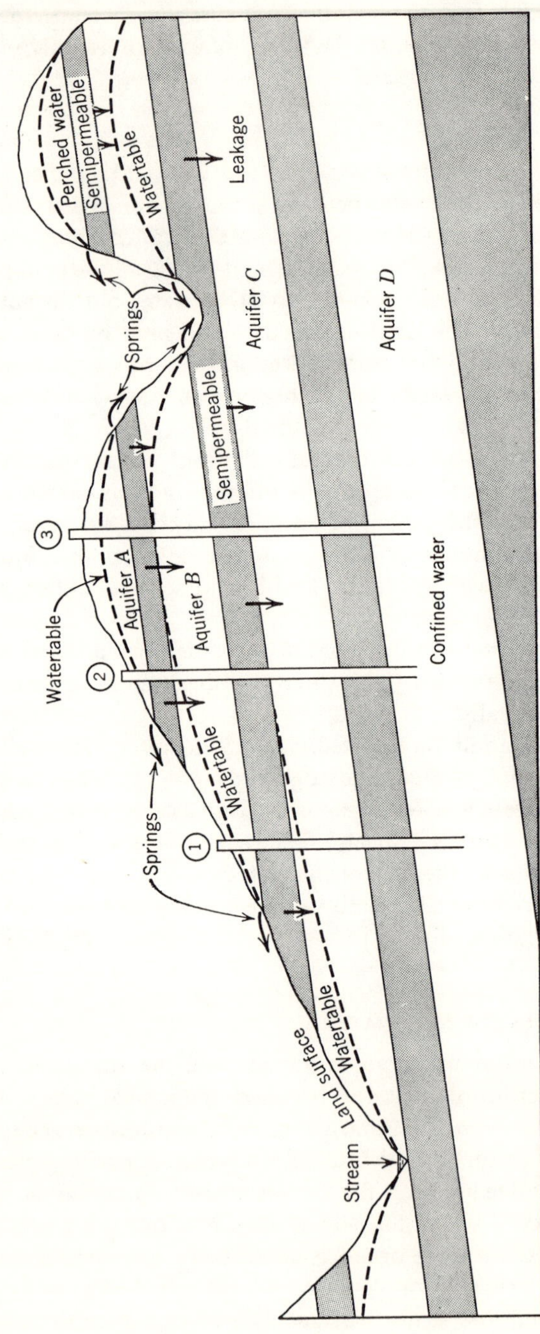


Figure 2.20 Confined, unconfined, and perched water in a simple stratigraphic sequence of sandstone and shale.

The distinction between confined water, semiconfined water, unconfined water, and perched water is generally a very difficult distinction to make. The geologic cross section in Figure 2.20 presents several situations commonly encountered within sedimentary rocks. If a single well were drilled at position 1 in the diagram, water in aquifer B would, without further information be classified as perched water and water in aquifer C would be unconfined. At point 2 the well would encounter perched water in aquifer A, unconfined water in aquifer B, and confined water in aquifers C and D. At point 3 all the aquifers except A would have confined water. Thus in a small area with a simple geologic structure, water within a single aquifer could be classified as perched, normal unconfined phreatic water, and confined water. In areas of more complex geology, the terms perched, confined, and unconfined water are difficult or impossible to apply.

Confined water is also called artesian water; this term, however, was first applied to water under sufficient pressure to produce flowing wells. In recent years artesian has been used more or less as a synonym for confined water.

The most spectacular confined water has been encountered in some of the widespread consolidated aquifers in Australia and central North America. Early wells into these aquifers encountered water with sufficient pressure to flow upward in columns more than 150 feet high, and natural discharges of more than 1000 gallons per minute (gpm) were not uncommon. Two wells drawing water from the Dakota sandstone in north-central United States had such strong flow that they supplied power for flour mills near the end of the nineteenth century. Unfortunately the head diminished rapidly and the water from most of these aquifers must be pumped at the present time.

The material overlying an aquifer may be semipermeable so that water is only semiconfined. Under these conditions high pressures do not develop and water will rarely have heads of more than 5 to 10 feet above the ground surface. Most artesian water within recent alluvial deposits is semiconfined.

### ELEMENTS OF AN ARTESIAN SYSTEM

The original concepts of artesian flow were based on a comparison with a closed conduit which had a source of water with a hydrostatic head above the discharge point (Figure 2.21). Most examples in textbooks show a synclinal basin with water in an intake area, or forebay, above the level of ground-water discharge (Figure 2.21). The projected water-level surface will slope slightly basinward from the water level in the forebay. The aquifer is confined by porous but almost impermeable layers called aquicludes, or by slightly permeable layers called aquitards. This classical structural picture is true of many large artesian systems. In nature, however, the variety is endless. A few of the common types are described as follows.

Figure 2.22*a* shows an artesian system of low head which is found in areas of stabilized sand dunes. Swampy areas form semiconfining layers in low depressions. Figure 2.22*b* is an aquifer formed by a fractured zone along a fault. The confining rock is dense and neither transmits nor stores water. Wells in this type of aquifer may encounter large pressures but are

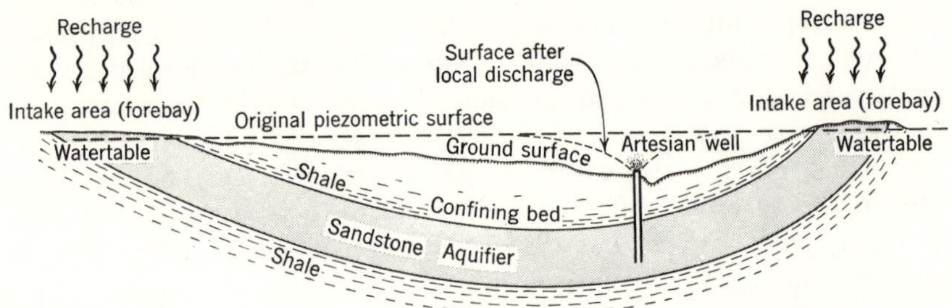


Figure 2.21 Elements of a classical artesian system within a syncline.

rarely dependable sources of large quantities of water. In the complex structure of Figure 2.22*c*, it is seen that an artesian aquifer system can exist in anticlines as well as synclines. Horizontally layered rocks are commonly artesian aquifers (Figure 2.22*d*). The intermixing of permeable and impermeable materials deposited along glacial margins gives rise to many confined aquifers and flowing wells as indicated in Figure 2.22*e*.

SOURCES OF ARTESIAN HEAD

The fact that the water elevation in the forebay is higher than the discharge point at the well accounts for the hydraulic head of a flowing well. This simple explanation, however, is only strictly true of local aquifers in consolidated rock such as the fault zone in Figure 2.22*b*. Larger aquifers have many indications of flexibility, which means that a more complicated mechanical model is necessary to explain the phenomena observed in most artesian systems.

If an artesian system were perfectly rigid and filled with a frictionless and incompressible fluid, the effects of well discharge would be observed as an instantaneous lowering of water levels in all the wells and in the forebay of the system. This is not observed in artesian systems. Commonly the effects of pumping are large enough to be measured only within ten miles of the well. Meinzer [14] was one of the first geologists to carry this

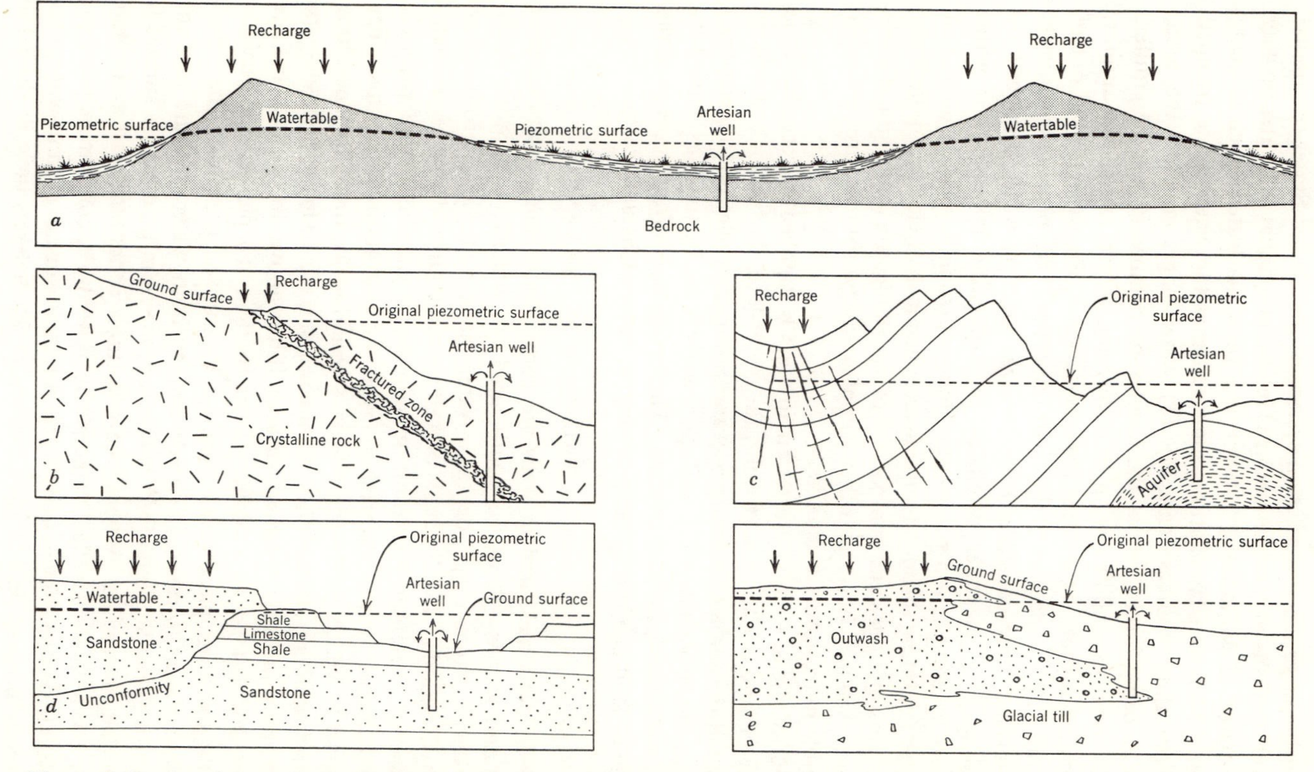


Figure 2.22 Artesian wells in (*a*) stabilized sand dunes, (*b*) crystalline rock, (*c*) complexly folded and fractured sedimentary rocks, (*d*) horizontal sedimentary rocks, and (*e*) glacial deposits.

line of reasoning to its logical conclusion. He noted that water levels of wells in the large Dakota artesian system of the United States responded only to the effects of local pumping. He also noted that the water flowing into an area of withdrawal is much less than the water which is being pumped. Meinzer's calculations indicated that the water being pumped was from three sources, namely: (1) water migrating into the area, (2) water being forced out of the aquifer by compaction of the aquifer, and (3) water expanding owing to a lowering of pressure in the aquifer. He concluded that of the three sources of water, the water forced out of the aquifer by compaction is the most important. Supplementary sources of water not fully considered by Meinzer are: (1) water displaced by expanding gas, (2) water displaced by expanding minerals, (3) water extruded by compaction of the aquicludes surrounding the aquifer, and (4) water migrating through temperature, chemical, or electrical potentials. The expansion of gas and compaction of aquicludes are important in some artesian systems, but the other supplementary sources are probably negligible in fresh-water aquifers.

Proof of the flexibility of artesian aquifers is supplied by records of the change of land elevation near wells, the water-level fluctuations caused by loading at the surface, quantity of water pumped in relation to water-level changes in the vicinity of a well, and water-level fluctuations caused by passage of seismic or tidal stresses. The best proof is from records of the change of land elevation near pumping wells (see Chapter 11).

## PRESENTATION OF WATER LEVEL DATA ON MAPS AND GRAPHS

Water levels measured in wells are conveniently studied by means of maps and graphs. Most frequently used are water-level contour maps, water-level change maps, depth-to-water maps, water-level profiles, and well hydrographs. If full details of well construction and aquifer geometry are known, water-level contour maps can be classified more precisely as piezometric maps, watertable maps, or potentiometric maps. The elevation to which water will rise in artesian wells, or wells penetrating confined aquifers, defines the piezometric surface. This surface may be either above or below the land surface; if it is above, the well casings are assumed to extend to a higher elevation than the piezometric surface. Strictly speaking, piezometric means a measure of pressure. The piezometric surface, however, is determined by both water pressure and the elevation of the aquifer. For this reason, the term piezometric surface is somewhat misleading. Where true watertable conditions exist and where water levels in wells reflect reliably the watertable, contour maps of the watertable can be drawn. The force potential causing ground-water flow is directly proportional to the elevation of water levels in wells drilled in both confined and unconfined aquifers (see Chapters 6 and 7). The term potentiometric map is best suited, therefore, to describe water-level maps which are representative of a single flow system within an aquifer.

The general direction of ground-water flow can be shown on watertable maps as well as on piezometric maps. A watertable map indicates the elevation of the watertable by means of contours. If the hydraulic gradient is less than 1 per cent and the transmissivity is more or less

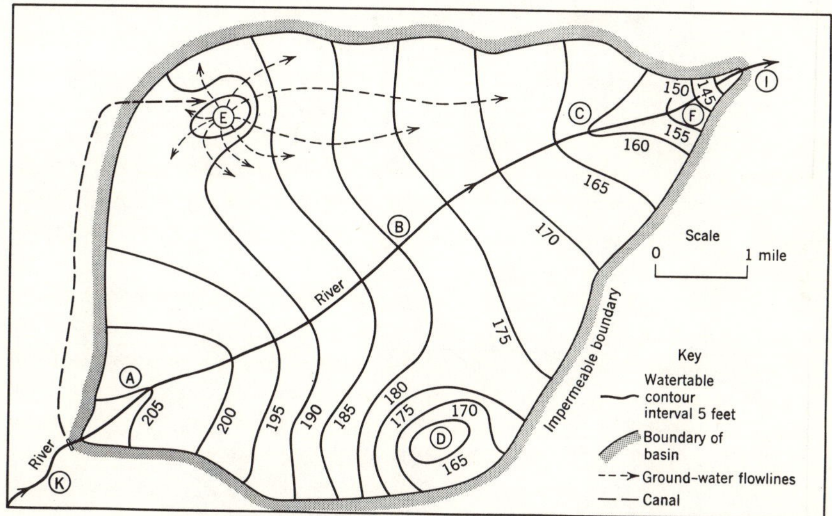


Figure 2.23 Contour map of the watertable in a small hypothetical ground-water basin. If the aquifer is homogeneous and isotropic and if the slope of the watertable is not large, the map can be used to construct a flow net; that is, a regular "square" net (see Chapter 7). A small number of flowlines have been drawn on the map. Excessive convergence of the flowlines suggests a changing transmissivity of the aquifer.

uniform, the watertable is also an accurate representation of the potentiometric surface of water in the aquifer. Inasmuch as flow in such an aquifer is almost horizontal, a flow net can be constructed using the watertable contours as equipotential lines and drawing flow lines perpendicular to the contours (Figure 2.23). The flow net can then be used to find centers of recharge or discharge and calculate quantity of ground-water flow (see Chapters 6 and 7). Figure 2.23 presents a number of watertable configurations related to common geologic or hydrologic causes.

Area A is an area of recharge within an alluvial fan where the surface is 80 feet above the watertable. Here the stream continually looses water to the permeable substrata. Streams with this relation to the watertable are called influent streams. At point B the water in the stream is at the same elevation as the watertable. The watertable contour is normal to the stream at this point because there is no flow from the stream and ground-water flowlines are therefore tangent to the direction of the stream. At C the surface of the stream is below the watertable, and the stream receives ground-water discharge. At C the stream is called an effluent stream.

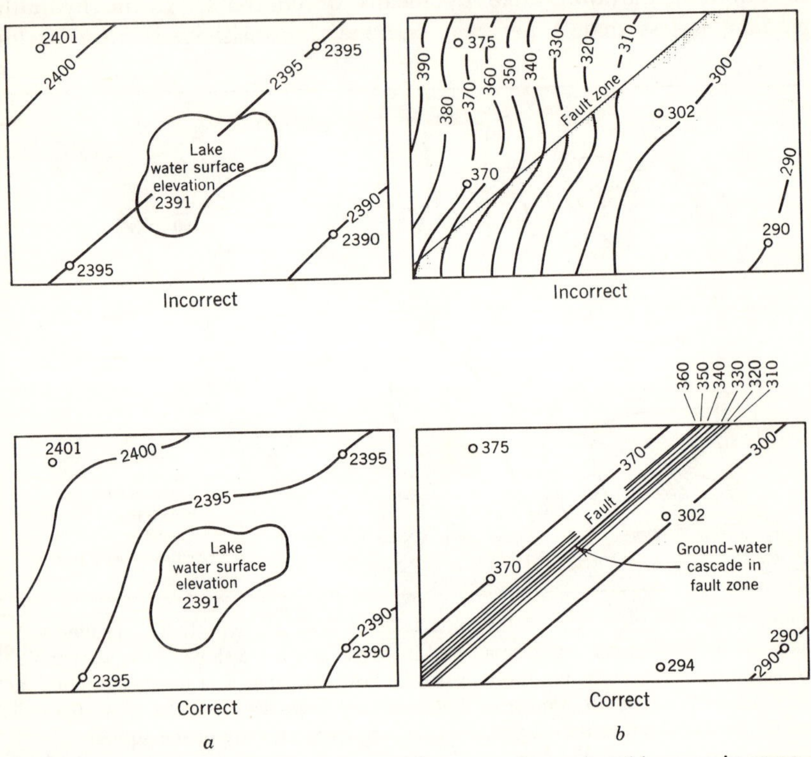


Figure 2.24 Common errors encountered in contouring watertable maps in areas of (*a*) topographic depressions occupied by lakes, and (*b*) fault zones.

At F the stream is still an effluent stream, but most of the ground water has already been discharged into the stream so the contours no longer bend sharply upstream. Point D is an area of heavy pumping in which the water has been lowered to 20 feet below the stream level at B. After a short period of time the pumping at D should make the contours shift so the river will be influent at B. Area E is an area of recharge in which surplus irrigation water has produced a ground-water mound 10 feet above the stream surface at B. The stream at K and I is flowing in an impervious channel. The difference between the discharges at K and I is equal to the water lost or gained within the ground-water basin.

Mistakes in constructing watertable maps are associated with purely mechanical extrapolation of contours between measured water levels. The watertable thus can be placed mistakenly above the land surface (Figure 2.24*a*), or obvious geologic structures are ignored (Figure 2.24*b*). Depth-to-water maps are useful in determining areas of potential water loss by evapotranspiration and also in determining the approximate depths needed for water wells. Thus in Figure 2.25, A designates an area in

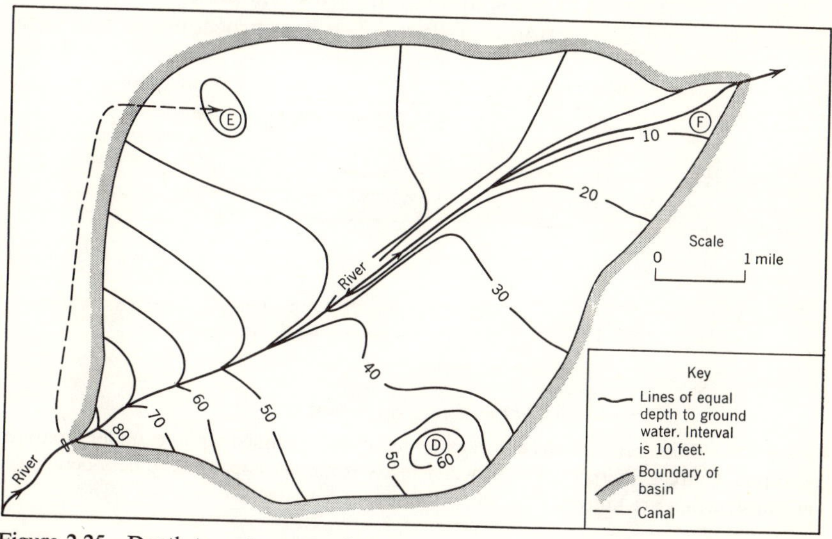


Figure 2.25 Depth-to-water map of the same ground-water basin as shown in Figure 2.23.

which ground water is lost through evapotranspiration. The general outline of areas D and E of Figure 2.23 also are reflected in the depth-to-water map.

Common mistakes in constructing piezometric maps are a failure to distinguish between the water levels of different aquifers and a failure to identify wells which have contact with more than one aquifer (Figure 2.26). If the area is one of complex stratigraphy or structure, the data should be interpreted with maximum use of geologic information. Commonly the geologic effects are indistinct until after the aquifer is developed. The initial condition of slow ground-water flow causes only a minimum change of water-level elevations across semipermeable boundaries. After development, however, the zones of small permeability cause marked differences in water-level elevations.

As a rule, piezometric surfaces are much smoother than watertables. Effects of aquifer discharge through wells or springs will be distributed rapidly through a large area. For example, measurable lowering of the piezometric surface around a pumping well may be propagated in less than 30 seconds to distances of more than one mile, whereas a similar lowering around a watertable well will take several months.

Water-level change maps are constructed by plotting the change of water levels in wells during a given span of time. If the study is of a short span of time, data from the same wells can be used. If, however, the time span is long, it is impossible in some areas to measure the same wells owing to their rather rapid destruction or failure. The best procedure in this case

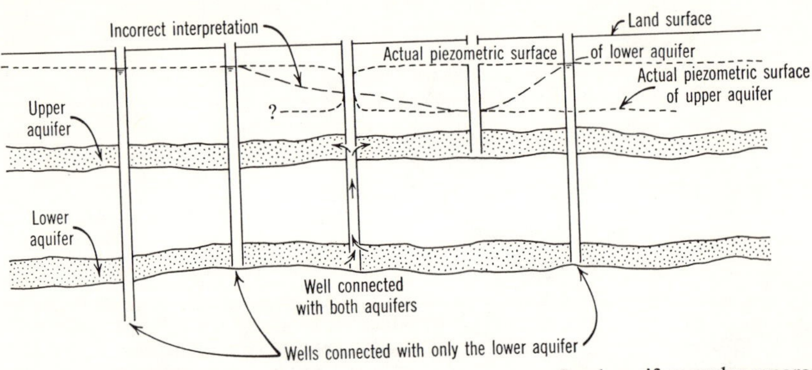


Figure 2.26 Observation wells in a region having two confined aquifers under separate pressures. Correct interpretation of water levels is almost impossible unless details of well construction are known.

is to draw on transparent paper two watertable maps of the years of interest. The maps are then superimposed and the water-level changes at contour intersections are recorded. The values can then be transferred to a separate map and lines of equal water-level change can be drawn (Figure 2.27).

The water-level change map can be used to calculate the changes in the volume of the saturated part of an unconfined aquifer. The total volume of dewatered sediment can then be divided into the volume of water pumped to obtain the water yielding character, or specific yield, of the dewatered sediments. Thus if $15 \times 10^6$ cubic feet of sediments are dewatered by pumping $5 \times 10^6$ cubic feet of water, the average specific yield would be 33 per cent.

Water-level change maps are also useful in measuring the local effects of recharge or discharge. Commonly the effects show as distinct anomalies which are difficult to detect by only the comparison of successive watertable maps. This is seen in Figure 2.27 in which the effects of recharge show only a deflection of the contours of the watertable, but as a well-defined circular area in the water-level change map.

Hydrographs of water levels in wells can be constructed from data obtained from individual measurements by chalked tape, electric probes, air lines, or the reflection of sound (Figure 2.28). Continuous records can be obtained also by mechanical or electrical devices (Figure 2.28). Some of the automatic recordings are adapted to direct use by high-speed data-processing equipment.

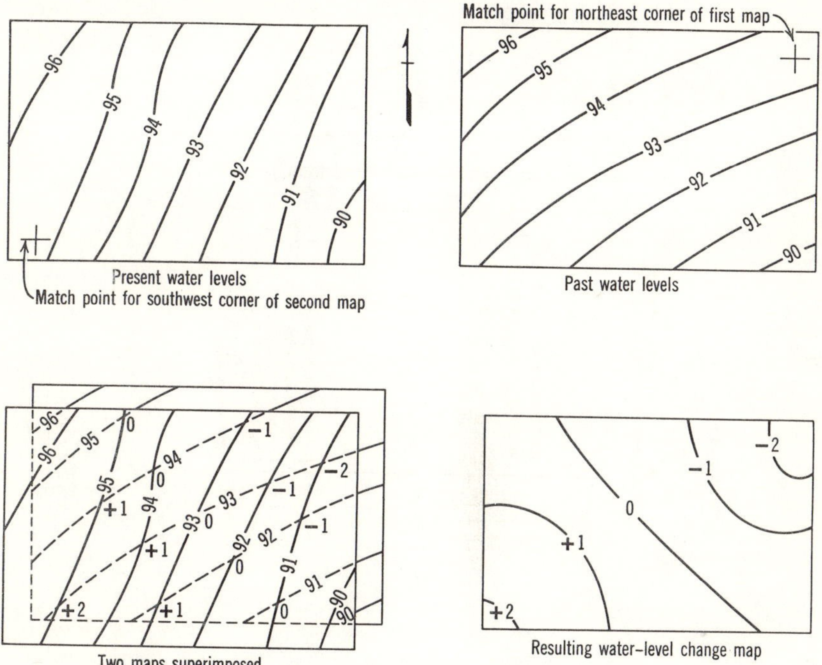


Figure 2.27 Construction of a water-level change map by superimposing water-level contour maps.

Well hydrographs are among the most diagnostic sources of hydrogeologic information. Long-term records will give some indication of the ultimate yield as well as the rate of replenishment of aquifers. A study of short-term fluctuations with amplitudes of less than a tenth of a foot may give information concerning the mechanical properties of material overlying the aquifer as well as the extent of the connection between the aquifer and the atmosphere. Records of large changes in water levels caused by pumping of ground water can be analyzed to obtain information on the geometry and water-yielding properties of aquifers. The response of water levels to ground-water pumping is discussed in detail in Chapter 7.

Most water-level fluctuations can be classed within four basic types: (1) fluctuations due to changes in ground-water storage, (2) fluctuations

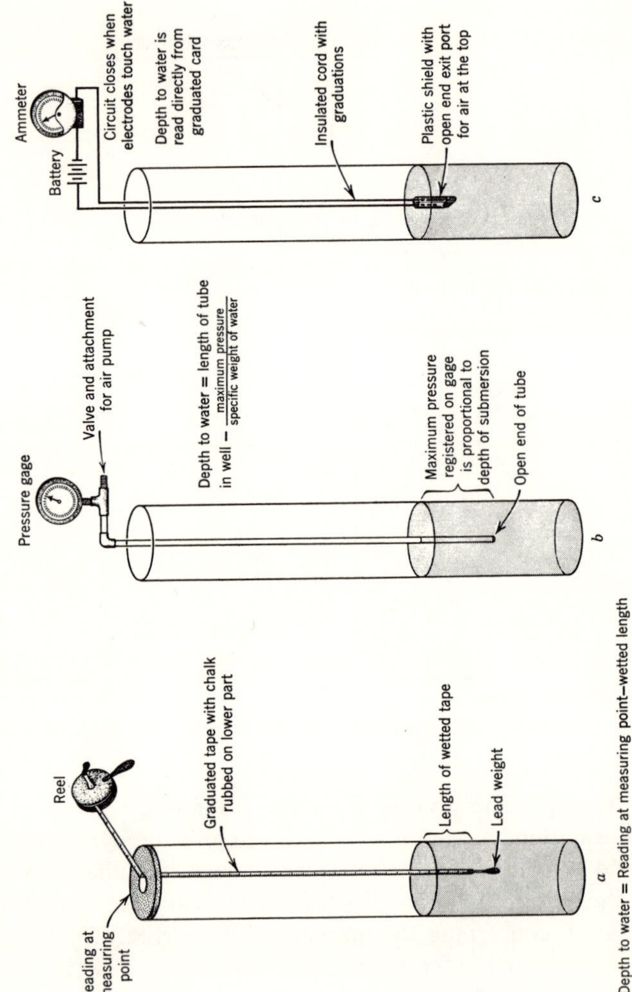
Reading at measuring point
Reel
Graduated tape with chalk rubbed on lower part
Length of wetted tape
Lead weight
a
Depth to water = Reading at measuring point–wetted length
Pressure gage
Valve and attachment for air pump
Depth to water = length of tube in well − maximum pressure / specific weight of water
Maximum pressure registered on gage is proportional to depth of submersion
Open end of tube
b
Battery
Ammeter
Circuit closes when electrodes touch water
Depth to water is read directly from graduated card
Insulated cord with graduations
Plastic shield with open end exit port for air at the top
c

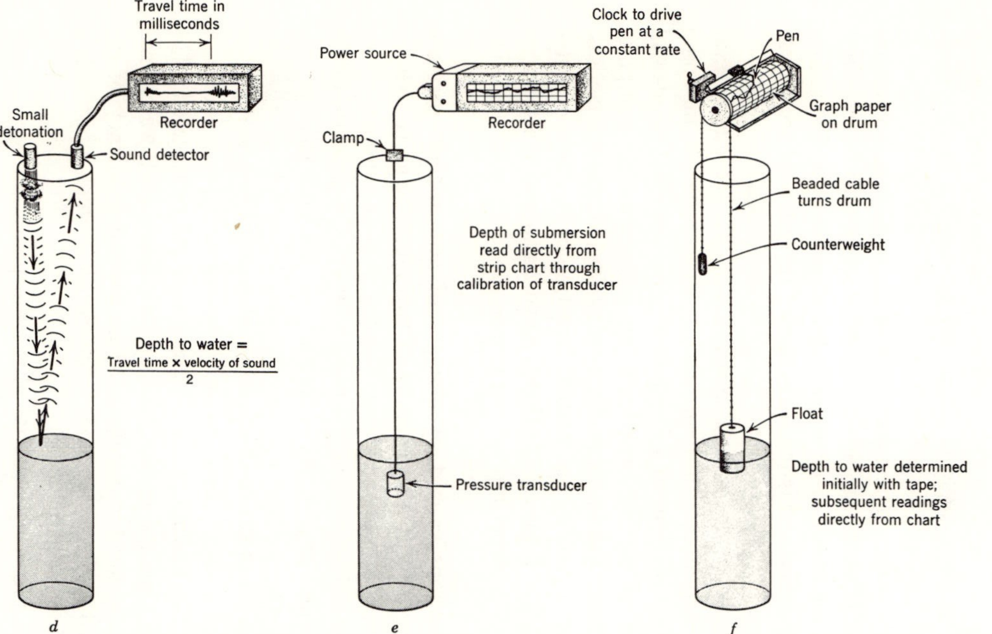


Figure 2.28 Various methods used to determine depth to water in wells. (*a*) Steel surveyor's tape that can be read with an accuracy of about 0.005 foot. (*b*) Air pressure gage, if functioning correctly, will yield values accurate to about 2 feet. (*c*) The electric probe is generally accurate to 0.1 foot, but if used with a surveyor's tape can yield readings accurate to 0.001 foot. (*d*) Sonic methods are useful only as a rough measure of depth to water and are accurate to about 10 feet if air temperature, and consequently velocity of sound, is known. (*e*) Various types of electrically actuated pressure transducers are available and can measure changes of submersion depth of less than 0.001 foot. (*f*) Mechanically actuated drum recorders are one of the oldest and most reliable types of water-level measuring devices. With proper gear ratios, water-level changes of less than 0.001 foot can be recorded.

caused by direct fluctuations of atmospheric pressure in contact with the water surface in wells, (3) fluctuations caused by deformation of aquifers, and (4) fluctuations caused by disturbances within the well. Minor fluctuations are also caused by chemical or thermal changes in and near wells [9,15].

Changes in storage account for most large fluctuations of water levels. Natural changes of storage, such as caused by recharge or spring flow, generally give rise to rather gradual changes in water levels. Near river channels, however, the increase in storage may be rather abrupt in response to flood flow. This will cause rapid rises in water levels near the streams. Figure 2.29*a* shows the effect of changing river stages on the fluctuation of the water level in a well. Figure 2.29*b* shows the effect of long-term storage changes caused by a natural discharge from the aquifer.

Rapid fluctuations of water levels through vertical distances of several feet are primarily in response to pumping [21,22]. The short-term pumping effects are shown as fluctuations of a few days length. Seasonal trends will have several months of rise followed by a fall in response to pumping. Seasonal trends are common in areas of pumping for irrigation purposes. Long-term trends are caused by the gradual depletion or recharge of water stored in an aquifer [22]. Figure 2.29*c* shows a hydrograph with a seasonal irrigation pumping cycle superimposed on a generally downward trending curve.

Water-level fluctuations caused by changes of atmospheric pressure on the water surface in wells are generally of two types. One is a short-term fluctuation caused by gusts of wind passing over the mouth of the well [17]. The increased velocity at the mouth lowers the pressure in the well and causes the water level to rise in the well. As the wind subsides the water level drops again.

Changes in barometric pressure also produce changes of greater duration in ground-water levels in wells [21]. As the atmospheric pressure increases the water levels in wells which tap confined aquifers will be lowered. The ratio of water-level changes in the well to the inverse of water-level changes in a water barometer is called the barometric efficiency of the aquifer. Some aquifers have barometric efficiencies which approach a maximum value of about 80 per cent.

The weight of the overlying material plus atmospheric pressure causes a pressure, $p_a$ on the aquifer. This pressure is resisted by the pressure $p_w$, of the water in the aquifer and the reactive pressure, $p_s$, of the solid grains in the aquifer. If the pressure on the aquifer is increased by $\Delta p_a$ by changes in barometric pressure, this increase is distributed partly on the grains and causes an increase in intergranular pressure, $\Delta p_s$, and it is distributed partly on the water and causes an increase in water pressure,

$\Delta p_w$, within the aquifer. Thus

$$p_a = p_w + p_s \tag{2.8}$$

and

$$\Delta p_a = \Delta p_w + \Delta p_s \tag{2.9}$$

When a well is present, an increase in atmospheric pressure of the magnitude of $\Delta p_a$ is transmitted directly to the aquifer through the water in the well. The water will thus be forced from the well into the aquifer. This movement, $\Delta h$, will continue until it is balanced by the counteracting pressure change; thus, in Figure 2.30

$$\Delta h\gamma + \Delta p_w = \Delta p_a \tag{2.10}$$

in which $\gamma$ is the specific weight of water. The foregoing equation assumes that the volume of water forced into the aquifer is negligible in comparison to the volume of water in the aquifer. For large wells in rocks of low permeability, this is certainly not true, but for most aquifers in unconsolidated sediments this assumption can be made. In general $\Delta p_w$ is small for rigid aquifers and large for unconsolidated silty aquifers. Thus the barometric efficiency is directly proportional to aquifer rigidity.

Fluctuations caused by the deformation of aquifers arise from several different forces. One of the most common sources of deformation is through a loading and unloading over coastal aquifers which is produced by fluctuations of the ocean level. Onshore winds during a storm may cause the ocean level to rise as much as two meters. This type of rise will be superimposed on the normal cyclic tidal fluctuations. If the ocean level rises, a greater load will be imposed on the aquifer so that

$$\Delta p_t = \Delta p_w + \Delta p_s \tag{2.11}$$

in which $\Delta p_t$ is the change in pressure on the aquifer caused by changes in the ocean level (Figure 2.31). If the aquifer is flexible, the value of $\Delta p_w$ will be large. The rise of water in the well will be equal to $\Delta h$, or

$$\Delta h\gamma = \Delta p_w \tag{2.12}$$

Jacob [7] has demonstrated the relation between tidal efficiency and barometric efficiency. If

$$B = \frac{\Delta h\gamma}{\Delta p_a} \tag{2.13}$$

and

$$C = \frac{\Delta h\gamma}{\Delta p_t} \tag{2.14}$$

then

$$C + B = 1 \tag{2.15}$$

in which C is the tidal efficiency at the coast and $B$ is the barometric efficiency at the coast. Thus tidal efficiency is a measure of the flexibility of an aquifer.

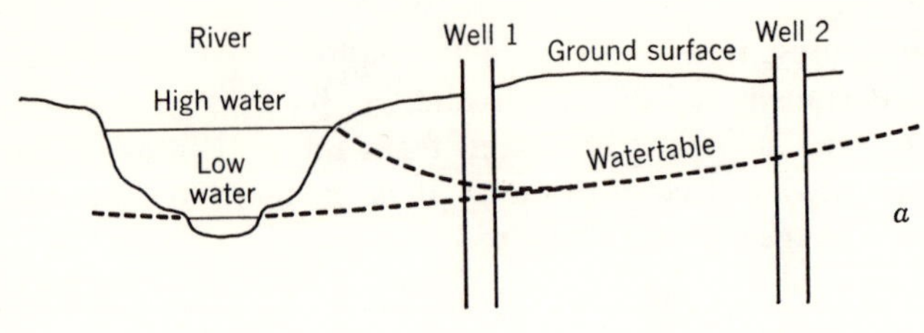


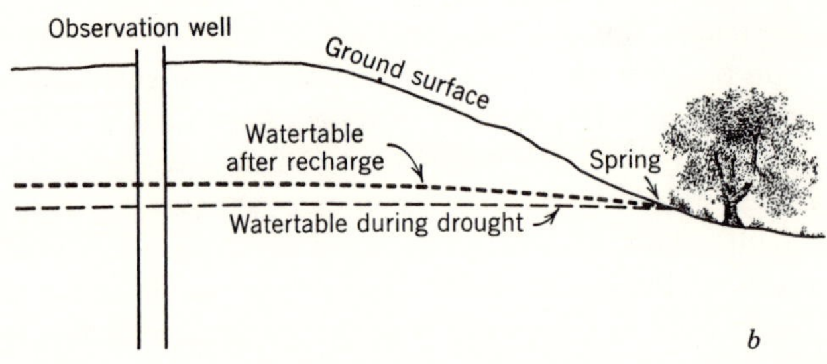


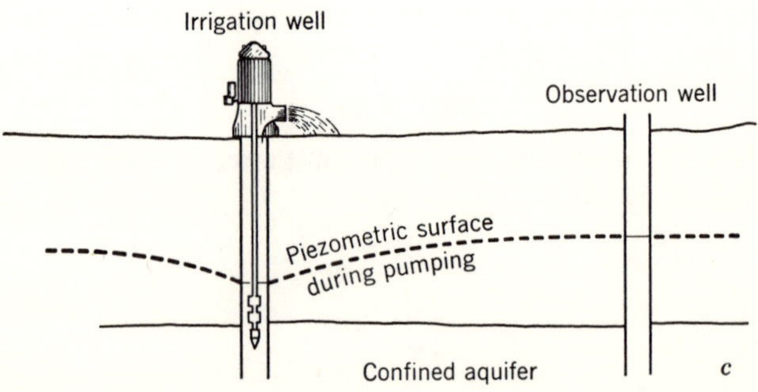


Figure 2.29 Water-level fluctuations caused by (*a*) changes in the water surface in a an irrigation well.

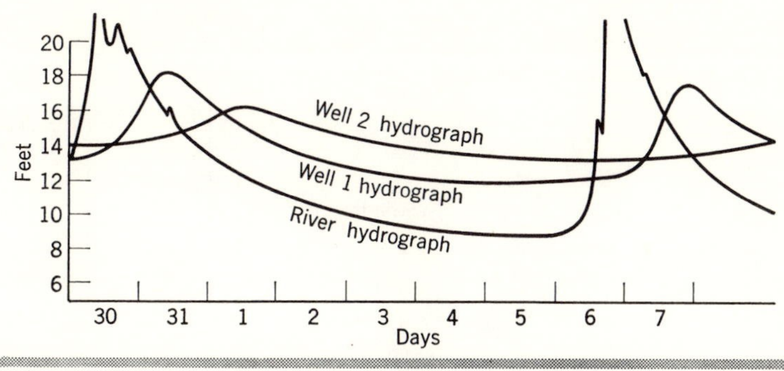


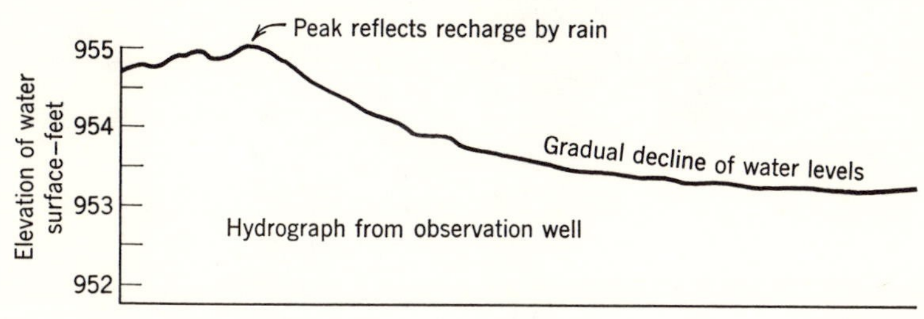


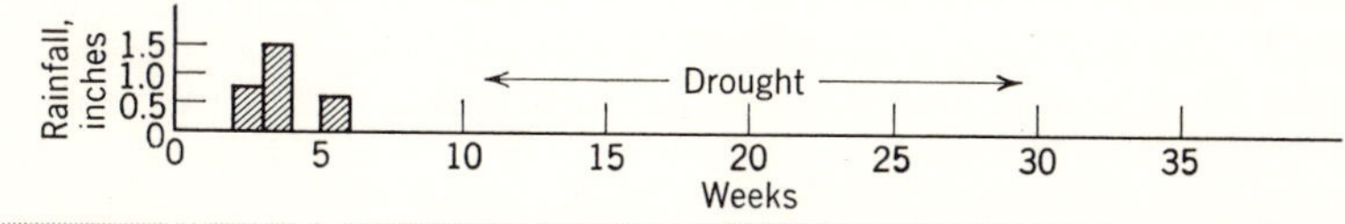


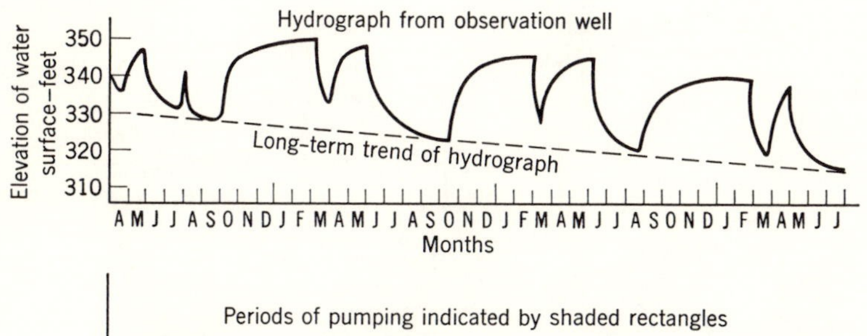


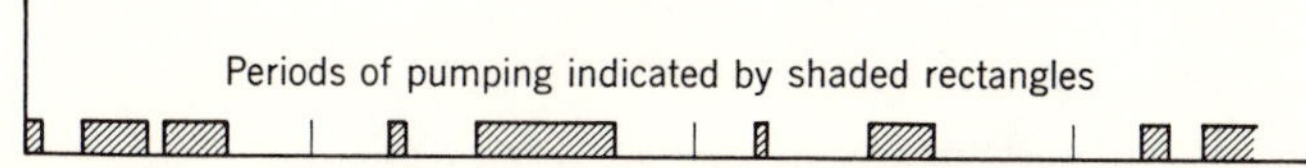


nearby river (*b*) gradual discharge of an aquifer during a drought, and (*c*) discharge of

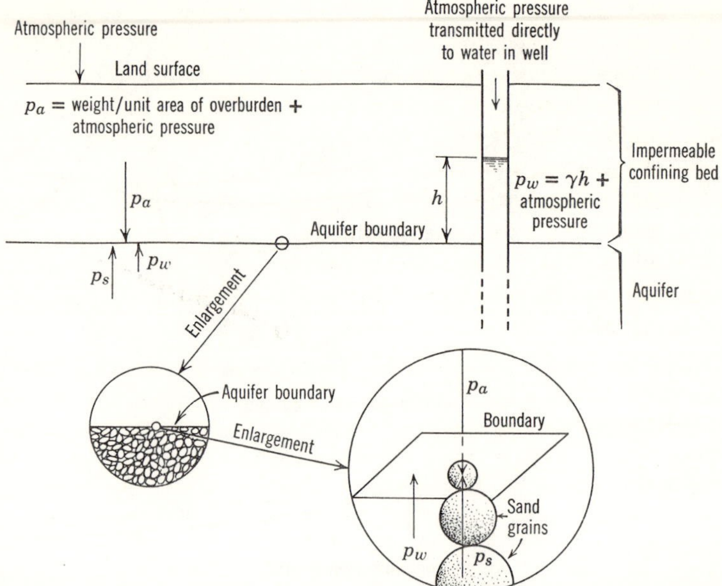


Figure 2.30 The effect of changes in atmospheric pressure on a confined aquifer. If the atmospheric pressure is increased by an amount equal to $+\Delta p_a$, then the water pressure will be increased by $+\Delta p_w$ and consequently, $+\Delta p_w = \gamma(-\Delta h) + \Delta p_a$; the negative $\Delta h$ indicates a drop in the water level since $\Delta p_a > \Delta p_w$. If the atmospheric pressure is decreased by an amount equal to $-\Delta p_a$, then the water pressure will be decreased by $-\Delta p_w$ and consequently, $-\Delta p_w = \gamma\Delta h - \Delta p_a$; the positive $\Delta h$ indicates a rise in the water level of the well.

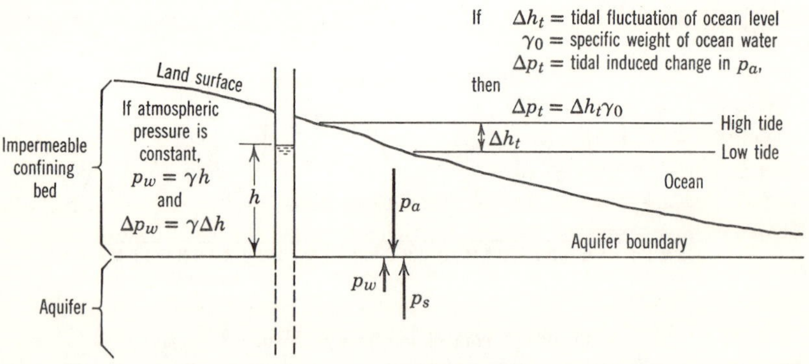


Figure 2.31 The effect of tidal fluctuations of the ocean surface on a confined aquifer. Pressure on aquifer, $p_a$, is produced by the combined effects of atmospheric pressure, pressure of ocean water, and weight of confining bed and overlying material. Since $\Delta p_w = \gamma\Delta h$ and $\Delta p_w$ is directly proportional to $\Delta h_t$, the algebraic sign (or direction of movement) of $\Delta h_t$ and $\Delta h$ will be the same. However, the change in water level of the well, $\Delta h$, will always be less than the change of ocean level, $\Delta h_t$, by the amount $\frac{\Delta p_s}{\gamma}$, provided that $\gamma \approx \gamma_0$.

Earth tides also cause small water-level fluctuations in some wells. The fluctuations of generally less than 0.1 foot amplitude show two minima each day which correspond with the moon's upper and lower culminations. The minima are produced by maximum tidal attraction which allows a slight dilation of the aquifer. The dilation in turn reduces the pressure and causes the water to drop in the well.

Nonperiodic fluctuations are caused by earthquakes, trains, earth-moving machinery, explosions, and other sources of temporary stresses in the aquifers [7,21]. These fluctuations are commonly less than 0.1 foot in amplitude but may be more than several feet in the case of extremely large earthquakes.

Water-level fluctuations caused by earthquakes include two types: one is elastic in nature, the other is nonelastic and is caused primarily by rearrangement of grain particles in or near the aquifer. Elastic deformation is propagated through rocks in all parts of the world in response to seismic movements in the crust. The small volume changes in the aquifer caused by the deformations will be reflected by water-level changes in wells which tap confined, yet moderately permeable, aquifers [2,3,5,29].

Permanent deformation of unconsolidated sediments is common within a one-hundred-kilometer radius of the epicenters of exceptionally large earthquakes and closer to the epicenters in smaller earthquakes [31]. The most spectacular effects are seen in recently deposited sand aquifers. Water may begin to flow from wells which tap these aquifers during and immediately after strong shocks. This effect is caused by the expulsion of water from aquifers which are being compacted by seismic vibrations. In many places where sand is at relatively shallow depths the hydraulic head produced is large enough to force a mixture of sand and water to the surface and form small sand mounds which have the shape of miniature volcanoes [23].

Water-level fluctuations of less than one foot are caused infrequently by disturbances within the well. Water cascading from leaking pipes, small animals falling into wells, gas bubbling through the water, and other disturbances may defy identification on hydrographs. Fortunately, the nonperiodic nature of the fluctuations together with their small amplitude enable the hydrogeologist to differentiate these miscellaneous movements from more significant parts of the hydrographs.

## 2.7 *Discharge of Ground Water*

The longest segment of the hydrologic cycle is completed when ground water is discharged at the surface or into bodies of surface water. The discharge may take place in a variety of ways of which springs, artificial discharge, and transpiration by plants are the most important. Locally, the water may come to the surface as diffuse discharge that evaporates directly from the soil surface or that seeps into rivers and lakes.

#### SPRINGS

Few manifestations of ground water have held more popular interest than springs. Springs in arid regions were largely responsible for the localization of ancient settlements. Religious tradition is filled with references to springs. Medical superstition of the past centuries was commonly interwoven with therapy based on mineral-spring or hot-spring treatments. People in general still consider spring water as something more or less magical. This belief is widely exploited by advertising firms, particularly in the sale of beverages. The public mind today also associates

*Table 2.2 Meinzer's Classification* [12] *of Spring Discharge*

| Magnitude | English Units | Metric Units |
|---|---|---|
| First | Greater than 100 $ft^3$/sec | Greater than 2.83 $m^3$/sec |
| Second | 10 to 100 $ft^3$/sec | 0.283 to 2.83 $m^3$/sec |
| Third | 1 to 10 $ft^3$/sec | 28.3 to 283 liters/sec |
| Fourth | 100 gal/min to 1 $ft^3$/sec | 6.31 to 28.3 liters/sec |
| Fifth | 10 to 100 gal/min | 0.631 to 6.31 liters/sec |
| Sixth | 1 to 10 gal/min | 63.1 to 631 ml/sec |
| Seventh | 1 pt/min to 1 gal/min | 7.9 to 63.1 ml/sec |
| Eighth | less than 1 pt/min | less than 7.9 ml/sec |

spring water with water of exceptional purity. This association is commonly not based on fact. People have been observed waiting patiently in line to obtain spring water which contains two to three times as much dissolved solids as the local public-water supplies. Also, spring water is generally more easily contaminated than water from properly constructed wells or from municipal supplies.

Springs are classified in a number of ways. The classifications can be based on magnitude of discharge, (Table 2.2), type of aquifer, chemical characteristics, water temperature, direction of water migration, relation to topography, and geologic structure. Clearly the number of variables is great enough so that several thousand distinctive types of springs could be described. The discussion which follows will attempt to cover basic principles as well as to describe a few representative types of springs. An exhaustive classification of springs will not be attempted.

Any natural surface discharge of water large enough to flow in a small rivulet can be called a spring. Discharge smaller than this is called surface seepage. Springs also discharge below the surfaces of oceans, lakes, and rivers. Subaqueous springs are commonly hard to detect unless their discharge is greater than several cubic feet per second.

The three principal variables that determine spring discharge are aquifer permeability, area contributing recharge to the aquifer, and quantity of recharge. A high permeability allows large volumes of water to be concentrated in a small area. Many aquifers have a considerable discharge in the form of springs, but their permeability is so low the water is forced to the surface over a large area. For example, the banks of an entire stream system may be lined with small individual seeps and springs with an aggregate discharge of several hundred cubic feet per second. The largest spring in the system, however, may not be more than one gallon per minute.

The area contributing water to the spring may range from less than a thousand square feet in an area having large amounts of infiltration to more than five thousand square miles in an arid region.

The quantity of water entering the ground as recharge may be as much as 10 feet per year in certain areas which have a combination of high rainfall and very permeable surface rock. Impervious rock or arid regions commonly have less than 0.1 inch of infiltration per year.

At most there are only a few hundred first-magnitude springs in the world. The relation between watershed area and infiltration is seen in Figure 2.32. The reason for the small number of first-magnitude springs can be readily appreciated inasmuch as a rare combination of large rainwater infiltration, large drainage area, and favorable geologic structure is needed. Some large springs are fed by water from rivers or lakes which seeps into permeable aquifers. This water commonly travels only a short distance to emerge again in large springs. A good example of this type of spring is found in the Rio Maule drainage area of central Chile. Here, a first-magnitude spring of roughly 1000 cubic feet per second emerges from beneath a lava flow which has dammed a lake. Recent construction of an artificial dam has raised the level of the lake and increased the discharge.

It can also be seen from Figure 2.32 that significantly large springs can issue from rather small catchment areas. For example, with a modest infiltration rate of only 2 centimeters per year, an area of less than 0.01 square kilometers can supply a spring capable of furnishing enough water for the domestic needs of an entire family. Figure 2.32 also illustrates the origin of springs found on some mountain peaks. Actually, the springs are always found below the crests of the peaks and have catchment areas of several tens, if not hundreds, of acres. With higher precipitation near the tops of some mountains, it is not at all unusual to have springs of several liters per minute originate from areas of less than 10 acres.

Most springs show measurable fluctuations of discharge in response to seasonal fluctuations of precipitation. Most springs of magnitude 8 are springs which flow only a short time following a period of precipitation.

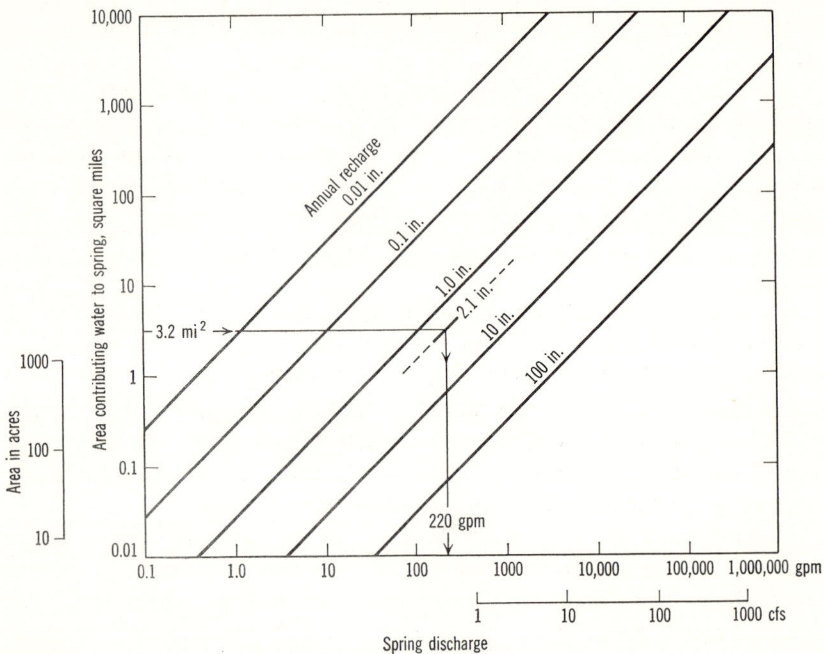


Figure 2.32 The relation between catchment area and spring discharge for various amounts of annual recharge.

Other springs which discharge from aquifers of large storage capacity may have only a very slight fluctuation of discharge. The variability of a spring can be measured by the following formula [12]:

$$V_a = \frac{Q_{\text{max}} - Q_{\text{min}}}{Q_{\text{md}}} 100 \tag{2.16}$$

in which $V_a$ is the percentage of variability, $Q_{\text{max}}$ is the maximum discharge, $Q_{\text{min}}$ is the minimum discharge, and $Q_{\text{md}}$ is the median discharge. The period of record used to calculate variability must be specified. In other words, a spring would be described as having a 30 per cent variability between June 1957 and September 1960. To simply state that a spring has a variability of 30 per cent does not indicate whether it applies to a two-week record or a five-year record.

Daily fluctuations of the discharge of small springs is commonly caused by the use of water by vegetation. The springs will flow vigorously between midnight and sunrise, but may be dry during the day. These springs will resume a steady discharge during the winter when transpiration almost ceases.

Almost all first-magnitude springs issue from lava, limestone, boulder, or gravel aquifers. Most first-magnitude springs also flow from cavern-like openings; however, some may originate from more diffuse discharge. Other common aquifers such as sandstone, conglomerate, and sand generally lack sufficient permeability to form first- or second-magnitude springs.

Small springs can be found in all types of rock. Such diverse rock types as loess, dolomite, graywacke, gypsum, and serpentine may give rise to small springs of magnitude 7 or 8. Even hard crystalline rock sometimes has small springs issuing from fault zones and joints. Shale may have very small springs which flow from joint planes and small lenses of silty or sandy rock.

If geologic materials were perfectly homogeneous, direct surface discharge of water would generally be in the form of diffuse seepage over relatively large areas. Thus where the topography is favorable the land surface would intersect the watertable (Figure 2.33*a*), and surface flow would result. This type of seepage is found in sand-dune areas, loess deposits, massive sandstone terrain, and other homogeneous types of rock and loose sediment.

A vertical or horizontal variation of permeability is the most common cause of the localization of springs. Small seasonal springs are generally associated with permeability changes in the weathered mantle. Sliderock deposits, soil horizons, and landslides help localize the flow of springs (Figure 2.33*b*). Vertical variations of permeability associated with layered sedimentary rocks are the cause of larger, more permanent, springs (Figure 2.33*c*).

Structural variability in rocks caused by earth movements produce many changes in permeability and thereby localize springs. If faults cross hard brittle rock, zones of greater permeability along the fault are formed. If faults cross unconsolidated rocks, the fault zone is usually less permeable than the surrounding rocks. Springs which arise from fault zones are illustrated in Figures 2.33*d* and 2.33*e*. Joints and joint systems commonly are responsible for small springs. This is particularly true of exfoliation joints in massive granitic rocks (Figure 2.33*f*).

Earth movements also cause tilting and folding which bring permeable or impermeable beds near the surface. Two common types of springs associated with folding are illustrated in Figures 2.33*g* and 2.33*h*.

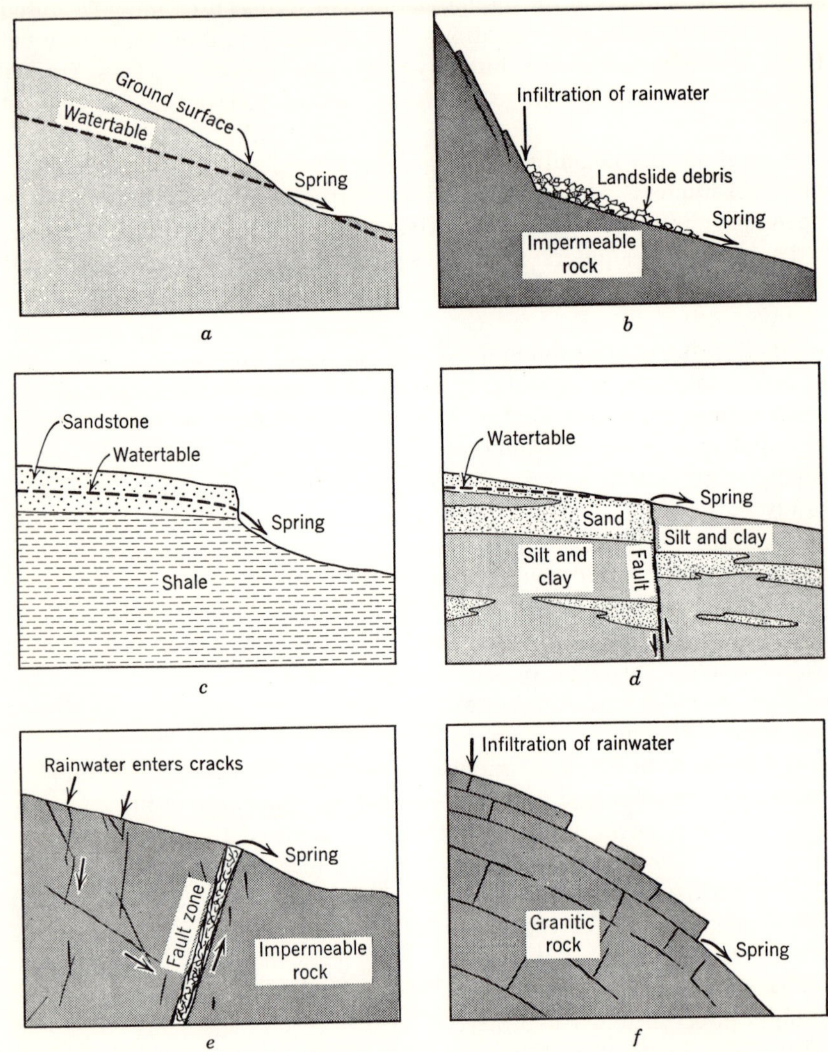


Many of the largest known springs originate from volcanic rocks or from gravel associated with the flows (Figure 2.33*i*). Dikes, sills, layers of tuff, and buried soils commonly control the location of springs in volcanic deposits (see Chapter 9).

A special type of spring comes from glacial meltwater discharged through subsurface channels within the ice. Some of the largest springs of the world may be of this type, but little data are available concerning

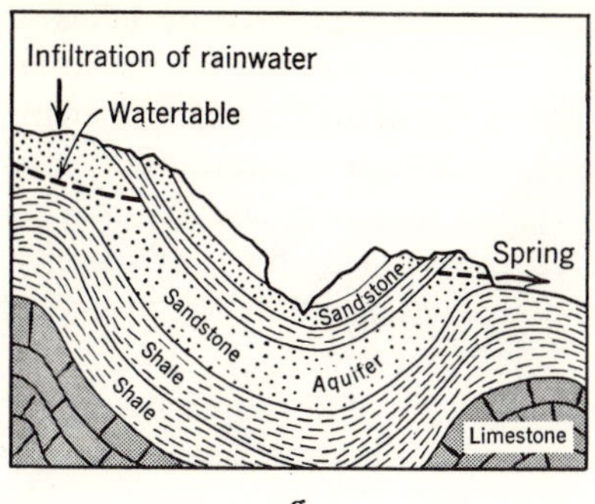


*g*

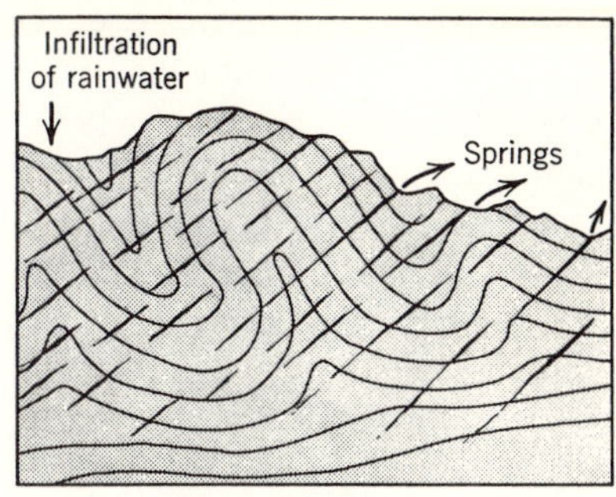


*h*

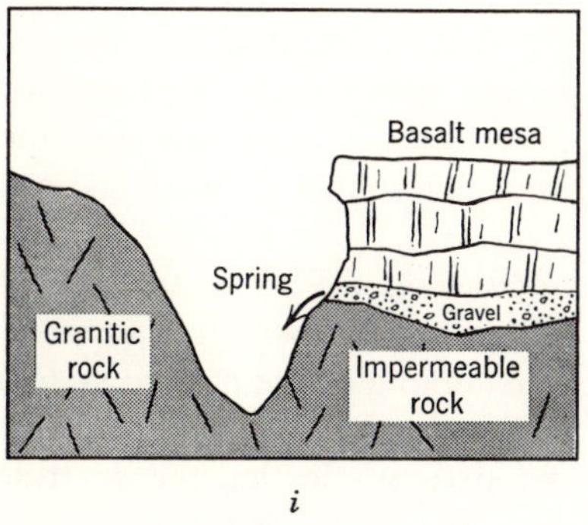


*i*

Figure 2.33 Springs localized by (*a*) a surface depression that intersects the watertable, (*b*) infiltration of rain water into coarse and permeable landslide rubble, (*c*) permeable sandstone overlying impermeable shale, (*d*) a fault that offsets impermeable beds against permeable beds in alluvium, (*e*) a fault that forms an open fractured zone in brittle rock, (*f*) sheet structure in granitic rock, (*g*) outcrop of an artesian aquifer, (*h*) dominant jointing in one direction, and (*i*) outcrop of permeable gravel and basalt overlying impermeable granitic rock.

discharges. Glacial springs are not usually classified as springs, but if glacial ice is considered as a rock, there is no reason to exclude these springs from a general geologic classification.

### ARTIFICIAL DISCHARGE

Discharge of ground water through wells is central to much of the discussion in the following chapters of this book. Although the social and economic aspects of ground-water utilization are beyond the strictly geological or even engineering training of the average hydrogeologist, he soon finds himself faced with problems of zoning regulations, conflicting water rights, conflicting desires of conservationists and engineers, lack of finances for needed well fields, and a myraid of other details only remotely related to scientific matters. The following discussion is a brief introduction to some of these problems. The reader is referred to writings by Thomas [25,26,27], Piper [18], Nace [16], and others [4,8,19] for extensive discussions of the legal, social, and economic problems encountered by the hydrogeologist.

The legal rights to surface water are most commonly acquired either by land ownership or by the appropriation of water sources located some distance from the point of use. Originally, if land bordered on lakes or streams, the owner had a riparian right to the unlimited use of the water. As water use increased, however, it became apparent that this doctrine was not adequate, particularly during times of drought. Judical decisions as well as statutes have modified the original common law in order to provide equitable water distribution for all users during times of scarcity. Also, the right to use water in a wasteful manner is generally not protected in modern laws.

In regions of extreme water scarcity the doctrine of appropriative rights has always prevailed. In simple terms, the doctrine states that the first person or group of persons to appropriate water has first right to the water. The appropriation can be of water on the owner's land or it can be of water several hundreds of miles away from the point of use. Successive appropriators have rights of less strength until the last appropriator may have a right only to surplus flood flow. The appropriative doctrine has been modified in most regions in a manner similar to modifications of the riparian doctrine.

The widespread attempt to apply legal concepts developed for surface water to ground-water problems has been somewhat less than successful. Part of the difficulty lies in a lack of understanding of the natural system of ground-water movement. For example, a futile effort is sometimes made to define the boundaries of "underground streams" so that the doctrine of riparian rights can be applied. Another difficulty stems from the large number of individuals or groups affected by ground-water utilization. Aquifers may extend beneath hundreds of thousands of individual property owners. Water taken from one part of an aquifer will ultimately affect to some extent the water levels in the entire aquifer. An equitable division of ground-water rights among such a large number of owners who represent a host of diverse needs is virtually impossible.

A modern tendency is to give district, state, provincial, or federal agencies the power to regulate the utilization of ground water. In the United States the power is generally vested in appropriate agencies within individual states. The nature of the laws vary widely from one state to another. In many, the state engineer is given the power to issue licenses for the construction of large-capacity wells. In this manner overproduction can be prevented provided that the state engineer's office is kept fully informed of the current ground-water situation in all areas of interest. In California, on the other hand, there has been little state regulation of ground-water production. Local water districts are formed, however, that have the power to regulate water use within well-defined hydrogeologic units. Some of these districts also tax owners of large wells in order to raise money for ground-water recharge operations.

The popular classification of all ground water as a renewable resource has raised some interesting social and economic problems in arid regions. Few individuals that purchase property in such regions realize that almost all extensive developments of ground water will eventually deplete the aquifers. If the development is limited to the amount of natural recharge, thus avoiding overproduction, small and inefficient mining, farming, or industrial operations would result. In contrast, extensive use of the water would allow efficient utilization of roads, equipment, manpower, and the like. Thus the removal of the water should be planned in a manner similar to the removal of coal, iron ore, or other nonrenewable resources. The construction of buildings and other structures involving large investments of money should be controlled by the expected life of the ground-water development.

It is often impossible to plan ground-water development in advance. Technical advice is deferred until serious water shortages develop. This is usually after cities and roads have been developed, small farms established, and the pattern of ground-water withdrawal fixed. Generally the economic investment is so large that the area cannot be abandoned. Expensive alternatives such as artificial recharge, legal control of pumping, importation of water, and sewage reclamation therefore, must, be employed.

## REFERENCES


1. Barnes, B. S., 1940, Discussion of analysis of runoff characteristics by O. H. Meyer: *Am. Soc. Civil Eng. Trans.*, v. 105, pp. 106.
2. Blanchard, F. B. and P. Byerly, 1935, A study of a well gauge as a seismograph: *Seismological Soc. Am. Bull.*, v. 25, pp. 313–321.
3. Da Costa, J. A., 1964, Effect of Hebgen Lake earthquake on water levels in wells in the United States: *U.S. Geol. Survey Prof. Paper* 435–O, pp. 167–178.
4. Deutsch, M., 1963, Ground-water contamination and legal controls in Michigan: *U.S. Geol. Survey Water-Supply Paper* 1691, 79 pp.
5. Grantz, A., and others, 1964, Alaska's Good Friday earthquake, March 27, 1964, a preliminary geologic evaluation: *U.S. Geol. Survey Circ.* 491, 35 pp.
6. Horton, R. E., 1933, The role of infiltration in the hydrologic cycle: *Am. Geophys. Union Trans.*, v. 14, pp. 446–460.
7. Jacob, C. E., 1940, On the flow of water in an elastic artesian aquifer: *Am. Geophys. Union Trans.*, v. 21, pp. 574–586.
8. Kazmann, R. G., 1958, Problems encountered in the utilization of ground-water reservoirs: *Am. Geophys. Union Trans.*, v., 39, pp. 94–99.
9. Kohout, F. A., 1961, Fluctuations of ground-water levels caused by dispersion of salts: *Jour. Geophys. Research*, v. 66, pp. 2429–2434.
10. Langbein, W. B., 1940, Some channel storage and unit hydrograph studies: *Am. Geophys. Union Trans.*, v. 21, pp. 620–627.
11. Lewis, D. C. and R. H. Burgy, 1964, The relationship between oak tree roots and

groundwater in fractured rock as determined by tritium tracing: *Jour. Geophys. Research.*, v. 69, pp. 2579–2588.

12. Meinzer, O. E., 1923, Outline of ground-water hydrology with definitions: *U.S. Geol. Survey Water-Supply Paper* 494, 71 pp.
13. —— 1927, Plants as indicators of ground water: *U.S. Geol. Survey Water-Supply Paper* 577, 95 pp.
14. —— 1928, Compressibility and elasticity of artesian aquifers: *Econ. Geol.* v. 23, pp. 263–291.
15. Meyer, A. F., 1960, Effect of temperature on ground-water levels: *Jour. Geophys. Research*, v. 65, pp. 1747–1752.
16. Nace, R. L., 1960, Water management, agriculture, and ground-water supplies: *U.S. Geol. Survey Circ.* 415, 12 pp.
17. Parker, G. G., and V. T. Stringfield, 1950, Effects of earthquakes, trains, tides, winds, and atmospheric pressure changes on water in the geologic formations of southern Florida: *Econ. Geol.*, v. 45, pp. 441–460.
18. Piper, A. M., 1960, Interpretation and current status of ground-water rights: *U.S. Geol. Survey Circ.* 432, 10 pp.
19. Robinove, C. J., 1963, What's happening to water?: *Smithsonian Institution Report for* 1962, pp. 375–389.
20. Robinson, T. W., 1958, Phreatophytes: *U.S. Geol. Survey Water-Supply Paper* 1423, 84 pp.
21. Russell, R. R., 1963, Ground-water levels in Illinois through 1961: *Illinois State Water Survey Report of Inv.* 45, 51 pp.
22. Sasman, R. T., and others, 1962, Water-level decline and pumpage during 1961 in deep wells in the Chicago Region, Illinois: *Illinois State Water Survey Circ.* 85, 32 pp.
23. Swenson, F. A., 1964, Ground-water phenomena associated with the Hebgen Lake Earthquake: *U.S. Geol. Survey Prof. Paper* 435-N, pp. 159–165.
24. Tanner, C. B., and W. L. Pelton, 1960, Potential evapotranspiration estimates by the approximate energy balance method of Penman: *Jour. Geophs. Research*, v. 65, pp. 3391–3413.
25. Thomas, H. E., 1951, *The conservation of ground water:* New York, McGraw-Hill Book Co., 327 pp.
26. —— 1955, Water rights in areas of ground-water mining: *U.S. Geol. Survey Circ.* 347, 16 pp.
27. —— 1961, Ground water and the law: *U.S. Geol. Survey Circ.* 446, 6 pp.
28. Thornthwaite, C. W., and J. R. Mather, 1955, The water budget and its use in irrigation, in Water, the yearbook of agriculture: *U.S. Dept. of Agriculture Yearbook of Agriculture* 1955, pp. 346–358.
29. Vorhis, R. C., 1964, Earthquake-induced water-level fluctuations from a well in Dawson County Georgia: *Seismological Soc. Am. Bull.*, v. 54, pp. 1023–1133.
30. University of California Sanitary Engineering Research Laboratory, 1955, An investigation of sewage spreading on five California soils: *Univ. California Sanitary Eng. Lab. Tech. Bull.* 12, 53 pp.
31. Zones, C. P., 1957, Changes in hydraulic conditions in the Dixie Valley areas, Nevada, after the earthquake of December 16, 1954: *Seismological Soc. Am. Bull.*, v. 47, pp. 387–396.

# *chapter 3*

# PHYSICAL AND CHEMICAL PROPERTIES OF WATER

## 3.1 *The Water Molecule*

Water is one of the most remarkable substances known. It is the only substance found in vast quantities in nature in three states: solid, liquid, and gaseous. Of the common liquids, it is the most universal solvent, the liquid with the highest surface tension, the highest dielectric constant, the greatest heat of vaporization, and, with the exception of ammonia, the highest heat of fusion. Unlike most other substances, water expands when it freezes under low pressure [6].

These peculiar properties are related to the special molecular structure of water. The chemical formula for water, $H_2O$, is deceptively simple. To begin with, the arrangement of hydrogen nuclei with respect to the electrons and the oxygen nucleus is not symmetrical. If the oxygen atom is pictured at the center of a tetrahedron, the centers of mass of the two hydrogen atoms would each occupy one corner, and the center of charge of two pairs of electrons would each occupy the other two corners (Figure 3.1). Thus, four electrons are as far from both the oxygen nucleus and the hydrogen nuclei as they can be and still be attracted to the oxygen nucleus. Of the other six electrons in the water molecule, four are in a position to form chemical bonds between the oxygen and the hydrogen nuclei, and the two remaining electrons stay close to the oxygen nucleus [4,19,22,32].

This unsymmetrical arrangement gives rise to an unbalanced electrical field which imparts a polar characteristic to the molecule. This arrangement also enables the molecules to join together by hydrogen bonding [26].

The spacing of hydrogen and oxygen atoms within the aggregations is thought to be roughly similar to the spacing of silica and oxygen atoms in quartz. This is particularly true of ice, but also to a lesser extent of liquid water where the aggregates, or arrays, are always in a state of rearrangement. When water is cooled, molecules group together in larger and

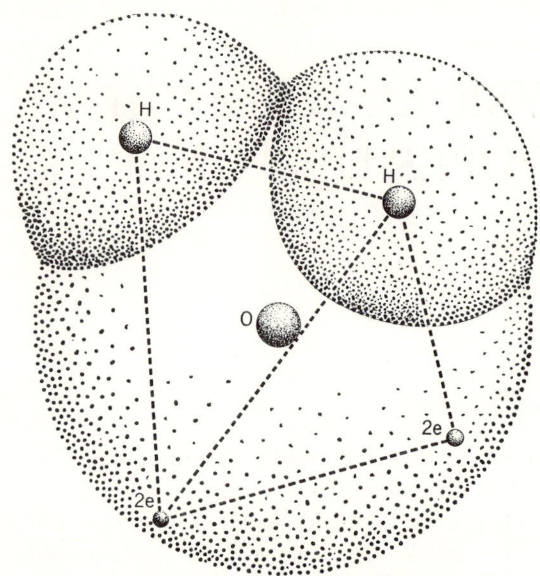


Figure 3.1 Interpenetrating hydrogen and oxygen atoms form water molecules in which the average position of the two hydrogen nuclei and each of two pairs of electrons are near the corners of a tetrahedron. The nucleus of the oxygen atom is in the center of the tetrahedron.

longer lasting aggregates until a maximum compactness, or density, is achieved at 4°C. At this temperature the structure is not rigid but includes many unattached molecules within the larger arrays of molecules. Further cooling causes a greater ordering of molecules so that the hitherto unattached molecules start taking their places in the open arrays, and, as a result, the density of the water decreases. When freezing is accomplished, a further ordering, and consequently a further reduction in density, takes place. In ice all the molecules are more or less rigidly held in place in the open arrays, which make up the crystals.

The fact that a relatively large amount of energy is needed to separate a single molecule from surrounding molecules explains the high values of both surface tension and heat of vaporization of water. The tendency to participate in hydrogen bonding and the polar character of the molecules explain the unusual solvent powers of water. A number of compounds such as sugars and alcohols are held in solution by hydrogen bonding. Other compounds which are highly ionized in water, sodium chloride for example, are held in solution because ions of opposite charge tend to be neutralized by clusters of oriented water molecules.

An additional complexity of water chemistry is the fact that there are natural variations in the masses but not in the charges of the nuclei of the hydrogen and oxygen atoms. Such variants of elements are called isotopes.

*Table 3.1 Relative Abundance of Hydrogen and Oxygen Isotopes*

| Isotope | Relative Abundance, per cent | Half-Life |
|---|---|---|
| $H^1$ | 99.984 | stable |
| $H^2$ | 0.016 | stable |
| $H^3$ | Trace | 12.4 years |
| $O^{14}$ | not known in nature | 76 seconds |
| $O^{15}$ | not known in nature | 2.1 minutes |
| $O^{16}$ | 99.76 | stable |
| $O^{17}$ | 0.04 | stable |
| $O^{18}$ | 0.20 | stable |
| $O^{19}$ | not known in nature | 29 seconds |

The two common isotopes in water are $H^1$ and $O^{16}$; these account for more than 99 per cent of all the atoms in pure water. Other isotopes are $H^2$, $H^3$, $O^{14}$, $O^{15}$ $O^{17}$, $O^{18}$, and $O^{19}$. Owing to the greater mass of most of these isotopes, they tend to be concentrated by processes of partial evaporation. Furthermore, $H^3$, $O^{14}$, $O^{15}$, and $O^{19}$ are radioactive; of these, $H^3$, or tritium, is the most common, being produced in the upper atmosphere by cosmic ray bombardment. It also has been produced in the past few years by nuclear explosions. From these and other facts about the isotopes, the partial history of some water can be inferred if an analysis of its isotopic composition has been made [3,8,9,21]. For example, surface water with an abundance of heavy isotopes would indicate a history of partial evaporation, produced perhaps by a long period of evaporation such as is experienced by the water in the Dead Sea, the Great Salt Lake, or other closed bodies of water. Ground water with a high tritium content would probably indicate that the water is rapidly circulating meteoric water, because the half-life of this isotope is only 12.4 years. Unfortunately, isotopic analyses are too expensive to be used for many routine water studies.

## 3.2 *Physical Properties of Water*

Among the various effects produced by heat, two of the most important are a change of temperature and a change of state. Both are important in the consideration of ground-water migration, accumulation, and utilization. The temperature of water commonly determines its usefulness for industry. The state of water determines its migration characteristics. Lenses of ice in the subsurface are essentially static, liquid water migrates slowly, usually under the influence of gravity, and vapor also may migrate, but through the influence of a concentration and/or temperature gradient.

The quantity of heat required to raise a unit mass of water one degree varies slightly from point to point along the thermometer scale. The mean calorie is defined as one one-hundredth part of the heat required to raise 1 gram of water from 0 to 100°C; this value is very close to the amount of heat needed to raise 1 gram of water from 15 to 16°C.

If pressure is constant, a change of state takes place at a constant temperature. The heat which is necessary for a change from solid to liquid is called the heat of fusion. At atmospheric pressure this value is 79.7 calories per gram for water. The heat necessary to change water from a liquid to a gaseous state is called heat of vaporization, its value for water at atmospheric pressure is 539.6 calories per gram.

The resistance of a liquid to flow is called viscosity. The common unit used for viscosity is the poise, or the more useful centipoise, which is one one-hundredth of a poise. The units of the poise are dyne-second per centimeter squared. Because the viscosity is due to intermolecular attraction, which in turn is reduced by thermal agitation, the value of viscosity is temperature dependent, being 1.0 centipoise for water at 20°C and 0.28 centipoise for water at 100°C. Viscosity as well as other important physical properties of water, which are also temperature dependent, are given in Figure 3.2.

## 3.3 *Dissociation*

In addition to water molecules, pure water also contains dissociated $H^+$ and $OH^-$ ions in very low concentrations. The symbol pH is used to designate the logarithm (base 10) of the reciprocal of the hydrogen-ion concentration. Thus, if there are $10^{-5}$ mole per liter of $H^+$, then the pH is 5.00. The pH of pure water at 25°C is 7.0.

When material goes into solution in water the pH is commonly changed owing to the fact that some of the new ions will combine with $H^+$ or $OH^-$ from the water and shift the chemical equilibrium. In the case of sodium chloride this shift is very small, but in the case of calcium carbonate the

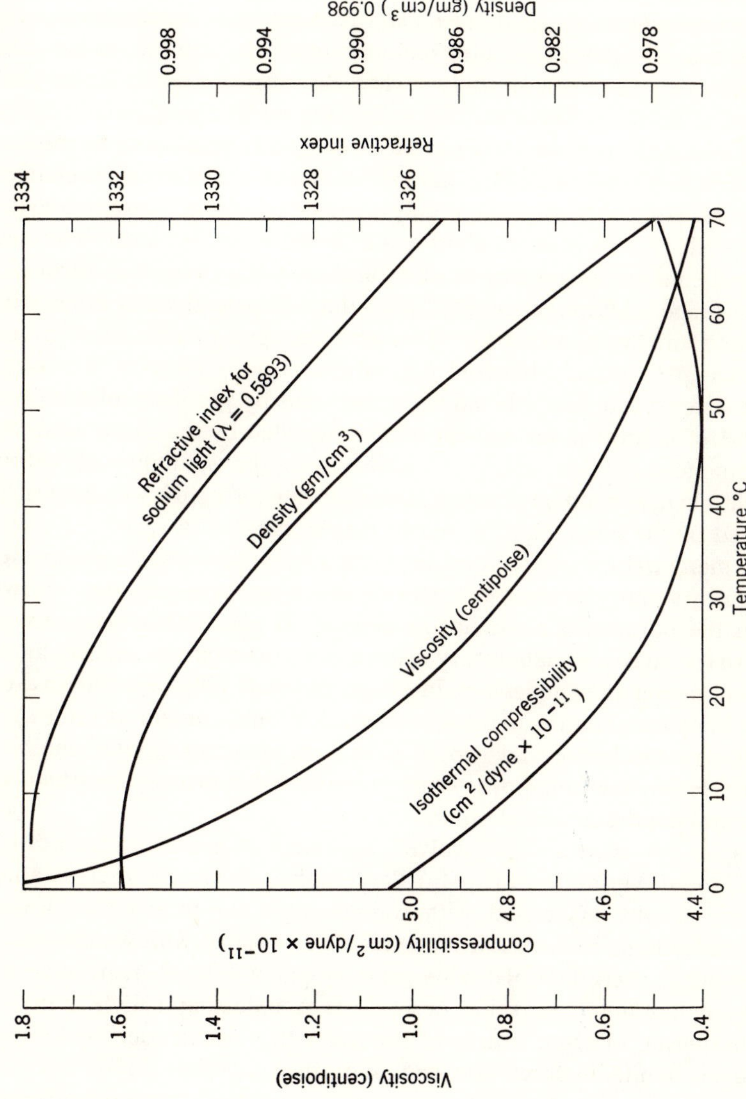


Figure 3.2 Some of the physical properties of water which vary with temperature [6,20,22].

shift is in the direction of fewer $H^+$ ions, or a basic reaction. Salts of aluminum and iron, which are rarely dissolved in concentrations large enough to affect greatly the pH of natural water, have an acid reaction [15].

Solutions whose pH tends to remain constant when small amounts of acid or base are added are called buffered solutions. This action is caused by acids or bases which are only slightly dissociated. When $H^+$ or $OH^-$ ions are added to buffered solutions, the ions are first used in shifting the chemical equilibrium so that the acid or base is converted to the salt. Until this has been completed, the pH of the solution remains relatively constant. Natural water commonly contains dissolved carbon dioxide gas and bicarbonate ions which form a buffered system with carbonic acid. This is the major reason why most natural pH values vary only through a limited range. Although recorded pH values of ground water range from 11.0 for alkali-spring water to 1.8 for acid hot-spring water, most ground water has pH values of between 5.0 and 8.0

The determination of pH values is restricted by problems of sampling. The pH of natural water is most often controlled by the carbon dioxide-bicarbonate-carbonate system. Inasmuch as the solubility of carbon dioxide changes with pressure and temperature, the pH will in turn be changed by the pumping of a well or the discharge of a spring. Furthermore, many pH determinations are made on samples several weeks after storage in a warm laboratory. During this time algal and other growth within the bottle will cause further changes of pH. Differences between field pH values and laboratory values are commonly as much as 0.5 or even greater in samples low in dissolved solids [29]. For this reason, water samples taken for pH determinations should contain as little air as possible in the bottle and should have a nonionizing growth inhibitor added. Field determinations of pH are essential if precise measurements of natural conditions are desired.

Despite the uncertainties involved, a number of generalizations can be made about variations of pH. Very high pH values, that is above 8.5, are usually associated with sodium-carbonate-bicarbonate waters. Moderately high pH values are commonly associated with waters high in bicarbonate. Very low pH values, that is below 4.0, are associated with waters containing free acids derived from oxidizing sulfide minerals, usually pyrite, or from waters in contact with volcanic gases containing hydrogen sulfide, hydrochloric acid, and other volatiles. Moderately low pH values may be associated with small amounts of mineral acids from sulfide sources or with organic acids from decaying vegetation. In general, water from clay-rich sediments has a lower pH than water from limestone.

## 3.4 *Methods of Expressing Analysis*

Water analyses most commonly deal with samples in which the total amount of dissolved solids constitutes only a small fraction of one per cent of the total weight of the sample. The analyses are, therefore, expressed more conveniently in parts per million (ppm) instead of percentages. One part per million means one part by weight of dissolved matter in a million parts by weight of solution, such as one kilogram of dissolved solids in one million kilograms of water or one ton of dissolved solids in one million tons of water. The measure is, therefore, independent of the units used. A similar measure which is, however, dependent on units, is milligrams per liter. This is the most commonly used unit in the laboratory inasmuch as the water sample is measured in fractions of a liter and chemical constituents are determined in milligrams. Parts per million and milligrams per liter are numerically almost the same if the concentration of dissolved solids is low and the specific gravity of the water is nearly 1.0. Grains per gallon is yet another measure in common use in English speaking countries. The following equations show the relations between the various units of measure.

$$\text{parts per million} = \frac{\text{milligrams per liter}}{\text{specific gravity of the water}}$$

1 part per hundred (percentage) = 10,000 ppm

1 grain per U.S. gallon = 17.12 milligrams per liter

1 grain per Imperial gallon = 14.3 milligrams per liter

A further unit of measure is convenient for many geochemical studies, this is equivalents per million (epm), or more exactly milligram equivalents per kilogram. If derived from milligrams per liter, the unit is called milligram equivalents per liter. Equivalents per million are calculated by dividing parts per million by the equivalent weight of the ion under consideration. This unit is helpful in picturing the true chemical character of the water and is one one-thousandth of the numerical value of the common chemical unit, normality. Inasmuch as the total equivalent weights of cations and anions in a solution must be the same, the sum of the equivalents per million can be used to check the accuracy and completeness of chemical analyses. The following examples will illustrate the nature and use of equivalents per million.

*a.* To convert 63 ppm $Mg^{2+}$ to epm:

Atomic weight Mg = 24.32 $\qquad$ Equivalent weight $= \dfrac{24.32}{2} = 12.16$

Valence = 2 $\qquad$ 63 ppm $Mg^{2+} = \dfrac{63}{12.16} = 5.19$ epm *b*. To convert 2.5 ppm $PO_4^{3-}$ to epm:

Atomic weight P = 30.97
Atomic weight O = 16.0
Molecular weight $PO_4$ = 94.97
Valence = 3

$$\text{Equivalent weight} = \frac{94.97}{3} = 31.66$$

$$2.5 \text{ ppm } PO_4^{3-} = \frac{2.5}{31.66} = 0.079 \text{ epm}$$

*c*. To check the following analysis, tabulate the epm of anions and cations and then add each column.

| Ion | ppm | epm cations | epm anions |
|---|---|---|---|
| $Ca^{2+}$ | 42 | 2.10 | |
| $Mg^{2+}$ | 27 | 2.22 | |
| $HCO_3^-$ | 196 | | 3.21 |
| $SO_4^{2-}$ | 15 | | 0.31 |
| $Cl^-$ | 72 | | 2.03 |
| $NO_3^-$ | 5 | | 0.08 |
| Total | | 4.32 | 5.63 |

Inasmuch as the cation and anion totals are not equal, the analysis is either incomplete or in error. The cation deficiency probably reflects undetected $Na^+$ and $K^+$ ions in the sample.

Silica, suspended material, and certain organic compounds found in water are not ionized. For this reason they are not expressed in equivalents per million.

Many older water analyses are given in terms of hypothetical combinations such as $CaCO_3$, $MgCl_2$, and NaCl which are present in the anhydrous residue of the water. This method may have some merit if studies are of the relation between water and rock types; nevertheless certain questionable assumptions must always be made concerning the relative solubilities of the compounds. Modern analyses are rarely reported in this manner. An exception is water hardness and alkalinity (see Chapter 4) which are reported as the hardness or alkalinity which would be produced if a certain amount of $CaCO_3$ were dissolved in water. This method is retained to avoid ambiguity, because more than one ion contributes to alkalinity and water hardness. The following example will show how to convert compounds expressed as parts per million to ions expressed as parts per million:

For compound $A_nB_m$,

$$\text{ppm ion A} = (\text{ppm compound } A_nB_m)\,\frac{n\,(\text{atomic weight A})}{(\text{molecular weight } A_nB_m)}$$

and consequently,

$$\text{ppm ion B} = (\text{ppm compound } A_nB_m)\,\frac{m\,(\text{atomic weight A})}{(\text{molecular weight } A_nB_m)}$$

EXAMPLE

Given that a certain water has 32 ppm $CaCl_2$, calculate the ppm of ionic Cl in the water. Since the atomic weight of Cl is 35.5 and of Ca is 40.0, then: $\text{ppm } Cl^- = (32)\,\frac{2\,(35.5)}{111} = 20.5.$

Chemical analyses of a large number of samples yield an unwieldy mass of data. Consequently, various types of graphs and maps are useful in summarizing the salient facts drawn from the analyses. One of the best, and also easiest, methods is to plot spacial variations of water quality on maps. For example, maps showing lines of equal chloride content (or isochlors) are common (Figure 3.3). Caution should be used in map representations not to group chemical data from hydrologically unrelated sources. Thus, in Figure 3.3 a locally high chloride content could indicate contamination of the shallow aquifer, or it could be simply from a deep well which bears little or no relation to the aquifer being studied.

Bar graphs of various types are used to represent chemical analyses of individual samples [15]. The graphs are most commonly of either equivalents per million or percentage of total equivalents per million (Figure 3.4). As the sums of equivalents per million of anions and cations are equal, the two columns are always the same length. Circular diagrams and radial coordinates (Figures 3.5 and 3.6) are also common. Analyses shown in Figure 3.7 are plotted on horizontal lines. The resulting diagrams, commonly known as Stiff [30] diagrams, are very useful if a rapid qualitative comparison of many analyses is desired.

The foregoing methods are limited because of the space required to represent a single analysis. A more efficient method is to plot the position of analyses with respect to various coordinates. Figure 3.8 shows analyses in a two-coordinate field and Figure 3.9 shows analyses in a multi-coordinate field. The second type is widely used and is known as a trilinear diagram. The trilinear diagram illustrates the various percentages of anions and cations in the two triangular fields and a combined position of all major ions in the diamond-shaped field. Percentages of anions and

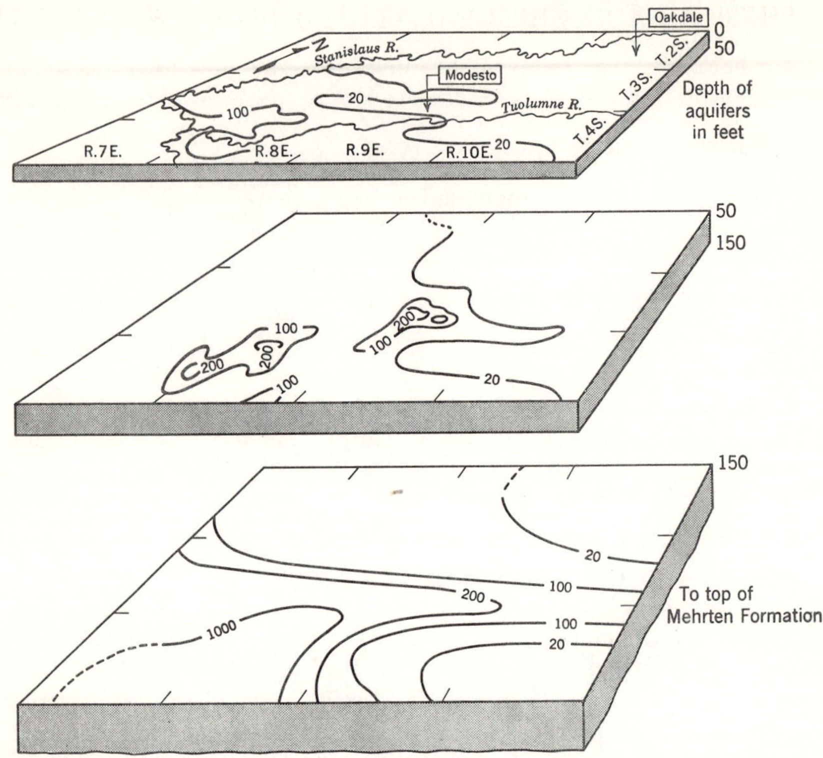


Figure 3.3 Change in water quality with depth in the northern San Joaquin Valley, California. Lines indicate equal chloride content in parts per million. (Data from Davis and Hall [5].)

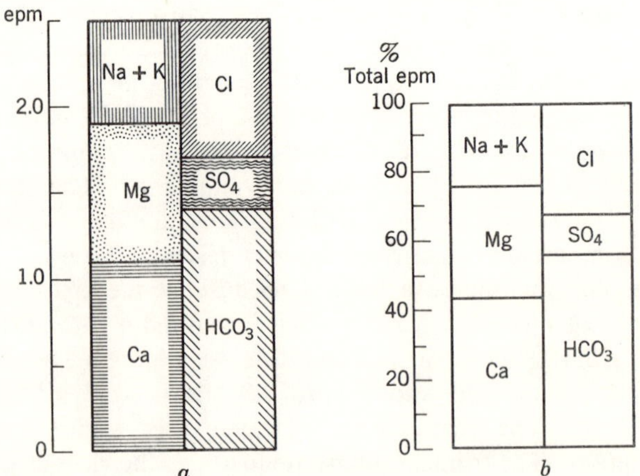


Figure 3.4 A chemical analysis of water represented by bar graphs of (*a*) total equivalents per million and of (*b*) percentage of total equivalents.

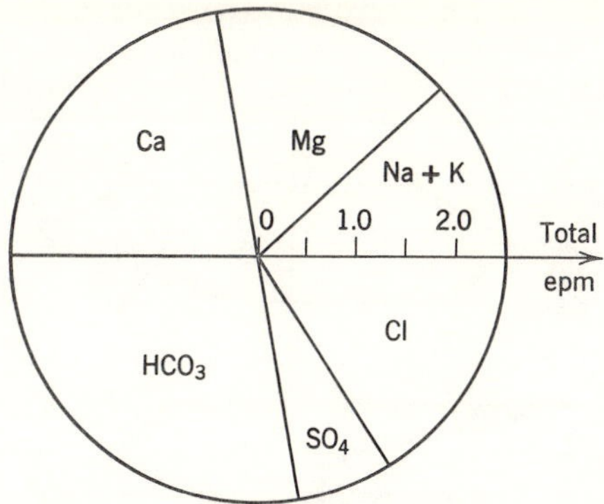


Figure 3.5 The same analysis as in Figure 3.4 shown on a circular diagram. Arcs of the circle are proportional to the percentage of each ion in equivalents and the radius is proportional to the total equivalents per million.

cations are based on total equivalents per million of the major ions. If only a relatively few analyses are plotted, it is also possible to represent the concentrations of the analyses with circles of various sizes.

Trilinear diagrams, along with several other types of diagrams, are a useful means of pointing up differences or similarities among waters. The trilinear diagrams also show the effects of mixing between waters, because mixtures of two different waters will plot on a straight line. In

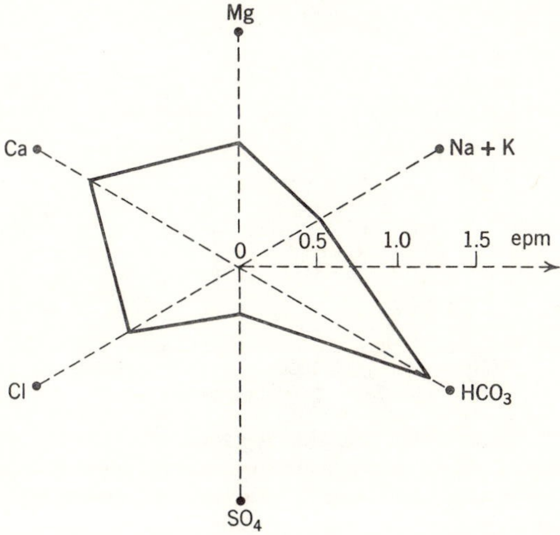


Figure 3.6 Radial coordinates used to show the same information as Figure 3.4*a*.

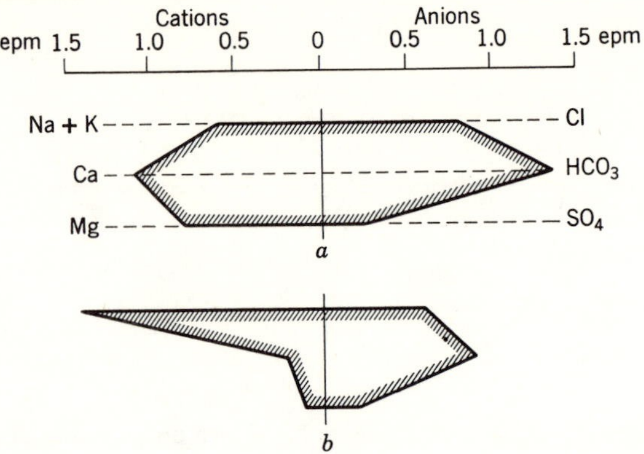


Figure 3.7 Diagram proposed by Stiff. Distinctive shapes facilitate rapid comparison of analyses. Diagram (*a*) represents same analysis as does Figure 3.4*a*.

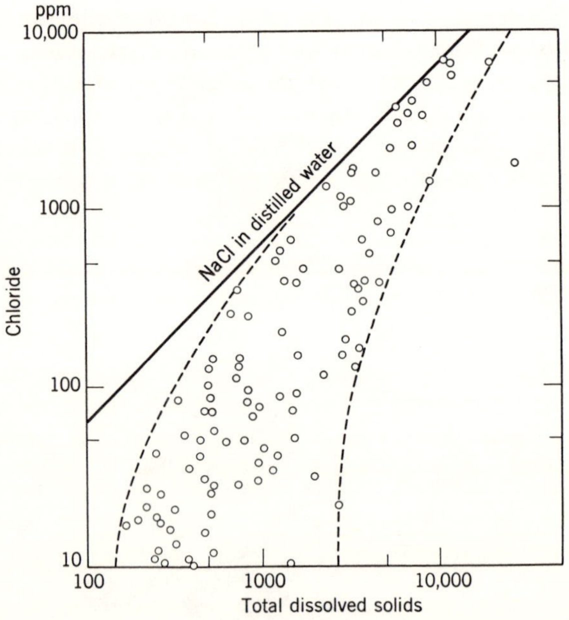


Figure 3.8 Two coordinate field showing total dissolved solids and chloride content of 100 analyses of ground water taken from various publications. Diagram illustrates the tendency of most ground water to approach the composition of NaCl brine as total dissolved solids increases.

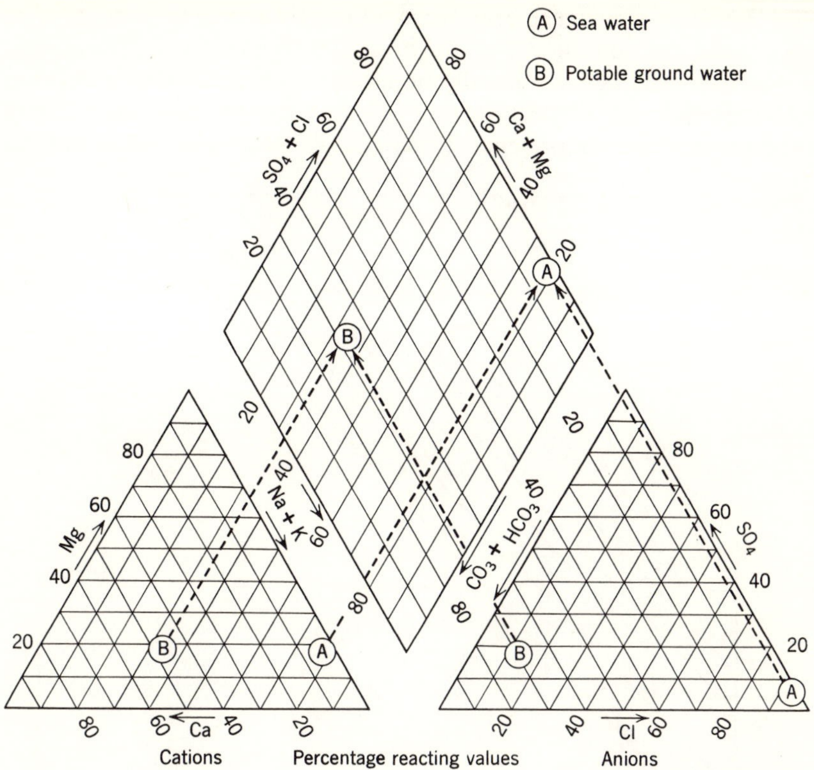


Figure 3.9 Trilinear diagram of the type proposed by Piper [27]. The positions of average potable ground water and sea water are indicated. Each analysis is represented by three points; two on the triangular fields and one on the combined diamond field.

addition, if two groups of data tend to converge along two straight lines to a common point in the field, a common source of some of the ions may be indicated. If the concentrations of various constituents are in proper proportions, mixing is also probable.

## 3.5 *Specific Electrical Conductance*

The ability of a cube one centimeter on a side to conduct an electrical current is called specific electrical conductance or electrical conductivity. Conductance is the reciprocal of resistance and is measured in mhos, which is, conveniently, the reverse spelling of ohm, the unit of resistance. Because the mho is usually too large a unit for fresh water, micromhos, or millionths of mhos, are used in most work with subsurface water.

The specific conductance of water is a function of temperature, type of ions present, and concentration of various ions (Figure 3.10). The specific conductance readings are usually adjusted to 25°C, so that variations in conductance are a function only of the concentration and type of dissolved constituents present. Inasmuch as measurements of specific

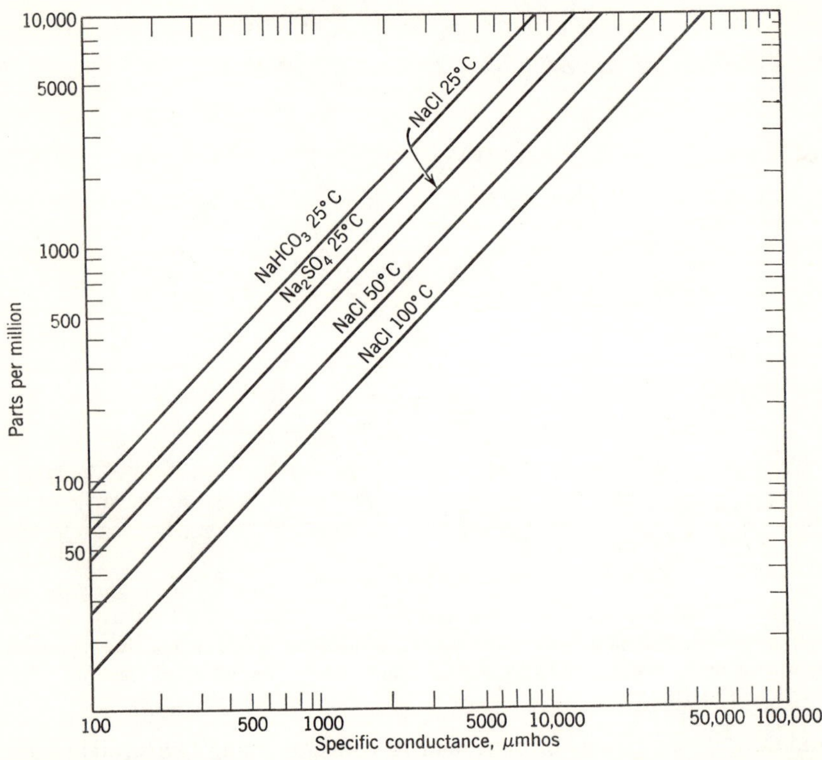


Figure 3.10 Specific conductivity of aqueous solutions of various compounds. The importance of temperature is illustrated by the three widely spaced curves for NaCl solutions.

conductance can be made very rapidly, they afford a quick method of estimating the chemical quality of water samples.

Of the common types of natural water, calcium bicarbonate and calcium sulfate water generally have the lowest conductance and sodium chloride water the highest conductance for a given total concentration of dissolved solids. For a very rough estimate of total dissolved solids in parts per million in fresh water, the specific conductance of the water in micromhos should be multiplied by 0.7. A more exact relation exists between equivalents per million and conductance in micromhos. For nearly pure water, the conductance divided by 100 gives the equivalents per million of the solution with an accuracy of about 5 per cent. For water with between 1.0 and 10.0 epm, the accuracy is about 15 per cent. The following empirical rules have been proposed [25] for relating epm and specific conductance, C:

*a.* When total epm, B, is less than 1.0, $C = 100B$ (2.2)

*b.* When B is between 1.0 and 3.0, $C = 12.27 + 86.38B + 0.835B^2$ (2.3)

*c.* When B is between 3.0 and 10.0, $C = B[95.5 - 5.54(\log B)]$ (2.4)

*d.* When B exceeds 10.0 and $HCO_3^-$ is the dominant anion, $C = 90.0B$ (2.5)

*e.* When B exceeds 10.0 and $Cl^-$ is the dominant anion, $C = 123B^{0.939}$ (2.6)

*f.* When B exceeds 10.0 and $SO_4^=$ is the dominant anion, $C = 101B^{0.949}$ (2.7)

Inasmuch as total epm taken from the anion sum commonly differs slightly from total epm from the cation sum, the value of B is taken as the mean of the two sums. The relations are valid only for values of B which are less than 1000 epm.

Pure water has a conductance of 0.055 micromhos at 25°C. Laboratory distilled water commonly has a conductance of 0.5 to 5.0 micromhos. Rainwater will usually range from about 5.0 to 30 micromhos, potable subsurface water ranges from 30 to 2000 micromhos, ocean water from 45,000 to 55,000 micromhos, and oil-field brines are commonly more than 100,000 micromhos.

## 3.6 *Mobility of the Elements in the Hydrosphere*

When igneous minerals first form they are adjusted to conditions of temperature and pressure not normally encountered at the surface of the ground. Therefore, when they are exposed to weathering their initial state of semiequilibrium with their environment is changed radically, and the physical and chemical behavior of the individual elements is considerably different from that within the igneous environment. The type of behavior of most interest to hydrogeology is the mobility of the elements. This is defined as the "tendency of an element to move in a given chemical environment" [14]. The movement can take place when the element is solid mineral matter, a gas, a melt, or in solution within a liquid. The average composition of igneous rocks, which are assumed to be the primary components of the earth's crust, is given in Table 3.2. A comparison of this composition with the composition of ground water, ocean water, and sedimentary carbonates gives a general indication of the mobility of the common elements in the hydrosphere.

Several striking contrasts can be seen in Table 3.2. Although aluminum and iron are the third and fourth most abundant elements in igneous rocks (oxygen, the most abundant element, is not shown in the table), their mobility in the hydrosphere is very low. Silica, the second most abundant element, is only moderately mobile; whereas chloride, a relatively scarce element in the crust, is very mobile and abundant in the hydrosphere. Calcium is moderately abundant and mobile in all environments. Sodium and potassium are also both mobile and abundant in all environments; however sodium is considerably more mobile than potassium even though both occur in roughly equal amounts in igneous rocks.

The mobility of an element in the hydrosphere depends on the solubility of its various compounds, the tendency of its ions to participate in ion exchange, and the extent to which organisms extract the element from the hydrosphere.

*Table 3.2 Composition of Two Rock Groups and Two Types of Water (parts per million)*

| Element | Igneous Rocks | Sedimentary Carbonates | Sea Water | Potable Ground Water |
|---|---|---|---|---|
| Si | 277,000 | 24,000 | 1 | 8 |
| Al | 81,000 | 4200 | 0.01 | 0.04 |
| Fe | 50,000 | 3800 | 0.01 | 0.07 |
| Ca | 36,300 | 302,000 | 400 | 45 |
| Na | 26,000 | 400 | 10,500 | 35 |
| K | 26,000 | 2700 | 380 | 2.5 |
| Mg | 21,000 | 47,000 | 1350 | 11 |
| Ti | 4400 | 400 | 0.001 | 0.001 |
| P | 1200 | 400 | 0.07 | 0.03 |
| Mn | 1000 | 1100 | 0.002 | 0.02 |
| F | 600 | 330 | 1.3 | 0.2 |
| S | 400 | 1200 | 885 | 14 |
| C | 320 | 115,000 | 28 | 40 |
| Cl | 200 | 150 | 19,000 | 16 |

* Data from various sources [12,14,15,28,31]

### SOLUBILITY

The solubility of a given element in water is controlled by variations of temperature, pressure, hydrogen and hydroxyl ion concentrations (pH), redox potential (Eh), and the relative concentrations of other substances in solution. In a natural environment these variables are related in such a complex manner that exact solubilities cannot be predicted. Chemical laws, however, which have been formulated from laboratory work also must apply to natural systems, regardless of their complexity. By using these well-established laws certain definite limits of natural solubilities can be established. Although these limits in some cases are broad, they help in the understanding of the concentrations of certain elements in the hydrosphere.

The solubility product is a concept of physical chemistry which can be useful in estimating equilibrium concentrations of certain ions. It is based on the principle that for a saturated solution of a poorly soluble compound, the product of the molar concentrations of the ions is a constant at any fixed temperature. If the surface area of the solid compound $A_nB_m$ is assumed to be constant, the solubility product $K_{sp}$ is

$$[A]^n[B]^m = K_{sp} \tag{2.8}$$

in which [A] and [B] are concentrations in moles per liter of ions A and B. As an example, find the amount of fluoride, $F^-$, which might be in equilibrium with 20 ppm $Ca^{2+}$ and the mineral fluorite, $CaF_2$.

### SOLUTION

$K_{sp}$ for 20°C estimated from data in chemical tables [see reference 20] = $3.5 \times 10^{-11}$

$$20 \text{ ppm} = 5 \times 10^{-4} \text{ mole/liter } Ca^{2+}$$
$$[Ca^{2+}][F^-]^2 = [5 \times 10^{-4}][F^-]^2 = 3.5 \times 10^{-11}$$
$$\text{Concentration of } F^- = 2.64 \times 10^{-4} \text{ mole/liter} \approx 5 \text{ ppm}$$

Thus, if these conditions are assumed to exist in ground water, a maximum of 5 ppm $F^-$ would be expected in solution. A greater concentration of $F^-$ would cause precipitation of $CaF_2$. The foregoing use of solubility product assumes that the activity coefficients are unity. Precise calculations should take into consideration the chemical activities of $Ca^{2+}$ and $F^-$ as influenced by all the constituents in the water. This is particularly important when concentrations exceed a few ppm. The interested reader, therefore, should consult appropriate references [7,10,16] before undertaking an analysis of chemical equilibria in natural water.

Temperature is important in controlling the solubility of gases; the lower the temperature the more gas can be held in solution. Of the gases, carbon dioxide is the most important. An increase in carbon dioxide will shift the carbon-dioxide-carbonate-bicarbonate system so that more material such as calcite can go into solution. The direct effect of temperature on the solubility of minerals is not as great as in the case of gases. Under normal ground-water conditions, slight temperature differences within a given region probably have only a slight effect on mineral solubilities.

The surface area exposed to water is an important factor in determining the rate of dissolution of minerals, inasmuch as dissolution is a surface phenomenon. This factor is particularly important in nonsaturated zones. through which infiltrating water passes more rapidly than in zones that are continuously saturated. Thus, if gravel is at the surface, water will percolate downward with little chemical modification; however, if silt of the same mineral composition is at the ground surface, the total surface area of the rock flower may be a thousand times as great as the gravel and the percolating water will undergo considerable modification.

The pressure normally encountered in ground water will have little direct effect on mineral solubilities. An increase in the partial pressure of gases in contact with water, however, will increase the solubility of gases which in turn will increase or decrease the mobility of some constituents. The presence of oxygen will decrease the mobility of iron; in contrast, carbon dioxide will increase the mobility of many constituents of which calcium and magnesium are the most important. It is possible that some deposits of minerals are localized by changes of pressure in actively circulating ground water.

The pH of water has a profound effect on the mobility of many of the elements. Only a few ions such as sodium, potassium, nitrate, and chloride remain in solution through the entire range of pH values found in normal ground water. Most metallic elements are soluble as cations in acid ground water but will precipitate as hydroxides or basic salts with an increase of pH. All but traces of ferric ions will be absent above a pH of 3. Abundant aluminum ions are absent above a pH of 5. Ferrous ions diminish rapidly as the pH increases above 6.0, and magnesium ions are mostly precipitated at a pH of more than 10.5.

The solubility of certain elements will depend on their oxidation state which is determined by the redox potential or oxidation-reduction potential, Eh, of the environment. The Eh is a measure of the energy needed to remove electrons from ions in a given chemical environment. In natural water the presence or absence of uncombined oxygen is one of the primary causes for variations of Eh. Two other oxidizing agents, chlorine and fluorine,* probably do not occur in significant amounts in normal ground water. The actual Eh at which oxidation or reduction takes place is also a function of pH. In general, if the solution is more alkaline, oxidation of Fe and Mn hydroxides takes place at a lower Eh. Thus ferrous iron can be stable in ground water at low pH values but will oxidize to ferric iron at high pH values and be precipitated as ferric hydroxide (Figure 3.11). The mobility of manganese, copper, vanadium, and uranium are also affected by the Eh of ground water, although other processes such as absorption may be more effective in controlling their distribution.

Stability fields shown in Figure 3.11 are for iron and pure water, but can be used to draw general conclusions concerning the occurrence of iron in fresh water. The diagram suggests that ferrous ions are the most common form of dissolved iron and that ferric hydroxide will precipitate in an oxidizing environment. The presence of significant amounts of sulfur and carbonate will shift the equilibrium fields of Figure 3.11 and will necessitate additional stability fields for the solids $FeCO_3$ and $FeS_2$. Hem [17,18] has calculated the effects of 100 ppm $HCO_3^-$ and 10 ppm $SO_4^{2-}$. A portion of his diagram is shown in Figure 3.12. The stability field for $FeCO_3$ is above a pH of 9.0 and is centered on an Eh value of $-0.3$ volt. Similar diagrams can be constructed for other concentrations of $HCO_3^-$ and $SO_4^{2-}$ by methods outlined by Garrels [10].

The effect of other material in a solution on the solubility of a given mineral is difficult to calculate for complex systems found in nature. An increase in the concentration of other ions may increase the solubility, but this is probably not too important in the dilute concentrations found in potable water. If, however, concentrations are near saturation, then the effects of other ions are large. Also, if one substance is being precipitated, some foreign ions are commonly trapped in or substituted within the structure of the newly formed substance. This process has been called coprecipitation. It is particularly effective in removing trace amounts of radium when barite is precipitated, copper when limonite is formed, and a large number of other trace elements when manganese oxides are formed [14].

### ION EXCHANGE

All minerals, even the stable silicates such as quartz, have surfaces with small unbalanced electrical charges. These surfaces attract water because it is a polar compound and also attract ions from water. A few minerals

* The gases chlorine and fluorine should not be confused with their ionic counterparts, fluoride and chloride, which are common in water.

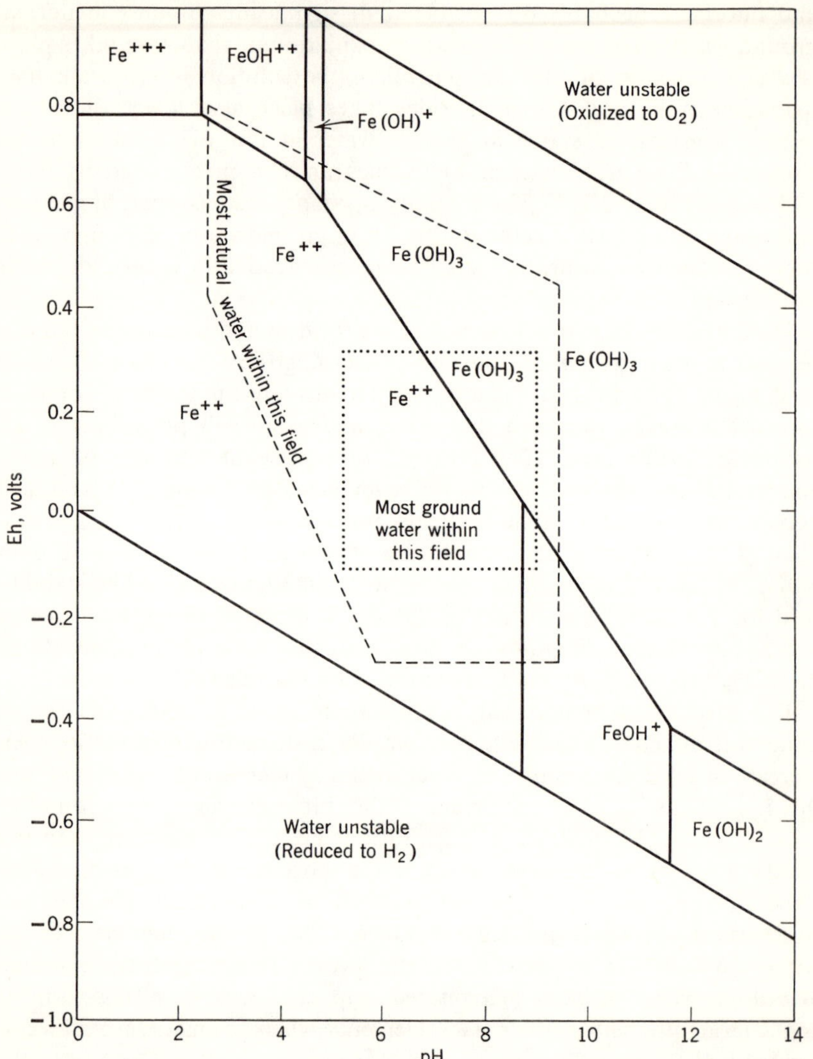


Figure 3.11 Stability fields for the aqueous ferric-ferrous system. (Hem and others [18].) Limits for natural water and ground water are from data by Baas Becking and others [1].

will have structures open enough to allow ions to exchange from the interior of the crystal. The ions held by the mineral surface can be displaced by other ions due to differences in the size or attraction of the ions themselves or differences in relative concentrations of ions [2,13,15]. Thus, most commonly, divalent ions will displace monovalent ions; however,

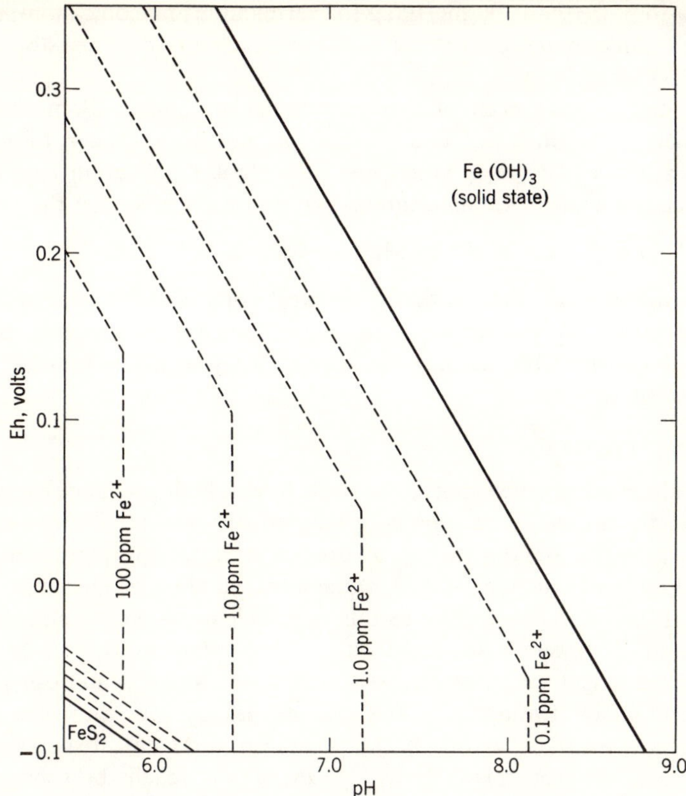


Figure 3.12 A portion of the stability fields for iron in water which has 100 ppm $HCO_3^-$ and 10 ppm $SO_4^{2-}$. (From a diagram by Hem [17].)

if monovalent ions are in greatest abundance they will tend to displace the divalent ions. This action of exchange is therefore reversible. As an example, calcium will normally be held more tightly than sodium, but if a solution with a high concentration of sodium ions, such as sea water, is passed through a substance containing adsorbed calcium, the sodium will tend to replace the calcium. The highest exchange capacities of common minerals are possessed by zeolites, vermiculite, and montmorillonite. Organic material such as humus and roots of living plants also has a high-exchange capacity.

Cation, or positive ion, exchange capacity is the amount of exchangeable cations measured in milliequivalents per gram or per 100 grams of solid material at a pH of 7.0. Finely ground igneous rock will have exchange capacities of 0.5 to 5 me/100 g. Clays will range from 3 to 15 me/100 g for kaolinite to more than 100 me/100 g for vermiculite and some montmorillonite. Organic material from soil will have exchange capacities between 100 and 500 me/100 g.

Most studies have been of cation-exchange processes. Some anions also undergo exchange but this process has not been studied fully. In general, anion exchange is much less important than cation exchange.

The relative ease of exchangeability of cations is given by the series:

$$Li^+ > Na^+ > K^+ > Mg^{2+} > Ca^{2+} > Sr^{2+} > Ba^{2+}$$

Lithium will thus be held with the weakest force and barium with the strongest force. This order is by no means fixed for all minerals and is subject to considerable natural variations. At low pH values $H^+$ can replace other cations.

### BIOLOGICAL FACTORS

It has already been mentioned that roots have a high-exchange capacity. This capacity is used by the plants to help adsorb mineral matter needed for plant growth. Besides carbon, hydrogen, and oxygen, plants contain large amounts of calcium, silicon, potassium, sulphur, magnesium, and sodium. Plants also concentrate certain minor elements such as strontium, molybdenum, copper, boron, and zinc. Thus, plant activity intercepts many of the dissolved constituents in soil water which would otherwise be leached by infiltrating rainwater and eventually join the other ions common in subsurface water. If the plants grow on flat areas, mineral matter mobilized from plant decay will eventually reach the subsurface water. In this case, the presence of plants increases the ions in ground water, because the activity of organic acids, carbon dioxide, and plant roots will accelerate chemical weathering in the soil horizon. If plants grow on steep slopes, the organic material may be removed rather rapidly by surface erosion and will not have an opportunity to contribute mineral matter to subsurface water.

Microorganisms are probably even more important than larger plants in altering the chemical character of water. The effectiveness of mirco-organisms to abstract elements such as copper, manganese, iron, calcium, and silica from lake and sea water is well known [18,22]. The low concentration of silica and the relatively low concentration of calcium in sea water is related to the extensive use of these elements in the hard parts of marine organisms. The use of iron and manganese by certain fresh-water bacteria is also extensively documented. Some of the bacteria even derive energy for life processes from the oxidation of iron and manganese.

One of the most interesting life processes is that of sulfate-reducing bacteria [11,22,24,33,34]. These bacteria utilize the oxygen from sulfate ions to oxidize hydrocarbons; in this process hydrogen sulfide is given off as a byproduct. Some types of sulfate-reducing bacteria are found in soil horizons. Other types are thought to thrive at moderately high temperatures under anaerobic conditions far below the surface of the ground. Live bacteria have been recovered from depths of more than 4000 feet in oil wells. Although bacteria can be introduced by drilling operations, the consensus is that the bacteria are actually living at these depths and have been possibly far removed from the surface environment for millions of years.

Recent Russian hydrogeochemical research has emphasized the importance of microorganisms, mainly bacteria, in the modification of the chemical characteristics of ground water [24]. The depth at which bacterial action can thrive is thought to be controlled by the temperature and permeability of the media. Temperatures above 100°C will probably kill most organisms, and impermeable rocks will not allow circulating fluids to supply nutrients to the organisms. The pH, Eh, and types of nutrients available will control the kind of biochemical activity which is possible at a given place in the subsurface. In addition to the important modifications in the soil horizon and during early diagenesis of sediments, many Russian workers believe that significant changes in water chemistry also take place at great depth through the biochemical reduction of $NO_3^-$ and $SO_4^{2-}$, or through the biochemical oxidation of propane, methane, and similar hydrocarbons. Some aquifers, which in the geologic past have been impermeable or have been subjected to temperatures above 100°C, are thought to have been recolonized by bacteria transported in ground water. Nutrients, chiefly from the decomposition of surface plants, are fed continuously by infiltrating surface water to the subsurface microorganisms. This concept certainly must be true of many limestone aquifers where extensive cavernous zones exist. Whether or not the microorganisms can be repopulated in almost all aquifers, as conceived of by some Russian hydrogeologists, remains to be demonstrated.

## REFERENCES


1. Baas Becking, L. G. M., I. R. Kaplan, and D. Moore, 1960, Limits of the natural environment in terms of pH and oxidation-reduction potentials: *Jour. Geol.*, v. 68, pp. 243–284.
2. Carroll, Dorothy, 1959, Ion exchange in clays and other minerals; *Geol. Soc. Am. Bull.*, v. 70, pp. 749–780.
3. Craig, Harmon, 1961, Isotopic variations in meteoric waters: *Sci.*, v. 133, pp. 1702–1703.
4. Davis, K. S., and J. A. Day, 1961, Water the mirror of science: Garden City, New York, Doubleday and Company, Inc., 195 pp.

5. Davis, S. N., and F. R. Hall, 1959, Water quality of eastern Stanislaus and northern Merced counties, California: *Stanford University Publications in Geol. Sciences*, v. 6, No. 1, 112 pp.
6. Dorsey, N. E., 1940, Properties of ordinary water-substance in all its phases; water-vapor, water, and all the ices: *Am. Chem. Soc. Monograph No.* 81: New York, Reinhold Publishing Corp., 673 pp.
7. Dutt, G. R. and K. K. Tanji, 1962, Predicting concentrations of solutes in water percolated through a column of soil: *Jour. Geophys. Research*, v. 67, pp. 3437–3439.
8. Epstein, S., and T. Mayeda, 1953, Variation of $O^{18}$ content of waters from natural sources: *Geochim. et Cosmochim. Acta*, vol. 4, pp. 213–224.
9. Friedman, Irving, 1953, Deuterium content of natural waters and other substances: *Geochim. et Cosmochim. Acta*, v. 1, pp. 33–48.
10. Garrels, R. M., 1960, Mineral equilibria at low temperatures and pressures: New York, Harpers and Brothers, 254 pp.
11. Ginter, R. L., 1934, Sulphate reduction in deep subsurface waters in W. E. Wrather and F. H. Lahee (editors), *Problems of petroleum geology:* Tulsa, Oklahoma, Am. Assoc. Petroleum Geologists, pp. 907–925.
12. Goldberg, E. D., 1963, The oceans as a chemical system in M. N. Hill (editor), *The sea*, v. 2, Composition of sea water: London, Interscience Publishers, pp. 3–25.
13. Grim, R. E., 1953, *Clay mineralogy:* New York, McGraw-Hill Book Co., 384 pp.
14. Hawkes, H. E., 1957, Principles of geochemical prospecting: *U.S. Geol. Survey Bull.* 1000 F, pp. 225–355.
15. Hem, J. D., 1959, Study and interpretation of the chemical characteristics of natural water: *U.S. Geol. Survey Water-Supply Paper* 1475, 269 pp.
16. —— 1961, Calculation and use of ion activity: *U.S. Geol. Survey Water-Supply Paper* 1535-C, 17 pp.
17. —— 1963, Some aspects of chemical equilibrium in ground water: *Ground Water*, v. 1, pp. 30–34.
18. —— and others, 1962, Chemistry of iron in natural water: *U.S. Geol. Survey Water-Supply Paper* 1459, 269 pp.
19. Hendricks, S. B., 1955, Necessary, convenient, commonplace in *Water*, *Yearbook of Agriculture:* U.S. Department of Agriculture, pp. 9–14.
20. Hodgman, C. D. (editor), 1963, *Handbook of chemistry and physics*, 44th ed. Cleveland, Ohio, Chemical Rubber Publishing Co., 3604 pp.
21. Horibe, Yoshio and Mituko Kobayakawa, 1960, Deuterium abundance of natural waters; *Geochim. et Cosmochim. Acta*, v. 20, pp. 273–283.
22. Hutchinson, G. E., 1957, A treatise on limnology, v. 1, *Geography, physics, and chemistry:* New York, John Wiley and Sons, 1010 pp.
23. Kavanau, J. L., 1964, *Water and solute-water interactions:* San Francisco, Holden-Day, Inc., 101 pp.
24. Kuznetsov, S. I. (editor), 1961, Geologic activity of microorganisms (Consultants Bureau Enterprises, Inc., English translation): Moscow, U.S.S.R. *Academy of Sci. Press, Trans. Inst. of Microbiology No.* 9, 112 pp.
25. Logan, John, 1961, Estimation of electrical conductivity from chemical analysis of natural waters: *Jour. Geophys. Research*, v. 66, pp. 2479–2483.
26. Pauling, Linus C., 1940, The nature of the chemical bond, and the structure of molecules and crystals: Ithaca, New York, Cornell University Press, 450 pp.
27. Piper, A. M., 1944, A graphic procedure in the geochemical interpretation of water analyses: *Am. Geophys. Union Trans.*, v. 25, pp. 914–923.

28. Rankama, Kalervo, and T. G. Sahama, 1950, *Geochemistry:* Chicago, Illinois, Chicago University Press, 912 pp.
29. Roberson, C. E., and others, 1963, Differences between field and laboratory determinations of pH, alkalinity, and specific conductivity of natural water: *U.S. Geol. Survey Prof. Paper* 475-C, pp. 212–215.
30. Stiff, H. A., Jr., 1951, The interpretation of chemical water analysis by means of patterns: *Jour. of Petrol. Tech.*, v. 3, No. 10, *Technical Note* 84, pp. 15–16.
31. Turekian, K. K. and K. H. Wedepohl, 1961, Distribution of the elements in some major units of the earth's crust: *Geol. Soc. Am. Bull.*, v. 72, pp. 175–192.
32. Wells, A. F., 1945, Structural inorganic chemistry: Oxford, University Press, 590 pp.
33. Zobell, C. E., 1958, Ecology of sulfate-reducing bacteria: *Producers Monthly*, v. 22, No. 7, pp. 12–29.
34. Zobell, C. E. and D. Q. Anderson, 1936, Vertical distribution of bacteria in marine sediments: *Am. Assoc. Petroleum Geologists Bull.*, v. 20, pp. 258–269.

# chapter 4

# WATER QUALITY

## 4.1 *Introduction*

The chemical and biological characteristics of water determine its usefulness for industry, agriculture, or the home. The study of water chemistry gives important indications of the geologic history of the enclosing rocks, the velocity and direction of water movement, and the presence of hidden ore deposits. Many large resorts have been located at springs having water of unusual chemical composition. Waters which are very high in dissolved solids may yield lithium, potassium, sodium chloride, and other chemicals in commercial quantities.

Despite its scientific and economic importance, water quality is poorly understood by many people who are otherwise well informed. Too often the natural variations in water chemistry are assumed to be entirely random or even enigmatical. It is true that in detail the chemistry of ground water can be exceedingly complex. Many of the gross characteristics of water chemistry, however, have been understood for eighty years or more. Recent advances in general geochemistry as well as in analytical techniques, moreover, have stimulated research in hydrogeochemistry so that it is one of the most rapidly expanding fields in the earth sciences.

The purpose of this chapter is to discuss the general occurrence of the various constituents in ground water and to mention briefly the relation of these constituents to water use. Particular attention is given to the general geologic controls and to the natural variations of the constituents. A discussion of the radionuclides in ground water is reserved for Chapter 5.

The number of major dissolved constituents in ground water is quite limited, and the natural variations are not as great as might be expected from a study of the complex mineral and organic material through which the water has passed. The most important constituents together with an indication of their natural variations in potable water are shown in Figures 4.1 and 4.2. Examples of waters of various compositions are given in Table 4.1.

*Table 4.1 Waters of Various Chemical Types*

| | 1<br>Snow | 2<br>Well | 3<br>Well | 4<br>Well | 5<br>Well | 6<br>Well | 7<br>River | 8<br>Acid Hot Spring | 9<br>Basic Spring |
|---|---|---|---|---|---|---|---|---|---|
| $SiO_2$ | 0.2 | 135 | 42 | 36 | 7.2 | 14 | 14 | 216 | 3970 |
| Al | 0.06 | ... | ... | 0.1 | ... | 0.0 | 0.02 | 56 | 1.8 |
| Fe | 0.02 | 0.06 | 0.03 | 0.0 | 0.33 | 9.0 | 0.01 | 33 | 0.00 |
| Mn | ... | ... | 0.05 | 0.0 | ... | 0.0 | ... | 3.3 | 0.00 |
| Ca | 0.9 | 5.2 | 0.4 | 173 | 288 | 696 | 9.6 | 185 | 7.3 |
| Mg | 0.2 | 1.0 | 0.5 | 32 | 41 | 204 | 13 | 52 | 2.6 |
| Na | 0.42 | 28 | 51 | 40 | 37 | 462 | 3.2 | 6.7 | 10,900 |
| K | 0.1 | 3.6 | 3.0 | 2.6 | ... | 9.1 | 0.5 | 24 | 135 |
| Li | ... | ... | ... | ... | ... | ... | ... | ... | 3.2 |
| $NH_4$ | ... | ... | ... | ... | ... | ... | ... | ... | 148 |
| $HCO_3$ | 4.0 | 50 | 120.0 | 218 | 267 | 72 | 89 | 0.0 | 0.0 |
| $CO_3$ | 0.0 | 0.0 | 0.0 | 0.0 | 0.0 | 0.0 | 0.0 | 0.0 | 5560 |
| OH | 0.0 | 0.0 | 0.0 | 0.0 | 0.0 | 0.0 | 0.0 | 0.0 | 1430 |
| $SO_4$ | 0.7 | 7.8 | 10 | 268 | 190 | 2130 | 5.0 | 1570 | 267 |
| Cl | 0.6 | 3.5 | 7.2 | 135 | 94 | 850 | 3.5 | 3.5 | 7180 |
| F | 0.0 | 12 | 0.0 | 0.1 | 0.3 | 1.5 | 0.1 | 1.1 | 3.0 |
| Br | ... | ... | ... | ... | ... | ... | ... | ... | 11 |
| I | ... | ... | ... | ... | ... | ... | ... | ... | 5.7 |
| $NO_3$ | 0.2 | 0.2 | 2.4 | 7.5 | 518 | 0.0 | 0.5 | 0.0 | 0.0 |
| $PO_4$ | ... | ... | ... | 0.0 | ... | 0.0 | 0.0 | ... | 4.3 |
| B | ... | 0.3 | ... | ... | ... | ... | ... | ... | 242 |
| Dissolved solids | 5.4 | 309 | 177 | 820 | 1306 | 4400 | 94 | ... | 31,200 |
| Conductance $\mu$mhos | 10 | 164 | 214 | ... | ... | 5610 | 141 | 4570 | 36,800 |
| pH | 6.0 | 7.6 | 7.9 | 7.4 | ... | 7.3 | 7.3 | 1.9 | 11.6 |
| Temperature, °F | ... | 107 | 78 | ... | ... | 73 | 63 | ... | 54 |
| $H_2S$ | ... | ... | ... | ... | ... | ... | ... | ... | 1000 |

1. Snow from near Mestersvig, Greenland (From *Air Force Survey in Geophysics* 127, 1961.)
2. Well water, Yellowstone National Park (From *U.S. Geol. Survey Water Supply Paper* 1475-F, 1962.)
3. Well water, Gulfport, Mississippi (From *U.S. Geol. Survey, Water Supply Paper* 1299, 1954.)
4. Well water, Santiago, Chile (From *Inst. Investigaciones Geol. Boletin* 1, 1958.)
5. Well water, Chase County, Kansas (From *State Geol. Survey Kansas*, **11**, 1951.)
6. Well water, near Carlsbad, New Mexico (From *U.S. Geol. Survey, T.E.I.*, 1962.)
7. Smith River, northern California (From *California Dept. Water Resources Bull.* 65-59, 1961.)
8. Hot spring, Sulfur Springs, New Mexico (From *U.S. Geol., Survey Water Supply Paper* 1473, 1959.)
9. Mineral Spring, near Mt. Shasta, California (From *Geochim. et Cosmochim. Acta*, **22**, 1961.)

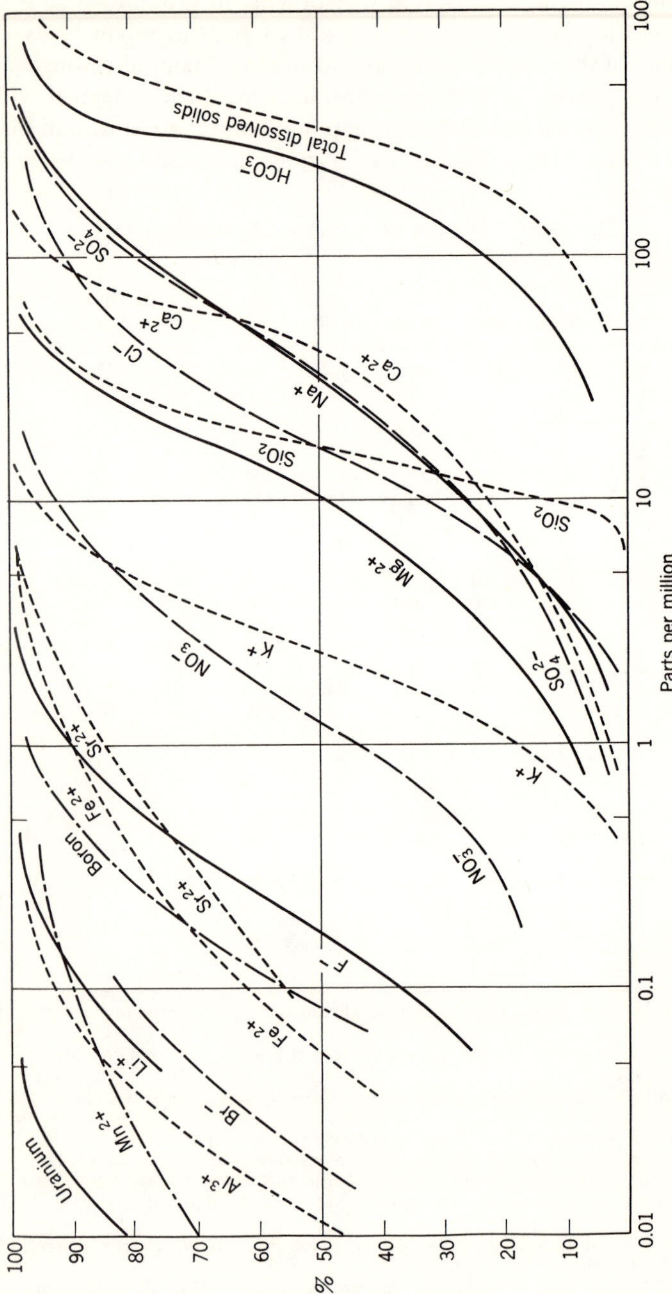


Figure 4.1 Cumulative curves showing the frequency distribution of various constituents in potable water. Data are mostly from the United States from various sources [5,7,13,28,29,39,43,49].

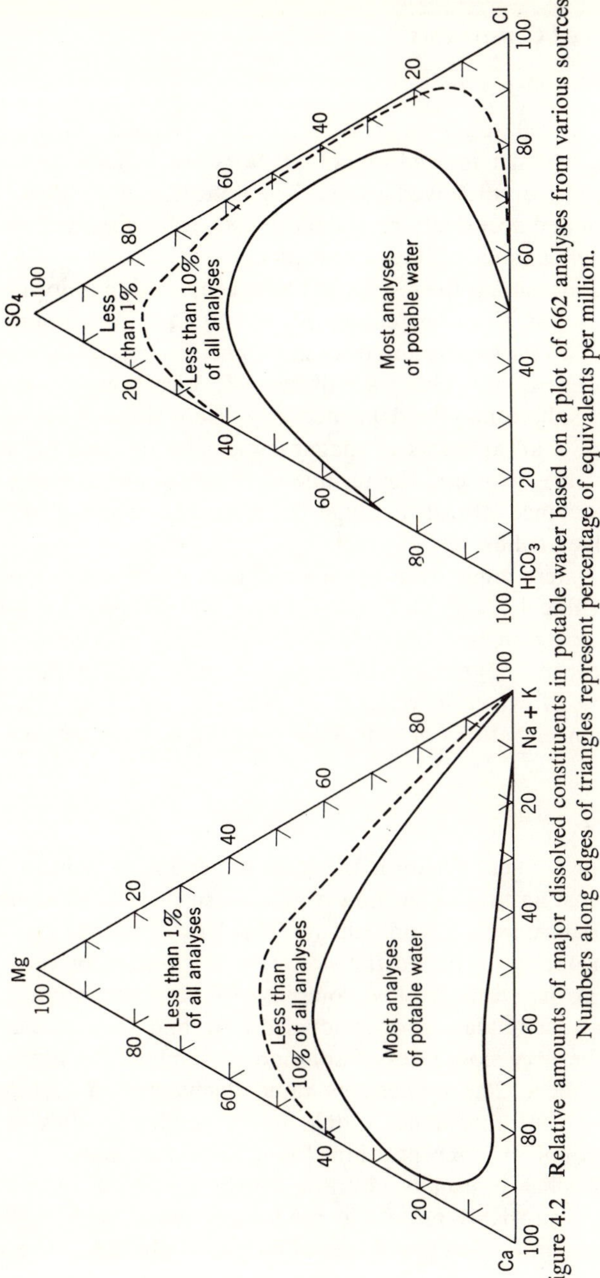


Figure 4.2 Relative amounts of major dissolved constituents in potable water based on a plot of 662 analyses from various sources. Numbers along edges of triangles represent percentage of equivalents per million.

## 4.2 *Dissolved Constituents*

### TOTAL DISSOLVED SOLIDS

Total dissolved solids in a water sample includes all solid material in solution, whether ionized or not. It does not include suspended sediment, colloids, or dissolved gases. Theoretically, if all dissolved solids were determined accurately by chemical tests, total dissolved solids would be the numerical sum of these constituents.

A related measure is the residue left after evaporation followed by drying in an oven at 180°C, or less commonly at 110°C, for one hour. The solid material left after evaporation does not coincide completely with material originally in solution. Gases are driven off; bicarbonate is converted to carbonate; sulfate may be deposited as gypsum which traps some of the water; and small amounts of magnesium, chloride, and nitrate may be volatilized. Nevertheless, the residue after evaporation, along with electrical conductance, affords a rough check on the accuracy of the sum of total dissolved solids.

Natural waters range from less than 10 ppm of dissolved solids for rain and snow Table 4.1, sample 1, to more than 300,000 ppm for some brines. Water for most domestic and industrial uses should be less than 1000 ppm, and water for most agricultural uses should be below 3000 ppm. The final classification of water in relation to potential use, however, should be based on concentrations of individual ions rather than on total dissolved solids.

### SILICA

Except for oxygen, silicon is the most abundant element in the earth's crust. In most natural water, however, it is usually only the fourth or fifth most abundant dissolved constituent. This lack of mobility in the hydrosphere is probably owing to the slow rate of solution of certain natural silicates and the relatively low solubility of silica compounds in water. The natural accumulation of residual quartz and various clay minerals attests to the very slow rates of solution of some of the more abundant sources of silica. The moderately rapid disintegration of other silicate minerals, on the other hand, should release sufficient silica in a soluble form to account for concentrations found in nautral waters.

Although silica is usually given as silicon dioxide in reports of water analyses, it is probably present as monomeric silicic acid, $H_4SiO_4$, in the normal temperature and pH ranges of natural water [25]. The silicic acid is essentially nonionized; silicate ions are thought to be present in appreciable amounts only above a pH of 9.

The solubility of amorphous silica ranges from about 50 ppm at 0°C to nearly 400 ppm at 100°C [25]. Subsurface water saturated with amorphous silica at temperatures normally encountered in temperate regions should contain from 90 to 110 ppm of silica. These high values are rarely found except in water from volcanic rocks and sediments derived from volcanic (Table 4.1 sample 2) or granitic rocks. Most commonly, ground water contains between 5 and 40 ppm silica. These lower concentrations probably reflect the low solubility of clay minerals, feldspars, quartz, and other common silicate minerals. The solubility of quartz is particularly low and would account for only 5 to 15 ppm $SiO_2$ in ground water [27,31]. Other silicate minerals have solubilities intermediate between quartz and amorphous silica. The pH of ground water will normally have little effect on the amount of silica which will go into solution [25]. Above a pH of 9.0, however, the solubility of silicates increases rapidly. The unusually high silica content of 3970 ppm shown in Table 4.1, Sample 9, is related to the exceptionally high pH of 11.6.

### IRON

Abundant sources of iron exist in the earth's crust. Some of the important minerals and mineral groups which may contain large amounts of iron are pyroxenes, amphiboles, magnetite, pyrite, biotite, and garnets. The weathering of these minerals releases large quantities of iron which usually are converted to the relatively insoluble and stable iron oxides. A small amount of iron is transported by surface and subsurface water in solution, but most transportation is probably by erosion and movement of solid particles including colloids and organic matter in surface water.

Most iron in solution is thought to be ionized; however, organic complexes may be important in surface water. If the pH of the water is below 3.0, the iron should occur in the ferric state ($Fe^{3+}$). Above this pH, ferric iron may be present as a complex ion. If the Eh is not too high, water contains ferrous iron [21]. The concentration of ferrous ions in ground water is probably limited by the solubility of ferrous carbonate, but it still ranges between 1 and 10 ppm if the pH is between 6.0 and 8.0, and if the bicarbonate concentration is relatively low.

When ground water containing ferrous ions comes in contact with the atmosphere, the following reaction can occur [19].

$$2Fe^{2+} + 4HCO_3^- + H_2O + \tfrac{1}{2}O_2 \rightleftharpoons 2Fe(OH)_3 + 4CO_2$$

Inasmuch as some contact with air almost always takes place during sampling and the amount of oxygen needed is small, this reaction will occur in most samples taken of natural water. Although the reaction tends to lower the pH somewhat, the solubility of ferric hydroxide is so low in the normal pH range that most of the iron will be precipitated.

### CALCIUM

Subsurface waters in contact with sedimentary rocks of marine origin derive most of their calcium from the solution of calcite, aragonite, dolomite, anhydrite, and gypsum. In igneous and metamorphic rocks, weathering also releases calcium from such minerals as apatite, wollastonite, fluorite, and various members of the feldspar, amphibole, and pyroxene groups. Inasmuch as calcium is both abundant in the earth's crust and extremely mobile in the hydrosphere, it is one of the most common ions in subsurface water.

Calcium carbonate is easily soluble in water provided that there is an abundant supply of $H^+$. The dissociation of carbonic acid, $H_2CO_3$, is one of the most important sources of $H^+$. The series of equations expressing the various reactions is as follows:

$$CO_2 + H_2O \rightleftharpoons H_2CO_3 \rightleftharpoons H^+ + HCO_3^-$$

$$HCO_3^- \rightleftharpoons H^+ + CO_3^{2-}$$

$$CaCO_3 + H^+ \rightleftharpoons Ca^{2+} + HCO_3^-$$

If abundant carbon dioxide is present, the dissociation will proceed only as far as the bicarbonate stage. If the pH is increased, however, then the ratio of carbonate ($CO_3^{2-}$) to bicarbonate ($HCO_3^-$) ions increases and calcium carbonate may be precipitated. In most natural waters the carbon dioxide in the system is the factor which is most important. If carbon dioxide is being added, solution continues; if it is removed, deposition will most likely occur.

Pure water at 23°C can dissolve only about 13 ppm of calcium carbonate, or about 5 ppm of calcium. Inasmuch as some carbon dioxide is almost always present, the actual solubility in natural water is much greater. Also, the presence of sodium and potassium salts increases the solubility of calcium carbonate. Surface water in equilibrium with atmospheric carbon dioxide can contain as much as 20 or 30 ppm of calcium at saturation. If the surface water is moderately high in dissolved solids, as much as 40 or 50 ppm may be in equilibrium with the carbon dioxide-bicarbonate-carbonate system. The concentration of carbon dioxide in the soil air can be from ten to more than one hundred times the concentration in the atmosphere. With the added carbon dioxide, soil water can probably contain 100 ppm or more of calcium. The presence of larger amounts of carbon dioxide from igneous activity, metamorphism of carbonates, or diagenesis of organic deposits may increase the solubility of calcium to 200 or 300 ppm in the presence of large quantities of bicarbonate. If the major anion in the water is sulfate ($SO_4^{2-}$), and the bicarbonate concentration is low, the solubility of calcium sulfate (gypsum) will tend to control the final calcium concentration in the water. About 600 ppm of calcium will be contained in water saturated with gypsum (Table 4.1, sample 6). The largest amounts of calcium are found in chloride brines. A few analyses of these brines indicate more than 50,000 ppm of calcium. These concentrations are below the saturation values for calcium chloride, so the concentrations are chemically reasonable. The complex problem of the influence of $HCO_3^-$, $Cl^-$, and $SO_4^{2-}$ on $Ca^{2+}$ concentrations has been investigated by Hall [18].

The mechanism for concentration of calcium chloride brines is not well understood. Sea water contains about five times as much magnesium as calcium, therfore if the brines are derived from sea water, a mechanism for removal of magnesium must operate. Most magnesium salts are more soluble than corresponding calcium salts. For this reason, processes of exchange and sorption have been postulated as mechanisms of concentration rather than precipitation of insoluble minerals. The diagenetic change from calcite to dolomite in marine sediments is one possible method of extracting magnesium ions from marine brines.

Concentrations of calcium in normal potable ground water generally range between 10 and 100 ppm. Calcium in these concentrations has no known effect on the health of humans or animals. Indeed, as much as 1000 ppm of calcium may be harmless. The widespread belief that calcium in water causes hardening of the arteries, kidney stones, and liver ailments is without factual support.

The most commonly noticed effect of calcium in water is its tendency to react with soap to form a precipitate called soap curd. Ions of magnesium, iron, manganese, copper, barium, and zinc also cause a similar difficulty, and, together with calcium, must be precipitated before soap can either cleanse or lather. The soap neutralizing power of these ions is called hardness. Inasmuch as all the ions except magnesium and calcium occur in only trace amounts, hardness is usually considered as the sum of the effects of only calcium and magnesium. Hardness is expressed in terms of the weight of calcium carbonate which would produce the hardness actually measured in the water.

### MAGNESIUM

The common sources of magnesium in the hydrosphere are dolomite in sedimentary rocks; olivine, biotite, hornblende, and augite in igneous rocks; and serpentine, talc, diopside, and tremolite in metamorphic rocks. In addition, most calcite contains some magnesium, so a solution of limestone commonly yields abundant magnesium as well as calcium.

The geomchemistry of magnesium is quite similar to that of calcium. The solubility of magnesium carbonate is also controlled by the presence of carbon dioxide. The solubility of pure magnesium carbonate, however, is much greater than that of calcium carbonate so that magnesium carbonate is not precipitated under ordinary temperatures and pressures which prevail in near-surface water. Under the influence of atmospheric carbon dioxide, about 190 ppm of magnesium will remain in solution. With the added influence of carbon dioxide in the soil, much larger amounts will be held in solution. Magnesium sulfate and magnesium chloride are both very soluble; consequently several thousand parts per million of magnesium can be held in solution in equilibrium with chloride and sulfate.

Despite the higher solubilities of most of its compounds, magnesium is generally found in lesser concentrations in natural waters that is calcium. This difference is probably owing to the slow dissolution of dolomite together with the greater abundance of calcium in the earth's crust. In contrast with most natural water, sea water contains about five times as much magnesium as calcium. The deficiency of calcium in sea water is undoubtedly owing to the preferential abstraction of calcium by plants and animals which use it for hard parts of aragonite and calcite. Some water from magnesium-rich rocks such as olivine basalt, serpentine, and dolomite may also contain two or three times as much magnesium as calcium.

Common concentrations of magnesium range from about 1 to 40 ppm. Water from rocks rich in magnesium may have as much as 100 ppm, but concentrations of more than 100 ppm are rarely encountered except in sea water and brines. Exceptionally low values of magnesium and calcium are found in some waters which have undergone natural softening by cation exchange (Table 4.1, sample 3). Most commonly, clays will exchange sodium, if available, for both magnesium and calcium ions.

### SODIUM

Sodium, unlike calcium, magnesium, and silica, is not found as an essential constituent of many of the common rock-forming minerals. The primary source of most sodium in natural water is from the release of soluble products during the weathering of plagioclase feldspars. In areas of evaporite deposits, the solution of halite is also important. Clay minerals may, under certain conditions, release large quantities of exchangeable sodium [11,32,33]. Less important sources of sodium in natural waters are the minerals nepheline, sodalite, stilbite, natrolite, jadeite, arfvedsonite, glaucophane, and aegirite. These minerals are locally abundant in some igneous and metamorphic rocks, but are quantitatively of minor importance in comparison with the feldspars.

Sodium salts are soluble and will not precipitate unless concentrations of several thousand parts per million are reached. The saturation point for NaCl is 264,000 ppm at 20°C, or about 105,000 pm $Na^+$. The saturation point for sodium nitrate is almost twice that of sodium chloride and for sodium bicarbonate it is about one third that of sodium chloride. Owing to the high solubility of its compounds, sodium is concentrated by evaporation in the ocean and in desert basins which lack outlets. The only common mechanism for removal of large amounts of sodium ions from natural water is through ion exchange, which operates if the sodium ions are in great abundance. The removal of sodium ions from sea water which has infiltrated into fresh-water aquifers has been ascribed by a number of workers to the action of ion exchange [32,34]. The conversion of calcium bicarbonate water to sodium bicarbonate water in many aquifers is also undoubtedly owing to ion exchange [3,11]. Thus, the process is reversible, the direction of exchange depending on the relative concentrations of ions.

All natural water contains measurable amounts of sodium. Actual concentrations range from about 0.2 ppm in some rain and snow to more than 100,000 ppm in brines in contact with salt beds. Areas of igneous and metamorphic rocks that are also in regions of moderate to high rainfall generally have waters with 1 to 20 ppm of sodium. Waters with total dissolved solids ranging from 1000 to 5000 ppm generally have more than 100 ppm of sodium. Exceptions are waters from gypsum beds and some water from limestone aquifers.

Although most modern analyses contain direct determinations of sodium older analyses were commonly based on the assumption that all undetermined cations were potassium and sodium. This was done by determining all common anions plus the two cations, calcium and magnesium; the excess of equivalents per million of anions over cations was assumed to be owing to undetermined potassium and sodium. The old method was unsatisfactory because all the errors of the analysis and of undetermined ions were included within the reported value for combined sodium and potassium. For this reason, older analyses which report values of combined sodium and potassium should be used with considerable caution.

### POTASSIUM

Common sources of potassium are the products formed by the weathering of orthoclase, microcline, biotite, leucite, and nepheline in igneous and metamorphic rocks. Waters percolating through evaporite deposits may contain very large quantities derived from the dissolution of sylvite and niter.

Although the abundance of potassium in the earth's crust is about the same as sodium, potassium is commonly less than one tenth the concentration of sodium in natural water. This relative immobility of potassium is owing, first, to the fact that potassium enters into the structure of certain clay and clay-like minerals during weathering, and, second, to the higher resistance to weathering of many potassium minerals in relation to the sodium minerals. Vegetation also concentrates potassium; this, however, accounts for the removal of only a small amount of potassium for most of the organic material is decomposed in place and relatively little is removed by erosion.

The solubilities of potassium salts are all high and generally similar in magnitude to the solubilities of sodium salts. For example, at 20°C, 255,000 ppm KCl, or about 133,000 ppm K, will be held in solution. Owing to this extreme solubility, potassium will not usually be removed from water except by sorption, ion exchange, or precipitation during evaporation.

All natural waters contain measurable amounts of potassium. Some snow and rainwater may contain as little as 0.1 ppm and some brines as much as 100,000 ppm. Most potable ground water, however, contains less than 10 ppm and commonly ranges between 1.0 and 5.0 ppm. An interesting feature of potassium is that dilute waters containing only 20 ppm total dissolved solids may contain about 2 ppm potassium. As the concentrations increase, however, the amount of potassium increases only slightly; thus water with a total dissolved solids concentration of 2000 ppm most likely will have less than 20 ppm potassium.

### BICARBONATE AND CARBONATE

The amount of a standard concentration of sulfuric acid which is needed to titrate a water sample to an endpoint of pH 4.5 is a measure of the alkalinity* of the water. Alkalinity is produced almost exclusively by bicarbonate and carbonate ions. Hydroxide, iron, and silicate may have some influence on alkalinity when the pH is above 9.0, and phosphate will affect alkalinity throughout normal ranges of pH values. The natural concentrations of phosphate, however, are almost insignificant in comparison with carbonate and bicarbonate ions; also, natural waters with a pH of more than 9.0 are rare. Alkalinity is, therefore, a reliable measure of carbonate and bicarbonate ions for most natural water.

* It should be noted that this usage of the word "alkalinity" is contrary to the common chemical usage in which only water with a pH of more than 7.0 is considered to be alkaline.

The dissociation of bicarbonate to carbonate ions is effective largely above a pH of 8.2. Below this pH most of the carbonate ions add hydrogen to become bicarbonate ions ($H^+ + CO_3^{2-} \rightleftharpoons HCO_3^-$), and the ratio of bicarbonate to carbonate ions increases to more than 100 to 1. The part of the alkalinity titration above a pH of 8.2 is, therefore, a measure of carbonate ions present and that below 8.2 is a measure of bicarbonate ions. Below a pH of 4.5, most of the bicarbonate ions are converted to carbonic acid molecules ($H^+ + HCO_3^- \rightleftharpoons H_2CO_3$). In titrating, a convenient color indicator for a pH of 8.2 is phenolphthalein and for a pH of 4.5, methyl orange. The two types of alkalinity are called carbonate and bicarbonate alkalinity respectively; however, they are also called, in industrial-water analyses, phenolphthalein and methyl-orange alkalinity, because chemicals causing alkalinity which have little relation to the carbonate bicarbonate system are commonly added to industrial water.

Most carbonate and bicarbonate ions in ground water are derived from the carbon dioxide in the atmosphere, carbon dioxide in the soil, and solution of carbonate rocks. Some ground waters and many oil-field waters probably obtain bicarbonate from the carbon dioxide generated by diagenesis of organic compounds [12]. Sodium bicarbonate water can be concentrated by evaporation in soils and desert basins, but if much calcium is present the bicarbonate will be taken out of the water through precipitation of calcium carbonate.

Ground water generally contains more than 10 ppm but less than 800 ppm bicarbonate. Concentrations between 50 and 400 ppm are most common. Only rarely will ground water have pH values of less than 4.5, causing bicarbonate to be converted to carbonic acid, or pH values of more than 8.2, so that the bicarbonate ions will dissociate to carbonate ions.

### ACIDS

The so called "mineral-acid acidity" of water is present when the pH of the water is less than 4.5. The acidity is determined by titrating the water with a 0.02N solution of sodium hydroxide until a pH of 4.5 is reached. Acidity can be caused by the presence of hydrochloric acid or sulfuric acid in areas of volcanic activity, the oxidation of sulfide deposits, the leaching of organic acids from decaying vegetation, or the hydrolysis of iron and aluminum. The oxidation of iron sulfide and the hydration of iron probably proceeds according to the following equations [19].

$$FeS_2 + H_2O + 7O \rightleftharpoons FeSO_4 + H_2SO_4$$

$$H_2SO_4 \rightleftharpoons 2H^+ + SO_4^{2-}$$

The ferrous sulfate $FeSO_4$ may be oxidized to ferric sulfate, $Fe_2(SO_4)_3$, which in solution yields:

$$Fe_2(SO_4)_3 \rightleftharpoons 2Fe^{3+} + 3SO_4^{2-}$$

$$Fe^{3+} + 3H_2O \rightleftharpoons Fe(OH)_3 + 3H^+$$

Thus the hydrogen-ion concentration is increased by both the dissociation of sulfuric acid and the hydrolysis of ferric ions. With present methods of water analysis, the exact nature of acidity is not easily determined. Acid waters, however, are comparatively rare in normal ground water.

Acidity is expressed as the acidity which would be produced by an equivalent amount of sulfuric acid even though acidity may be caused by other factors. Values for acidity rarely exceed a few parts per million, but may be as high as 900 ppm in acid hot-spring water.

### SULFATE

Despite a relatively large amount of sulfur, mostly in the form of sulfates, in water and in sedimentary rocks, sulfur is only a minor constituent of igneous rocks. Early in the history of the hydrosphere most sulfates probably originated from the oxidation of sulfides from igneous rocks and volcanic sources, but at present sulfates are largely recycled from the atmosphere [23] and from the solution of sulfate minerals in sedimentary rocks. Sedimentary rocks, particularly organic shales, may also yield large amounts of sulfates through the oxidation of marcasite and pyrite. All atmospheric precipitation contains sulfate, which, although commonly in absolute concentrations of less than 2 ppm, is one of the major dissolved constituents of rain and snow. The sulfate in the atmosphere is derived from dust particles containing sulfate minerals, from the oxidation of sulfur dioxide gas, and from the oxidation of hydrogen sulfide gas. The hydrogen sulfide in the atmosphere is derived from the decomposition of organic material and locally from volcanic emanations. One of the most prolific sources of hydrogen sulfide gas is probably from the bacterial reduction of organic material in tidal mud flats. Sulfur dioxide is discharged into the atmosphere from volcanic sources, and, more important, from the burning of coal and oil and smelting of ores so that precipitation near these sources is commonly high in sulfate [23].

Most sulfate compounds are readily soluble in water. Of the more common compounds found in nature, barium sulfate is the least soluble. The amount of barium available for precipitation of the sulfate ions, however, is quite small, so the removal of sulfate by this process is probably of only minor significance. Strontium, which is more abundant in water than barium, will precipitate all but about 60 ppm of sulfate from distilled water. The absence of large amounts of secondary celestite ($SrSO_4$) in most aquifers suggests that this process is also of secondary importance. One of the most effective natural process for the removal of sulfate from water appears to be the reduction of sulfate by bacteria. As mentioned in Chapter 3, certain bacteria, collectively known as sulfur bacteria, use sulphides and sulfates in their life cycles. The sulfate-reducing bacteria derive energy from the oxidation of organic compounds and in the process obtain oxygen from the sulfate ions in subsurface water. The resulting reduction of sulfate ions produces hydrogen sulfide gas as a by-product. Most of the gas will remain in the subsurface water which will retain several hundred parts per million $HS^-$ ions and $H_2S$ gas. If iron is present in the water under moderately reducing conditions, iron sulfide may be precipitated, thus removing both iron and sulfide from the water. If the sulfate reduction is accomplished by soil bacteria, much of the hydrogen sulfide may escape directly to the atmosphere.

Concentrations of sulfate from less than 0.2 ppm to more than 100,000 ppm are found in nature. The lowest concentrations of sulfate are in rainwater, snow, and subsurface waters subject to sulfate reduction. The highest concentrations are in magnesium sulfate brines. Ground water from igneous and metamorphic rocks, or from sediments derived from them, generally contain less than 100 ppm and may contain less than 1 ppm if sulfate-reducing bacteria are active in the soil through which recharge water has percolated.

### CHLORIDE

Chloride is a minor constituent of the earth's crust, but a major dissolved constituent of most natural water. Sodalite and apatite are the only common minerals in igneous and metamorphic rocks which contain chloride as an essential constituent; although micas, hornblende, and natural glass may also contain significant amounts. Liquid inclusions in rocks and minerals are another source of chloride from igneous rocks. The foregoing sources are generally thought to be insufficient to account for the amount of chloride which must have been contributed to the oceans since they were first formed. The chloride in evaporite deposits and in ocean water at present may be as much as 100 times the amount of chloride which would be expected on the basis of the volume of rocks which must have been weathered to produce the amount of sodium present. It is probable that the small, but more or less continuous, contribution of chloride from volcanic gases may account for much of the chloride which has accumulated in sea water.

Most chloride in ground water comes from four different sources. First, chloride from ancient sea water entrapped in sediments; second, solution of halite and related minerals in evaporite deposits; third, concentration by evaporation, of chloride contributed by rain or snow; and fourth, solution of dry fallout from the atmosphere, particularly in arid regions. A locally important source may be from volcanic water in hot spring systems. Quantitatively speaking, the most important source of chloride in near-surface water appears to be chloride transported in the atmosphere and carried to earth by rain or snow. The chloride content of most rain near coastal areas ranges between 3 and 6 ppm. This diminishes rapidly to 1.0 ppm or less 100 miles inland, and to 0.3 ppm or less 500 miles inland [24]. Thus it is inferred that most chloride in coastal areas is from oceanic sources [14,23,52]. The rain and snow condense around minute dried particles of ocean spray which are continually being carried aloft from turbulent areas on the ocean surface. As air masses migrate inland, dust particles from terrestrial sources become more important as nuclei of condensation, and the ratio of chloride to other ions in rain decreases accordingly. If, however, dust is rich in chlorides, such as is true near desert playas and cities which use salt to control street icing, the chloride content of rain may be locally high even in the interior of continents.

All chloride salts are highly soluble, so chloride is rarely removed from water by precipitation except under the influence of freezing or evaporation. Chloride is also relatively free from effects of exchange, adsorption, and biological activity. Thus, if water once takes chloride into solution, it is difficult to remove the chloride through natural processes.

Chloride concentrations found in natural water vary between about 0.1 ppm in arctic snow to 150,000 ppm in brines. Continental rain and snow may contain from 1.0 to 3.0 ppm, but probably average less than 1.0 ppm. Shallow ground water in regions of heavy precipitation generally contains less than 30 ppm of chloride. Concentrations of 1000 ppm or more are common in ground water from arid regions.

### NITRATE

Although igneous rocks contain small amounts of soluble nitrate or ammonia, most nitrate in natural water comes from organic sources or from industrial and agricultural chemicals [45]. An additional minor source is from nitric oxides produced by lightning discharges. Nitrogen is an essential constituent of protein in all living organisms. When organic material decomposes through bacterial action, the complex proteins change through amino acids to ammonia, nitrites, and, finally to nitrates. Some of the nitrates produced may be leached by percolating water and eventually reach the ground water: however, most nitrate is probably used by plants soon after it is released by bacterial action. Certain plants, such as alfalfa and peas, have bacteria living on root nodules which fix nitrogen gas from the atmosphere. The nitrates thus formed are commonly in excess of the plant's needs, so a surplus is available for leaching.

Nitrate compounds are highly soluble, so nitrate is taken out of natural water only through activity of organisms or through evaporation. Common nitrate concentrations in water range from 0.1 to 0.3 in rainwater to as much as 600 ppm in ground water from areas influenced by excessive applications of nitrate fertilizer or runoff from barnyards (Table 4.1, sample 5). Normal ground water contains only from 0.1 to 10.0 ppm nitrate [13,28,29].

#### MINOR AND SECONDARY ELEMENTS

Besides the more abundant secondary elements already discussed, there are a number of additional minor and secondary elements of considerable importance. Most of these elements are listed, along with some trace elements, in Table 4.2. Unfortunately, information concerning many of the minor and trace constituents is scarce. For example, based on crustal abundance and known chemical properties, rubidium should be present in most water in concentrations of from 0.0001 to 0.1 ppm. Information concerning rubidium in fresh water, however, is almost entirely lacking.

Minute amounts of dissolved constituents are of interest for at least three important reasons. First, many of the elements are of potential importance in the interpretation of the geologic history of water [16,32,37, 38,47,48,50]. Second, anomalously large amounts of certain metals in water may indicate the presence of ore deposits [7,26]. Third, many of the elements affect the health of plants and animals, even though they are in very small concentrations [1,8,22,30,43].

Bromide is chemically similar to chloride in natural water but has a much lower abundance. Most water contains only about 1 ppm bromide for each 300 ppm of chloride. Bromide in natural concentrations is not known to affect the health of plants or animals. It may be, however, a useful ion to study if the source of water is of interest. White [47,49] has suggested that water from a volcanic source may have a smaller amount of bromide in relation to chloride than is found in other water.

Iodide, along with chloride and bromide, are ions of the halogen elements. The geochemistry of iodide is somewhat different, however, than the more abundant ions, chloride and bromide. Iodine is concentrated by plants and animals so that sea water is very low in this element. After burial, organic material will decompose and release the iodide to the interstitial water. The ratio between iodide and chloride in ocean water is only 0.000003, whereas in most oil-field brines it is about 0.001 [49]. This contrast may be useful to differentiate between ancient sea water and newly intruded sea water in coastal aquifers.

*Table 4.2 Dissolved Solids in Potable Water—a Tentative Classification of Abundance*

| | |
|---|---|
| **Major Constituents (1.0 to 1000 ppm)** | |
| Sodium | Bicarbonate |
| Calcium | Sulfate |
| Magnesium | Chloride |
| Silica | |
| **Secondary Constituents (0.01 to 10.0 ppm)** | |
| Iron | Carbonate |
| Strontium | Nitrate |
| Potassium | Fluoride |
| Boron | |
| **Minor Constituents (0.0001 to 0.1 ppm)** | |
| Antimony* | Lead |
| Aluminum | Lithium |
| Arsenic | Manganese |
| Barium | Molybdenum |
| Bromide | Nickel |
| Cadmium* | Phosphate |
| Chromium* | Rubidium* |
| Cobalt | Selenium |
| Copper | Titanium* |
| Germanium* | Uranium |
| Iodide | Vanadium |
| | Zinc |
| **Trace Constituents (generally less than 0.001 ppm)** | |
| Beryllium | Ruthenium* |
| Bismuth | Scandium* |
| Cerium* | Silver |
| Cesium | Thallium* |
| Gallium | Thorium* |
| Gold | Tin |
| Indium | Tungsten* |
| Lanthanum | Ytterbium |
| Niobium* | Yttrium* |
| Platinum | Zirconium* |
| Radium | |

Major sources used in compiling this list are given in the references at the end of this chapter [4,5,6,7,8,10,22,26,28,29,39,40,43,49,53]. Elements marked with an asterisk (*) are those which occupy an uncertain position in the list.

Fluorine, also of the halogen group, is quite different from other elements in the group in its geochemistry. Unlike chlorine, bromine, and iodine, many of the compounds of fluorine have a low solubility. Natural concentrations of fluoride commonly range from about 0.01 to 10.0 ppm. A few water samples have been analyzed which have more than 10.0 ppm (Table 4.1, sample 2); the highest reported is 67 ppm from the Union of South Africa. The natural concentration of fluoride appears to be limited by the solubility of fluorite ($CaF_2$), which is about 9 ppm fluoride in pure water. It has been observed, however, that waters high in calcium do not contain more than about 1 ppm of fluoride [19].

Boron is an element which is essential to plant growth, but also is injurious if present in large quantities. Sensitivities of plants vary widely. Citrus trees may be damaged by as little as 0.5 ppm, but alfalfa will tolerate more than 10.0 ppm if soil drainage is good. Normal ground-water concentrations range from about 0.01 ppm to 1.0 ppm. Highest concentrations are known from volcanic hot springs and oil-field brines. These commonly exceed 10 ppm, but rarely exceed 100 ppm.

Boron is probably important as an indication of water origin. Ocean water and surface water concentrated by evaporation have a boron chloride ratio of about 0.0002, but oil-field brines may be as high as 0.02, and volcanic hot springs as high as 0.1 [46,47].

The geochemistry of manganese is very similar to that of iron [20]. In natural waters its concentration is less than one half that of iron. A small number of acid waters have been analyzed which contain more than 1 ppm, but most waters contain less than 0.2 ppm. Like iron, manganese in water promotes the growth of certain bacteria and stains bathroom fixtures and laundry [17]. Although manganese is essential for healthy plant growth, the manganese which may be contributed by irrigation water and rain is probably insignificant in comparison with manganese released through weathering of minerals.

Aluminum is one of the most abundant elements in the earth's crust, but it is also one of the least mobile in the hydrosphere. Waters within the pH range of 5.0 to 9.0 have less than 1.0 ppm aluminum. Normal ground water probably ranges between 0.005 and 0.3 ppm. Acid waters may have as much as 100 ppm, but lower values are more typical (Table 4.1, sample 8). Aluminum in water is of considerable interest to geologists who study the weathering of rocks and the formation of ore deposits. For those interested in the utilization of water, however, aluminum is of minor importance. Despite widespread pseudoscientific literature to the contrary, aluminum is not a poison.

Strontium is chemically similar to calcium and occurs in minerals in structural positions normally occupied by calcium. Its natural concentration in water is probably limited by ion exchange with calcium-rich clays. In a few waters its concentration may be limited by the solubility of strontium sulfate which is about 132 ppm in pure water at 20°C, or equivalent to 63 ppm $Sr^{2+}$. Strontium occurs in most ground water in concentrations ranging between 0.01 and 1.0 ppm. The maximum amount reported is 2730 ppm in a calcium chloride brine, and the maximum reported in potable water is 52 ppm.

Lithium is relatively rare in the earth's crust, being concentrated principally in granitic rocks. Most compounds of lithium are soluble, so once it is dissolved lithium tends to remain in solution. Concentrations of lithium in ground water generally range between 0.001 and 0.5 ppm, but may exceed 5 ppm in brines.

The natural concentrations of arsenic, copper, zinc, lead, nickel, and uranium have been measured by several workers interested in using water analyses to aid in prospecting for ore deposits. These ions, nevertheless, are not usually reported in routine chemical analyses. In general, waters with a low pH or a high temperature will contain the largest concentrations of these metals. Some mine waters and many industrial wastes contain harmful amounts of arsenic and lead, and if mixed with ground water may render it unfit for consumption by humans or livestock. Only rarely are copper, zinc, and uranium found in harmful concentrations.

Selenium has been studied widely in areas affected by selenium poisoning in western United States. Most of the difficulty has arisen from livestock feeding on plants that concentrate selenium in their leaves. Although cases of possible selenium poisoning through water are rare, it is an important potential source of harm. Most of the areas of high-selenium soils are in the outcrop region of upper Cretaceous marine shales. Here, some drainage water has been reported to exceed 1.0 ppm of selenium. Normal ground water in the same region generally contains less than 0.1 ppm.

Although more analytical evidence is needed, circumstantial evidence based on crustal abundances and solubilities of compounds suggests that cobalt, barium, rubidium, and titanium occur in most waters in concentrations of more than 0.0001 ppm. Cobalt is essential for animal growth, but most cobalt is taken up through food, and water-born cobalt is probably of only minor importance. Natural concentrations of barium, rubidium, and titanium have not been reported to be significant in terms of water use.

### TRACE CONSTITUENTS

All elements are soluble in water to at least a small degree, even though natural concentrations are so small that they may be difficult to measure.

Of the trace constituents listed in Table 4.2, mercury, cadmium, gallium, cesium, indium, bismuth, and silver may occur in concentrations of between 0.0001 and 0.001 ppm. Radium, however, is known to average much less than $10^{-10}$ ppm, and gold, beryllium, and platinum probably occur in concentrations of less than $10^{-5}$ ppm.

Considerable research must be undertaken before the importance of trace constituents is understood. It may be possible to locate ore deposits of mercury, silver, and bismuth by finding waters with anomalously high concentrations of these metals. Some of the trace elements may be important from the standpoint of public health, but only the radioactive isotopes have been studied extensively in this regard.

### GASES IN GROUND WATER

Analyses of gases which are dissolved in ground water are rarely made. Gases are, however, of great importance in hydrogeology. Gases coming out of solution will form bubbles which tend to obstruct water movement in aquifers, particularly near wells. Certain dissolved gases, such as oxygen and carbon dioxide, will alter water chemistry. Other gases affect the use of water. Methane coming out of solution may accumulate and present a fire or explosion hazard. Dissolved oxygen will promote the corrosion of metals. Hydrogen sulfide in concentrations of more than 1.0 ppm will render the water unfit for consumption because of the objectionable odor. Hydrogen sulfide also promotes the growth of certain bacteria which clog well screens and pipes.

The solubility of most gases in water is directly proportional to pressure and inversely proportional to temperature. The effect of temperature is seen from data in Table 4.3.

Dissolved gas probably occurs in concentrations of 1.0 to 100.0 ppm in most ground water. Higher concentrations are possible under certain conditions. Some wells produce large amounts of methane, hydrogen sulfide, and carbon dioxide. Water associated with these gases is probably saturated with respect to the gases which are under pressures of several atmospheres. Also, water percolating through decaying organic material may contain large amounts of dissolved ammonia which, however, probably oxidizes rapidly to nitrate. Ammonia is also associated with some hot-spring waters [36].

Dissolved gases in rain and in cascading streams are in near equilibrium with atmospheric gases (Table 4.3). If this surface water infiltrates directly into highly permeable aquifers, the dissolved gases will be retained in their original concentrations. More commonly, however, water infiltrates through soil and underlying material before reaching aquifers. Oxidation of minerals, microbiological activity, and diagenesis of organic matter will deplete the oxygen in the percolating water [42]. Temperature increases will cause gas to be released from the water, or, if recharge water cools as it enters the ground, more gas may go into solution. Methane, hydrogen sulfide, ammonia, and other gases may be enriched in the recharge zone, so new equilibria with gases in solution will be established. Nitrogen may be depleted by certain bacteria. As a consequence, the gas content of water as it first enters aquifers in most areas will be considerably different than the gas content of surface water.

*Table 4.3 Solubility of Gases in Water*

| Temperature, °C | $N_2$ | $O_2$ | Ar | $CO_2$ |
|---|---|---|---|---|
| *(ppm by weight in equilibrium with the atmosphere at 760 mm)* | | | | |
| 0 | 23 | 15 | 0.9 | 1 |
| 10 | 19 | 11 | 0.7 | 0.3 |
| 20 | 16 | 9 | 0.6 | 0.6 |
| 30 | 14 | 8 | 0.5 | 0.4 |
| 50 | 11 | 6 | 0.4 | 0.3 |
| (ppm by weight in equilibrium with given gas at 760 mm) | | | | |
| | $N_2$ | $O_2$ | $H_2S$ | $CO_2$ |
| 0 | 30 | 70 | 7000 | 3400 |
| 10 | 23 | 54 | 5200 | 2300 |
| 20 | 19 | 44 | 3900 | 1700 |
| 30 | 16 | 37 | 3000 | 1300 |
| 50 | 12 | 27 | 1900 | 760 |

Once infiltrating water reaches a fully saturated zone, the gas content of the water should remain relatively constant, particularly if the saturated zone is under artesian pressure. Slight changes in gas content, however, will take place owing to the slow oxidation of minerals and possibly owing to microbiological activity at depth [42].

Concentrations of inert gases such as helium, argon, and neon will be subject to the slightest changes of the various gases in ground water. Sugisaki [41] has used this fact to determine the original temperature of ground water at the time it percolated from a river into a permeable artesian aquifer. Inasmuch as a high argon content indicated low temperature and a low argon content indicated a high temperature at the time of recharge, bands of winter and summer water could be located down gradient from the recharge area and the rate of water movement could be inferred (Figure 4.3).

Given enough time, ground water is enriched in helium and, to a lesser extent, in argon. Uranium and thorium will produce helium through radioactive decay. Potassium-40 will produce argon, also through radioactive decay. Rapidly circulating ground water will continually remove the small quantity of helium and argon which is released from rocks in contact with the water. Slowly moving ground water in large artesian basins and nearly static connate water, however, should show significant increases in at least helium and possibly also in argon. Judging by the

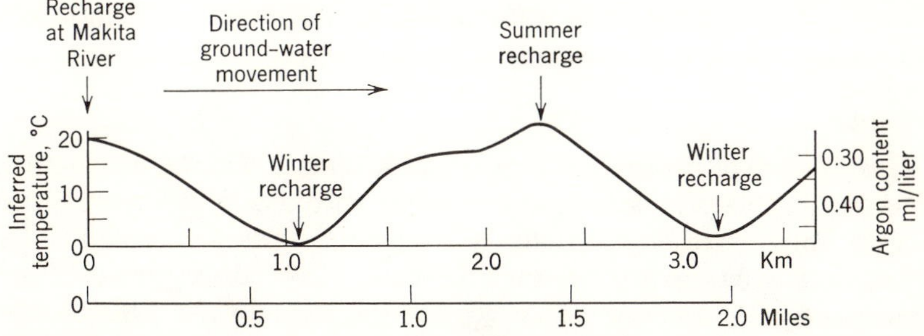


Figure 4.3 Argon in ground water in the vicinity of Takada, Japan. Temperature at time of recharge is inferred from argon content. Profile is parallel with direction of ground-water movement, so distance between temperature minima indicates water has moved more than 2 kilometers in one year. (Redrawn from Sugisaki [41].)

relative abundance of helium in the atmosphere [23] as compared with natural gas deposits [2,54], the ratio of helium/argon should increase from about 0.0005 in surface water to as much as 1.0 in some saline water in contact with natural-gas deposits.

Artificially injected helium has proved to be a satisfactory ground-water tracer [9]. It is not biologically harmful, and it is inert and so does not react chemically with the aquifer. The need for special analytic equipment, however, has discouraged widespread use of helium as a tracer.

## 4.3 *Suspended Material*

### INORGANIC

Several types of suspended inorganic material may be found in well water. Small particles of ferric oxide may come from rusty casings and pump columns, or, if in flocs, it may be from the oxidation and precipitation of dissolved iron in the water. Far more common are suspensions of silt and fine sand, particularly in wells which are pumped at high discharge rates. Suspensions of clay are also encountered in some wells drawing water from limestone aquifers. Turbidity often occurs following rains when rapidly flowing water carries clay from fissures and caves into the wells and in some places indirectly from the surface soil into the wells.

Concentrations of suspended sediment rarely exceed 500 ppm and most commonly do not exceed 5 ppm. Even concentrations of 5 ppm, nevertheless, may cause trouble. For example, a well discharging 1000 gallons per minute (gpm) and having 5 ppm suspended solids will produce almost a ton of solids each month. Unless removed at the well, much of this material may lodge in pipes and tanks in the distribution system and greatly increase maintenance costs. The removal of sand, silt, and clay from the water-bearing zones in the subsurface may also cause subsidence and damage to the well and surrounding structures.

#### ORGANIC

Several types of bacteria may thrive inside wells and in the surrounding water-bearing zones. Pathogenic organisms, fortunately, are rarely found in well water. Most organically contaminated wells are in regions of limestone aquifers where caverns and smaller solution openings afford direct connection between the surface and the ground water. Most other cases of contamination are caused by poor well construction which allows surface drainage to enter the well. Diseases which have been known to be spread through ground water are typhoid, cholera, amoebic dysentery, and infectious hepatitis.

Soluble organic complexes may enter shallow wells and cause water to have a color and sometimes a disagreeable taste. Total amount of organic material exclusive of gases is rarely more than 15 ppm in well water.

## 4.4 *Classification of Water*

#### TOTAL DISSOLVED SOLIDS

The simplest classification of water is based on the total concentration of dissolved solids. The classification used in this book is given in Table 4.4. This is similar to a classification suggested by Gorrell [15].

*Table 4.4 Types of Water*

| Name | Concentration of Total Dissolved Solids in Parts per Million. |
|---|---|
| Fresh water | 0–1000 |
| Brackish water | 1000–10,000 |
| Salty water | 10,000–100,000 |
| Brine | more than 100,000 |

DOMINANT IONS

The foregoing classification can be combined with a classification based on the dominant ions present in the water. Most classifications of this type use a percentage of equivalents per million of the anions and cations. Figure 4.4 represents one possible system of nomenclature in which water represented by point *A* would be called a calcium bicarbonate water and

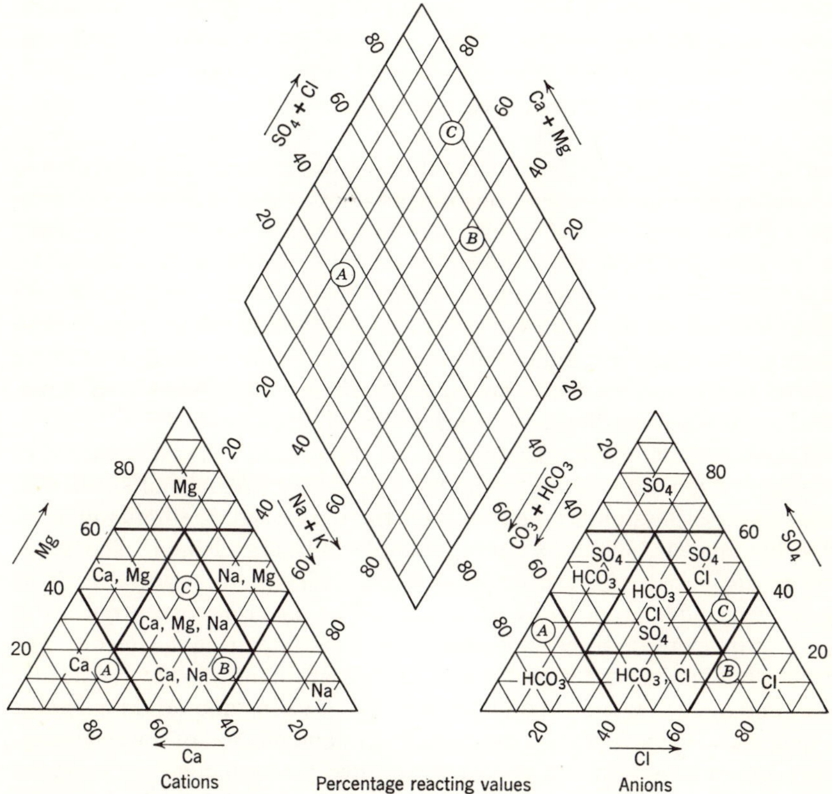


Figure 4.4 Water classification based on the percentage of equivalents per million. Point *A* represents calcium bicarbonate water; point *B*, calcium, sodium, chloride water; point *C*, sodium, calcium, magnesium, chloride, sulfate water.

point *B* would represent calcium, sodium, chloride water. Point *C* would represent sodium, calcium, magnesium, chloride, and sulfate water. If the total dissolved solids concentration in *B* were 7000 ppm, it could be called brackish calcium sodium chloride water, thus combining the sense of total dissolved solids with an indication of relative proportions of the various ions. A rigid classification of this type, however, is not commonly too useful, because normally the entire analysis is reported together with a graphical presentation, as a much more accurate understanding can be gained by reference to diagrams or tables than by awkward word descriptions.

### WATER USE

A classification of water according to possible uses is more important. All such classifications, however, should be used with caution and should never be applied too rigidly. As an example, ocean water is clearly unsuited for most domestic, agricultural, and industrial uses. Nevertheless, if fresh water is too expensive, sea water can be used for washing, cooling, recreation, and fire protection. A second example could be drawn from agricultural water classifications. Alfalfa grown in sandy soil can tolerate concentrations of dissolved constituents many times greater than citrus trees grown in clayey soil. Thus, soil drainage and vegetation type need to be considered. A third example can be taken from drinking water standards. In some parts of Australia, North Africa, and other arid regions many people drink water with total dissolved solids of more than twice the generally recommended limit. As far as can be determined, most of the people have suffered no injurious effects. An abbreviated classification of water quality for various uses is given in Table 4.5.

Drinking water standards are based on two criteria: (1) the presence of objectionable tastes, odors, or colors, and (2) the presence of substances with adverse physiological effects. Of the ions listed, the limits of lead, fluoride, arsenic, sulfate, sodium, chloride, nitrate, selenium, and chromium are based on adverse physiological effects. Substances such as hydrogen sulfide, iron, and manganese can be taken in concentrations of 5 to 10 times those listed in Table 4.5 without impairing the health. The exact optimum limits for all the ions are actually controlled by the health, size, and age of the individual as well as by his eating and drinking habits. Climate also has an effect on the liquid intake and method of waste elimination, so limits in Table 4.5 should be adjusted for local conditions and for specific individuals according to advice of medical authorities in the region of interest.

Water for general household use includes water for baths, laundry, dishes, flowers, and cooking. Limits are established to prevent objectionable odors, stains, and deposits as well as to prevent waste of soap owing to high hardness values. Good water as defined in Table 4.5 will rarely provoke complaints from homeowners and tenants, but poor water will be only barely acceptable for many uses.

Irrigation-water criteria are dependent on the types of plants, amount of irrigation water used, soil, and climate [35]. Tables 4.6 and 4.7 list

*Table 4.5 Quality Criteria for Various Uses. Numbers are maximum recommended concentrations in parts per million.*
(Abstracted from various sources [19,30,35,43,44,51])

| | Drinking | General Household Use | | Irrigation | | Food Processing | Boiler water | |
|---|---|---|---|---|---|---|---|---|
| | | Good | Poor | Good | Poor | | High Pressure | Low Pressure |
| Antimony | 0.05 | ... | ... | ... | ... | 0.05 | ... | ... |
| Arsenic | 0.05[a] | ... | ... | ... | ... | 0.05 | ... | ... |
| Barium | 1.00[a] | ... | ... | ... | ... | 1.00 | ... | ... |
| Bicarbonate | 500 | 150 | 500 | 200 | 500 | 300 | 5 | 50 |
| Boron | 20 | ... | ... | 0.3 | 3.0 | ... | ... | ... |
| Cadmium | 0.01[a] | ... | ... | ... | ... | 0.01 | ... | ... |
| Calcium | 200 | 40 | 100 | ... | ... | 80 | 1 | 40 |
| Chloride | 250 | ... | ... | 100 | 300 | 300 | ... | ... |
| Chromium | 0.05[a] | ... | ... | ... | ... | 0.05 | ... | ... |
| Copper | 1.0 | 0.5 | 3.0 | ... | ... | 3.0 | ... | ... |
| Cyanide | 0.2[a] | ... | ... | ... | ... | 0.2 | ... | ... |
| Fluoride | 1.5 | ... | ... | ... | ... | 1.5 | ... | ... |
| Hydrogen sulfide | 1.0 | 0.05 | 2.0 | ... | ... | 0.5 | 0 | 5 |
| Iron | 1.0 | 0.2 | 0.5 | ... | ... | 0.2 | ... | ... |
| Lead | 0.05[a] | ... | ... | ... | ... | 0.05 | ... | ... |
| Magnesium | 125 | 20 | 100 | ... | ... | 40 | 1 | 20 |
| Manganese | 0.05 | 0.05 | 0.3 | ... | ... | 0.1 | ... | ... |
| Nitrate | 20 | ... | ... | ... | ... | 20 | ... | ... |
| Phenol | 0.001 | ... | ... | ... | ... | 0.001 | ... | ... |
| Selenium | 0.01[a] | ... | ... | ... | ... | 0.01 | ... | ... |
| Silica | ... | 10 | 50 | ... | ... | 50 | 1 | 30 |
| Silver | 0.05[a] | ... | ... | ... | ... | 0.05 | ... | ... |
| Sodium | 200 | 100 | 300 | 50 | 300 | 300 | ... | 50 |
| Sulfate | 250 | 100 | 300 | 200 | 500 | ... | ... | ... |
| Synthetic detergents | 0.5 | 0.2 | 1.0 | ... | ... | 0.5 | 0 | 0 |
| Total solids | 1500 | 300 | 2000 | 500 | 3000 | 1000 | 100 | 2000 |
| Zinc | 5 | ... | ... | ... | ... | 5 | ... | ... |

[a] Mandatory limits of the U.S. Public Health Service for water used on interstate public transportation facilities.

the relative tolerances of plants to boron and salt. If a crop receives most of its water from irrigation, the tolerance for poor water will be increased by the use of larger volumes of water. This is owing to the fact that the surplus water will leach salts from the soil and thus prevent a dangerous build up of soil-moisture salinity. In general, a clayey soil will cause most difficulty with water quality because drainage is poor and the opportunity for leaching of excess salts is thereby lessened. If plants are struggling against adverse climatic conditions, they will be susceptible to injury by poor irrigation water. Also, plants in a hot, dry climate will abstract more moisture and thereby concentrate dissolved solids in the soil moisture

*Table 4.6 Tolerance of Plants to Boron*
[Listed in order of increasing tolerance]
(From Richards [35])

| | Sensitive | Semitolerant | Tolerant |
|---|---|---|---|
| Excellent water: | Less than 0.3 ppm | Less than 0.7 | Less than 1.0 |
| Unsuitable water: | More than 1.3 ppm | More than 2.5 | More than 3.8 |
| | Lemon | Bean (Lima) | Carrot |
| | Grapefruit | Pepper | Lettuce |
| | Avocado | Pumpkin | Cabbage |
| | Orange | Oat | Turnip |
| | Apricot | Corn (Maize) | Onion |
| | Peach | Wheat | Alfalfa |
| | Cherry | Barley | Beet |
| | Persimmon | Olive | Sugar beet |
| | Grape | Radish | Date palm |
| | Apple | Tomato | Asparagus |
| | Pear | Cotton | |
| | Artichoke | Sunflower | |
| | Walnut | | |

*Table 4.7 Relative Tolerance of Crop Plants to Salt*
(From Richards [35])

| Low Tolerance | Medium Tolerance | High Tolerance |
|---|---|---|
| Pear | Grape | Date Palm |
| Apple | Olive | Beets |
| Orange | Fig | Asparagus |
| Almond | Pomegranate | Spinach |
| Apricot | Tomato | Bermuda grass |
| Peach | Cabbage | Barley |
| Lemon | Cauliflower | Cotton |
| Avocado | Lettuce | |
| Radish | Corn | |
| Celery | Carrot | |
| Green beans | Onions | |
| Clover | Alfalfa | |
| | Wheat | |
| | Rye | |
| | Oats | |
| | Sunflower | |

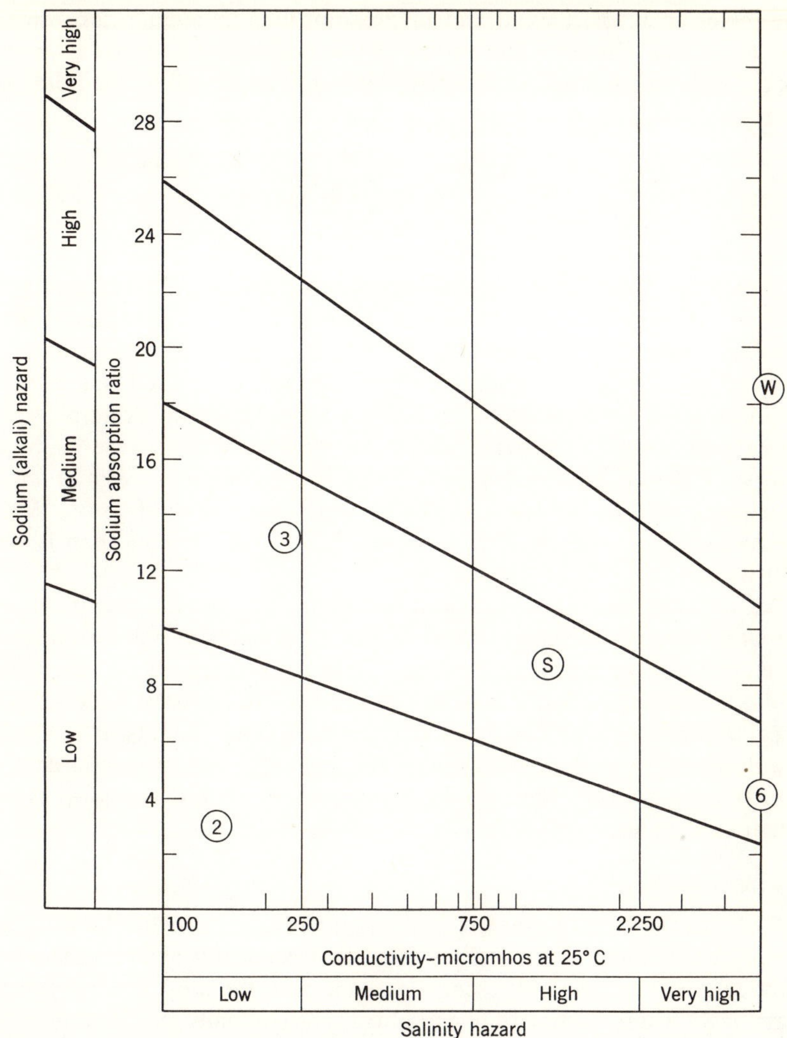


Figure 4.5 Diagram for the classification of irrigation waters. Numbers are those of analyses in Table 4.1. The letter W represents sea water diluted to 1/10 of its original concentration with distilled water. The letter S represents sea water diluted to 1/50 of its original concentration. Diagram modified from Wilcox [51].

faster than in a cool, moist climate. The salinity hazard can be estimated by measuring the electrical conductivity of the water (Figure 4.5).

Besides potential dangers from high salinity and boron, a sodium hazard sometimes exists. The two principal effects of sodium are a reduction in soil permeability and a hardening of the soil. Both effects are caused by the replacement of calcium and magnesium ions by sodium ions on the soil clays and colloids. The extent of this replacement can be estimated by the sodium adsorption ratio (SAR) which is expressed by the following formula:

$$SAR = \frac{Na}{\sqrt{\frac{Ca + Mg}{2}}}$$

in which Na, Ca, and Mg represent concentrations in milliequivalents per liter of the respective ions. The relative hazard of various values of SAR are indicated in Figure 4.5.

Food processing is a complex operation and each food has its own requirements. General limits in Table 4.5 are based on drinking water limits plus special consideration for constituents which cause stains or odors. Calcium and magnesium limits are rather low because many vegetables tend to harden if boiled in water high in these ions. Water moderately high in these ions, however, is better for baking than is very soft water.

Most boiler water is conditioned before use by chemical treatment and sometimes by ion exchange or distillation. The limits of boiler-feed water given in Table 4.5 indicate the extent of treatment necessary for this water. Although some raw water may be suitable for low-pressure boilers, high-pressure boilers require water of the greatest purity. Limits in Table 4.5 are designed to prevent both corrosion and scale. Most standards also specify a pH of more than 8.0 for low-pressure and 9.0 for high-pressure boiler-feed water.

### WATER ORIGIN

The classification of greatest geologic interest is a genetic classification of subsurface water. Eventually a somewhat detailed genetic classification may be possible; at the present, however, even the most general genetic classifications are filled with uncertainty. The common genetic terms used in modern classifications are marine, meteoric, connate, metamorphic, magmatic, volcanic, plutonic, and juvenile. The following definitions are mostly adapted from White [46,47].

*Marine water*, or sea water, is ocean water which has recently invaded rocks and unconsolidated sediments which are in contact with the ocean.

*Meteoric water* is water that has been recently involved in atmospheric circulation.

*Connate water* is water that has been out of contact with the atmosphere for long periods of time, commonly measured in millions of years.

Contrary to some definitions, connate water is not restricted to water which has been virtually static since the burial of the surrounding rocks. Some connate water undoubtedly has migrated many miles since burial.

*Metamorphic water* is water that is, or that has been, associated with rocks during their metamorphism.

*Magmatic water* is water that is in, or is derived from, a magma.

*Volcanic water* is water that is in, or is derived from, a magma at shallow depths.

*Plutonic water* is water that is in, or is derived from, a magma at considerable depths, probably in the order of several kilometers.

*Juvenile water* is water that has never been part of the hydrosphere.

The genetic classification given by White is based on evidence from the geologic environment as well as from important chemical considerations. A summary of his criteria for the differentiation of meteoric, marine, connate, metamorphic, and magmatic waters is given in Table 4.8.

*Table 4.8 Tentative Criteria for Recognition of Major Types of Ground Water*
(White [46,47,49])

| Type | Chemical Composition | Isotopic Composition | Temperature |
|---|---|---|---|
| Meteoric Water | Controlled by surface water and bedrock | Same as or close to surface water | Near normal |
| Marine Water | Very similar to sea water. May have more Ca than sea water | Same as or close to sea water | Near normal |
| Connate Water (chloride type) | Enriched in I, B, $SiO_2$, combined N, Ca, and low in $SO_4$ and Mg relative to sea water | $H^2/H^1 \leqq$ sea water<br>$O^{18}/O^{16} >$ sea water | Normal to slightly thermal |
| Metamorphic Water | Little known. High in combined $CO_2$ and B, low in Cl relative to ocean. Moderately high in I | Little known<br>$H^2/H^1 \leqq$ sea water<br>$O^{18}/O^{16} >$ sea water | Normal to moderately thermal |
| Magmatic Water (Volcanic sodium chloride type) | Relatively high in Li, F, $SiO_2$, B, S, $CO_2$; low in I, Br, Ca, Mg, combined N (?) | Little known.<br>$H^2/H^1 <$ sea water<br>$O^{18}/O^{16} >$ sea water | Strongly thermal |

Although White [47] recognized that connate water could be fresh water, his discussion was confined largely to saline waters of marine origin.

## REFERENCES


1. Aldrich, D. G., A. P. Vanselow, and G. R. Bradford, 1951, Lithium toxicity in citrus: *Soil Sci.*, v. 71, pp. 291–295.
2. Anderson, C. C., and H. H. Hinson, 1951, Helium-bearing natural gases of the United States: *U.S. Bur. Mines Bull.* 486, 141 pp.
3. Back, William, 1960, Origin of hydrochemical facies of ground water in the Atlantic Coastal Plain: *Internat. Geol. Cong.*, 21 Session, Part 1, *Geochemical Cycles*, pp. 87–95.
4. Blanchard, R. L., G. W. Leddicotte, and D. W. Moeller, 1959, Water analysis by neutron activation: *Am. Water Works Assoc. Jour.*, v. 51, pp. 967–980.
5. Bradford, G. R., 1963, Lithium survey of California's water resources: *Soil Sci.*, v. 96, pp. 77–81.
6. Braidech, M. M., and F. H. Emery, 1935, The spectrographic determination of minor chemical constituents in various water supplies in the United States: *Am. Water Works Assoc. Jour.*, v. 27, pp. 557–580.
7. Brown, Philip M., 1958, The relation of phosphorites to ground water in Beaufort County, North Carolina: *Econ. Geol.*, v. 53, pp. 85–101.
8. Cannon, H. L., 1963, The biogeochemistry of vanadium: *Soil Sci.*, v. 96, pp. 196–204.
9. Carter, R. C., and others, 1959, Helium as a ground-water tracer: *Jour. Geophys. Research*, v. 64, pp. 2433–2439.
10. Durum, W. H., and Joseph Haffty, 1961, Occurrence of minor elements in water: *U.S. Geol. Survey Circ.* 445, 11 pp.
11. Foster, M. D., 1942, Base exchange and sulfate reduction in salty ground waters along Atlantic and Gulf coasts: *Am. Assoc. Petroleum Geologists Bull.*, v. 26, pp. 838–851.
12. ——, 1950, The origin of high sodium bicarbonate waters in the Atlantic and Gulf coastal plains: *Geochim. et Cosmochim. Acta*, v. 1, pp. 33–48.
13. George, W. O., and W. W. Hastings, 1951, Nitrate in the ground water of Texas: *Am. Geophys. Union Trans.*, v. 32, pp. 450–456.
14. Gorham, Eville, 1961, Factors influencing supply of major ions to inland waters with special reference to the atmosphere: *Geol. Soc. Am. Bull.*, v. 72, pp. 795–840.
15. Gorrell, H. A., 1958, Classification of formation waters based on sodium chloride content: *Am. Assoc. Petroleum Geologists Bull.*, v. 42, pp. 2513.
16. Grantz, Arthur, and others, 1962, Saline springs Copper River Lowland, Alaska: *Am. Assoc. Petroleum Geologists Bull.*, v. 46, pp. 1990–2002.
17. Griffin, A. E., 1958, Problems caused by manganese in water supplies: *Am. Water Works Assoc. Jour.*, v. 50, pp. 1386–1388.
18. Hall, F. R., 1963, Calculated chemical composition of some sulfate-bearing waters: *Int. Assoc. Sci. Hydrology Pub. No.* 64, pp. 7–15.
19. Hem, J. D., 1959, Study and interpretation of the chemical characteristics of natural water: *U.S. Geol. Survey Water-Supply Paper* 1473, 269 pp.
20. ——, 1963, Chemical equilibria and rates of manganese oxidation: *U.S. Geol. Survey Water-Supply Paper* 1667-A, 64 pp.

21. Hem, J. D., and W. H. Cropper, 1959, Survey of ferrous-ferric chemical equilibria and redox potentials: *U.S. Geol. Survey Water-supply Paper* 1459-A, 31 pp.
22. Hutchinson, G. E., 1957, A treatise on limnology, v. 1, *Geography, physics, and chemistry:* New York, John Wiley and Sons, 1010 pp.
23. Junge, C. E., 1963, Air Chemistry and Radioactivity: New York, Academic Press, 382 pp.
24. Junge, C. E., and P. E. Gustafson, 1957, On the distribution of sea salt over the United States and its removal by precipitation: *Tellus*, v. 9, pp. 164–173.
25. Krauskopf, K. B., 1956, Dissolution and precipitation of silica at low temperatures: *Geochim. et Cosmochim. Acta*, v. 10, pp. 1–26.
26. Leutwein, F., and L. Weise, 1962, Hydrogeochemische Untersuchungen an erzgebirgischen Gruben- und Oberflächenwässern: *Geochim. et Cosmochim. Acta*, v. 26, pp. 133–1348.
27. Lier, J. A. van, 1959, *The solubility of quartz:* Utrecht, Kemink en Zoon, 54 pp.
28. Lohr, E. W., and S. K. Love, 1954*a*, The industrial utility of public water supplies in the United States, 1952, Part 1, States east of the Mississippi River: *U.S. Geol. Survey Water-Supply Paper* 1229, 639 pp.
29. Lohr, E. W., and S. K. Love, 1954*b*, The industrial utility of public water supplies in the United States, 1952, Part. 2, States west of the Mississippi River: *U.S. Geol. Survey Water-Supply Paper* 1300, 462 pp.
30. McKee, J. E., and others, 1957, Water quality criteria: *California Water Pollution Control Board Pub. No.* 3, 512 pp., and 164 pp. In addendum.
31. Morey, G. W., R. O. Fournier, and J. J. Rowe, 1962, The solubility of quartz in water in the temperature interval from 25 to 300°C: *Geochim. et Cosmochim. Acta*, v. 26, pp. 1029–1043.
32. Piper, A. M., A. A. Garrett, and others, 1953, Native and contaminated waters in the Long Beach-Santa Ana area, California: *U.S. Geol. Survey Water-Supply Paper* 1136, 320 pp.
33. Renick, B. C., 1925, Base exchange in ground water by silicates as illustrated in Montana: *U.S. Geol. Survey Water-Supply Paper* 520-D, pp. 53–72.
34. Revelle, Roger, 1941, Criteria for recognition of sea water in ground water: *Am. Geophys. Union Trans.*, v. 22, pp. 593–597.
35. Richards, L. A. (editor), 1954, Diagnosis and improvement of saline and alkali soils: *U.S. Dept. of Agriculture Handbook No.* 60, 160 pp.
36. Roberson, C. E., and H. C. Whitehead, 1961, Ammoniated thermal waters of Lake and Colusa Counties: *U.S. Geol. Survey Water-Supply Paper* 1535-A, 11 pp.
37. Schoeller, H., 1955, Geochimie des eaux souterraines, application aux eaux des gisements de petrole: Paris, *Revue de L'Institut Francais de Petrole et An. de Conbustibles Liquides*, 213 pp.
38. Schofield, J. C., 1955, Methods of distinguishing sea-ground-water from hydrothermal water: *New Zealand Jour. Sci. and Technology*, v. 37, pp. 597–602.
39. Skougstad, M. W., and C. A. Horr, 1963, Occurrence and distribution of strontium in natural water: *U.S. Geol. Survey Water-Supply Paper* 1496-D, pp. 55–97.
40. Sugawara, H., H. Naito, and S. Yamada, 1956, Geochemistry of vanadium in natural waters: Nagoya University (Japan), *Jour. Earth Sci.*, v. 4, pp. 44–61.
41. Sugisaki, R., 1961, Measurement of effective flow velocity of ground water by means of dissolved gases: *Am. Jour. Sci.*, v. 259, pp. 144–153.
42. Sugisaki, R., 1962, Geochemical study of ground water: Nagoya University (Japan), *Jour. Earth Sci.*, v. 10, pp. 1–33.
43. Taylor, F. B., 1963, Significance of trace elements in public, finished water supplies: *Am. Water Works Assoc. Jour.*, v. 55, pp. 619–623.

44. U.S. Department of Health, Education, and Welfare, 1962, Public Health Service drinking water standards: *U.S. Public Health Service Pub. No.* 956, 61 pp.
45. Waring, F. G., 1949, Significance of nitrate in water supplies: *Am. Water Works Assoc. Jour.*, v, 41, pp. 147.
46. White, D. E., 1957*a*, Thermal waters of volcanic origin: *Geol. Soc. Am. Bull.*, v. 68, pp. 1637–1658.
47. White, D. E., 1957*b*, Magmatic, connate, and metamorphic waters: *Geol. Soc. Am. Bull.*, v. 68, pp. 1659–1682.
48. White, D. E., E. T. Anderson, and D. K. Grubbs, 1963, Geothermal brine well: mile-deep drill hole may tap ore-bearing magmatic water and rocks undergoing metamorphism: *Sci.*, v. 139, pp. 919–922.
49. White, D. E., J. D. Hem, and G. A. Waring, 1963, Chemical composition of subsurface waters, Chapter F, in *Data of geochemistry*, 6th Ed.: *U.S. Geol. Survey Prof. Paper* 440-F, 67 pp.
50. White, D. E., and C. E. Roberson, 1962, Sulphur Bank, California, a major hot-spring quicksilver deposit: *Geol. Soc. Am. Bull.*, Buddington volume, pp. 397–428.
51. Wilcox, L. V., 1955, Classification and use of irrigation waters: *U.S. Dept. Agr. Circ. No.* 969, 19 pp.
52. Yaalon, D. H., 1963, On the origin and accumulation of salts in groundwater and in soils of Israel: *Research Council of Israel Bull.*, v. 11G, No. 3, pp. 105–131.
53. Young, R. S., 1957, The geochemistry of cobalt: *Geochim. et Cosmochim. Acta*, v. 13, pp. 28–41.
54. Zartman, R. E., G. J. Wasserburg, and J. H. Reynolds, 1961, Helium, argon, and carbon in some natural gases: *Jour. Geophys. Research*, v. 66, pp. 277–306.

# chapter 5

# RADIONUCLIDES IN GROUND WATER

## 5.1 *Introduction*

Modern concepts of atomic structure were first developed at the close of the nineteenth century. Less than 50 years later the awesome energy available within the nucleus of the atom was utilized for explosions as well as for controlled reactions. Although new hazards exist with this utilization, the potential benefits to humanity are almost too numerous to mention. Within the field of water supply, radioactive material provides the means of dating and tracing both surface and ground water. In addition, there is some hope that nuclear explosives can be used to modify water-bearing characteristics of rocks. Converting saline ground water to fresh water may also be done in conjunction with the operation of nuclear-power plants.

Dangers of radioactive contamination are perhaps the most restricting aspect of the practical utilization of nuclear energy. Burns caused by near contact with the skin and deaths from internal accumulations of small amounts of radium gave warning to early workers of the potential dangers of radioactive substances. Today, however, despite occasional accidents [11], the handling and processing of radioactive materials has one of the best safety records of any major industrial activity. This achievement has been possible only through a full awareness of the hazards of radioactivity.

Unstable nuclides that tend to change spontaneously to other species of nuclides through various decay reactions are called radioactive nuclides, or radionuclides. Nuclides with the same number of protons but with varying masses are isotopes of a single element. Many of the elements have more than ten isotopes. Most of these are radioactive and have been produced artificially.

Nuclear-decay reactions produce alpha ($\alpha$), beta ($\beta$), and gamma ($\gamma$) radiation. Alpha rays are made of positively charged helium ions. Owing to the charge and the relatively large size of the particles, they have very limited powers of penetration. Almost all alpha particles can be stopped by a sheet of paper. Even though the range is small, they are highly ionizing and cause serious internal damage to sensitive parts of the body if radionuclides that emit alpha particles are ingested. In contrast to alpha particles, gamma radiation can penetrate several feet of soil and rock and can completely penetrate the human body. Like X-rays, gamma rays are electromagnetic radiations. Gamma rays, however, have a higher frequency and possess much higher energies. Biological damage from gamma radiation is very similar to X-ray damage. Beta particles are electrons, most of which are the negative species, negatrons. A few radionuclides emit positrons. Beta particles have a greater penetrating power than alpha particles, but far less penetrating power than gamma radiation. Owing to a lesser ability to ionize, beta particles are not as damaging to living cells as are alpha particles.

## **5.2** *Units of Measurement*

The quantity of gamma or X-ray radiation is measured by the amount of ionization produced. The roentgen, which is the basic unit of measure, is defined as the quantity of X-ray or gamma radiation which will give rise to $2.08 \times 10^9$ ion pairs per cubic centimeter of dry air at standard temperature and pressure [17,21,46]. This is equivalent to the release of about 88 ergs of energy. Another commonly used unit is the roentgen-absorbed-dose, or rad, which is equivalent to 100 ergs of energy absorbed per gram of soft tissue (unless otherwise specified) from any type of ionizing radiation. A third unit, roentgen-equivalent-man, or rem, is the absorption by the body of any type of ionizing radiation that produces an effect equivalent to the absorption by man of one roentgen of X-ray or gamma radiation. The values of radiation in rads are measured instrumentally. They are converted to rems by using a multiplying factor which takes into account the relative biological effectiveness of the radiation [44,46]. For example, if all the radiation is gamma radiation, then the measured radiation in rads will equal numerically the biological effect in rems. For alpha radiation, the numerical value in rads must be multiplied by twenty in order to obtain rems. Hazards from various types of radiation are most commonly expressed today in rems, which gives a direct indication of the possible harmful biological effects.

The quantity of radionuclides present is measured indirectly by measuring the number of decay reactions during a given length of time. The basic unit of measure used in the curie, which was originally defined as the number of decay reactions per unit time of one gram of radium. Inasmuch as several isotopes of radium exist, each with a different disintegration rate, a more exact quantity of $3.7 \times 10^{10}$ disintegrations per second has been adopted for the curie. This unit is too large for many practical problems, so millicuries (mc), microcuries ($\mu$c), and micro-microcuries ($\mu\mu$c) are commonly used. A micro-microcurie is also called a picocurie (pc). Inasmuch as the curie refers to radiated energy or matter, it is possible to specify a certain number of curies without identifying the radionuclide that gives rise to the activity. Thus, many analyses give the micro-microcuries of alpha or beta activity per liter or milliliter of water without specific reference to nuclear species.

## 5.3 *Permissible Doses of Radiation*

Radiation escaping from work areas into residential or other areas not directly connected with the production and utilization of the radiation should not exceed 0.5 rem to the body in one year [17,67]. This is probably somewhat greater than, but still the same order of magnitude as, the radiation effects received from medical and natural sources by the average individual [34,59]. In areas of natural concentrations of radioactive minerals, however, the background radiation received by an individual can be higher than 2.0 rems per year [34]. Somatic effects of radiation are not immediately apparent in man below short-term exposures of about 20 rems. Changes in the blood-forming system become marked above 150 to 200 rems. Above 1000 rems the effects of short-term exposure are generally fatal. Greater tolerances for radiation exist if it is received in small increments over a period of many months or years [21].

Although physical damage is not immediately apparent from small amounts of radiation, it is generally thought that even the smallest increases in absorbed radiation will increase the likelihood of cataracts, leukemia, tumors, a general shortening of life, and retarded development of children which were *in utero* at the time of exposure [20,21,24,34,61]. Genetic effects from nuclear radiation have also been produced in biological experiments and undoubtedly cause a certain percentage of the observed mutations in humans [34].

Small amounts of radiation which are received by the body as the result of the ingestion of radionuclides are of greater importance in hydrogeologic investigations than the radiation received from external sources. Ideally, water used for drinking or the production and preparation of food

*Table 5.1 Radionuclides of Interest in Hydrogeology*
(Federal Register, November 17, 1960)

| Element | Mass Number of Radioisotope | Half-Life years, y; days, d; hours, h | Radiation | MPC above Natural Background μc/ml in Solution in Water [67] |
|---|---|---|---|---|
| Barium | 131 | 13 d | $\gamma$ | $2 \times 10^{-4}$ |
| | 140 | 12.8 d | $\beta^-, \gamma$ | $3 \times 10^{-5}$ |
| Bromine | 82 | 36 h | $\beta^-, \gamma$ | $3 \times 10^{-4}$ |
| Calcium | 45 | 153 d | $\beta^-$ | $9 \times 10^{-6}$ |
| Carbon | 14 | 5600 y | $\beta^-$ | $8 \times 10^{-4}$ |
| Cerium | 144 | 290 d | $\beta^-, \gamma$ | $1 \times 10^{-5}$ |
| Cesium | 135 | $2.9 \times 10^6$ y | $\beta^-$ | $1 \times 10^{-4}$ |
| | 137 | 33 y | $\beta^-, \gamma$ | $2 \times 10^{-5}$ |
| Chlorine | 36 | $4 \times 10^5$ y | $\beta^-$ | $8 \times 10^{-5}$ |
| Chromium | 51 | 27.8 d | $\gamma$ | $2 \times 10^{-3}$ |
| Cobalt | 57 | 270 d | $\beta^+, \gamma$ | $5 \times 10^{-4}$ |
| | 60 | 5.3 y | $\beta^-, \gamma$ | $5 \times 10^{-5}$ |
| Hydrogen | 3 | 12.4 y | $\beta^-$ | $3 \times 10^{-3}$ |
| Iodine | 129 | $1.72 \times 10^7$ y | $\beta^-, \gamma$ | $4 \times 10^{-7}$ |
| | 131 | 8.04 d | $\beta^-, \gamma$ | $2 \times 10^{-6}$ |
| Phosphorus | 32 | 14.3 d | $\beta^-$ | $2 \times 10^{-5}$ |
| Plutonium | 238 | 92 y | $\alpha, \gamma$ | $5 \times 10^{-6}$ |
| | 239 | $2.4 \times 10^4$ y | $\alpha, \gamma$ | $5 \times 10^{-6}$ |
| | 240 | 6580 y | $\alpha$ | $5 \times 10^{-6}$ |
| | 242 | $5 \times 10^5$ y | $\alpha$ | $5 \times 10^{-6}$ |
| Radium | 226[a] | 1620 y | $\alpha, \gamma$ | $1 \times 10^{-8}$ |
| | 228 | 6.7 y | $\beta^-$ | $3 \times 10^{-8}$ |
| Radon | 222 | 3.83 d | $\alpha$ | A gas |
| Rubidium | 86 | 18.7 d | $\beta^-, \gamma$ | $7 \times 10^{-5}$ |
| | 87 | $6 \times 10^{10}$ y | $\beta^-$ | $1 \times 10^{-4}$ |
| Ruthenium | 103 | 40 d | $\beta^-, \gamma$ | $8 \times 10^{-5}$ |
| | 106 | 1 y | $\beta^-$ | $1 \times 10^{-5}$ |
| Sodium | 22 | 2.6 y | $\beta^+, \gamma$ | $4 \times 10^{-5}$ |
| Strontium | 89 | 51 d | $\beta^-$ | $1 \times 10^{-5}$ |
| | 90[a] | 29 y | $\beta^-$ | $1 \times 10^{-7}$ |
| Sulfur | 35 | 88 d | $\beta^-$ | $6 \times 10^{-5}$ |
| Uranium | 235 | $7.1 \times 10^8$ y | $\alpha$ | $3 \times 10^{-5}$ |
| | 238 | $4.5 \times 10^9$ y | $\alpha, \gamma$ | $4 \times 10^{-5}$ |
| Zinc | 65 | 245 d | $\beta^+, \gamma$ | $1 \times 10^{-4}$ |

[a] The U.S. Public Health Service has adopted the following limits in their recommendations of drinking water standards.

| | |
|---|---|
| Radium 226 | $3 \times 10^{-9}$ μc/ml |
| Strontium 90 | $1 \times 10^{-8}$ μc/ml |

should not contain radioactive material either in solution or suspension. Normal methods of water purification will remove some of the radioactive material. The hazards from radioactive material in water, however, are usually so low that significant biological damage cannot be detected. As a consequence, special processing to remove radionuclides from water has been undertaken only when unusual amounts of radioactive material have been introduced into the water as a result of reactor operations.

Standards giving the maximum permissible concentration, or MPC, of radionuclides in water have been established by federal and state agencies. Table 5.1 gives the MPC of various important radionuclides. Inasmuch as the term MPC might suggest incorrectly that there is a limit below which immunity is insured but above which harmful effects will always result, the title radioactive concentrations guide, or RCG, has been used by the Federal Radiation Council for their latest list. The expression MPC is, nevertheless, still widely used.

Determination of MPC values is a complicated problem [44]. The most difficult task is to determine an acceptable risk for the general public [31]. Certainly nobody will want to expose himself needlessly; nevertheless, occupational and travel hazards are accepted because of the benefits involved. The same philosophy has been applied to the use of atomic energy. Those evaluating radioactive hazards generally recommend that radiation exposure be kept as low as possible and be no greater than the amounts received by many populations from their natural environment. All MPC and RCG values for the general population are set low enough so that biological effects are so slight or infrequent that they are impossible to detect. Limits set for workers exposed regularly to small amounts of radiation are higher than for the general public. Even these higher values are not known to produce detectable damage, provided that the age of the worker and the accumulated radiation dose is taken into account in setting individual limits.

Even if an acceptable limit of radiation dosage is determined, there still remains the task of calculating the effects of the individual radionuclides. Besides the fact that different species emit different types of radiation, and at various rates (Table 5.1), the energy of alpha and gamma radiation is unique for each radionuclide. In addition, certain elements will tend to accumulate in different parts of the body. Radioisotopes of these elements will then concentrate their damage in restricted areas. For example, plutonium, radium, and strontium will accumulate in the bones, iodine in the thyroid glands, and lead in the kidneys [44]. Other elements, such as carbon and hydrogen, tend to pass through the body after a relatively short period of time. The resulting differences in biological hazards are illustrated by the wide contrast in MPC values between $Ra^{228}$ and $H^3$

(Table 5.1). Both have half-lives of a few years and emit low energy beta radiation, but their residence time in the body is vastly different. Most $H^3$ will pass through the body in a matter of a few weeks; $Ra^{228}$, on the other hand, will remain fixed once it has accumulated in the bone.

## 5.4 *Natural Radionuclides in Ground Water*

The largest number of natural radionuclides are the so-called primordial radionuclides, which, by virtue of their exceptionally long half-lives, are thought to have been present in the primitive material which first accumulated to form the earth. Some of these radionuclides decay to daughter products that are also radioactive. The most abundant primordial radionuclides are $K^{40}$, $Rb^{87}$, $Th^{232}$, $U^{235}$, and $U^{238}$. Of the daughter products, $Rn^{222}$ and $Ra^{226}$, which come from $U^{238}$, are the most important from the standpoint of hydrogeology. A very small quantity of radioactive material is also formed from natural fission and neutron activation from the fission.

A second important group of natural radionuclides originates from cosmic-ray activation of the stable nuclides $N^{14}$, $O^{16}$, and $Ar^{40}$, and to a lesser extent from various nuclides in meteoritic dust and terrestrial water and soil [25]. Most of these activation products have half-lives of less than one million years and are, therefore, confined largely to water, air, living matter, and deposits that are young geologically. Products of cosmic-ray activation that have been studied most extensively are $H^3$ and $C^{14}$.

### URANIUM

Uranium is present in trace or minor amounts in all ground water. The solubility of many uranium compounds in water is high enough to suggest that several ppm could occur in solution [64]. Actual amounts found generally vary from 0.05 to 10.0 parts per billion (ppb). The median of more than 500 analyses of potable ground water is roughly 1.5 ppb [2,3,14,28,35,58]. Ground water moving through rocks rich in uranium generally has more than 200 ppb, with a maximum reported of 18 ppm from uranium-rich sandstone in the Colorado Plateau area of the United States and 90 ppm from the Soviet Union [49,64].

Like many of the elements found in minor and trace amounts, uranium concentrations probably vary widely in the same aquifer in response to local changes in pH, Eh, and temperature [41]. Barker and Scott [2] have found that ground water high in $HCO_3$ is also somewhat higher in uranium (Figure 5.1). They explained this relation by assuming that a soluble uranyl-carbonate complex is formed as follows:

$$HCO_3^- \rightleftharpoons H^+ + CO_3^{-2}$$
$$UO_2^{2+} + 2CO_3^{-2} + 2H_2O \rightleftharpoons UO_2(CO_3)_2(H_2O)_2^{-2}$$
$$UO_2^{2+} + 3CO_3^{-2} \rightleftharpoons UO_2(CO_3)_3^{-4}$$

If this hypothesis is correct, any increase in the amount of bicarbonate will shift the equilibrium and favor the formation of the two uranyl-carbonate complexes. Under certain circumstances, the nonionized sulfate

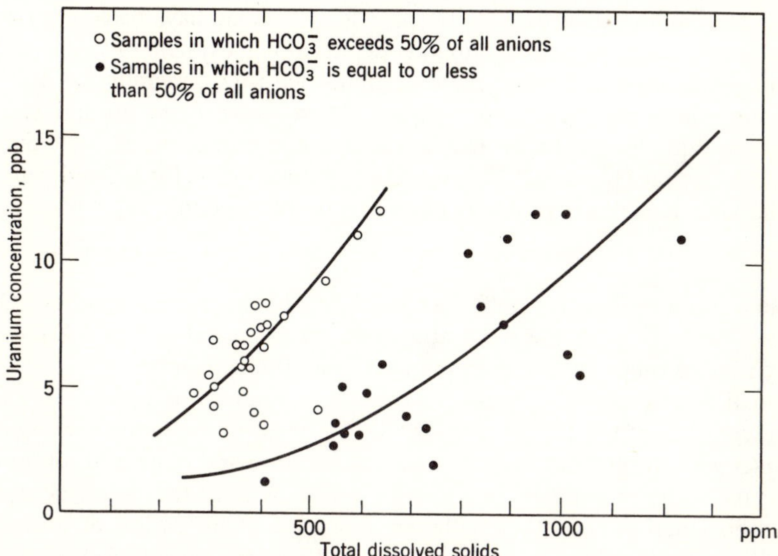


Figure 5.1 Relation between uranium concentration and dissolved solids content of water from the Ogallala Formation of the Llano Estacado, Texas. (Redrawn from Baker and Scott [2].)

complex, $UO_2SO_4$, may also occur in appreciable amounts in ground water [41,64].

The MPC of $4 \times 10^{-5}$ $\mu c$/ml of uranium, or equivalent to about 120 ppm uranium, is not exceeded in natural water. It is interesting to note that in the case of the common uranium isotope, $U^{238}$, the danger from chemical toxicity which causes liver damage is greater than the danger from radiation [54].

### RADIUM

The most abundant isotope of radium, $Ra^{226}$, is one of the decay products of $U^{238}$. The only other radium isotope with a half-life of more than a few days is $Ra^{228}$. It is one of the decay products of $Th^{232}$. Radium is geochemically similar to barium, and as such should be readily sorbed by clays, and colloids. Radium is also known to coprecipitate with barium, calcium, and manganese in nature. The relative immobility of radium, which is inferred from its chemistry, in the hydrosphere has also been used to explain a general deficiency of radium as compared with uranium in ground water of northern Texas [2].

The common isotope of radium, $Ra^{226}$, is the most toxic of all inorganic material. The concentration limit of $3 \times 10^{-9}$ $\mu c/ml$ set by the U.S. Public Health Service for drinking water is exceeded in many well waters [64]. Maximum values of $1 \times 10^{-7}$ to $7 \times 10^{-7}$ $\mu c/ml$ have been reported from water which may rise from great depths along a system of faults [43]. As might be expected, water from uranium-rich rock generally has $Ra^{226}$ concentrations above $3 \times 10^{-9}$ $\mu c/ml$. Most water from normal sedimentary aquifers probably has a radium concentration of less than $1 \times 10^{-9}$ $\mu c/ml$ [39,58,64]. The median radium content for ground water in the United States is probably close to $3 \times 10^{-10}$ $\mu c/ml$ [58].

### RADON

Radon is a radioactive noble gas which has 12 short-lived isotopes. Of these, $Rn^{222}$ is by far the most abundant, having a half-life of 3.8 days, which is five times greater than the next most stable isotope.

Small quantities of $Rn^{222}$ are produced continuously in aquifers by the disintegration of the parent nuclide, $Ra^{226}$. Most of the radon coming in contact with the ground water will go into solution and remain in solution until the water is exposed to the atmosphere or until the radon decays. The short half-life together with the slow rate of migration of ground water allows the radon in solution to be in approximate equilibrium with the trace amounts of radium in the local rocks. Once the ground water discharges into lakes, rivers, or artificial bodies of water, the supply of radon is greatly reduced because of the limited amount of radium in contact with the water. Even more important, gases in the water will approach equilibrium with gases in the atmosphere. Inasmuch as the atmosphere contains only traces of radon, most of the radon in the water will pass into the atmosphere. Arndt and Kuroda [1] reported that spring water in Arkansas lost 41 per cent of its radon after flowing only 4 feet as a surface stream. Rogers [53] found similar rapid declines of radon in turbulent mountain streams in Utah. He was able to take numerous samples of stream water and to locate places where ground water entered the streams by noting the position of anomalously high concentrations of radon.

Ground water with the highest radon content is most commonly associated with uranium deposits which are in turn high in radium. Also, rapidly ascending ground water along faults, and hot-spring water may contain unusually large amounts of radon. Normal ground water contains from less than $1 \times 10^{-7}$ $\mu$c/ml to about $3 \times 10^{-5}$ $\mu$c/ml radon. The median concentration for ground water may be close to $2 \times 10^{-6}$ $\mu$c/ml. Maximum values reported for various regions generally range from about $5 \times 10^{-5}$ to $4 \times 10^{-4}$ $\mu$c/ml [1, 5, 27, 43,53].

Radon dissolved in water has not been reported to be a health hazard. In gaseous form, however, it will produce undesired radiation exposure to the lungs, and has been of concern to miners working in improperly ventilated uranium mines.

### POTASSIUM 40

About 0.012 per cent of potassium in nature is the radioisotope $K^{40}$. Normal potable water contains about 2.5 ppm potassium, or $3 \times 10^{-4}$ mg/1 of $K^{40}$. This concentration is equivalent to about 2 $\mu\mu$c/1 or $2 \times 10^{-9}$ $\mu$c/ml.

Although $K^{40}$ may account for 5 to 50 per cent of the natural beta activity in potable water, it has not been a major consideration in determining the biological hazard of water. It will be taken into the body and will be concentrated somewhat in the gonads, but its low specific activity and the lack of large amounts in the water do not make it one of the more hazardous radionuclides.

### TRITIUM

The heavy isotope of hydrogen, $H^3$ or tritium, is produced continuously by cosmic-ray activation of nitrogen [25]. Before 1952, the tritium content of most rain ranged from about 1 to 10 tritium units (a tritium unit, TU, is a concentration of one atom of $H^3$ for every $10^{18}$ total hydrogen atoms; 1 TU produces roughly $3.2 \times 10^{-3}$ $\mu\mu$c/ml activity). Testing of thermonuclear devices since 1952 has greatly increased the $H^3$ content of rain and surface water. The first tests in the Ivy series in 1952 contributed a small amount in comparison with the Castle series in 1954 and several other test series since 1954. Peak values in the Northern Hemisphere generally have reached several hundred tritium units, but range widely both with time and with geographic position. A maximum of 2937 TU was reported from Ottawa in March of 1954 [36]. Variations are related to the time of testing, type of test, location of the test, mechanics of individual storms, regional wind patterns, and percentage of the precipitation that is derived from recondensation of old ocean water in areas of upwelling ocean currents.

Data are not extensive enough to calculate general averages of tritium that might be found in water which has recently infiltrated into the ground.

Several measurements suggest that water entering the ground since 1954 should have had initially more than 10 TU and in many areas more than 20 TU [18,29,36,37]. Local peak values would, of course, be almost the same as the rain, or several hundred tritium units. Dispersion of the infiltrating water will reduce the peak values as the ground water moves through the aquifer.

Tritium can be used both to date and to trace water [12,13,18,31,36,62, 66]. The relatively short half-life of tritium, 12.4 years, means that within 40 to 60 years the tritium content of pre-1954 water will be below most methods of detection. Within about 90 years the tritium content of water originating as rain during the period of testing of thermonuclear weapons should be below the limits of detection.

The tritium method of dating ground water is only approximate owing to the uncertainty of the original tritium content of the infiltrating rain, the extent of mixing by dispersion in the aquifer, and the possibility of mixing through the discharge of multiple aquifers into wells during sampling. Although only approximate, useful applications of dating have been made to ground-water problems in various parts of the world. For example, tritium concentrations of 3.7 to 5.0 TU in shallow ground water from wadi gravels in the Arabian peninsula suggested that the water sampled in 1958 was about 10 years old, inasmuch as the values were too low to be associated with post-1954 rain but high enough to indicate that less than one half life of tritium had elapsed [62]. Libby reported several applications of tritium dating in which the water could be identified as water which had recently originated from rain or water which had resided out of contact with the atmosphere for more than a decade or so [36]. Commonly, tritium dating can be used to solve pressing problems in hydrogeology. Water from behind a dike complex on Molokai Island, Hawaii, was found to be, in part, water which is relatively old. This fact indicated that the zone could be used for holdover storage of wet-season water. If the water had been less than a year old, it would have indicated a very rapid recharge and outflow in the system, thus suggesting that the zone could not have been used for storage [36].

Artificial tritium is almost an ideal tracer [30,31]. It is actually part of the water and as such is not selectively sorbed by the aquifer material; it can be detected in relatively small amounts; its half-life is long enough to be used for experiments lasting several years; its biological hazard is lower than most other radionuclides, and it is not present in large amounts in the original water.

Natural tritium can be used also as a ground-water tracer. In a study in New Jersey, samples of water were taken for tritium analyses at varying depths from a series of wells which were more or less perpendicular to the regional flow of the ground water [12]. Although local stratification exists, the material as a whole is moderately homogeneous, but probably quite anisotropic. Results of the measurements are shown in Figure 5.2. The figure illustrates the fact that there is layering of the water and that water below 30 meters is essentially free of tritium. River-water samples indicated further that the low flow in the river is sustained by inflow of the lower tritium free water.

### CHLORINE-36

A small amount of $Cl^{36}$ is produced continuously by cosmic-ray activation of atmospheric $Ar^{40}$. The small amount of sorption of chlorides plus the long half-life make $Cl^{36}$ a potentially important radionuclide with which to date water. Unfortunately, the exceedingly low pre-1954 natural concentrations are almost impossible to detect in ground water. Even with the large increase in $Cl^{36}$ from the activation of sea water by thermonuclear explosions, water samples of several thousand liters are necessary in order to recover sufficient $Cl^{36}$ for study [55]. Despite this limitation, future work may find that the water related to the period of testing of thermonuclear devices can be detected as it moves in the subsurface.

### CARBON-14

The dating of organic material by its $C^{14}$ content is a well-established procedure. The method depends on the fact that while living, carbon within organic material is in near equilibrium with the carbon in the atmosphere. The ratio between radiocarbon, $C^{14}$, and stable carbon is thus the same as the atmosphere. Once the organism dies, atmospheric carbon is no longer added and the radiocarbon decays so that the ratio of $C^{14}$ to $C^{12}$ and $C^{13}$ decreases with time. The ratio then can be used to date the ancient organic material being tested. Most rainwater and small bodies of surface water are also in near equilibrium with the $C^{14}$ in the atmosphere. If this water is buried in the subsurface, then the ratio of radiocarbon to stable carbon would decrease with time. The $C^{14}$ content could then be used to date subsurface water by removing all the carbon [15,19,45,62], most of which is in the carbonate ion, and counting the activity of the carbon. Unfortunately, the carbonate in most ground water comes from various sources, of which the atmosphere may, under some circumstances, be only of secondary importance. Water percolating long distances from the surface to the watertable may encounter large amounts of $CO_2$ in the soil atmosphere which originates from the decomposition of ancient organic material buried in the subsurface. Solution of carbonate

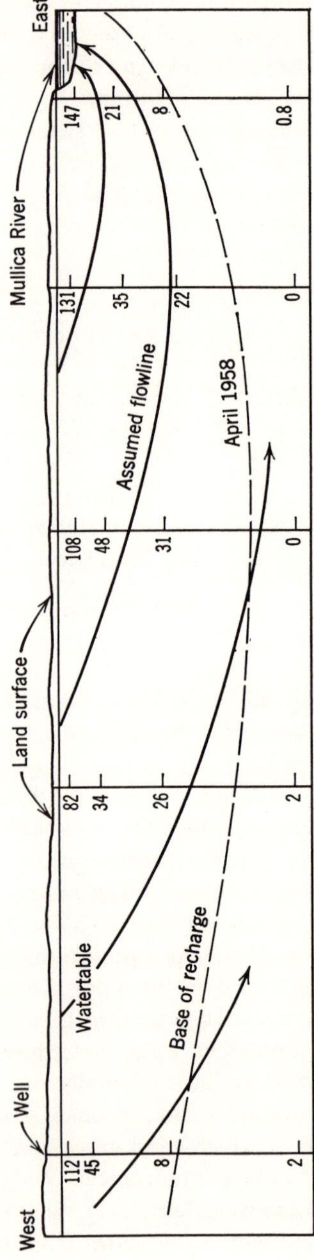


Figure 5.2 Cross section of part of Wharton Tract well field in New Jersey. The line labeled "base of recharge" is based on tritium analyses. Numbers beside wells indicate tritium units of samples from various depths. (Redrawn from a diagram by Carlston, Thatcher, and Rhodehamel [12].)

minerals and possibly some anion exchange will contribute carbonate to the water which may have been out of contact with the atmosphere for millions of years and will thus contain no measurable $C^{14}$. Dates of ground water based on the radiocarbon method will, therefore, indicate ages that are in general older than the water. Preliminary work by Munnich and Vogel [45] indicates that recent ground water has 85 per cent of the $C^{14}$ content of recent material. If this is generally true, it means that $C^{14}$ dating of water may be more accurate than would be suggested by general geologic reasoning. Moderately accurate dating may be possible for water from desert regions where the water percolates through sandy aquifers free of carbonate and organic matter. Dates of water in the Arabian peninsula and of water from the United Arab Republic suggest that much of the deeply buried water may be more than 10,000 years old and some of its may be more than 30,000 years old [15,62].

Natural carbon of recent biological origin has an activity of about 16 disintegrations per minute per gram which means that water containing 150 ppm $HCO_3$ which derives all carbon from the atmosphere or the decay of recent organic material has an activity of $2.1 \times 10^{-10}$ $\mu c/ml$, which is far below the maximum permissible concentration for drinking water. Unlike $H^3$ and $Cl^{36}$, the total amount of $C^{14}$ in the atmosphere has increased only a few per cent above the pre-1954 level as a result of the testing of nuclear devices [9].

## 5.5 *Man-Made Concentrations of Radionuclides*

Artificial radioactivity arises from the bombardment of matter with subatomic particles. If bombardment results in the production of elements of lower atomic numbers, the process is called fission. A fission chain is produced when neutrons from fission induce other fission reactions. Plutonium-239 and uranium-235 are able to sustain a fission chain and are essential components of fission explosives and reactors. The basic reaction is:

Neutron + uranium-235 (or plutonium-239) $\rightarrow$ fission fragments + 2 or 3 neutrons + energy (about 200 million electron volts which is equivalent to about $1.6 \times 10^{-6}$ erg [21]).

Most of the fission fragments have mass numbers between 80 and 110 and between 125 and 155. Out of some 80 nuclides produced, $Sr^{89}$, $Sr^{90}$, $Ru^{106}$, $I^{131}$, $Cs^{137}$, and $Ce^{144}$ are probably the most important from the standpoint of biological hazards.

Nuclear reactions whereby lighter elements are brought together to form heavier elements are called fusion reactions. Fusion reactions produced by thermonuclear explosions [21] are given as follows:

$$H^2 + H^2 \rightarrow He^3 + n + \text{energy}$$
$$H^2 + H^2 \rightarrow H^3 + H^1 + \text{energy}$$
$$H^3 + H^2 \rightarrow He^4 + n + \text{energy}$$
$$H^3 + H^3 \rightarrow He^4 + 2n + \text{energy}$$

in which n stands for a neutron produced. Inasmuch as the fusion reactions have been initiated by the heat of fission explosions, fission products have been also produced by the testing of thermonuclear devices.

Excess neutrons from both the fission and fusion reactions will bombard the environment and produce new radioactive nuclides. Of those produced, $Fe^{55}$, $Ca^{45}$, $Cl^{36}$, $Co^{60}$, $H^3$, $P^{32}$, $S^{35}$, and $C^{14}$ are probably most important in relation to the safety of water supplies.

The environment is likely to be contaminated through the refinement and processing of uranium ore, initial production of nuclear fuels and explosives, reprocessing used reactor elements, discharge of cooling water which has been exposed to nuclear activation, escape of volatile material from evaporation and burning, dispersion of products of nuclear explosions [40], and the release of radionuclides used in science and medicine. Of the various sources of radioactive contaminants, these associated with reactor operations and nuclear explosions have been of most concern.

### ORE PROCESSING

Radium associated with waste water from uranium mills has posed a serious contamination problem in surface of the uranium mining districts of the Colorado Plateau in the United States. Most problems have been solved by filtration and water processing [6,65]. In at least one place tailing water from a mill is disposed of in a well which feeds the water. into a thick sandstone more than 1000 feet below the surface. The sandstone was originally saturated with saline water which had little or no economic value. The sandstone is isolated from overlying potable water by impermeable evaporate and shale beds.

### REACTOR WASTES

The safe disposal of wastes from reactor operations and fuel reprocessing is one of the major problems in the widespread utilization of nuclear power [6,16,22]. Although individual estimates vary, it is likely that by the year 2000 several tens of millions of curies of $Sr^{90}$ and $Cs^{137}$ will be produced each month as by-products of the atomic energy industry in the United States alone [16]. The magnitude of the disposal problem can be realized from the fact that only a single curie of $Sr^{90}$, if dissolved in water, could render $10^{11}$ liters unacceptable as drinking water according to U.S. Public Health Service standards. It is easily seen that simple dilution in even the largest lakes and rivers will not be sufficient to dispose of the waste. Indeed, the oceans would not be large enough, even if they were evenly mixed, because many of the hazardous radionuclides will be concentrated by marine life.

Disposal practices depend on the radioactivity of the waste, the general chemical character of the waste, and the physical environment in the area of disposal. The radioactivity of liquid waste is broadly referred to as low level if it has fractions of a microcurie per gallon; intermediate level if it has less than a few curies per gallon but greater than a microcurie per gallon; and high level if it it has more than a few curies per gallon [6]. Some waste initially contains several hundred curies per gallon.

Low-level wastes result from a number of nuclear-energy operations. Minute amounts of radioactive dust in wash water, volatile radionuclides in steam condensate, and activation of impurities in cooling water accounts for much of the large volume of low-level waste. Some installations dispose of more than a million gallons of low-level waste each day. For example, almost 500 million gallons were injected into a single well in the National Reactor Testing Station in Idaho during 1958 [56]. A much greater amount has been disposed of in natural depressions at Hanford, Washington. Most of the water disposed of in the depressions is cooling water which contains less than $10^{-5}$ $\mu$c/ml gross beta activity. Steam condensate at Hanford which may contain from $10^{-13}$ to $10^{-1}$ $\mu$c/ml has been disposed of in cribs which are box-like tiber structures in the ground (Figure 5.3) [38]. Water high in chemicals that will adversely affect the ion sorption has been disposed of in trenches which are used only for a limited length of time.

Total activity disposed of by spreading, cribs, and trenches at Hanford has been almost 3 million curies of mixed fission wastes. Most of the radionuclides are sorbed on the soil particles [51]. A small quantity will move downward with the percolating water. Extensive studies of the soil have been made to determine the capacity of the soil to receive the radionuclides. In addition, more than 600 wells have been installed to study the natural circulation of the ground water and to monitor the migration of the radionuclides in the water [7,10]. At the disposal area, the temperature of the water as well as analyses of the more mobile radionuclides give a good indication of the movement of the waste water. The artificial recharge has altered the natural movement of the water and has created a large ground-water mound (Figures 5.4 and 5.5) [7]. The resulting gradients plus the local variations in permeability control the direction of contaminant migration.

At Hanford the most mobile radionuclides which have been reported are $Ru^{103}$ and $Ru^{106}$. Inasmuch as $Ru^{106}$ has a half-life about ten times as long as $Ru^{103}$, it is much more significant as a potential contaminant. Next in order of mobility of the radionuclides reported is $Sr^{90}$. The least mobile is $Cs^{137}$, which is very strongly sorbed on the clay surfaces (Figure 5.3).

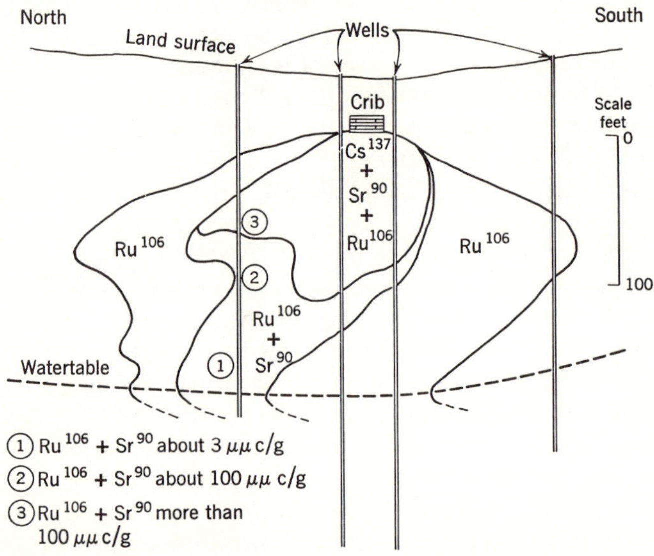


Figure 5.3 Downward movement of radionuclides from crib used for waste disposal at Hanford, Washington, January 1956. Two outer wells were drilled after use of crib was discontinued. (Diagram from Linderoth and Pearce [38].)

The different geologic conditions and possibly different chemical nature of the wastes cause a different mobility of the contaminants at the Savannah River Plant in South Carolina [50]. Here wastes are sent to large ponds or basins excavated into the Tuscaloosa Formation which at this place is sand, silt, and clay. The clays are mostly kaolinitic. Here, the most mobile constituent traced is tritium. The next most mobile is $Sr^{90}$. Samples recovered from perched ground water adjacent to the pits have had $H^3$, $Sr^{89}$, $Sr^{90}$, $Ce^{141}$, $Ce^{144}$, $Cs^{137}$, $Ru^{103}$, $Ru^{106}$, and Zr-$Nb^{95}$. Sorption plus decay of the radionuclides with short half-lives has prevented the movement of all but the $H^3$ and $Sr^{90}$ to the deeper ground water. $Pu^{239}$ has not been observed to leave the seepage basins. Pattern of movement from three basins is shown in Figure 5.6.

Most high-level wastes come from the reprocessing of reactor fuels. In addition to the radioactivity, most of these wastes have troublesome

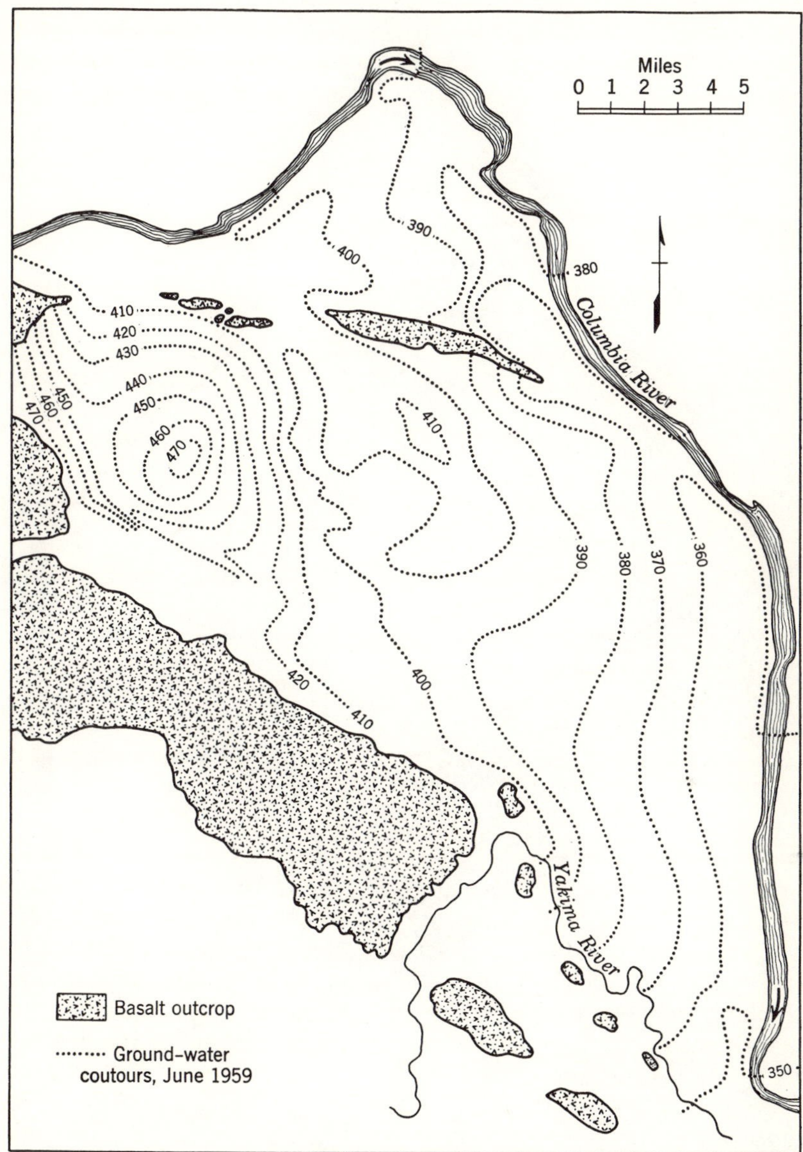


Figure 5.4 Elevation of water levels in wells at Hanford, Washington. (Redrawn from a diagram by Bierschenk [7].)

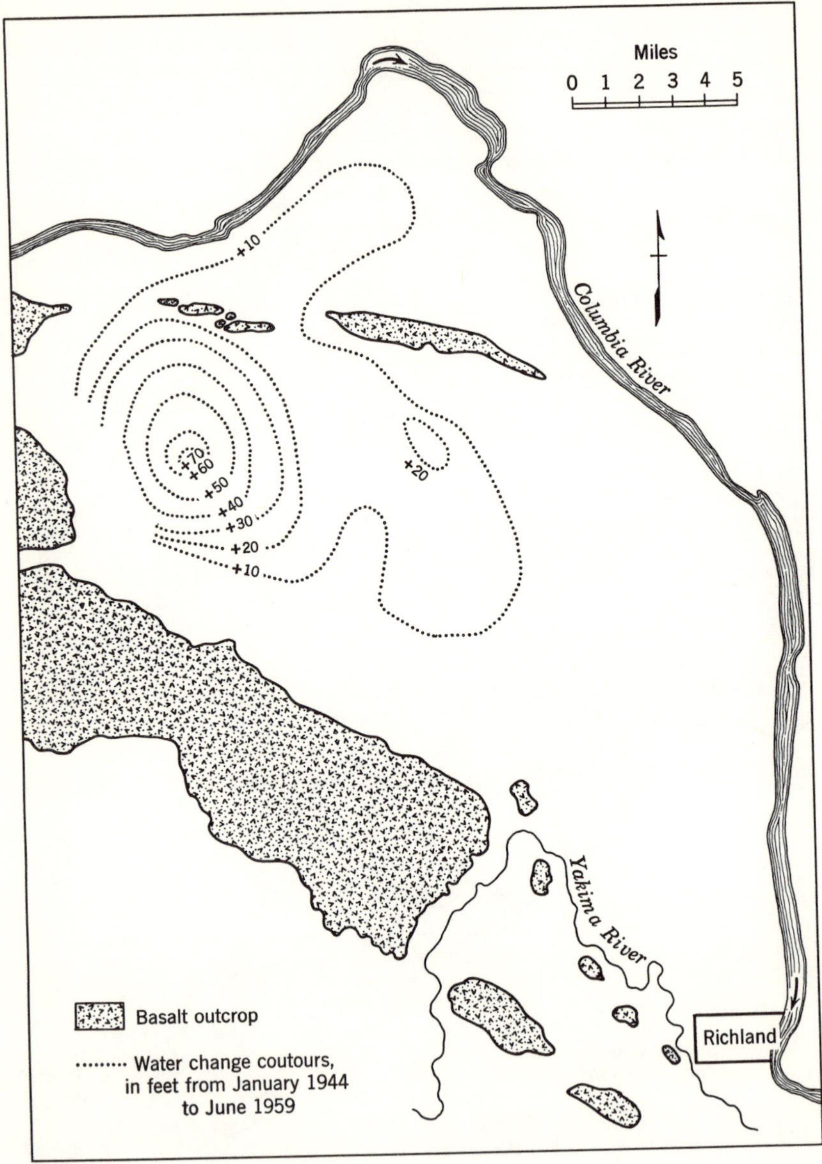


Figure 5.5 Water-level change caused by the disposal of waste water at Hanford, Washington. (Data from Bierschenk [7].)

chemical properties. One of the common types of waste is an acid waste high in $NO_3$ and Al. Most of the high-level wastes are presently stored at or near the surface in tanks. A number of schemes are under study for the eventual disposal of these wastes [6,16,52]. They are accumulating very rapidly and are a major expense in the atomic energy program. Storage in mines, specially excavated cavities, injected grout, and as injected liquids are some of the possibilities. All plans for terrestrial storage have stressed the need for careful hydrogeologic studies to insure that the wastes will not contaminate existing or future water supplies.

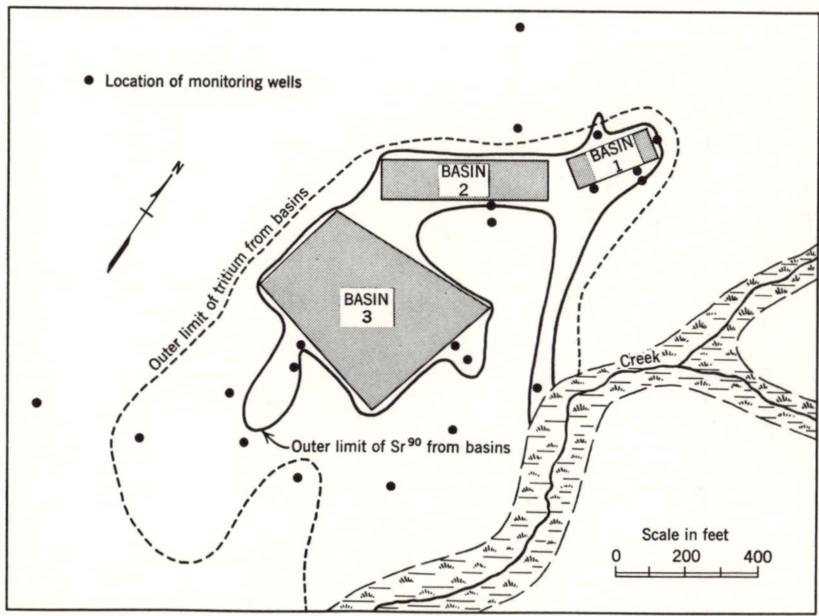


Figure 5.6 $H^3$ and $Sr^{90}$ in ground water as a result of waste-water disposal at Savannah River Plant, South Carolina, January-February 1962. (From Reichert [50].)

All reactors under the supervision of the U.S. Atomic Energy Commission are required to have a complete hazard evaluation before installation. An important part of this evaluation is a study of the local ground water [8,33,47,48]. Although most accidents at reactor stations do not involve large amounts of material [11] and only a few cases of radioactive material leaving the reactor building are known, complete safety of the ground water is desired even in the case of the worst accident which might be the result of reactor or natural disaster. Normally, there is an almost negligible amount of radioactivity discharged from the operation of a reactor, since all the spent fuel is shipped to localities where the fuel is reprocessed.

### NUCLEAR EXPLOSIONS

The only radionuclides from nuclear explosions which have been reported from ground water are $H^3$ and $Cl^{36}$. Undoubtedly some $C^{14}$ which

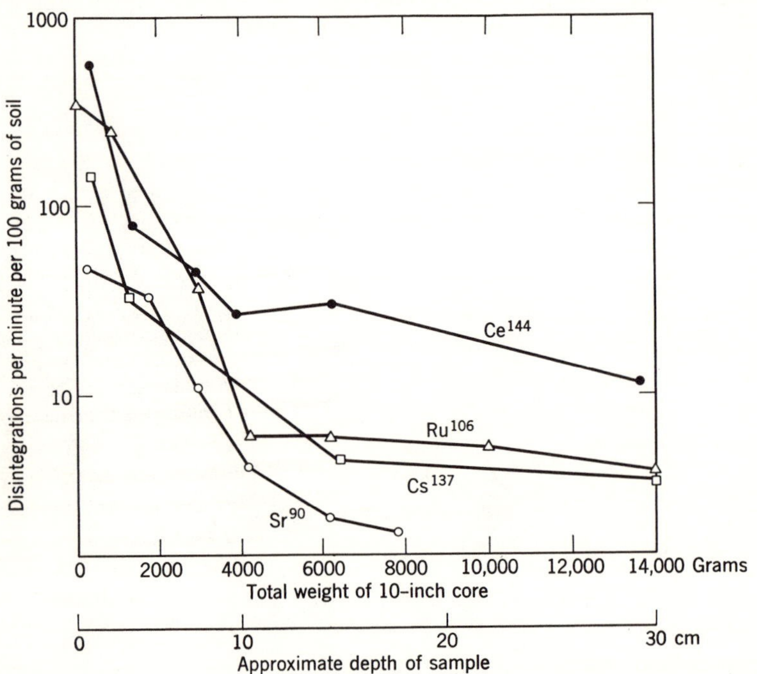


Figure 5.7 Variation of radionuclides with depth in Keyport soil of New Jersey. The soil is a sandy clay. All radionuclides are from worldwide fallout from weapons testing before the summer of 1960. (Data are from Walton [69].)

originated from nuclear tests has also entered the subsurface. The reported concentrations of these radionuclides are far below the values considered as significant hazards. The more hazardous radionuclides such as $Sr^{90}$ and $Cs^{137}$ are sorbed in the soil and are moved downward very slowly by infiltrating ground water. For example, extensive sampling in New Jersey [69] indicated more than 90 per cent of the activity from these nuclides was still in the upper 9 inches of most of the soils in 1960 (Figure 5.7). The greatest penetration was measured in sand which presumably had the lowest ion-exchange capacity. Even in the sand, however, the surface concentration was roughly five times the concentration at a depth of 12 inches. The data from New Jersey suggest that $Sr^{90}$ may reach shallow watertables in sand dunes within regions of high rainfall within a few decades. Downward movement in most desert soils probably has been negligible [63]. Aerial distribution of $Sr^{90}$ as a result of tests is only a small fraction of that needed to create water having $Sr^{90}$ concentrations above MPC, particularly if the amount of $Sr^{90}$ which will be sorbed on the solid particles in the soil is taken into account.

Subsurface nuclear detonations in or near saturated and permeable rocks have not been described in the literature. Such detonations are visualized in the ultimate utilization of nuclear explosives for peaceful purposes. Arguments by Higgins [26] and Batzel [4] suggest that under some circumstances the long-term contamination hazards to regional aquifers could be negligible. Each case must be judged individually, however, because of the variations of device, rock type, water quality, and regional hydrology. Especially important is the fact that cavernous limestone, coarse gravel, some volcanic rocks, and highly fractured rocks may not have sufficient sorptive capacity to delay the movement of the radionuclides enough to insure safe conditions in the aquifer. Stead [60] has shown that tritium may become a troublesome contaminant if large nuclear-fusion explosions are near aquifers. Under certain circumstances several million curies of $H^3$ could enter into a ground water system from a single detonation.

### RADIONUCLIDES AS ARTIFICIAL TRACERS

The greatest advantage of artificially introduced radioactive tracers comes from the ability to detect very small quantities of the tracer in the water. Thus, the tracers can be used without causing extensive chemical modification of the aquifers and without involving the injection of highly concentrated chemicals that tend to separate from the water by gravity differentiation. Unfortunately, however, there are certain drawbacks to all tracers and to radioactive tracers in particular. The greatest drawback is the hazard involved, even with a relatively safe tracer such as tritium. A second difficulty is that special equipment is needed to detect the tracers; alpha and beta emitters are particularly difficult to detect in small quantities. Ideally a tracer should be easily detected, be used in small quantities, not change the hydraulic characteristics of the aquifer, be inexpensive to purchase and utilize, not be sorbed by the media, have a useful half-life, have low toxicity, and not normally be present in aquifers in large quantities. If tritium were easier to detect, it would fulfill almost all the requirements quite well.

Other radionuclides that have been used for artificial tracers are $I^{131}$,

$Cr^{51}$, $Co^{60}$, $Rb^{86}$, $Ru^{103}$, and $Br^{82}$ [23,30,32]. Iodine is quite strongly sorbed on clays. The complex ion, $Co(CN)_6^{-3}$, nevertheless, has been found very satisfactory. It travels with the water and is easy to detect. Chromium can be combined in an organic complex, ethylene diamine tetraacetic acid, which will not be affected greatly by sorption. Its half-life of about 28 days is ideal for tracer tests lasting less than a few months. Bromine-82 is easily detected by its gamma emission, but its half-life of only 36 hours limits its use for many tests.

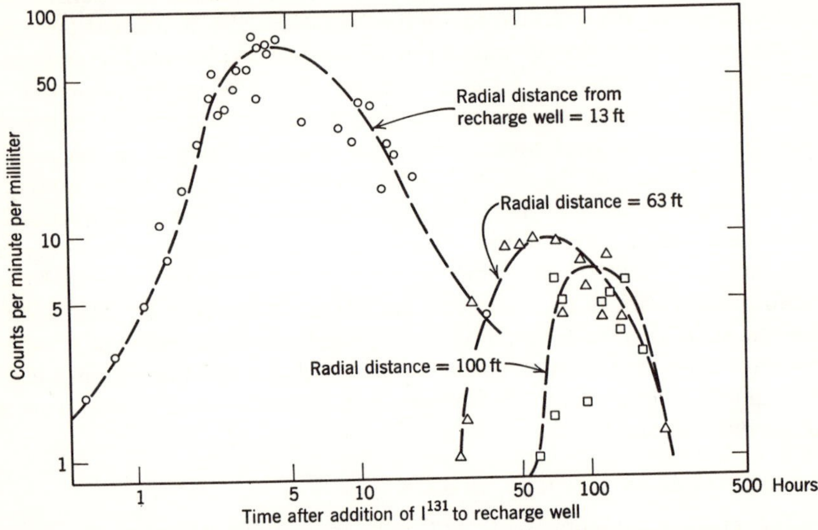


Figure 5.8 Variations of $I^{131}$ concentrations with time at various distances from an injection well. (From Kaufman and Orlob [32].)

The major factors that control the rate of migration of radionuclides in aquifers are the type and concentration of the radionuclide, the velocity of the water, the ionic species and their concentration in the water, the sorptive characteristics of the media, and the porosity of the aquifer. The media will tend to sorb ionized radionuclides from the water. They will be dislodged by other stable ions in the ground water that are competing for the space occupied by the radionuclides. Thus, the higher the ion concentration, the faster the radionuclide will move by successive displacements. An approximate expression for the rate of movement of radionuclides within one dimensional ground-water flow [26,42] is as follows:

$$V_i = \frac{V_w}{1 + K_d \dfrac{\rho_b}{n}}$$

in which $V_i$ is the velocity of the ionic species,
$V_w$ is the velocity of the water,
$\rho_b$ is the bulk density of the media,
$n$ is the porosity,
and $K_d$ is the distribution coefficient of the media in relation to the ionic species and the other ions in the water.

If both velocity values are in the same units and the density is measured in g/cm³, then $K_d$ has the units of cm³/g. Physically, the distribution

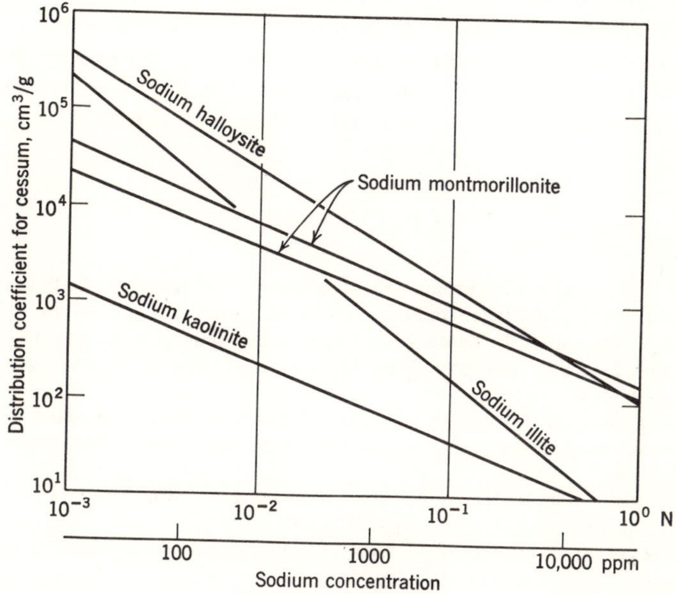


Figure 5.9 Distribution coefficients of Cs on various clay minerals as a function of sodium-ion concentration. The values are valid only for small concentrations of Cs, these are generally less that $10^{-8}$ N (1.3 ppb). (Data are from Wahlberg and Fishman [68].)

coefficient can be thought of as a measure of the distribution of an ion species between the water, or liquid phase, and the solid phase which tends to sorb the ion. A high distribution coefficient indicates a strong tendency for sorption.

The distribution coefficient is a variable that is uniquely determined for only a single set of physical and chemical conditions. In very low concentrations, however, the distribution coefficient is principally a function of the media and the concentrations of the competing ions. The influence of different concentrations of sodium and different minerals for small amounts of cesium is shown in Figure 5.9. The distribution coefficient for trace amounts of strontium in sandy aquifers with no clays or organic material but with water saturated with various cations ranges from about 0.5 to 50 $cm^3/g$ [57]. The distribution coefficient for cesium under similar circumstances will range from about 1 to 500 $cm^3/g$. Coarse quartz sand will have the lowest and fine calcite sand will have the highest distribution coefficients [57].

Inasmuch as $Sr^{90}$ and $Cs^{137}$ are two of the most troublesome contaminants produced by nuclear fission, the foregoing information is of great practical importance. Using reasonable values for the porosity and density of granular aquifers, it can be seen that even though contamination by $Sr^{90}$ and $Cs^{137}$ occurs, both radionuclides will travel with only a small fraction of the velocity of the ground water. Furthermore, the velocity of $Cs^{137}$ will almost always be less than $Sr^{90}$ in potable water.

## REFERENCES

1. Arndt, R. H., and P. K. Kuroda, 1953, Radioactivity of rivers and lakes in parts of Garland and Hot Springs Counties, Arkansas: *Econ. Geol.*, v. 48, pp. 551–567.
2. Barker, F. B., and R. C. Scott, 1958, Uranium and radium in the ground water of the Llano Estacado, Texas and New Mexico: *Am. Geophys. Union Trans.*, v. 39, pp. 459–466.
3. Barker, F. B., and R. C. Scott, 1961, Uranium and radium in ground water from igneous terranes of the Pacific Northwest: *U.S. Geol. Survey Prof. Paper* 424-B, pp. 298–299.
4. Batzel, R. E., 1960, Radioactivity associated with underground nuclear explosions: *Jour. Geophys. Research*, v. 65, pp. 2897–2902.
5. Belin, R. E., 1959, Radon in the New Zealand geothermal regions: *Geochim. et Cosmochim. Acta*, v. 16, pp. 181–191.
6. Belter, W. G., 1963, Waste management activities of the (U.S.) Atomic Energy Commission: *Ground Water*, v. 1, pp. 17–24.
7. Bierschenk, W. H., 1961, Observational and field aspects of ground-water flow at Hanford (Washington), in Ground disposal of radioactive wastes; *Sanitary Engineering Research Laboratory, University of California, Berkeley, Conference Proceedings*, pp. 147–156.
8. Bowen, B. M., and others, 1960, Geological factors affecting ground disposal of liquid radioactive wastes into crystalline rocks at the Georgia Nuclear Laboratory site: *Twenty-First Int. Geol. Congress Section* 20, pp. 32–48.
9. Broecker, W. S., and Alan Walton, 1959, Radiocarbon from nuclear tests: *Sci.*, v. 130, pp. 309–314.
10. Brown, D. J., and J. R. Raymond, 1962, Radiologic monitoring of ground water at the Hanford Project: *Jour. Am. Water Works Assoc.*, v. 54, pp. 1201–1212.
11. Buchanan, J. R., and L. J. King, 1963, Accidents in nuclear energy operations: *Nuclear Safety*, v. 4, pp. 147–155.
12. Carlston, C. W., L. L. Thatcher, and E. C. Rhodehamel, 1960, Tritium as a hydrologic tool, the Wharton Tract study: *Internat. Assoc. Sci. Hydrol. Publ. No.* 52, pp. 503–512.

13. Clebsch, Alfred, Jr., 1961, Tritium-age of ground water at the Nevada Test Site, Nye County, Nevada: *U.S. Geol. Survey Prof. Paper* 424-C, pp. 122–125.
14. Cohen, Philip, 1961, An evaluation of uranium as a tool for studying the hydrogeochemistry of the Truckee Meadows area, Nevada: *Jour. Geophys. Research*, v. 66, pp. 4199–4206.
15. Degens, E. T., 1961, Diagenesis of subsurface waters from the Libyan Desert (abstract): *Geol. Soc. American Special Paper* 68, p. 160.
16. deLaguna, Wallace, 1962, Engineering geology of radioactive waste disposal: *Geol. Soc. America, Reviews in Engineering Geology*, v. 1, pp. 129–160.
17. Department of the Navy, 1962, Radioactivity in water supply and waste water systems: *U.S. Bureau of Yards and Docks, NAVDOCKS* MO-218, 47 pp., Appendices.
18. Eriksson, Erik, 1962, Radioactivity in hydrology, in H. Israel, and A. T. Krebs (editors), *Nuclear radiation in geophysics:* New York, Academic Press, pp. 47–60.
19. Feltz, H. R., and B. B. Hanshaw, 1963, Preparation of water sample for Carbon-14 dating: *U.S. Geol. Survey Circ.* 480, 3 pp.
20. Glass, Bently, 1957, The genetic hazards of nuclear radiations: *Sci.*, v. 126, pp. 241–246.
21. Glasstone, Samuel (editor), 1962, *The effects of nuclear weapons:* Washington, D.C., U.S. Department of Defense and U.S. Atomic Energy Commission, 730 pp.
22. Glueckauf, E. (editor), 1961, *Atomic energy waste: Its nature, use and disposal:* London, Butterworths, 420 pp.
23. Halevy, E., and A. Nir, 1962, The determination of aquifer parameters with the aid of radioactive tracers: *Jour. Geophs. Research.*, v. 67, pp. 2403–2409.
24. Harley, J. H., 1957, Radiation hazards, in N. I. Sax (editor), *Dangerous properties of industrial materials:* New York, Reinhold Publishing Corp., pp. 73–127.
25. Haxel, O., and G. Schumann, 1962, Erzeugung radioactiver Kernarten durch die kosmische Strahlung, in H. Israel, and A. T. Krebs (editors), *Nuclear radiation in geophysics:* New York, Academic Press, pp. 97–135.
26. Higgins, G. H., 1959, Evaluation of the ground-water contamination hazard from underground nuclear explosions: *Jour. Geophys. Research*, v. 64, pp. 1509–1519.
27. Jourain, G., 1960, Moyens et résultats d'étude de la radioactivité due au radon dans les eaux naturelles: *Geochim. et Cosmochim. Acta.* v. 20, pp. 51–82.
28. Judson, Sheldon, and J. K. Osmond, 1955, Radioactivity in ground and surface water: *Am. Jour. Sci.*, v. 253, pp. 104–116.
29. Junge, C. E., 1963, Air chemistry and radioactivity: New York, Academic Press, 382 pp.
30. Kaufman, W. J., 1960, The use of radioactive tracers in hydrologic studies: University of California Proceedings of Conference on Water Research, *Water Resources Center Report* 2, pp. 6–14.
31. ——, 1961, Tritium as a ground water tracer: *Am. Soc. Civil. Eng. Trans. Paper* 3203, pp. 436–446.
32. Kaufman, W. J., and G. T. Orlob, 1956, An evaluation of ground-water tracers: *Am. Geophys. Union Trans.*, v. 37, pp. 297–306.
33. Keech, C. F., 1962, Geology and hydrology of the site of the Hallam nuclear power facility, Nebraska: *U.S. Geol.. Survey Bull.* 1133B, 51 pp.
34. Krebs, A. T., and N. G. Stewart, 1962, Biological aspects, in H. Israel, and A. T. Krebs (editors), *Nuclear radiation in geophysics:* New York, Academic Press, pp. 241–294.

35. Landis, E. R., 1960, Uranium content of ground and surface waters in a part of the central Great Plains: *U.S. Geol. Survey Bull.* 1087-G, pp. 223–258.
36. Libby, W. F., 1961, Tritium geophysics: *Jour. Geophys. Research*, v. 66, pp. 3767–3782.
37. —— 1963, Moratorium tritium geophysics: *Jour. Geophys. Research*, v. 68, pp. 4485–4494.
38. Linderoth, C. E., and D. W. Pearce, 1961, Operating practices and experiences at Hanford (Washington), in Ground disposal of radioactive wastes: Sanitary Engineering Research Laboratory, *University of California, Berkeley, Conference Proceedings*, pp. 7–16.
39. Lucas, H. F., Jr., and F. H. Ilcewicz, 1958, Natural radium 226 content of Illinois water supplies: *Am. Water Works Assoc. Jour.*, v. 50, pp. 1523–1532.
40. Machata, L., 1963, Worldwide radioactive fallout from nuclear tests, Part 1: *Nuclear Safety*, v. 4, pp. 103–111.
41. McKelvey, V. E., D. L. Everhart, and R. M. Garrels, 1955, Origin of uranium deposits: *Econ. Geol. Fiftieth Anniversary Volume*, pp. 464–533.
42. Mayer, S. W., and E. R. Tompkins, 1947, Ion exchange as a separations method: A theoretical analysis of the column separations process: *Jour. Am. Chem. Soc.*, v. 69, pp. 2866–2874.
43. Mazor, E., 1962, Radon and radium content of some Israeli water sources and a hypothesis on underground reservoirs of brines, oils, and gases in the Rift Valley: *Geochim. et. Cosmochim. Acta*, v. 26, pp. 765–786.
44. Morgan, K. Z., 1963, Permissible exposure to ionizing radiation: *Sci.*, v. 139, pp. 565–571.
45. Munnich, K. O., and J. C. Vogel, 1960, $C^{14}$ determination of deep ground-waters: *International Assoc. Sci. Hydrology, General Assembly of Helsinki Pub.* 52, pp. 537–541.
46. National Bureau of Standards, 1962, Radiation quantities and units, International Commission on Radiological Units and Measures (ICRU) *Report* 10*a: U.S. National Bureau of Standards Handbook* 84, 8 pp.
47. Norris, S. E., and A. M. Spieker, 1961, Geology and hydrology of the Piqua area, Ohio: *U.S. Geol. Survey Bull.* 1133A, 33 pp.
48. Norvitch, R. F., and others, 1963, Geology and hydrology of the Elk River, Minnesota nuclear-reactor site: *U.S. Geol. Survey Bull.* 1133C, 25 pp.
49. Phoenix, D. A., 1959, Occurrence and chemical character of ground water in the Morrison Formation, in Geochemistry and mineralogy of the Colorado Plateau uranium ores: *U.S. Geol. Survey Prof. Paper* 320, pp. 55–64.
50. Reichert, S. O., 1962, Radionuclides in ground-water at the Savannah River Plant waste disposal facilities: *Jour. Geophys. Research*, v. 67, pp. 4363–4374.
51. Robinson, B. P., 1962, Ion-exchange minerals and disposal of radioactive wastes—a survey of literature: *U.S. Geol. Survey Water-Supply Paper* 1616, 132 pp.
52. Roedder, Edwin, 1959, Problems in the disposal of acid aluminum nitrate high-level radioactive waste solutions by injection into deep-lying permeable formations: *U.S. Geol. Survey Bull.* 1088, 65 pp.
53. Rogers, A. S. 1958, Physical behavior and geologic control of radon in mountain streams: *U.S. Geol. Survey Bull.* 1052E, pp. 187–211.
54. Sax, N. I. (editor), 1957, *Dangerous properties of industrial materials:* New York, Reinhold Publishing Corp., 1467 pp.
55. Schaffer, O. A., S. O. Thompson, and N. L. Lark, 1960, Chlorine-36 radioactivity in rain: *Jour. Geophys. Research*, v. 65, pp. 4013–4016.
56. Schmalz, B. L., 1961, Operating practices, experiences, and problems at the National

Reactor Testing Station (Idaho), in Ground disposal of radioactive wastes: *Sanitary Eng. Research Laboratory, University of California, Berkeley, Conference Proceedings*, pp. 17–33.

57. Schroeder, M. C., and A. R. Jennings, 1963, Laboratory studies of the radioactive contamination of aquifers: *University of California, Lawrence Radiation Lab. Pub.* UCRL-13074, 51 pp. plus 66 pp. Appendices.
58. Scott, R. C., and F. B. Barker, 1958, *Radium and uranium in ground-water in the United States:* Second United Nations International Conference on Peaceful Uses of Atomic Energy, 10 pp.
59. Solon, L. R., and others, 1960, Investigations of natural environmental radiation: Sci., v. 131, pp. 903–906.
60. Stead, F. W., 1963, Tritium in ground water around large underground fusion explosions: *Sci.*, v. 142, pp. 1163–1165.
61. Sternglass, E. J., 1963, Cancer: Relation of prenatal radiation to development of the disease in childhood: *Sci.*, v. 140, pp. 1102–1104.
62. Thatcher, L., M. Rubin, and G. F. Brown, 1961, Dating desert ground water: *Sci.* v. 134, pp. 105–106.
63. Thornthwaite, C. W., and others, 1960, Movement of radiostrotium in soils: *Sci.*, v. 131, pp. 1015–1018.
64. Tokarev, A. N., and A. V. Shcherbakov, 1956, *Radiohydrogeology: Moscow, State Publ. of Sci.—Tech. Literature on Geol. and Conservation of Nat. Resources* (English Translation by U.S. Atomic Energy Commission, AEC-tr-4100), 346 pp.
65. Tsivoglov, E. C., 1963, Research for the control of radioactive pollutants: *Jour. Water Pollution Control Federation*, v. 35, pp. 242–259.
66. von Buttlar, H., 1959, Ground-water studies in New Mexico using tritium as a tracer, Part II: *Jour. Geophys. Research*, v. 64, pp. 1031–1038.
67. U.S. Atomic Energy Commission, 1960, Rules and regulations, Title 10—*Atomic energy: Federal Register*, November 17, 1960, 11 pp.
68. Wahlberg, J. S., and M. J. Fishman, 1962, Adsorption of cesium on clay minerals: *U.S. Geol. Survey Bull.* 1140A, 30 pp.
69. Walton, Alan, 1963, The distribution in soils of radioactivity from weapons tests: *Jour. Geophys., Research.*, v. 68, pp. 1485–1496.

# chapter 6

# ELEMENTARY THEORY OF GROUND-WATER FLOW

## 6.1 *Darcy's Law*

Ground water moves from levels of higher energy to levels of lower energy, whereby its energy is essentially the result of elevation and pressure. Kinetic energy, proportional to the square of the velocity, is neglected because ground-water velocities are very small, at least in laminar flow. While flowing, ground water experiences a loss in energy due to friction against the walls of the granular medium along its seepage path. This loss per unit length of distance travelled, or hydraulic gradient, is simply proportional to the velocity of ground water for laminar flow in sandy aquifers or seepage through earth embankments. When the proportionality of hydraulic gradient and ground-water velocity is expressed by a mathematical equation, a linear law of flow, called Darcy's law, arises.

The similarity between ground-water flow and laminar pipe flow was recognized by Darcy [3] and Dupuit [9]. Experiments with pipe flow were first conducted by Hagen in 1839 and almost simultaneously by Poiseuille in 1841, after whom the law for laminar flow in pipes was named. Darcy also was oriented toward experimental work, and in particular he investigated the friction factor of the pipe flow formulas. In 1856 Darcy ran an experiment on a vertical pipe of cross-sectional area $A$ filled with sand, under conditions simulated by Figure 6.1 [17]. From his investigations of the flow through horizontally stratified beds of sand, Darcy concluded that the flow rate $Q$ was proportional to the energy loss, inversely proportional to the length of the flow path, and proportional to a coefficient $K$, depending on the nature of the sand. Darcy's law may be expressed as

$$Q = KA\frac{(h_1 - h_2)}{dl}$$

$$= -KA\frac{dh}{dl} = KAS^* \tag{6.1}$$

in which

$$h = z + \frac{p}{\gamma} + \text{arbitrary constant} \tag{6.2}$$

In these equations $h$ is the energy per unit weight of fluid or hydraulic head in the case of water, $z$ is the elevation above an arbitrary datum

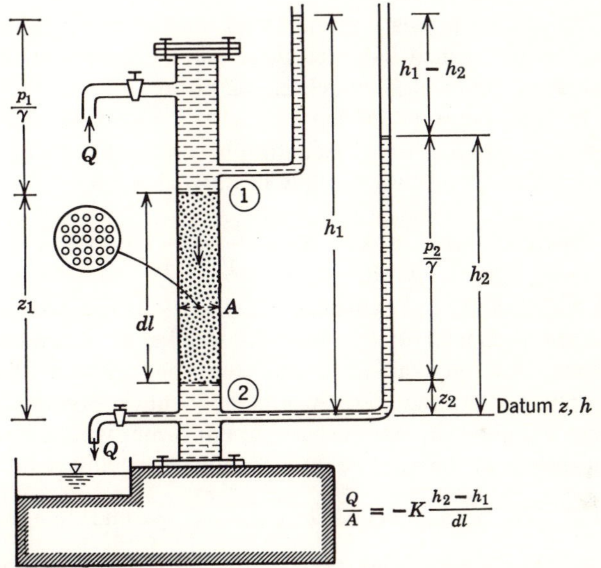


Figure 6.1 Apparatus to demonstrate Darcy's law. (After Hubbert.)

plane, $p$ is the pressure sustained by the fluid in the pores of the medium, and $\gamma$ is the specific weight of the fluid. When the fluid is water, the hydraulic gradient $S^*$ is defined as

$$S^* = -\frac{dh}{dl} \tag{6.3}$$

The subscripts 1 and 2 in Equation 6.1. refer to the value of $h$ at elevations $z_1$ and $z_2$.

Ground-water flow could be treated microscopically by the laws of hydrodynamics if the granular skeleton of the porous medium were a simple geometrical assembly of prismatic, unconnected tubes. The seepage path, far from being a prismatic channel, however, is tortuous, branching into a multitude of tributaries. In its original form, Darcy's law avoids the insurmountable difficulties of the hydrodynamic microscopic picture by introducing a doubly averaging macroscopic concept.

1. It considers a fictitious flow velocity, the Darcy velocity or specific discharge,** through a given cross section of a porous medium rather than the true velocity between the grains, as is evident from Equation 6.1:

$$V = Q/A = -K\,dh/dl$$

or

$$V = KS^* \tag{6.4}$$

2. It considers average hydraulic values rather than local hydrodynamic values of this velocity.

These simplifying concepts were introduced by the nature of Darcy's experiment, which only permitted the measurement of average hydraulic values in the sand-filled cylindrical pipe.

In 1857 Dupuit's [9] familiarity with pipe flow and open-channel flow led to his derivation of Equation 6.4 in the following way. Dupuit used the equation for uniform flow in open channel:

$$S^* = \frac{P}{A}(\alpha V + \beta V^2) \tag{6.5}$$

in which $S^*$ is the hydraulic gradient, $V$ is the average velocity, $P$ is the wetted perimeter of the cross-sectional area $A$, and $\alpha$ and $\beta$ are coefficients. He then assumed that the channel was filled with sand and, therefore, that the water would flow much slower than in an ordinary open channel. In fact, he pointed out, water would flow through the grain interstices as through an infinite number of tiny parallel pipes, and if the sand were homogeneous, Equation 6.5 would be applicable with $\frac{P}{A}\alpha = 1/K$, characteristic constant of the medium. Furthermore, the term in $V^2$ would be neglected because of the small velocities of ground-water flow. Thus Dupuit arrived at Equation 6.4 and then described Darcy's [3] experiment of the previous year (1856) in support of his formula. He stated,† "The fundamental equation $V = KS^*$ may therefore be considered as an experimental result. If we have derived it first from the equations of uniform flow, it is only to indicate the similarity which exists between the two kinds of flow."

---

** The Russian literature uses the term "seepage velocity" (see reference 26, p. 11), but this is not compatible with U.S. nomenclature for soil mechanics. (see reference 35, p. 100).

† For original text and symbols, see reference 5.

## 6.2 *Major Subdivisions of Ground-Water Flow*

The subject of ground-water flow may be divided into several compartments, according to the dimensional character of the flow, the time dependency of the flow, the boundaries of the flow region or domain, and the properties of the medium and of the fluid.

### DIMENSIONAL CHARACTER

All ground-water flow in nature is to a certain extent three-dimensional. It is practically impossible to solve a natural three-dimensional ground-water flow problem unless symmetry features of the problem allow us to reduce the number of dimensions involved by one or two. Fortunately this can be done in the majority of all problems of ground-water flow. A typical example is the flow to a well penetrating a pervious stratum of uniform thickness and confined between two impervious strata.

### TIME DEPENDENCY

Ground-water flow may be evaluated quantitatively by the knowledge of the velocity, pressure, density, temperature, and viscosity of water percolating through a geologic formation. These characteristics of the water are commonly the unknown variables of the problem and may vary in place, from point to point in the formation, and in time (that is, assume different values as time goes on). If the unknown or dependent variables are functions of the space variables or independent variables $x$, $y$, $z$ only, the flow is steady. On the other hand, if the unknown are also functions of time, the flow is unsteady or time dependent. Steady flow may be conceived of as a limit case of unsteady flow, as time goes to infinity, or else as the average of unsteady flow over a given period.

### BOUNDARIES OF FLOW REGION OR DOMAIN

Ground-water flow is confined when the boundaries or bounding surfaces of the medium (that is, the space made up by the water-filled pores) through which the water percolates are fixed in space for different states of flow. Ground-water flow is unconfined, on the other hand, when it possesses a free surface, the position of which varies with the state of flow. Unconfined flow is sometimes referred to as flow with a water-table.

### PROPERTIES OF THE MEDIUM AND OF THE FLUID

A medium is called isotropic if its properties at any point are the same in all directions emanating from that point. It is called anisotropic if, on the other hand, some properties are affected by the choice of direction at a point. The medium is of heterogeneous composition if its properties or conditions of isotropy or anisotropy vary from point to point in the medium; it is homogeneous if its properties, isotropic or anisotropic conditions are constant over the medium. A medium therefore can be isotropic and heterogeneous at the same time, as for example when its permeability has no preference to orientation, yet varies in space.

A fluid is homogeneous essentially when it is single phased. A mixture of completely miscible salt water and fresh water is treated as homogeneous, although the density of the mixture may vary from point to point. A dispersed mixture of oil and water, however, is heterogeneous. Oil and water may be treated as immiscible for some purposes, exhibiting a liquid-liquid interface on both sides of which the flow may be homogeneous.

The combination of properties of the medium and of the fluid in such characteristics as the hydraulic conductivity (see Section 6.4) leads to heterogeneous flow if either medium or fluid is heterogeneous. This textbook is restricted to homogeneous flow.

## 6.3 *Rock and Soil Composition*

Rock and soil form the porous medium in which water is collected by water-bearing zones and through which water flows under the influences of various forces. This porous medium has a solid matrix or skeleton, an assembly of solid mineral grains separated and surrounded by voids, pores or interstices which may be filled with water, gases, or organic matter.

### POROSITY

A given volume $\mathcal{V}_0$ of porous medium contains $\mathcal{V}_s$ of solids and $\mathcal{V}_v$ of voids. The porosity $n$ is defined as

$$n = \frac{\mathcal{V}_v}{\mathcal{V}_0} \tag{6.6}$$

Porosity of consolidated materials depends on the degree of cementation, the state of solution and fracturing of the rock; porosity of unconsolidated materials depends on the packing of the grains, their shape, arrangement, and size distribution. Small grains will fit into the openings left between grains of large diameter and thus a medium with nonuniform size distribution will have a smaller porosity than one in which the grains are well sorted. The effect of packing upon porosity may be evaluated numerically in the case of grains of uniform spherical shape. It has been shown by Graton and Fraser [13] that the loosest packing is that of a

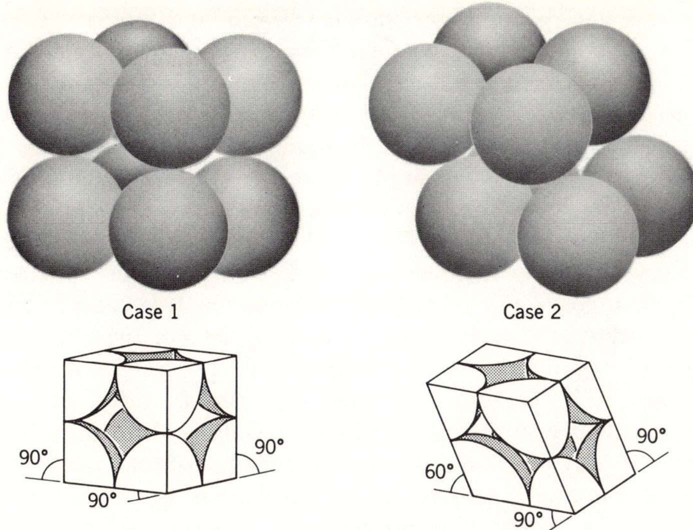


Figure 6.2 Packing of spherical grains. Unit cells of cubic (Case 1) and rhombohedral (Case 2) packing. (After Graton and Fraser.)

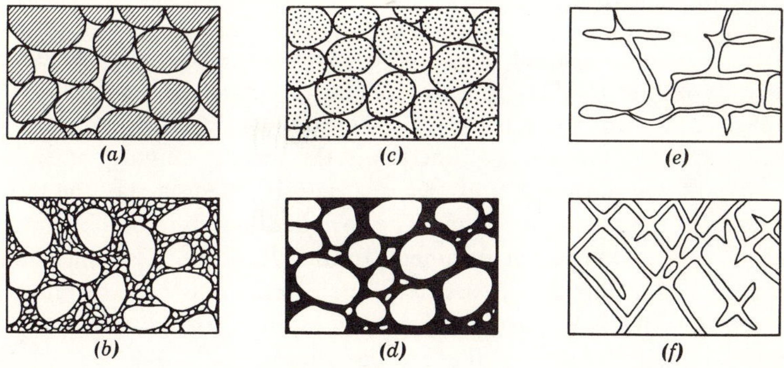


Figure 6.3 Rock interstices and the relation of rock texture to porosity. Diagram showing several types of rock interstices and the relation of rock texture to porosity. (*a*) Well-sorted sedimentary deposit having high porosity; (*b*) poorly sorted sedimentary deposit having low porosity; (*c*) well-sorted sedimentary deposit consisting of pebbles that are themselves porous, so that the deposit as a whole has a very high porosity; (*d*) well-sorted sedimentary deposit whose porosity has been diminished by the deposition of mineral matter in the interstices; (*e*) rock rendered porous by solution; (*f*) rock rendered porous by fracturing. (After Meinzer, *U.S. Geological Survey Water-Supply Paper* 489, 1923, Fig. 1, p. 3.)

cubical array of spheres, for which $n = 0.476$, and the tightest packing is that of a rhombohedral array, with $n = 0.26$ (Figure 6.2). Intermediate arrays, of course, have porosities lying between these limits. Figure 6.3 shows several types of rock interstices and the relation of rock texture to porosity. Porosity is a measure of the water-bearing capacity of a medium, and, as may be expected, it also plays a role in the capability of a medium to transmit water. This capability is expressed by the fluid (hydraulic, in the case of water) conductivity of the formation which is defined in Section 6.4. The relationship between porosity and hydraulic conductivity, however, is not a simple one and other factors besides porosity affect the value of hydraulic conductivity. Thus, pore size, of the same order of magnitude as grain size, is far more important than porosity for the water-transmitting capability of a medium. Sands with relatively large rounded or angular grains may have smaller porosity than clays which are composed of tiny plate-like particles with a large specific surface causing high-molecular forces between the clay and water particles. In spite of their smaller porosity, sandy materials are pervious and good aquifers, whereas clay forms aquicludes or at best aquitards. Methods to measure porosity may be found in a number of textbooks [25, 29].

## 6.4 *Hydraulic Conductivity K*

Both Darcy and Dupuit failed to recognize the fact that $K$ depends on properties of the fluid as well as on characteristics of the medium. A comparison of Equation 6.4 with a similar equation for laminar pipe flow would suggest that $K$ is a function of the specific weight $\gamma$ of the fluid, of its dynamic viscosity $\mu$, and of a characteristic length, say the average pore size $d$ of the porous medium. The correct dimensions of the functional relationship remain to be determined. In this relationship, a dimensionless constant or shape factor $C$ occurs, which takes into account effects of stratification, packing, arrangement of grains, size distribution, and porosity. It may be verified that the equation

$$K = Cd^2 \frac{\gamma}{\mu} \tag{6.7}$$

is dimensionally correct. A variety of names have been given to $K$, such as effective permeability, coefficient of permeability, seepage coefficient, and hydraulic conductivity. The latter term, in view of the analogy with thermal and electrical conductivity, is gaining widespread use and is adopted throughout this book. The name intrinsic permeability for

$$k = Cd^2 \tag{6.8}$$

is now generally accepted as being characteristic of the medium alone.

As is evident from Equation 6.4, $K$ has the dimensions of a velocity. It may therefore be expressed in a variety of consistent units, generally different from one discipline to the other (soil mechanics, petroleum engineering, hydrogeology) and from continent to continent (Europe vs the United States). In soil mechanics, cgs units are universally used, whereas water-supply studies in this country are carried out with $K$ in gallons per day and per square foot (gpd/ft$^2$).

The darcy unit, which is adopted in petroleum engineering, has been preempted as a measure of the intrinsic permeability $k$ because of an incomplete definition of Darcy's law. Indeed, for horizontal flow, Darcy's law may be written as

$$V = Q/A = -(k/\mu)(dp/dl)$$

or, in absolute value,

$$k = \frac{\mu \dfrac{Q}{A}}{dp/dl} \tag{6.9}$$

From Equation 6.9, the darcy has been defined as

$$1 \text{ darcy} = \frac{\dfrac{1 \text{ centipoise} \times 1 \text{ cm}^3/\text{sec}}{1 \text{ cm}^2}}{1 \text{ atmosphere/cm}}$$

The darcy unit has the dimensions of $[L]^2$. It may be expressed as such by replacing in the foregoing formula:

$$1 \text{ centipoise} = 0.01 \text{ poise} = 0.01 \frac{\text{dyne sec}}{\text{cm}^2}$$

$$1 \text{ atmosphere} = 1.0132 \times 10^6 \frac{\text{dynes}}{\text{cm}^2}$$

This leads to

$$1 \text{ darcy} = 0.987 \times 10^{-8} \text{ cm}^2$$

or

$$1 \text{ darcy} = 1.062 \times 10^{-11} \text{ ft}^2$$

Because Equation 6.7 may be written as

$$K = kg/\nu \tag{6.10}$$

in which $\nu = \mu/\rho$ is the kinematic viscosity of the fluid and $\rho$ is the density of the fluid, it is easy to convert from darcy units to gpd/ft$^2$. Indeed, the knowledge of the temperature of the fluid allows $\nu$ to be read from tables of physical properties of fluids [15]. The U.S. Geological Survey uses the meinzer unit to measure the hydraulic conductivity, honoring the late O. E. Meinzer. The meinzer unit is defined as the flow of water in gallons per day through a cross-sectional area of 1 square foot under a hydraulic gradient of 1 at a temperature of 60°F.

A second coefficient, the field coefficient of hydraulic conductivity, is also used by the Water Resources Division of the U.S. Geological Survey. This unit is defined as the flow of water in gallons per day through a cross section of aquifer 1 foot thick and 1 mile wide under a hydraulic gradient of 1 foot per mile at field temperature [38].

In Equation 6.7 it is assumed that both porous material and water are chemically and mechanically stable. This may never be true. Ion exchange on clay and colloid surfaces will cause changes of mineral volume which in turn will change the pore size and shape. Extreme changes in pressure will cause dilatation or compaction of aquifers. Moderate to high groundwater velocities will move colloids and small clay particles. Also, all water movement will facilitate solution or deposition of dissolved constituents. Relatively small changes in pressure or temperature may cause gases to come out of solution and clog pore space, thus reducing the hydraulic conductivity.

*Table 6.1 Average Values of K and k*

| Soil Class | $K$, cm/sec | $k$, darcys | $K$, gpd/ft$^2$ |
|---|---|---|---|
| Gravel | $1$–$10^2$ | $10^3$–$10^5$ | $10^4$–$10^6$ |
| Clean sands (good aquifers) | $10^{-3}$–$1$ | $1$–$10^3$ | $10$–$10^4$ |
| Clayey sands, fine sands (poor aquifers) | $10^{-6}$–$10^{-3}$ | $10^{-3}$–$1$ | $10^{-2}$–$10$ |

*Table 6.2 Representative Values for k and K*

| Geologic Classification | | Darcys, $k$ | Meinzers, $K$ |
|---|---|---|---|
| Argillaceous limestone | 2% porosity | $1.0 \times 10^{-4}$ | $1.80 \times 10^{-3}$ |
| Limestone | 16% porosity | $1.4 \times 10^{-1}$ | 2.50 |
| Sandstone, silty | 12% porosity | $2.6 \times 10^{-3}$ | $4.74 \times 10^{-2}$ |
| Sandstone, coarse | 12% porosity | 1.1 | 19.90 |
| Sandstone | 29% porosity | 2.4 | 43.60 |
| Very fine sand | well sorted | 9.9 | $18.00 \times 10$ |
| Medium sand | very well sorted | $2.6 \times 10^2$ | $4.60 \times 10^3$ |
| Coarse sand | very well sorted | $3.1 \times 10^3$ | $5.80 \times 10^4$ |
| Gravel | very well sorted | $4.3 \times 10^4$ | $7.88 \times 10^5$ |
| Montmorillonite clay[a] | | $10^{-5}$ | $10^{-4}$ |
| Kaolinite clay[a] | | $10^{-3}$ | $10^{-2}$ |

[a] For the clays only the order of magnitude is indicated.

*Table 6.3 Equivalence between K and k Values*

| | |
|---|---|
| 1 darcy | $= 9.87 \times 10^{-9}$ cm$^2$ $= 1.062 \times 10^{-11}$ ft$^2$ |
| $10^{-10}$ cm$^2$ | $= 1.012 \times 10^{-2}$ darcys |
| 0.1 cm/day | $= 1.15 \times 10^{-6}$ cm/sec $\approx 1.18 \times 10^{-11}$ cm$^2$ for water at 20°C |
| 1.0 cm/sec | $\approx 1.02 \times 10^{-5}$ cm$^2$ for water at 20°C |
| 1 darcy | $\approx$ 18.2 meinzer units for water at 60°F |
| 1 meinzer | $= 0.134$ ft/day $= 4.72 \times 10^{-5}$ cm/sec $\approx 5.5 \times 10^{-2}$ darcys for water at 60°F |

Tables 6.1, 6.2, and 6.3 contain useful data about $K$ and $k$. In Table 6.3, $\approx$ means "equivalent to."

$K$ may be determined by laboratory methods as well as by field measurements. In the laboratory, the hydraulic conductivity may be determined by means of so-called permeameters [34] in which small samples taken at different points of the aquifer are submitted to flow under constant head or variable head. These samples are generally disturbed, and, moreover, even if they were undisturbed they might not be representative of the average $K$ of the aquifer. More significant for the hydrogeologist is the determination of $K$ through a well-pumping test in the field.

Many researchers have tried to investigate the various factors which influence the intrinsic permeability and hence the hydraulic conductivity, both in experimental and analytical work. Among others, the influence of porosity was studied by Kozeny [22], that of grain size by Hazen [16], and that of void ratio by Zunker [39]. Other important contributions were made by Slichter [31,32,14], Fair and Hatch [11], Rose [27,28], and Bakhmeteff and Feodoroff [1].

The influence of the fluid properties on the value of the hydraulic conductivity is evident from Equation 6.7. Hence, hydraulic conductivity may be measured by using different fluids. It is possible, at least in principle, to measure the hydraulic conductivity using a gas, say air, and to compute the intrinsic permeability $k$. Once $k$ is known, it suffices to multiply $k$ by $g/\nu$ to find the hydraulic conductivity for any fluid. Klinkenberg [21], however, has shown that the permeability of a porous medium to a liquid and to a gas is not the same. Owing to the slip phenomenon, whereby the velocity of a gas layer in the immediate vicinity of the surface of the grains has a finite (instead of zero) velocity, the permeability to a gas is higher than to a liquid. Klinkenberg derived the equation

$$k_a = k\left(1 + \frac{b}{\bar{p}}\right) \tag{6.11}$$

where $k_a$ is the apparent permeability and $k$ is the intrinsic permeability when gas is used in the measurement, $b$ is a constant, and $\bar{p}$ is the mean pressure over the sample. From this equation it follows that the intrinsic permeability may be obtained by an extrapolation of the gas-permeability data for low pressure to those for an infinitely high mean pressure.

Also, samples containing clay and silt change hydraulic conductivity in response to the type of fluid used. In general, permeabilities determined with air have the highest values, those determined with brines are next, and those determined with distilled water are the lowest. The difference between air and water values is explained by the fact that hydration of clays and other minerals causes swelling of the grains and hence clogging of pores. Air permeabilities and water permeabilities may differ by a factor of more than 100 in clay-rich sediments. Brine permeabilities and distilled water permeabilities may also differ by a factor of 100. The latter difference is due also to the effects of partial hydration. The water molecules on the clay surface tend to be in osmotic equilibrium with the surrounding pore fluids. When the pore fluid is a brine, only a few water molecules cluster about the clay particles and the effective volume of the clay particle is small. With a decrease in ion concentration in the pore fluid, a new equilibrium will be established in which a much larger cluster of water molecules will surround each clay particle. Thus the effective volume of the clay is increased with a resultant decrease in pore space and permeability.

## 6.5 *Pore Pressure and Intergranular Stress; Soil-Moisture Tension; Piezometric Head; Watertable*

The pressure experienced by the water in the voids of a porous medium such as in the pores of the saturated soil of Figure 6.4 is called pore pressure. It is measured as pressure head $h_p$ in a point $P$ by the height of water in a piezometer inserted in $P$, as indicated in Figure 6.4. This height is measured with reference to point $P$, and is counted positive above $P$ and negative below $P$. In Figure 6.4 the pore pressure in $P_1$ is positive and equal to $\gamma h_p$, whereas in Figure 6.5 the pore pressure in $P_2$ is negative and equal to $\gamma h_t$ or $-\gamma h_p$. In the second case, where $P_2$ is situated in the unsaturated zone of the soil, the negative pore pressure is often called soil-moisture suction or soil-moisture tension and the piezometer is called a tensiometer.

Pressures used in this textbook are gage pressures unless it is stated explicitly that absolute pressures are adopted. Gage pressure and absolute pressure are interrelated by the equality

$$\text{Absolute pressure} = \text{local atmospheric pressure} + \text{gage pressure}$$

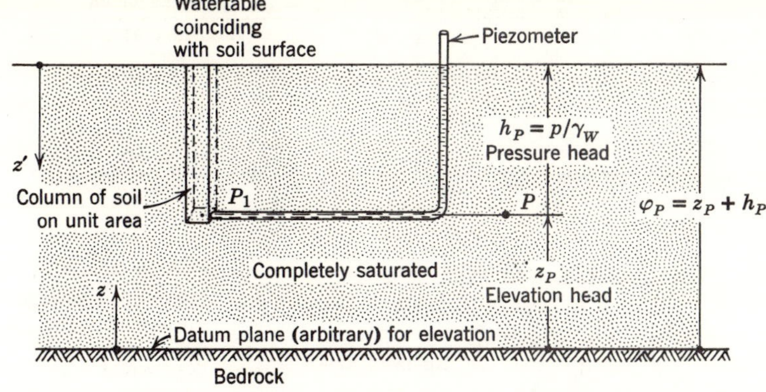


Figure 6.4 Piezometric potential in the saturated zone.

Positive gage pressure is called pore pressure; negative gage pressure is called tension or vacuum. Absolute pressure is always positive (Figure 6.6). In Figure 6.6 this definition of the watertable is implied: it is a phreatic surface at atmospheric pressure, or at zero gage pressure.

The elevation $z$ of a point $P$ above an arbitrary datum plane is called elevation head, positive above the datum plane and negative below the datum plane. The sum of elevation head and pressure head is called piezometric head or piezometric potential $\Phi$. The concepts of head and potential are developed in Section 6.6.

The total pressure or stress in a point $P$ of a porous medium is somewhat artificially defined as a macroscopic pressure [35], namely as the weight of the overburden above $P$ per unit area. Thus it is possible to say that the total stress (pressure) or combined stress (pressure) is equal to the sum of

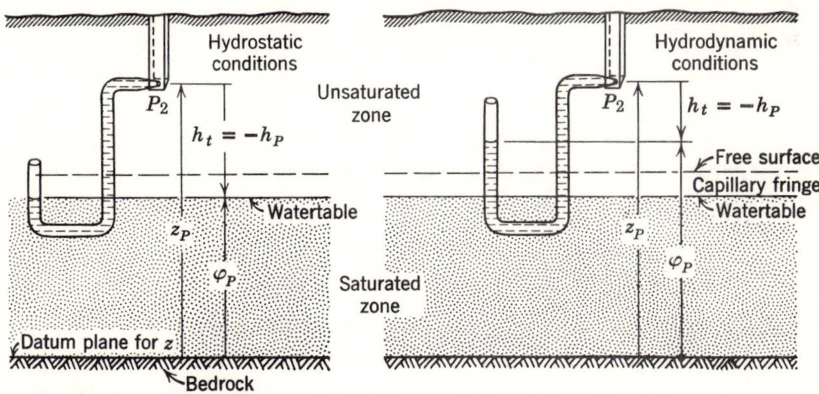


Figure 6.5 Piezometric potential in the unsaturated zone $\varphi_P = z_P + h_t = z_P - h_P$. In hydrostatic conditions, $\varphi_P = \varphi$ at watertable. In hydrodynamic conditions, $\varphi_P > \varphi$ at watertable.

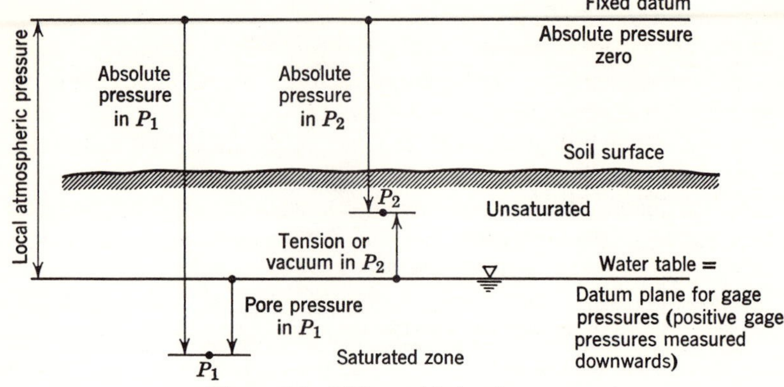


Figure 6.6 Different kinds of pressure.

the pore pressure and the intergranular stress, whereby intergranular stress is understood to be the stress in the granular skeleton. It would be difficult to define intergranular stress otherwise than as the difference between total stress and pore pressure, because the intergranular forces are transmitted from grain to grain contact and it would be hard to define the surface over which the intergranular stress acts. However, the difference definition is completely satisfactory for our purposes. In Figure 6.4, the intergranular stress in $P_1$ is equal to the overburden per unit area minus the pore pressure; that is, the intergranular stress is the buoyant overburden per unit area, considering the fact that the medium is completely saturated.

Similar considerations may be made when $P$ is in the unsaturated zone as $P_2$ in Figure 6.5. Here the intergranular stress in $P_2$ exceeds the total stress in that point by the amount $\gamma_w h_t$. In hydrostatic conditions, no water flows from $P_2$ to the watertable and this is expressed by the equality of piezometric potential between $P_2$ and the watertable. Therefore $h_t$ assumes a large value as indicated in Figure 6.5. When water flows from $P_2$ to the watertable, the flow conditions are hydrodynamic and therefore the piezometric potential in $P_2$ must be higher than that of the watertable. Hence $h_t$ in absolute value is smaller than in hydrostatic conditions.

## 6.6 *The Concept of Potential*

In Equation 6.2, $h$ is the energy per unit weight of fluid, also called the piezometric head or hydraulic head in the case of water. It is a scalar quantity, that is, at any point it can be expressed and defined solely by a number, in contrast to a vectorial quantity which additionally requires a direction. A scalar quantity preserves this property if it is multiplied by another scalar quantity.

In ground-water flow, the products $Kh$ and $gh$, in which $K$ and $g$ are constants, are called $\Phi$ and $\Phi^*$, and are respectively the velocity potential and the force potential. The concept of force potential, developed by Hubbert [18], is more general than the concept of velocity potential, which is strictly valid only for media with constant $K$, characterized by a fluid of constant density and viscosity and by an intrinsic permeability which is constant in space. This is the case in most simplified hydrogeologic applications and the continued use of the velocity potential may be defended for this reason alone.

To understand fully the meaning of these potentials, it is necessary to introduce a few definitions concerning vectors. A vector may be represented graphically by a line segment of a well-defined length and orientation. In three dimensions it may be expressed as the sum of its three components in the $x$, $y$, and $z$ directions, each multiplied by a unit vector in the corresponding direction. Thus a vector $\mathbf{A}$ may be written as

$$\mathbf{A} = A_x\mathbf{u}_x + A_y\mathbf{u}_y + A_z\mathbf{u}_z \tag{6.12}$$

in which $A_x$, $A_y$, and $A_z$ are the measures of the components. A special kind of vector that occurs in the present theory is that derived from a scalar quantity such as the potential $\Phi$ (or $\Phi^*$), namely the gradient of $\Phi$, or, abbreviated, grad $\Phi$. The meaning of the gradient vector is illustrated by the multiplication of both sides of Equation 6.4 by a unit vector $\mathbf{u}_l$ in the $l$ direction. Thus

$$V\mathbf{u}_l = -K\frac{dh}{dl}\mathbf{u}_l$$

or

$$\mathbf{V} = -K \text{ grad } h \tag{6.13}$$

because in unidirectional flow, gradient and directional derivative are overlapping concepts. Equation 6.13 is the vectorial form of Darcy's law and is valid in two and three dimensions as well as in one dimension. The physical interpretation of the gradient in two and three dimensions is as follows:

In three dimensions, $h(x, y, z) = m$ may be represented by a surface as sketched in Figure 6.7. If the parameter $m$ assumes an increment $\Delta m$, then $h$ assumes an increment $\Delta h$ and $h + \Delta h = m + \Delta m$ is represented by another surface. Let $\Delta n$ be the intercept between the two surfaces along the normal $n$ in point $P$ of $h(x, y, z) = m$ perpendicular to $h$. The gradient of $h$ is then defined as grad $h = \lim\limits_{\Delta n \to 0} \dfrac{\Delta h}{\Delta n}\mathbf{u}_n$ where $\mathbf{u}_n$ is the unit normal. It is the directional derivative of $h$ in the direction $\mathbf{n}$ normal to the surface $h(x, y, z) = m$ in the point $P$. The special feature of the gradient is that it represents the maximal value of the directional derivative. To see this, the directional derivative $D$ in any other direction, say $\mathbf{r}$, is taken as

$$\underset{\mathbf{r}}{Dh} = \lim_{\Delta\mathbf{r}\to 0} \frac{\Delta h}{\Delta r}\mathbf{u}_r \tag{6.14}$$

But since $|\Delta r| > |\Delta n|$, the latter limit is smaller than the gradient. From Figure 6.7, it follows that the component of the gradient in any direction

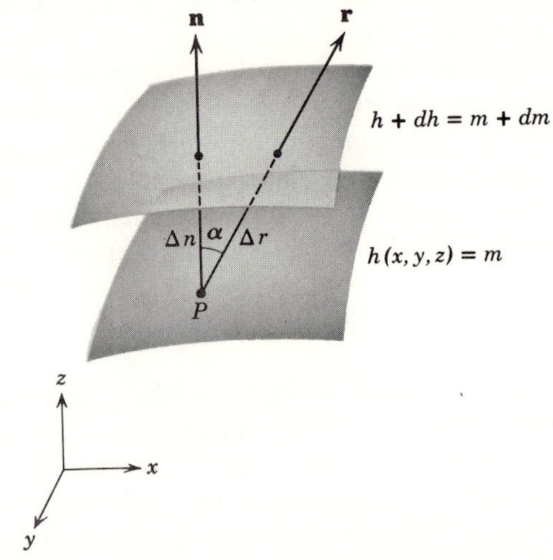


Figure 6.7 Physical interpretation of grad $h(x, y, z)$.

is the directional derivative in that direction. Indeed, $\Delta n = \Delta r \cos\alpha$, and if this is inserted in Equation 6.14 it follows that

$$(\text{grad } h)\cos\alpha = \underset{\mathbf{r}}{Dh}$$

The surfaces $h(x, y, z) = m$, for various values of the parameter $m$, are called level surfaces.

If these level surfaces are intersected with a plane $z$ = constant, contour curves $h(x, y) = m'$ are generated. A well-known case is that of topographic contours, where $h$ is simply the elevation of different points of the earth's surface (Figure 6.8). The gradient again could be defined exactly as in the three-dimensional case. Here it would be oriented along the path of steepest descent perpendicular to $h = m'$ in $P$.

Any vector may be decomposed into its components in a coordinate system. If $\mathbf{u}_x$, $\mathbf{u}_y$, $\mathbf{u}_z$ are unit vectors along the $x$-, $y$-, and $z$-axes, decomposition of grad $h$ renders:

$$\text{grad } h = (\text{grad } h)_x\mathbf{u}_x + (\text{grad } h)_y\mathbf{u}_y + (\text{grad } h)_z\mathbf{u}_z$$

But according to what has just been stated, the components of the gradient in any direction are the directional derivatives in that direction. Therefore $(\text{grad } h)_x = \frac{\partial h}{\partial x}$, and so forth, and

$$\text{grad } h = \frac{\partial h}{\partial x}\mathbf{u}_x + \frac{\partial h}{\partial y}\mathbf{u}_y + \frac{\partial h}{\partial z}\mathbf{u}_z \tag{6.15}$$

With these elements of vector-calculus, it is possible to understand Hubbert's concept of force potential $\Phi^*$, defined as

$$\Phi^* = g\int_{z_0}^{z} dz + \int_{p_0}^{p} \frac{dp}{\rho} \tag{6.16}$$

in which the kinetic energy per unit mass, say $v^2/2$, has been neglected and in which the density $\rho$ must be a function of only the pressure $p$ in

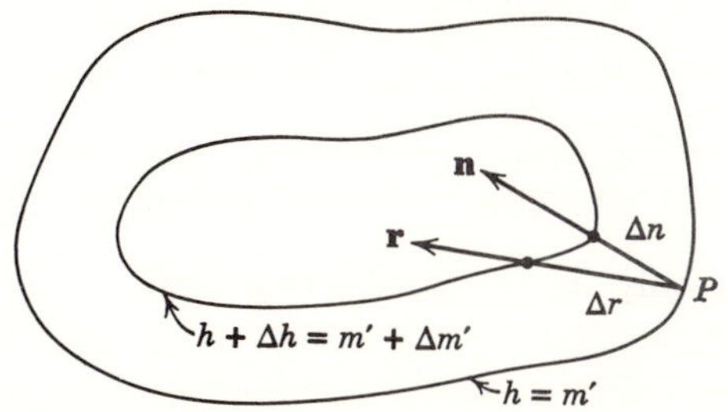


Figure 6.8 Physical interpretation of grad $h(x, y)$.

order to have a uniquely defined $\Phi^*$. This is the case for water, with $\rho = \text{constant}$, so that the second integration can be performed. For arbitrary but constant values of the lower limits of integration, $z_0$ and $p_0$, $\Phi^*$ for water becomes

$$\Phi^* = g\left(z + \frac{p}{\gamma}\right) + \text{constant} = gh \tag{6.17}$$

$\Phi^*$ is called a force potential because grad $\Phi^*$ represents a force per unit mass of fluid. If the gradient of Equation 6.16 is taken, the result is

$$\text{grad } \Phi^* = g \text{ grad } z + \frac{1}{\rho} \text{grad } p \tag{6.18}$$

Equation 6.18 may be rewritten conveniently as

$$-\text{grad } \Phi^* = \mathbf{g} - \frac{1}{\rho} \text{grad } p \tag{6.19}$$

because grad $z = \mathbf{u}_z$, according to Equation 6.15 which is general and valid for any potential, hence for $h = z$. Furthermore, $g\bar{u}_z = -\mathbf{g}$, $\mathbf{u}_z$ being oriented upward against the direction of gravity.

Equation 6.19 is a force equation per unit mass of fluid. The left side of this equation states that the unit mass of fluid is subject to a force oriented from surfaces of higher $\Phi^*$ to surfaces of lower $\Phi^*$ and perpendicular to these surfaces. The right side of the equation states that this force has a gravity component and a component caused by the gradient of the fluid pressure. In Equation 6.19 only the direction of $\mathbf{g}$ is fixed, but the vectors $\left(-\frac{1}{\rho}\operatorname{grad} p\right)$ and $(-\operatorname{grad}\Phi^*)$ may have arbitrary directions (Figure 6.9).

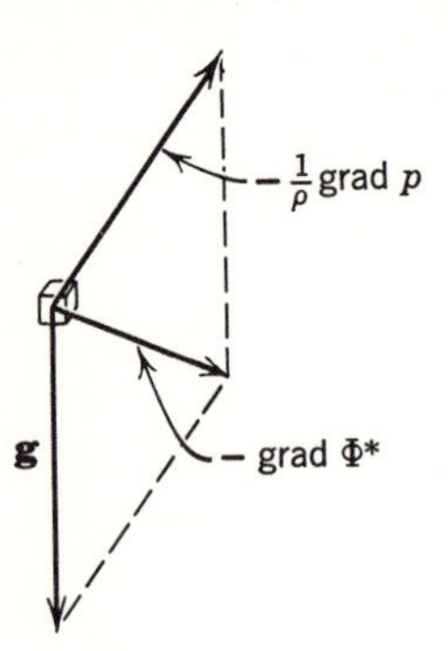


Figure 6.9 Graphical representation of forces on unit mass of fluid (Eq. 6.19).

The magnitude of $-\operatorname{grad}\Phi^*$ in most cases of ground-water flow is only a small fraction of $\mathbf{g}$ so that $\left(-\frac{1}{\rho}\operatorname{grad} p\right)$ becomes nearly vertical, which in effect means that the surfaces of constant pressure are nearly horizontal. As may be seen from Figure 6.9, only in the special case that $-\frac{1}{\rho}\operatorname{grad} p = -\mathbf{g}$ do we find that

$$\operatorname{grad}\Phi^* = 0$$

and this means that there is hydrostatic equilibrium. The vectors $-\operatorname{grad} p$ generate a vector field everywhere normal to the isobaric surfaces $p =$ constant. The vectors $-\operatorname{grad}\Phi^*$ generate a family of field lines or a vector field everywhere normal to the equipotential surfaces $\Phi^* =$ constant. How may the latter vector field be related to the streamlines? Let it be recalled that a streamline is a line tangent in each of its points to the velocity vector of a particle placed in these points. The relationship asked for is Darcy's law, which according to Hubbert [17], may be written as

$$\mathbf{V} = -\frac{K}{g}\operatorname{grad}\Phi^* \tag{6.20}$$

If $K$ is constant, as for a homogeneous and isotropic medium and for water, then $\mathbf{V}$ will have the same direction as $-\operatorname{grad}\Phi^*$, that is, streamlines are normal to surfaces or lines of $\Phi^* =$ constant. But for anisotropic media where $K$ has preferential properties in given directions, flowlines will in general be oblique to the direction of $-\operatorname{grad}\Phi^*$ and hence not be normal to equipotential surfaces or equipotential lines.

The concept of velocity potential arises when the product of $K$ and $h$ is formed, say

$$\Phi = Kh \tag{6.21}$$

The gradient of $\Phi$ has the dimensions of a velocity, hence the name velocity potential. If $K$ is not a function of the space variables, and only then, the equation

$$\mathbf{V} = -\text{grad}\,\Phi \tag{6.21*}$$

is a correct expression of Darcy's law. Indeed, if the fluid is water, $\Phi$ may be expressed in terms of $\Phi^*$ of Equation 6.17 as

$$\Phi = \frac{K}{g}\Phi^* + \text{constant} \tag{6.22}$$

and therefore $$-\text{grad}\,\Phi = -\frac{K}{g}\,\text{grad}\,\Phi^* - \frac{\Phi}{g}\,\text{grad}\,K \tag{6.23}$$

Equation 6.21* reduces to Equation 6.20 if and only if grad $K = 0$, or if $K$ is not a function of $x$, $y$, or $z$.

The advantage of the velocity potential $\Phi$ lies in the fact that it may be combined in a particular way with a stream function $\Psi$, having the same dimensions $[L]^2/[T]$. In those civil engineering problems in which $K$ may be treated as a constant, $\Psi$ may be expressed in discharge units per unit width of porous medium.

As is shown in Section 6.11, orthogonality of streamlines and equipotential lines in two-dimensional problems facilitates the construction of flownets, a graphical method to solve Laplace's equation. From hydrodynamics it is known that the satisfaction of Laplace's equation by a potential function is associated with irrotational flow. The fact that a ground-water potential has been derived seems to be in apparent contradiction with the predominance of viscous resistance forces in ground-water movement and the invariably rotational character of viscous flow. The paradox is explained because only the specific discharge has been derived from a potential. In the individual porous channels, the flow is truly viscous and perfectly analogous to Poiseuille flow. It is assumed, however, that when a great number of channels are assembled the rotations in the individual interstices balance out and that the average resultant flow is irrotational [19].

A discussion of the differential form of Darcy's law which is helpful in the understanding of the different potential concepts is given by Jones [20].

## 6.7 *Range of Validity of Darcy's Law*

In pipeflow, the transition from laminar to turbulent flow is characterized by well-known values of the Reynolds number which expresses the ratio of inertial to viscous forces. Thus there is a lower critical number, around 2,100, below which the flow in pipes is always laminar. By analogy with pipeflow a Reynolds number has been established in flow through porous media, namely

$$N_R = \frac{VD}{\nu} \tag{6.24}$$

in which $V$ is the specific discharge, $\nu$ is the kinematic viscosity of the fluid, and $D$ is a characteristic length. For $D$ some have used the average grain diameter and others the effective diameter of the particles, $d_{10}$ (equal to the diameter of the aperture of the sieve through which 10 per cent by weight of the soil sample is screened). It may be demonstrated that inertia forces are neglected in Darcy's law. If the flow at low velocities is laminar, it is expected to become turbulent at higher velocities. Turbulent flow, if the analogy with pipeflow has to hold, would require a nonlinear, near-quadratic or quadratic relationship between velocity and head loss. Hence, Darcy's law would no longer be valid. Experiments conducted by Lindquist [18,23], however, have shown that digressions from Darcy's law occur, even in the laminar flow regimen, when inertial forces become effective. Lindquist defined a special value of the Reynolds number $N_R{}^*$ at which digression from Darcy's law starts because the inertia forces become important. He found $N_R{}^*$ to be of the order of 4 in the case of a medium of uniform grain size, with diameters ranging from 1 to 5 millimeters and with a porosity $n$ of 38 per cent. He also found that for this medium the upper limit of $N_R$, above which there is always turbulence (called $N_{R,\text{crit}}$), was greater than 180.

These experiments were confirmed by Schneebeli [30] who used the same definition for $N_R$ as did Lindquist, and also the same value for $D$, namely the average grain diameter. For a medium of spheres of uniform diameter $d = 27$ mm, $k = 6.13 \times 10^{-3}$ cm², $n = 39$ per cent and water at 20°C, Schneebeli found $N_R{}^*$ to be of the order of 5, whereas $N_{R,\text{crit}}$ was of the order of 60. For a medium of crushed stone of equivalent diameter $d_e = 37$ mm, $k_e = 11.5 \times 10^{-3}$ cm², $n = 47$ per cent and water at 20°C, he found $N_R{}^*$ to be of the order of 2, and $N_{R,\text{crit}}$ of the order of 60 as in the case of the spheres. The equivalent diameter $d_e$ used in $N_R$ for $D$ is defined by

$$\frac{d_e^{\,2}}{d^2} = \frac{k_e}{k} \tag{6.25}$$

where $k$ and $d$ are the intrinsic permeability and grain size of a medium of uniform spheres and $k_e$ is the intrinsic permeability of the medium of crushed stone. Actually, Equation 6.25 may be rewritten as

$$\frac{d_e^2}{k_e} = \frac{d^2}{k} = \frac{1}{C}$$

in which $C$ is the constant of Equation 6.8. Here, to compare the two tests, Schneebeli used the values of his first test for $K$ and $d$. From these and other similar experiments it can be concluded that there are transition zones first from laminar flow where resistance (viscous) forces are predominant to laminar flow where inertial forces govern, and then finally to turbulent flow [37]. In the case of a typical discharging well, departures from laminar flow are to be expected in the immediate vicinity of the well, where the velocities of the water are maximal. A study of turbulent flow near well screens has been made by Engelund [10]. Considerable departures from Darcy's law probably exist if clay and colloidal material are abundant in the medium [33].

## 6.8 *General Equation for Ground-Water Flow*

The derivation of this equation is based on the principle of conservation of mass. In the present text [8], some changes are made in the original version given by Jacob [19]. This version was widely used until recently [7,14].

Consider an elemental parallelepipedum of porous medium, Figure 6.10*a*, completely saturated with fluid of density $\rho$. Let the vectorial specific discharge $V(x, y, z, t)$, in the center $P$ of the volume element, have the components $u(x, y, z, t)$, $v(x, y, z, t)$, $w(x, y, z, t)$. In order to find the flow rate through the sides of the volume element, the specific discharge at these sides must be referred to that at the center $P$ or to its components. This is done by an expansion of the functions $u(x, y, z, t)$, $v(x, y, z, t)$, $w(x, y, z, t)$ in a Taylor series about the values of $u$, $v$, $w$, in $P$. Since these functions contain more than one variable, it is necessary to introduce the concept of a partial derivative. This derivative differs from the ordinary derivative in that only one of the variables is considered as such in the process of differentiation, and the others are kept constant. For example, to find the value of $u$ through the plane at $x + \Delta x/2$. Taylor's expansion gives:

$$u\left(x + \frac{\Delta x}{2}, y, z, t\right) = u(x, y, z, t) + \frac{\Delta x}{2}\frac{\partial u}{\partial x} + \frac{1}{2}\left(\frac{\Delta x}{2}\right)^2 \frac{\partial^2 u}{\partial x^2} + \cdots$$

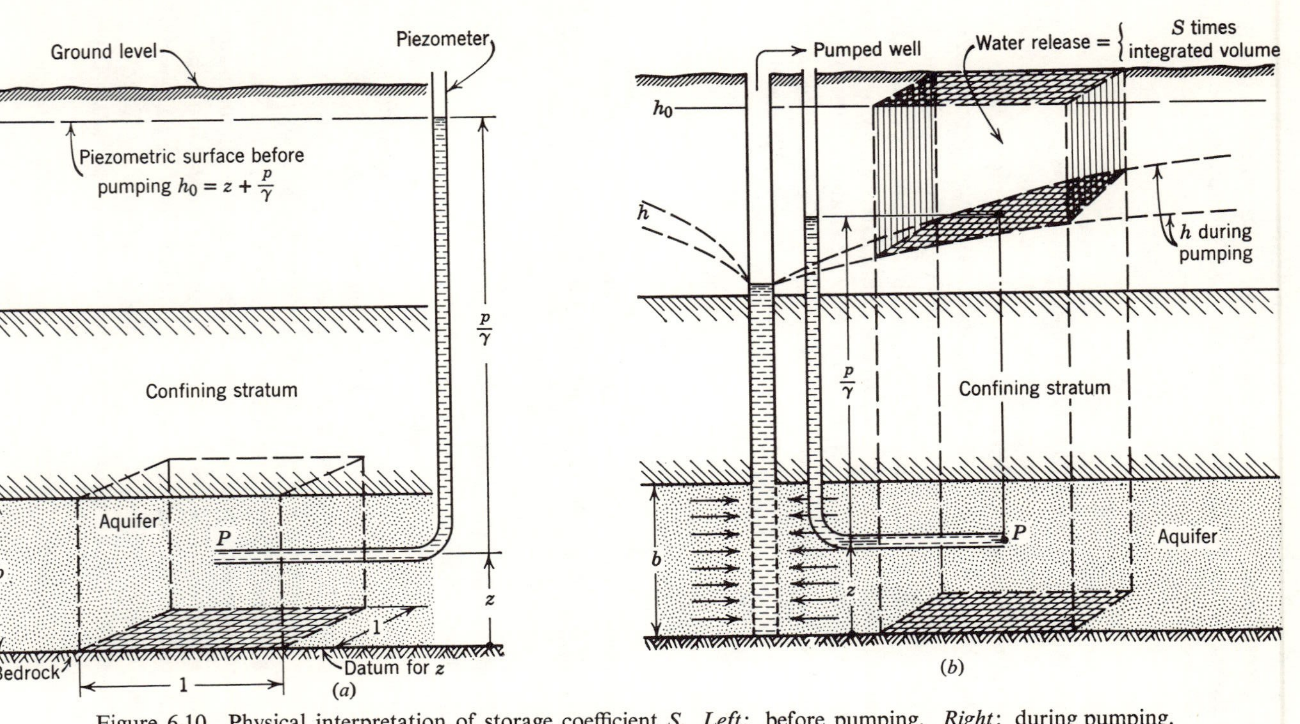


Figure 6.10 Physical interpretation of storage coefficient $S$. *Left*: before pumping. *Right*: during pumping.

Note that the derivatives of all orders are evaluated at the point $P$. It is common practice to break off the series after the linear term in $\Delta x$ and to neglect the higher order terms in $\Delta x$. The mass flowrate through an elemental area $\Delta A$ is expressed as $(\rho V_n \Delta A)$, where $V_n$ is the normal component of the velocity to the area $\Delta A$. Since $\rho$ may be a function $\rho(x, y, z, t)$, one has to consider the products $\rho u$, $\rho v$, $\rho w$ to compute the mass flow rate through the sides of the volume element. This is accomplished, for the flow in the $x$-direction for example, by replacing $u$ by $\rho u$ in the above Taylor series expansion. The terms that are kept in the different Taylor series are written in Figure 6.10*a*. The principle of conservation of mass for the volume element requires:

Mass inflow rate = mass outflow rate + change of mass storage in time

The contributions for the mass inflow rate are:
In the

$$x\text{-direction:} \quad \left[\rho u - \frac{\partial(\rho u)}{\partial x}\frac{\Delta x}{2}\right]\Delta y\,\Delta z$$

$$y\text{-direction:} \quad \left[\rho v - \frac{\partial(\rho v)}{\partial y}\frac{\Delta y}{2}\right]\Delta x\,\Delta z$$

$$z\text{-direction:} \quad \left[\rho w - \frac{\partial(\rho w)}{\partial z}\frac{\Delta z}{2}\right]\Delta x\,\Delta y$$

These contributions must be added to obtain the total mass inflow rate. Similarly, the contributions for the mass outflow rate are:
In the

$$x\text{-direction:} \quad \left[\rho u + \frac{\partial(\rho u)}{\partial x}\frac{\Delta x}{2}\right]\Delta y\,\Delta z$$

$$y\text{-direction:} \quad \left[\rho v + \frac{\partial(\rho v)}{\partial y}\frac{\Delta y}{2}\right]\Delta x\,\Delta z$$

$$z\text{-direction:} \quad \left[\rho w + \frac{\partial(\rho w)}{\partial z}\frac{\Delta z}{2}\right]\Delta x\,\Delta y$$

These contributions must be added to obtain the total mass outflow rate. From this it follows that

Mass inflow rate − mass outflow rate

$$= -\left[\frac{\partial(\rho u)}{\partial x} + \frac{\partial(\rho v)}{\partial y} + \frac{\partial(\rho w)}{\partial z}\right]\Delta x\,\Delta y\,\Delta z$$

This must be equal to the change of mass storage in time. The mass of fluid $\Delta M$ stored in the volume element $\Delta \mathcal{V}_0$ is

$$\Delta M = n\rho\,\Delta x\,\Delta y\,\Delta z$$

where $n$ is the porosity defined in Section 6.3. In the considerations of the change of $\Delta M$ in time, $n$ may vary due to vertical compression or expansion of the medium, and $\rho$ may change in time as well as in place. We assume that there is no change in the dimensions $\Delta x\,\Delta y\,\Delta z$ of the control volume of Figure 6.10*a*. Therefore

$$\frac{\partial(\Delta M)}{\partial t} = \left[\rho \frac{\partial n}{\partial t} + n \frac{\partial \rho}{\partial t}\right] \Delta x\,\Delta y\,\Delta z \tag{6.26}$$

It remains now to express the two terms in the right member of Equation 6.26 in terms of the compressibility $\alpha$ of the aquifer, the compressibility $\beta$ of the fluid, and the pore pressure $p$.

1° $$\rho \frac{\partial n}{\partial t}$$

Changes in porosity are essentially due to vertical compression or expansion of the medium, whereas variations of the lateral dimensions of aquifers are negligible because of the constraints of these aquifers by their surroundings. Thus the change of porosity $n$ of an elemental volume $\Delta x'\,\Delta y'\,\Delta z'$ may be computed as follows: The volume of solid grains $\Delta V_s = (1 - n)\,\Delta x'\,\Delta y'\,\Delta z'$ may be considered as constant because the compressibility of the individual grains is considerably smaller than that of their skeleton as considered above and is also smaller than the compressibility of water. The total derivative of this quantity is zero, or

$$d(\Delta V_s) = d[(1 - n)\,\Delta x'\,\Delta y'\,\Delta z'] = 0$$

Again $\Delta x'$ and $\Delta y'$, the lateral dimensions of the volume element, do not change in comparison to the change in the vertical dimension $\Delta z'$. Therefore $\Delta x'\,\Delta y'$ in the above total derivative is treated as a constant, and only two terms remain:

$$\Delta z'\,d(1 - n) + (1 - n)\,d(\Delta z') = 0$$

or

$$dn = \frac{1 - n}{\Delta z'}\,d(\Delta z')$$

and

$$\frac{\partial n}{\partial t} = \frac{1 - n}{\Delta z'}\frac{\partial(\Delta z')}{\partial t} \tag{6.27}$$

To find $\dfrac{\partial(\Delta z')}{\partial t}$, the concept of vertical compressibility $\alpha$ of the granular skeleton of the medium, treated as a continuum, is introduced; $\alpha = 1/E_s$, where $E_s$ is the bulk modulus of elasticity of this skeleton. The stress $\sigma_z$

on the intergranular skeleton in the vertical direction is called intergranular pressure or stress. By definition

$$E_s = -\frac{d\sigma_z}{\dfrac{d(\Delta z')}{\Delta z'}} = \frac{1}{\alpha}$$

so that

$$d(\Delta z') = -\alpha\,\Delta z'\,d\sigma_z$$

and

$$\frac{\partial(\Delta z')}{\partial t} = -\alpha(\Delta z')\frac{\partial \sigma_z}{\partial t}$$

If this value of $\dfrac{\partial(\Delta z')}{\partial t}$ is introduced in Equation 6.27, we find

$$\frac{\partial n}{\partial t} = -(1-n)\alpha\frac{\partial \sigma_z}{\partial t} \tag{6.28}$$

2°

$$n\frac{\partial \rho}{\partial t}$$

To introduce the compressibility $\beta$ of the fluid or the reciprocal of its bulk modulus of elasticity,

$$\beta = -\frac{\dfrac{d(\Delta V'_v)}{\Delta V'_v}}{dp}$$

the equation of conservation of mass is written as

$$\rho\,\Delta V'_v = \rho_0\,\Delta V'_{v_0} = \text{constant}$$

in which $\rho_0$, $\Delta V'_{v_0}$ are constant reference values of density and elemental volume of fluid. Total differentiation of this equation gives

$$\rho d(\Delta V'_r) + (\Delta V'_v)\,d\rho = 0$$

or

$$-\rho(\Delta V'_v)\beta\,dp + (\Delta V'_v)\,d\rho = 0 \tag{6.29}$$

and

$$\rho\beta\frac{\partial p}{\partial t} = \frac{\partial \rho}{\partial t}. \tag{6.30}$$

Here $p$ is the pressure experienced by the water in the pores and is called pore pressure or neutral pressure.

At any depth intergranular pressure and pore pressure add to render the total or combined pressure. The combined pressure is numerically equal to the weight of all the matter that rests above per unit area if arching effects of the overlying strata are neglected. It follows that

$$\sigma_z + p = \text{constant} \tag{6.31}$$

and

$$d\sigma_z = -dp$$

With the help of Equations 6.28, 6.30, and 6.31, the final expression of Equation 6.26 becomes

$$\frac{\partial(\Delta M)}{\partial t} = [\rho(1-n)\alpha + n\rho\beta]\,\Delta x\,\Delta y\,\Delta z\,\frac{\partial p}{\partial t}$$

and the final continuity equation becomes

$$-\left[\frac{\partial(\rho u)}{\partial x} + \frac{\partial(\rho v)}{\partial y} + \frac{\partial(\rho w)}{\partial z}\right] = \rho[(1-n)\alpha + n\beta]\frac{\partial p}{\partial t} \tag{6.32}$$

To derive the equation for ground-water flow in homogeneous isotropic medium, Equation 6.32 is transformed as follows. The expansion of its left side leads to

$$-\rho\left(\frac{\partial u}{\partial x} + \frac{\partial v}{\partial y} + \frac{\partial w}{\partial z}\right) - \left(u\frac{\partial \rho}{\partial x} + v\frac{\partial \rho}{\partial y} + w\frac{\partial \rho}{\partial z}\right)$$

The second term of this expression is in general quite small in comparison with the first [21] and may therefore be neglected, especially for low-angle flow with $\partial h/\partial z$ very small. Indeed, from Equation 6.16

$$h = z + \frac{1}{g}\int\frac{dp}{\rho} + \text{constant} \tag{6.16*}$$

The following partial derivatives follow

$$\frac{\partial h}{\partial x} = \frac{1}{\rho g}\frac{\partial p}{\partial x}$$

$$\frac{\partial h}{\partial y} = \frac{1}{\rho g}\frac{\partial p}{\partial y}$$

$$\frac{\partial h}{\partial z} = 1 + \frac{1}{\rho g}\frac{\partial p}{\partial z}$$

or after rearrangement

$$\frac{\partial p}{\partial x} = \rho g\frac{\partial h}{\partial x}$$

$$\frac{\partial p}{\partial y} = \rho g\frac{\partial h}{\partial y}$$

$$\frac{\partial p}{\partial z} = \rho g\left(\frac{\partial h}{\partial z} - 1\right)$$

They may be inserted in the partial derivatives derived from

$$d\rho = \rho\beta\, dp \tag{6.29}$$

to give

$$\frac{\partial \rho}{\partial x} = \beta\rho^2 g \frac{\partial h}{\partial x}$$

$$\frac{\partial \rho}{\partial y} = \beta\rho^2 g \frac{\partial h}{\partial y}$$

$$\frac{\partial \rho}{\partial z} = \beta\rho^2 g \left(\frac{\partial h}{\partial z} - 1\right)$$

Furthermore, Darcy's law (Equation 6.13) may be expressed as

$$u = -K \frac{\partial h}{\partial x}$$

$$v = -K \frac{\partial h}{\partial y} \tag{6.13}$$

$$w = -K \frac{\partial h}{\partial z}$$

Also, from Equation (6.16*)

$$\frac{\partial p}{\partial t} = \rho g \frac{\partial h}{\partial t}$$

Equation 6.32 may now be rewritten as

$$K\rho\left(\frac{\partial^2 h}{\partial x^2} + \frac{\partial^2 h}{\partial y^2} + \frac{\partial^2 h}{\partial z^2}\right) + 2K\beta\rho^2 g\left[\left(\frac{\partial h}{\partial x}\right)^2 + \left(\frac{\partial h}{\partial y}\right)^2 + \left(\frac{\partial h}{\partial z}\right)^2 - \frac{\partial h}{\partial z}\right]$$

$$= \rho^2 g[(1 - n)\alpha + n\beta]\frac{\partial h}{\partial t} \tag{6.33}$$

The quadratic terms of the left side of Equation 6.33 may be neglected, and it can be shown [8] that the term in $\partial h/\partial z$ is small for low-angle flow. The first term may be rewritten as $K\rho\nabla^2 h$ where

$$\nabla^2 = \frac{\partial^2}{\partial x^2} + \frac{\partial^2}{\partial y^2} + \frac{\partial^2}{\partial z^2}$$

is the Laplacean operator. The final form of Equation 6.33 is

$$\nabla^2 h - 2g\beta\rho \frac{\partial h}{\partial z} = \frac{S_s}{K}\frac{\partial h}{\partial t} \tag{6.34}$$

in which

$$S_s = \rho g[(1 - n)\alpha + n\beta] \tag{6.35}$$

is the specific storage. It has the dimensions $1/L$, as may be seen from Equation 6.34, and therefore it may be conceived of as the amount of water in storage that is released from a unit volume of aquifer per unit decline of head. Its two parts may be interpreted as:

$(1 - n)\rho g\alpha$ = water in storage released due to the compression of the intergranular skeleton per unit volume and per unit decline of head

$\rho g n\beta$ = water in storage released due to the expansion of the water per unit volume and per unit decline of head

It may indeed be observed that $\beta > 0$ for a decrease in pore pressure and a resulting expansion of the water. In the special case of a confined aquifer of thickness $b$, integration of Equation 6.34 leads to the value [8] of the storage coefficient

$$S = \rho g(\alpha + n\beta) \tag{6.36}$$

The transmissivity $T$ of the aquifer

$$T = Kb \qquad (6.36^*)$$

and the hydraulic head

$$\bar{h}(x, y, t) = \frac{1}{b}\int_0^b h(x, y, z, t)\, dz \tag{6.37}$$

may be introduced, and Equation 6.34 becomes

$$\nabla^2 \bar{h} = \frac{S}{T}\frac{\partial \bar{h}}{\partial t} \tag{6.38}$$

where $S$ and $T$ are called the formation constants of the confined aquifer. $S$ is dimensionless and therefore may be interpreted as the amount of water in storage released from a column of aquifer with unit cross section under a unit decline of head. In the computation of $S$, the compressibility of water cannot be neglected because the term $\rho g n\beta$ may be of the same order of magnitude as the term $\rho g\alpha$ [19]. The significance of $S$ may be visualized by means of Figure 6.10*b*. The decrease of head $\bar{h}$ in a given point of a confined aquifer is generally a result of water withdrawal through pumpage and the accompanying decrease in pore pressure.

Equations 6.34 and 6.38 are the general equations for unsteady flow respectively in an unconfined and in a confined aquifer.

In fact, Equation 6.34 for the unconfined aquifer may be simplified still further. The compressibility of a sand is relatively important only when the sand is completely saturated with water and confined between impervious strata. But when the flow is unconfined, the compressibility of the sand and of the water are relatively unimportant compared to unsteady perturbations or vertical displacements of the free surface which affect the flow pattern. For unconfined flow, the term $\frac{S_s}{K}\frac{\partial h}{\partial t}$ may be neglected in Equation 6.34 and if the term in $\frac{\partial h}{\partial z}$ can be neglected formally the so-called Laplace equation

$$\nabla^2 h = 0 \tag{6.39}$$

is obtained both for steady and unsteady unconfined flow. Treatment of problems of the latter type [4] is beyond the scope of this book.

Laplace's equation is also the equation for confined steady flow as follows immediately from Equation 6.38 when $\bar{h}$ is independent of $t$. The foregoing derivations of Equation 6.39 were made on a hydrodynamic basis. It is shown in the next paragraph how the hydraulic theory modifies Equation 6.39 for steady unconfined flow through the Dupuit assumptions.

The formation constants $S$ and $T$ are generally determined from pumping tests.

## 6.9 *The Dupuit Assumptions in Unconfined Flow*

The solution of Equation 6.39 for unconfined flow is made difficult because the location of the free surface is generally not known and because the boundary condition along this free surface is quadratic in terms of the derivatives of the head. Dupuit's simplifying assumptions permitted Forchheimer [5,24,12] to derive an alternate equation for Equation 6.39 in which the free-surface boundary condition is already satisfied, albeit in an approximate manner.

Dupuit made his assumptions first in the derivation of the flowrate $Q$ towards a well in the center of a circular island (Figure 6.11). In writing the formula for $Q$ as

$$Q = 2\pi r z_f K \frac{dz_f}{dr} \tag{6.40}$$

Dupuit implied that (1) The flow is horizontal, or the equipotential surfaces are cylinders coaxial with the well; (2) the velocity is uniform over the depth of flow $z_f(r)$; and (3) the velocity at the free surface may be expressed as $v = -K\frac{\partial h}{\partial r_i} = K\frac{\partial h}{\partial r}$ instead of $v = -K\frac{\partial h}{\partial s}$. Here $r_i$ denotes the radius oriented towards the well whereas $r$ is oriented away from the well, and $s$ denotes the flow path.

Because of the first implication, the head $h(r)$ may be expressed by the height of the free surface $z_f(r)$ above the impervious bedrock.

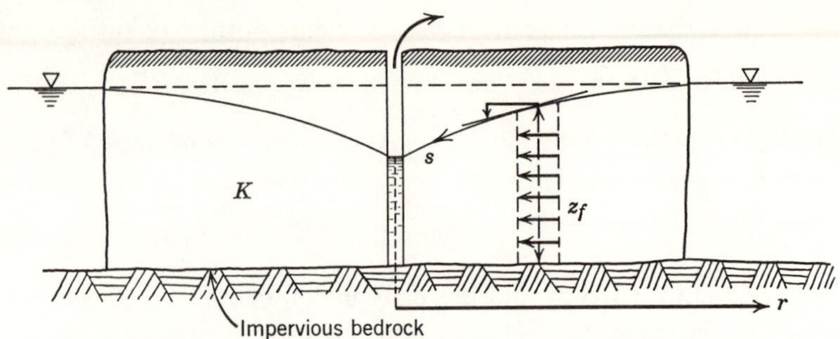


Figure 6.11 Dupuit's assumptions in unconfined flow.

Forchheimer considered a prism of porous medium as sketched in Figure 6.12; the bottom rests on impervious bedrock and the top surface is the watertable. The datum plane for $h$ is the bedrock. As a consequence of Dupuit's assumptions, $h(x, y)$ is the height of the free surface above the datum plane as well as the head in any point of the vertical dropped from a point of the free surface. Dupuit's assumptions are reasonable for mild curvatures of the free surface.

For flow in the $x$-direction, $h$ in the section through $x$ and parallel to $y$ may also be taken as the average value of the head in that section. The flowrate through the elemental area at $x$, say $h\,\Delta y$, may be expressed as

$$Q(x) = -K\frac{\partial h}{\partial x}\,h\,\Delta y \tag{6.41}$$

The flowrate through the elemental area at $x + \Delta x$ may be expressed as

$$Q(x + \Delta x) = -K\frac{\partial h}{\partial x}\,h\,\Delta y + \Delta x\,\frac{\partial}{\partial x}\left(-K\frac{\partial h}{\partial x}\,h\,\Delta y\right) + \cdots \tag{6.42}$$

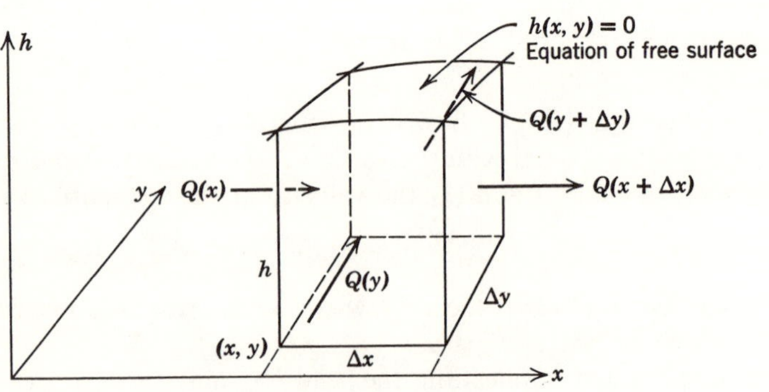


Figure 6.12 Continuity for unconfined aquifer.

For constant $K$, the difference between inflowrate and outflowrate is

$$\Delta x\,\Delta y K\,\frac{\partial}{\partial x}\left(h\,\frac{\partial h}{\partial x}\right) = \Delta x\,\Delta y K\,\frac{\partial}{\partial x}\left(\frac{1}{2}\,\frac{\partial h^2}{\partial x}\right)$$

In a similar manner, the difference between inflowrate and outflowrate in the $y$-direction may be found to be

$$K\,\Delta x\,\Delta y\,\frac{\partial}{\partial y}\left(\frac{1}{2}\,\frac{\partial h^2}{\partial y}\right)$$

By hypothesis there is no flow in the vertical direction. According to the principle of continuity, the difference between inflowrate and outflowrate must be equal to the change of water volume inside the prism. This change is zero if there are neither sources nor sinks inside the prism. Hence

$$\Delta x\,\Delta y\,\frac{K}{2}\left[\frac{\partial^2 h^2}{\partial y^2} + \frac{\partial^2 h^2}{\partial y^2}\right] = 0$$

and

$$\nabla^2 h^2 = 0 \tag{6.43}$$

In case recharge to the watertable takes place through percolation of infiltrated water, Equation 6.43 may be modified without difficulty. If the rate of recharge per unit area is expressed as $W$ (dimensions $L/T$), the rate of recharge to the prism of Figure 6.12 is $W\,\Delta x\,\Delta y$ and conservation of mass requires that

$$\Delta x\,\Delta y\,\frac{K}{2}\left[\frac{\partial^2 h^2}{\partial x^2} + \frac{\partial^2 h^2}{\partial y^2}\right] + W\Delta x\,\Delta y = 0$$

or

$$\nabla^2 h^2 + \frac{2W}{K} = 0 \tag{6.44}$$

**EXAMPLE**

Free surface flow through rectangular earth embankments (Figure 6.13) without replenishment of the watertable. Equation 6.43 reduces to

$$\frac{d^2 h^2}{dx^2} = 0$$

The integration of this ordinary differential equation renders

$$h^2 = Ax + B \tag{6.45}$$

in which $A$ and $B$ are constants of integration which remain to be determined. The constant $A$ is extracted from Equation 6.45 by differentiation

$$A = 2h\,\frac{dh}{dx}$$

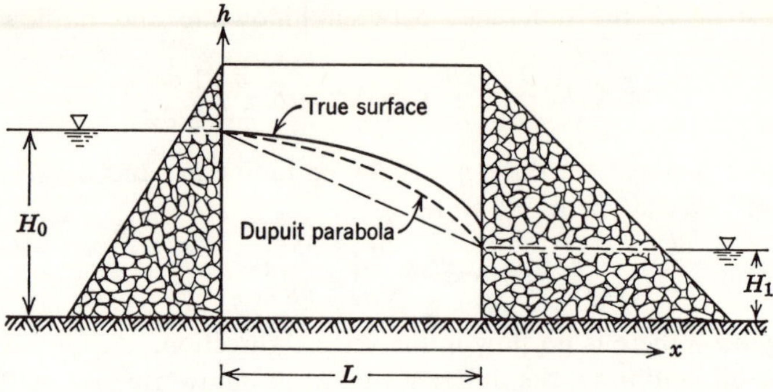


Figure 6.13 Dupuit-flow through dam.

Another expression of $h\,dh/dx$ is found in the flow rate $Q$ through any vertical cross section, based on Dupuit's assumptions

$$Q = -Kh\frac{dh}{dx}$$

Therefore

$$A = -\frac{2Q}{K}$$

The constant $B$ is determined by the condition that for $x = 0$, $h$ assumes the value $H_0$. Therefore

$$B = H_0^2$$

The equation of the free surface is

$$h^2 = -\frac{2Q}{K}x + H_0^2 \tag{6.46}$$

a parabola. If the existence of a seepage surface at the exit is ignored then $Q$ is determined from Equation 6.46 and from $h = H_1$ for $x = L$,

$$Q = \frac{K}{2L}(H_0^2 - H_1^2) \tag{6.47}$$

It may be proved that the flowrate expressed by Equation 6.47 is exact, independent of the existence of the seepage surface. This proof was given by Charny [6,26].

## 6.10 *Capillary Forces*

The tendency of water molecules to adhere to surfaces of solids combined with surface tension of water gives rise to capillary forces. Surface tension is due to molecular attraction forces in the surface of a liquid when exposed to a fluid with which it does not mix (water and air are immiscible) and as a result of which the free surface tends to assume a minimum area. It is a skin effect that is expressed as a force per unit length of contour made by the free liquid surface with its solid boundaries. When a circular tube of small diameter $2r$, wetted by the liquid to be tested, is partly immersed in a vessel containing the liquid, the liquid in the tube will rise to a height $h_c$ above the free liquid surface in the vessel.

The formulas for capillary rise given in most textbooks make certain simplifying assumptions concerning the physics and geometry of the water and tube. First, a finite time is assumed during which the capillary rise takes place. Actually the viscosity of the water in the tube exerts a restraining force which causes the upward movement of the capillary water to diminish exponentially with time, so that an infinite length of time is needed to reach a theoretically stable condition. Second, an angle of contact is assumed between the water surface and the solid. The water surface, nevertheless, approaches the solid surface as a continuously curving surface, and the so-called angle of contact is formed by a line tangent to that surface at the point of contact. Inasmuch as most natural materials have irregular surfaces in minute detail, the concept of a microscopic angle of contact is somewhat artificial as applied to problems in hydrogeology.

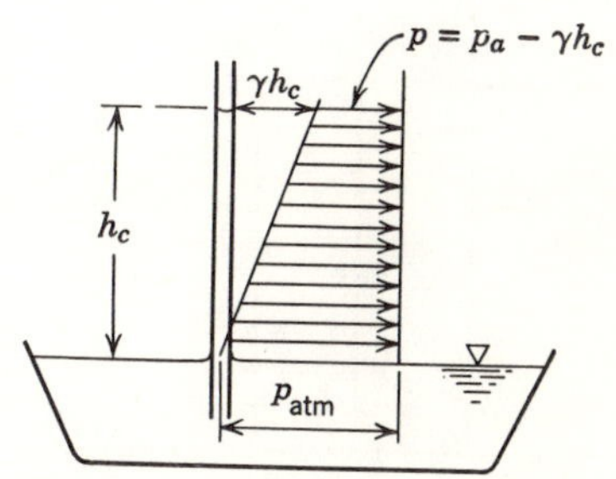


Figure 6.14 Rise in a capillary tube.

If surface tension is designated by $\sigma$, the maximum capillary rise by $h_c$, the tube radius by $r$, the angle of contact by $\theta$, the water density by $\rho$, and the acceleration due to gravity by $g$, then the vertical component of the upward force on the inside of a tube is $\sigma\, 2\pi r \cos\theta$ and the downward pull is the weight of the water, or $\pi r^2 h_c \rho g$ (Figure 6.14). If these forces are equated, the following expression results for $h_c$

$$h_c = \frac{2\sigma \cos\theta}{r\gamma} \tag{6.48}$$

If for water the approximate values $\sigma = 75$ dynes/cm, $\cos\theta \approx 1.0$, and $\gamma = 1$ gramweight/cm³ are used in the presence of solids such as glass and quartz at about 20°C, Equation 6.48 may be simplified for most purposes to

$$h_c = \frac{0.153}{r} \tag{6.49}$$

in which $h_c$ and $r$ are measured in centimeters. For a glass tube of radius 0.01 millimeter, the theoretical rise of water will be slightly more than $1\frac{1}{2}$ meters. The openings in many fine-grained unconsolidated sediments average about this radius, and the maximum capillary rises observed in these sediments range from about 1 to 2 meters. If the material is a sand, the capillary rise will be between 0.1 and 1.0 meter. In gravel the capillary rise will be less than 0.1 meter. The actual capillary rise will be a function of water temperature, mineral composition, orientation, shape, and packing of the component grains. In fine-grained clay the capillary rise is on the average from 2 to 4 meters. The permeability of such clay is, nevertheless, so small that the length of time needed for the rise is measured in terms of many years. The phenomenon observed in Figure 6.14 also occurs when the open end of a pipe filled with sand is immersed in water (Figure 6.15). The pressure at the bottom $AB$ is atmospheric so that in the pipe pore pressures lower than atmospheric prevail. All water within the capillary zone between $AB$ and $CD$ is under tension (negative pressure, if atmospheric pressure is assumed to be zero). Actually the granular skeleton is under pressure in the same way as the wall of the capillary tube of Figure 6.14 is under vertical compression. In the capillary zone the intergranular pressure is larger than the combined pressure, due to the weight of the overlying strata, by an amount equal to the capillary tension in the water. In $CD$, the pore pressure is minimum, namely

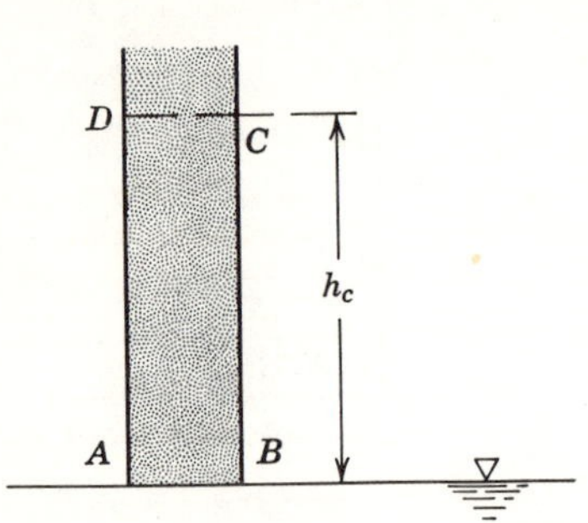

Figure 6.15 Capillary rise in a pipe filled with sand.

$$p = p_a - \gamma h_c \tag{6.50}$$

This condition must be satisfied at the free surface of ground-water flow, according to some Russian authors [26]. Averjanov [26] used a modified form of Equation 6.50, namely

$$p = p_a - \alpha\gamma h_c$$

where $\alpha$ is a number smaller than one, depending on the saturation of the soil. Averjanov proposed to take $\alpha = 0.3$ for most cases, because the capillary rise in laboratory tests is exaggerated in comparison with the actual capillary rise in the field [36].

## 6.11 *Flow Nets. Graphical Solution of* $\nabla^2 h = 0$

When the flow of ground water is steady or time independent, the fundamental equation results from Equation 6.34 as a first approximation

$$\nabla^2 h = 0 \tag{6.51}$$

This equation is called Laplace's equation and must be solved in a given flow region when certain conditions on $h$ are specified along the boundaries of the region. Analytical solutions of Equation 6.51 may be found with relative ease only when the geometry of the flow region is relatively simple and when the analytical conditions imposed along the boundaries of the region are also simple.

To make a solution of Equation 6.51 possible, it is mandatory that the value of $h$ or its normal derivative $\partial h/\partial n$ [$n$ indicates the normal to the boundary of the flow region] or a linear combination of both, be known along the entire boundary containing the flow region. It is sufficient to know $h$ along some parts of this boundary and to have the value of $\partial h/\partial n$, or a linear combination of both, along the remaining parts.

The geometry of the region is simple when the region is rectangular or circular, the boundary conditions are simple when $h =$ constant and when $\partial h/\partial n =$ constant. In general these restrictions are not compatible with the majority of ground-water flow problems and analytical solutions become impossible or unwieldy. Approximate graphical and semi-graphical methods, relaxation techniques and a variety of model studies are substituted for analytical methods when the latter become too complex. A widely used graphical method consists of the fitting of a flow net to the boundary conditions. This method has definite advantages in speed of execution and range of applicability, especially when geometrically complicated boundaries occur.

The construction of a flow net to solve Laplace's equation graphically is relatively easy in problems with fixed boundaries and for confined flows, whereas it is slightly more complicated for free surface flows or unconfined flows where the location of the watertable is not known beforehand. A flow net is a two-dimensional graph composed of two families of curves of a special nature: flowlines or streamlines which indicate how water travels and equipotential lines which join points of the same potential. Its use is therefore limited to the investigation of two-dimensional cross sections of a porous medium which are representative of the main flow and to the analysis of three-dimensional problems with either axial of radial symmetry. The first use of a flow net (to study seepage underneath a concrete dam) was made by Forchheimer [12].

BOUNDARY CONDITIONS

In general, four different types of boundary conditions occur in the construction of a flow net. Each type is represented in Figure 6.16, which illustrates the seepage through an earth embankment caused by a difference in head between headwater and tailwater. The inflow surface $AB$ is a line of constant head $h$ or constant potential $\Phi$, for $K$ constant. The impervious surface $BC$ is a streamline, characterized by $\partial h/\partial n = 0$, or by the fact that along this line $\Psi = \text{constant}$. The function $\Psi$ is the stream function and has the same dimensions as $\Phi$, namely $L^2/T$. This means that it may be expressed in discharge units per unit width of a porous medium measured perpendicularly to the plane of flow.

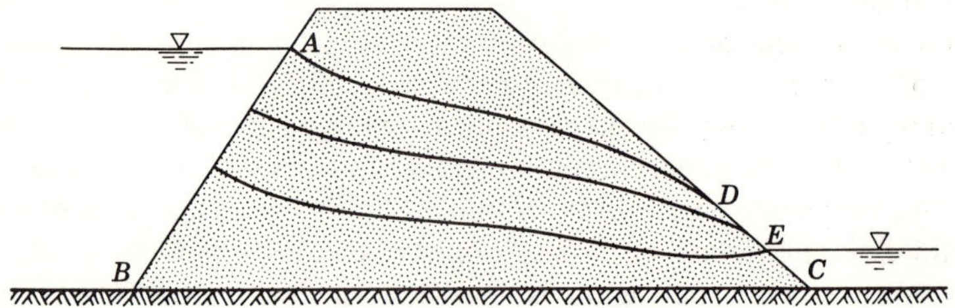


Figure 6.16 Free surface flow.

The top flowline $AD$ is at atmospheric pressure and is often referred to as representing the free surface. In the absence of replenishment to or evaporation from the free surface, $AD$ is a streamline. In general, the equation of the free surface is not available and is one of the unknowns of the problem.

The outflow surface of the dam has the property of being at constant potential between $E$ and $C$ and of being at atmospheric pressure between $D$ and $E$. The segment $DE$ represents a seepage surface and is neither a streamline nor a constant potential line. Along $DE$, water leaves the porous medium to enter the free space and to trickle down along the outflow surface.

CONSTRUCTION OF FLOW NET

In Section 6.6 it is shown that streamlines are orthogonal to equipotential lines in a homogeneous and isotropic medium for which Darcy's law is valid. A flow net may be conceived of as a grid by two families of mutually orthogonal lines established in such a way that, as a rule, streamlines terminate upon potential lines delineating in part the flow region, and vice versa. The exception to this rule occurs when a seepage surface is present. The construction of a flow net is limited to ground-water movement which satisfies the criteria of steadiness and homogeneity, and for which Darcy's law is valid.

There is no unique way to construct a flow net because the number of streamlines and equipotential lines to choose from is extremely large, as both $\Psi$ and $\Phi$ are continuously varying functions. Only a few representative lines of each family are picked. The ratio of the number of streamtubes to that of equipotential drops, however, is a constant for each problem as will follow from the formula for the seepage flowrate. It follows that the flow net becomes uniquely determined once the number of streamlines or equipotential lines is imposed. Casagrande [2] advises the limiting of the number of flow tubes to four or five. Actually fractional flow tubes or potential drops may be used, and experienced students may prefer to use them. In a beginner's attempt however, fractional flow tubes should be avoided. Only by coincidence will both the numbers of flow tubes and potential drops be an integer.

A cross section of a spillway or overflow weir with a single cutoff wall halfway underneath the base is shown in Figure 6.17. This hydraulic structure is supposedly very long in the direction perpendicular to the paper so that the flow may be treated as two dimensional. This assumption is acceptable at a reasonable distance, say 1.5 times the transversal dimension of the structure, away from the abutments of the weir. To draw this flow net, it was decided to use four streamtubes. It is easily discovered that the fluid-filled porous medium has symmetry about the axis $CE$, so that one half of the flow net must be the mirror image of the other half. The first line to draw is the upper flowline $UU$. Its first position is a trial one but it should be born in mind at this stage that there must be room for three more stream tubes. They could be laid out tentatively in the vicinity of $CE$ and an attempt could be made to construct curvilinear squares (why squares and not rectangles are used will be explained) bounded by the equipotential line 3 and 4 and the proposed streamlines. If this trial is successful, a significant step has been made towards the correct completion of the entire net.

The next problem is the proximity of the flowline near the tip $C$ of the cut-off wall. Actually this point is a singular point in the flow net. In hydrodynamics it is proved that for potential flow about a corner forming an obtuse angle the velocity near the corner point tends to infinity. Infinite velocities would, of course, not occur in ground-water flow because microscopically, as was pointed out before, the flow is not irrotational. Large velocities, however, may be expected in the vicinity of the tip where Darcy's law would break down. Hence the flowline may pass very close to the tip. The squares around the tip are likewise singular.

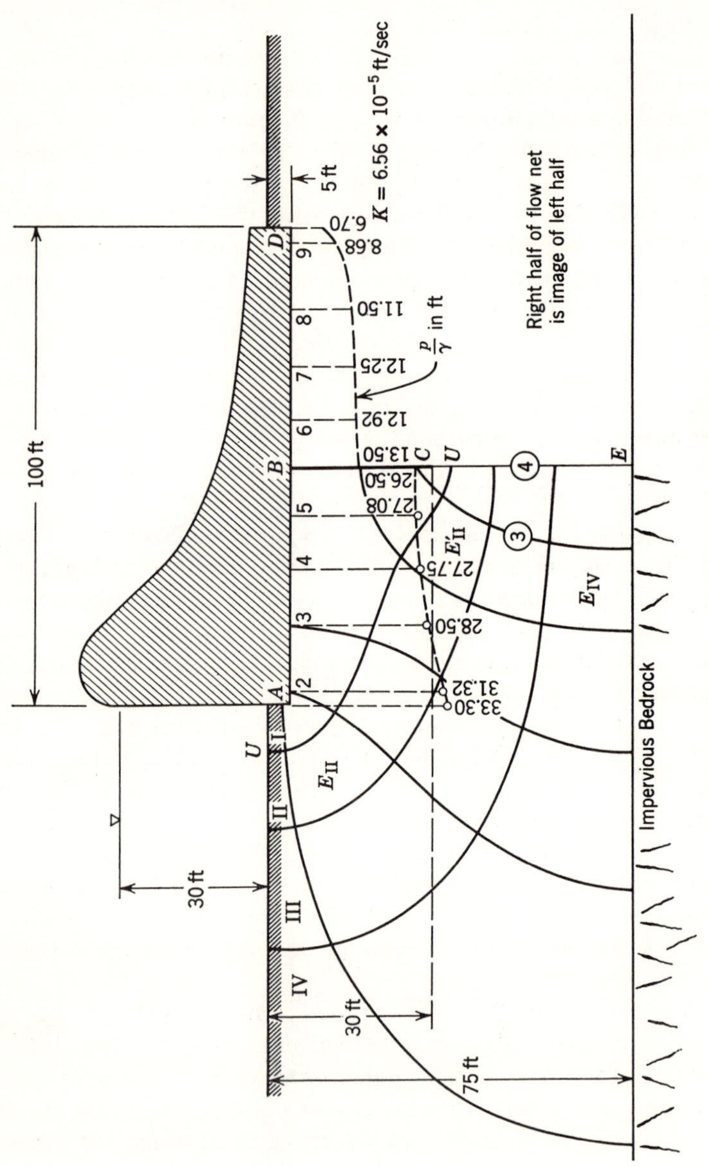


Figure 6.17 Seepage under overflow weir with single cut-off wall.

Some hints on how to start the construction of a flow net have just been given. Its completion may involve relocation, by trial and error, of parts of the first set of streamlines.

*Seepage Rate.* The computation of the seepage rate or flow underneath the hydraulic structure of Figure 6.17 shows the effect of a choice of squares or rectangles with constant width-length ratios on the properties of the flow net [35]. If the streamtubes of Figure 6.17 are labeled I, II, III, IV, it suffices to consider three curvilinear elements $E_{\text{II}}$, $E'_{\text{II}}$, and $E_{\text{IV}}$ and to write down the discharge through these elements:

$$E_{\text{II}} \quad \therefore \quad \Delta q_{\text{II}} = K\,\Delta h_{\text{II}}\left(\frac{\Delta b}{\Delta l}\right)_{\text{II}}$$

$$E'_{\text{II}} \quad \therefore \quad \Delta q'_{\text{II}} = K\,\Delta h'_{\text{II}}\left(\frac{\Delta b}{\Delta l}\right)'_{\text{II}}$$

$$E_{\text{IV}} \quad \therefore \quad \Delta q_{\text{IV}} = K\,\Delta h_{\text{IV}}\left(\frac{\Delta b}{\Delta l}\right)_{\text{IV}}$$

Provided that $\dfrac{\Delta b}{\Delta l}$ = constant, it follows from these three relationships that $\Delta q$ = constant and $\Delta h$ = constant for the three elements. Since the choice of these elements was arbitrary and since it is possible to cover the entire net by appropriate successive choices, it follows that $\Delta q$ = constant and $\Delta h$ = constant for all curvilinear elements. For the construction of the flow net, the simplest constant for $\dfrac{\Delta b}{\Delta l}$ is one. In this case, if

$n_s$ = number of stream tubes,

$n_d$ = number of equipotential drops $= \dfrac{H_t}{\Delta h}$

the flow rate $Q$ per unit width of hydraulic structure may be expressed as

$$Q = n_s\,\Delta q = n_s K\,\Delta h$$

or

$$Q = \frac{n_s}{n_d} K H_t \tag{6.52}$$

The flow net makes it possible to determine the pore pressure and the velocity in the porous medium from the knowledge of $h$. Actually discrete average values of $h$ are obtained for every curvilinear square instead of a continuously varying $h$ which would be the result of the analytical solution. Discrete values of $h$ for the same quality of the flow net are the more accurate the smaller the subdivision of elements is carried out. To find

$h$ at any point, advantage is taken of the fact that the total head is lost along any streamline and over the total length of the streamline between headwater and tailwater. The head at the origin of the streamline is determined after the datum plane for elevation is chosen. The loss between this point and the point where the values of pressure and velocity are sought is computed in terms of the number of equipotential drops between the two points and the loss per drop, say $H_t/n_d$. After $h$ is computed, the values of $p$ follow immediately from the knowledge of $z$ and the definition of $h = z + p/\gamma$.

*Table 6.4*

| Points | $h$, ft | $p/\gamma$, ft |
|---|---|---|
| *A* | 28.30 | 33.30 |
| 2 | 26.32 | 31.32 |
| 3 | 23.50 | 28.50 |
| 4 | 22.75 | 27.75 |
| 5 | 22.08 | 27.08 |
| *B* left | 21.50 | 26.50 |
| *B* right | 8.50 | 13.50 |
| 6 | 7.92 | 12.92 |
| 7 | 7.25 | 12.25 |
| 8 | 6.50 | 11.50 |
| 9 | 3.68 | 8.68 |
| *D* | 1.70 | 6.70 |

NUMERICAL EXAMPLE

Figure 6.17

$$n_s = 4, \quad n_d = 10.6, \quad K = 2 \times 10^{-3}\ \text{cm/sec} = 6.56 \times 10^{-5}\ \text{ft/sec}$$

$$H_t = 30\ \text{ft}, \quad \Delta h = 30\ \text{ft}/10.6 = 2.83\ \text{ft}$$

$$Q = \frac{4}{10.6} \times 6.56 \times 10^{-5}\ \text{ft/sec} \times 30\ \text{ft} = 7.43 \times 10^{-4}\ \text{cfs per ft}$$

Uplift pressures on the base are computed in points 1 to 6 and are represented graphically. The pressure distribution is, of course, anti-symmetrical about the center $B$. Since at the base $z = -5$ ft, it follows from the definition of $h$ that $p = \gamma(h + 5)$.

The final results are summarized in Table 6.4.

OTHER EXAMPLES OF FLOW NETS

Figures 6.18 and 6.19 show flow nets and pressure distributions respectively for the flow below a dam on a stratum of infinite thickness and

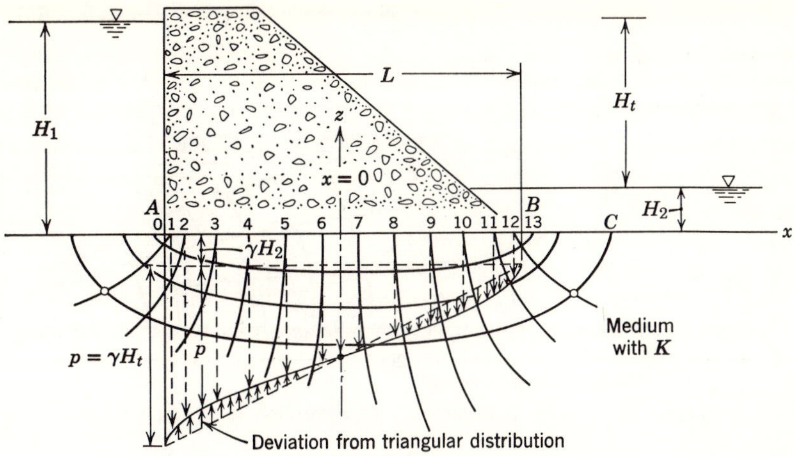


Figure 6.18 Dam on infinitely thick stratum.

for the flow about a sheetpile in a stratum of infinite thickness [7]. Figure 6.20 shows the flow through an earth embankment, exhibiting a free surface. The length of the seepage surface $CD$ is determined by a formula [2,7,35], whereas the location of the upper flowline is determined by trial and error. Its position must be such that its points of intersection with the lines of equipotential are spaced an even $\Delta z$ apart.

Flow nets may also be constructed for anisotropic media, for which the

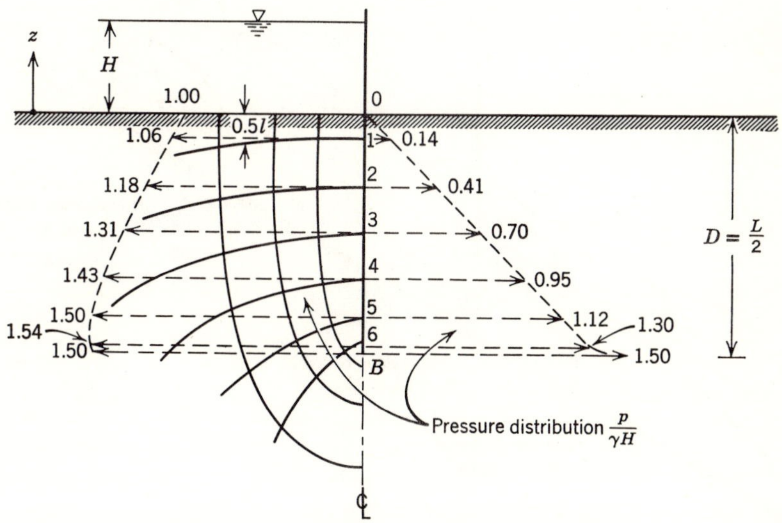


Figure 6.19 Sheet pile in infinitely deep stratum.

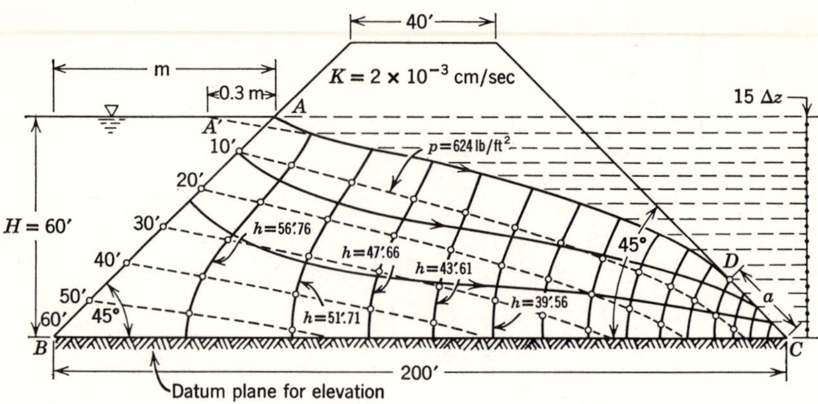


Figure 6.20 Numerical example. Flow through earth embankment.

hydraulic conductivity in one direction is greater than in another direction. This is the case in most stratified sediments, in which the flow proceeds more easily along the planes of deposition than across them. Figure 6.21 shows the flow about a sheetpile in a medium which has a hydraulic conductivity in the $x$-direction nine times larger than in the $z$-direction.

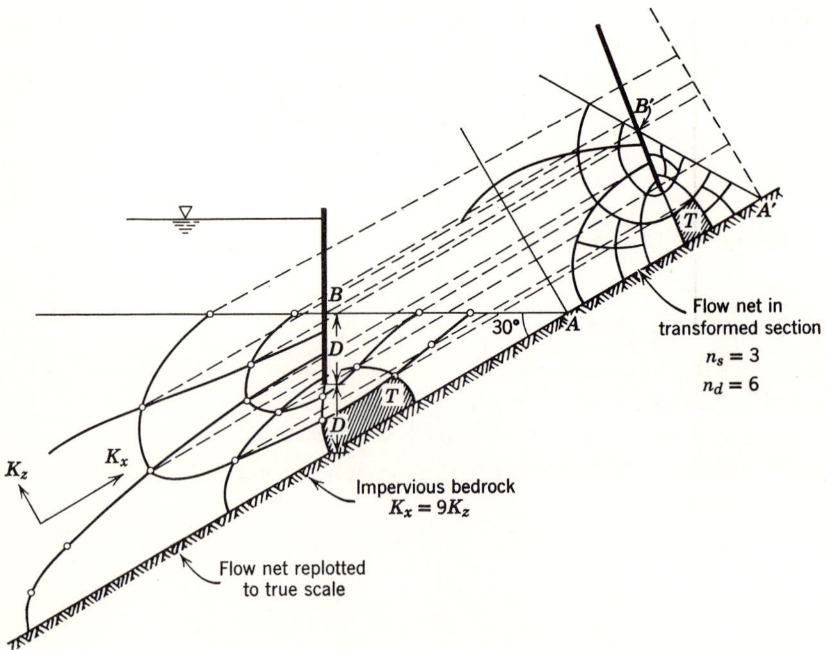


Figure 6.21 Flow net in anisotropic medium.

To obtain the flow net, a change of variable

$$x_t = x\sqrt{\frac{K_z}{K_x}} \tag{6.53}$$

was made and an equilateral flow net was traced in the transformed section and then replotted to true scale.

When streamlines from a medium of a given hydraulic conductivity $K_1$ cross the boundary of a medium with different hydraulic conductivity $K_2$,

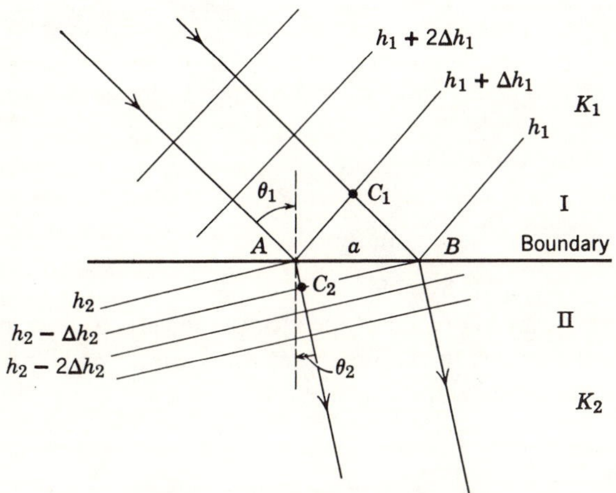


Figure 6.22 Refraction of streamlines.

they are refracted in a manner similar to optical refraction of light rays (Figure 6.22). The refraction follows the law

$$\frac{K_1}{\tan\theta_1} = \frac{K_2}{\tan\theta_2} \tag{6.54}$$

This relationship shows that the flow of Figure 6.22 proceeds from a coarse- to a fine-grained medium. The flow elements of Figure 6.22 are squares in medium I, but rectangles in medium II. If a flow net with squares on both sides of the boundary and like flow quantities is required, the head drop $\Delta h_1$ and $\Delta h_2$ in the media must satisfy

$$\frac{\Delta h_2}{\Delta h_1} = \frac{K_1}{K_2} \tag{6.55}$$

A flow net in a medium consisting of two strata of different hydraulic conductivity is sketched in Figure 6.23. The equipotential lines are

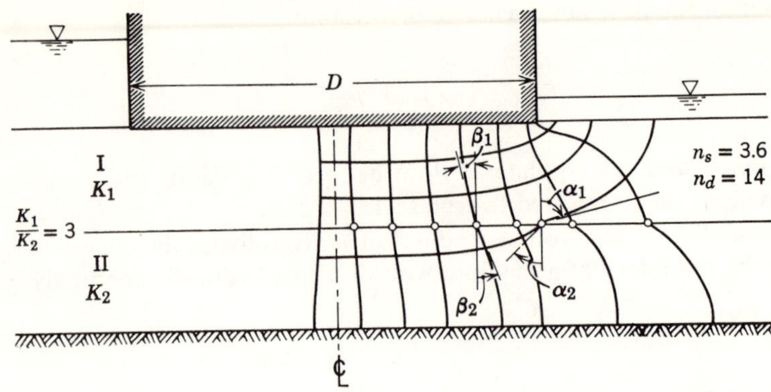


Figure 6.23 Flow net in medium with different hydraulic conductivity.

refracted according to the law

$$\frac{\tan \beta_1}{\tan \beta_2} = \frac{K_2}{K_1} \tag{6.56}$$

in which $\alpha_1 + \beta_1 = \alpha_2 + \beta_2 = 90°$.

Flow nets may not only be developed for plane vertical flow, but also for plane horizontal flow, as in the flow towards wells. For plane horizontal flow, however, analytical solutions are relatively easy to obtain.

## REFERENCES

1. Bakhmeteff, B. A., and N. V. Feodoroff, 1937, Flow through granular media: *Jour. Appl. Mech.*, v. 4A, p. 97—discussion, v. 5A, also 1937, pp. 86–90.
2. Casagrande, A., 1937 and 1940, Seepage through dams: *Jour. New England Water Works Assoc.*, June 1937: also, *Jour. of Boston Soc. Civ. Engrs.*, 1940, pp. 295–337.
3. Darcy, H., 1856, Les fontaines publiques de la ville de Dijon: Paris, V. Dalmont, 674 pp.
4. De Wiest, R. J. M., 1960, Unsteady flow through an underdrained earth dam: *Jour. Fluid Mechanics*, v. 8, pp. 1–9.
5. ——, 1964, History of the Dupuit-Forchheimer assumptions in ground-water flow. Paper presented at the Annual Winter Meeting of the Am. Soc., of Agric. Engrs., December, New Orleans.
6. ——, 1963, Russian contributions to the theory of ground-water flow: *Ground Water Journal of the NWWA*, v. 1, no. 1, January, pp. 44–48.
7. ——, 1965, *Geohydrology:* Chapter 5, New York, John Wiley and Sons, 366 pp.
8. ——, 1966, On the Storage Coefficient and the Equations of Ground-Water Flow, *Jour. Geophys. Research*, v. 71, No. 4, Febr. 15.
9. Dupuit, J., 1863, *Etudes théoriques et pratiques sur le mouvement des eaux dans les canaux découverts et à travers les terrains perméables* (2nd ed.): Paris, Dunod, 30 pp.
10. Engelund, F., 1953, *Trans. Dan. Acad. Techn. Sci.*, no. 3, p. 105.

11. Fair, G. M., and L. P. Hatch, 1935, Fundamental factors governing the streamline flow of water through sand: *Jour. Am. Water Works Assoc.*, v. 25, pp. 1151–1565.
12. Forchheimer, Ph. 1914, *Hydraulik:* Leipzig, B. G. Teubner, 566 pp.
13. Graton, L. C., and H. J. Fraser, 1935, Systematic packing of spheres with particular relation to porosity and permeability, and experimental study of the porosity and permeability of clastic sediments: *Jour. Geol.*, v. 43, p. 785.
14. Hantush, M. S. 1964, *Advances in Hydroscience*, vol. 1, pp. 282–430. New York, Academic Press.
15. Hatschek, E., 1928, *The viscosity of liquids:* Princeton, D. Van Nostrand.
16. Hazen, A., 1911, Discussion of "dams on sand foundations," by A. C. Koenig: *Trans. Am. Soc. Civ. Engrs.* v. 73, p. 199.
17. Hubbert, M. K., 1953, Entrapment of petroleum under hydrodynamic conditions: *Bull. Am. Assoc. Petroleum Geologists* v. 37, pp. 1954–2026.
18. ——, 1940, The theory of ground-water motion; *Jour. Geol.*, November, December, p. 319.
19. Jacob, C. E., 1950, Flow of ground water: *Engineering Hydraulics*, edited by H. Rouse, New York, John Wiley and Sons, pp. 321–386.
20. Jones, K. R., 1962, On the differential form of Darcy's Law: *Jour. Geophys. Research*, v. 67, no. 2, pp. 731–732.
21. Klinkenberg, L. J., 1941, The permeability of porous media to liquids and gases: *Drilling and Production Practice*, New York, American Petroleum Institute, pp. 200–214.
22. Kozeny, J., 1927, *Wasserkraft und Wasserwirtschaft:* v. 22, p. 86.
23. Lindquist, E., 1935, On the flow of water through porous soil: Stockholm, Premier Congrès des Grands Barrages, pp. 81–101.
24. Muskat, M., 1937 and 1946, *The flow of homogeneous fluids through porous media* (1st ed.) New York, McGraw-Hill Book Company, 1937, 737 pp, 2nd Printing, Ann Arbor, Edwards Brothers, 1946.
25. Pirson, S. J., 1958, *Oil reservoir engineering* (2nd ed.): New York, McGraw-Hill, Book Co., pp. 30–44.
26. Polubarinova-Kochina, P. Ya., 1962, *The theory of ground water movement:* Princeton, Princeton University Press, 613 pp. Charny's proof, pp. 281–283.
27. Rose, H. E., 1945, An investigation into the laws of flow of fluids through beds of granular materials: *Proc. Inst. Mech. Engrs.*, v. 153, pp. 141–148.
28. ——, 1945, On the resistance coefficient—Reynolds number relationship for fluid flow through a bed of granular material: *Proc. Inst. Mech. Engrs.* v. 153, pp. 154–168.
29. Scheidegger, A. E., 1960, *The physics of flow through porous media*, New York, The Macmillan Co., pp. 8–12.
30. Schneebeli, G., 1955, Expériences sur la limite de validité de la loi de Darcy et l'apparition de la turbulence dans un écoulement de filtration: *La Houille Blanche*, v. 10, no. 2, pp. 141–149.
31. Slichter, C. S., 1899, Theoretical investigation of the motion of ground waters: Washington, D.C., *U.S. Geol. Survey Nineteenth Ann. Report*, part 2, pp. 295–384.
32. ——, 1902, *The motions of underground waters:* Washington, D.C., *U.S. Geol. Survey Water-Supply Paper* 67, p. 106.
33. Swartzendruber, D., 1962, Non-Darcy flow behaviour in liquid saturated porous media: *Jour. Geophys. Research*, 67, December, pp. 5205–5213.
34. Taylor, D. W., 1948, *Fundamentals of soil mechanics:* New York, John Wiley and Sons, pp. 101–104.

35. ——, 1948, *Fundamentals of soil mechanics:* New York, John Wiley and Sons, pp. 126–128.
36. Van Everdingen, R. O., and B. K. Bhattacharya, 1963, Data for ground-water model studies, *Geological Survey of Canada, Paper* 63, 31 pp.
37. Ward, J. C., 1964, Turbulent flow in porous media: *Journal of the Hydraulics Division, Am. Soc. Civ. Engrs.*, September, pp. 1–12.
38. Wenzel, L. K., 1942, Methods for determining permeability of water-bearing materials: *U.S. Geol. Survey Water-Supply Paper* 887, pp. 7–11.
39. Zunker, F., 1930, Das Verhalten des Bodens zum Wasser: v. 6, of *Handbuch der Bodenlehre* by E. Blanck, Berlin, J. Springer, p. 152.

# chapter 7

# APPLICATIONS OF GROUND-WATER FLOW

The principles of ground-water flow explained in Chapter 6 are applied in Part A of this chapter to problems of flow to and from wells. Simple boundary conditions make it possible to solve Equations 6.38 and 6.39 without the use of advanced mathematical techniques.

The subject of well flow has been thoroughly treated in some of the existing textbooks and has been intensively investigated by research workers. Hence, there is some repetition of material that has become classic in the study of ground-water flow whereas on the other hand only the most idealized boundary conditions are treated. Nevertheless, practical problems have been solved satisfactorily with the equations which have been derived for these artificial boundary conditions even though the solutions are only a first approximation of the true situation in nature.

Part B deals with two-phase flow, in particular with the concepts of potential as developed by Hubbert and Lusczynski. Part C describes a few analogs used in the experimental investigation of ground-water flow.

## PART A
## WELL FLOW

### 7.1 *Steady Radial Flow to a Well*

#### CONFINED FLOW

Assume that a well is pumped at a constant discharge $Q$ (Figure 7.1) and that the boundary conditions are such that the resulting lowering of the piezometric head $h$ has the same distribution in any vertical section through the axis of the well. This means that the flow has radial symmetry (that is, it is independent of the angle $\theta$ in polar coordinates) and that the head must be constant along the perimeter of any circle concentric with the well. Actually, such requirements are only met by a well centered on a circular island and penetrating a homogeneous and isotropic aquifer. If complete penetration of the aquifer by the well is assumed, then the flow is everywhere parallel to the bedrock and follows streamlines $s'$ in a direction opposite to the rays $r$ emanating from the center of the well,

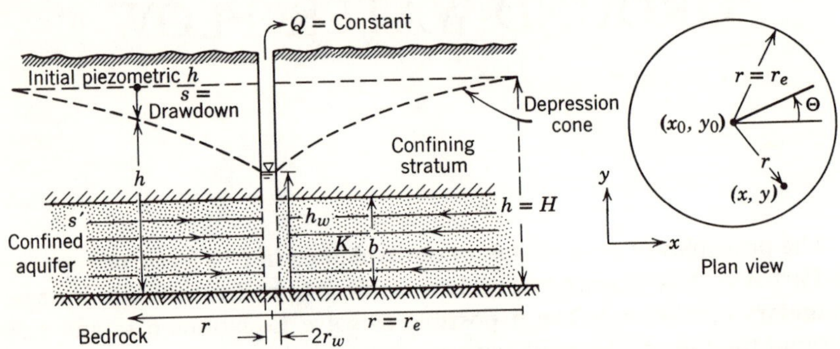


Figure 7.1 Radial flow to a well completely penetrating a confined aquifer.

taken as the origin of plane polar coordinates. Hence $Q$ equals the flow-rate through a cylinder with radius $r$ and height $b$, thickness of the aquifer, so that

$$Q = -K \frac{\partial h}{\partial s'} 2\pi r b = K \frac{dh}{dr} 2\pi r b \tag{7.1}$$

Equation 7.1 may be integrated at once after separation of the variables. Since it is a first-order differential equation in one independent variable, it needs only one boundary condition, namely

$$h = H \quad \text{for} \quad r = r_e \tag{7.2}$$

Equation 7.1 may be integrated according to condition 7.2 as follows:

$$\int_h^H dh = \frac{Q}{2\pi K b} \int_r^{r_e} \frac{dr}{r}$$

or

$$H - h = \frac{Q}{2\pi K b} \log_e \frac{r_e}{r} \tag{7.3}$$

The difference $H - h$ between the elevation of the initial piezometric surface and the elevation of this surface after $Q$ was pumped is called the drawdown. It is commonly designated by the symbol $s$. Because in well problems one is more interested in the drawdown $s$ than in the absolute value of $h$, and because the aquifer is confined, no special attention has been given to the choice of the datum plane for elevation. Note that in Figure 7.1, to avoid confusion, the symbol $s'$ has been used to indicate the streamlines. The product $Kb = T$ is the transmissivity of the aquifer [see Equation 6.37]. It has the dimensions $[L^2/T]$ and can be expressed in U.S. practical units of gallons per day and per feet (gpd/ft).

The integration of Equation 7.1 may also be written as

$$h = \frac{Q}{2\pi Kb} \log_e r + \text{constant} \tag{7.4}$$

This formula avoids the concept of cylindrical boundaries at constant head, coaxial with the well and at a distance $r_e$ from the center of the well. The radius $r_e$ corresponds to the rather ill-defined radius of influence [38], which determines a so-called circle of influence. The circle of influence is nothing but the vertical projection of a cylinder at constant head, not affected by the pumping of the well.

From Equation 7.4 it is clear that $h$ increases indefinitely with an increase in $r$. For extensive natural aquifers, however, values of $r$ can be very large but values of $h$ are relatively small and are commonly limited by the elevation at which recharge enters the system. Thus Equation 7.4 cannot be applicable except within reasonable distances of a well. Stated in another way, steady radial flow to a well, which is an implied condition for Equation 7.4, is only achieved near the well, and steady radial flow in large aquifers that approach infinite lateral dimensions is both theoretically and physically impossible.

Equations 7.3 and 7.4 also hold when the center of the well is located in the point $(x_0, y_0)$ of the coordinate system, but of course $r^2 = (x - y_0)^2 + (y - y_0)^2$. Equation 7.4 becomes

$$h(x, y) = \frac{Q}{4\pi Kb} \log_e [(x - x_0)^2 + (y - y_0)^2] + C \tag{7.5}$$

Equation 7.5 may be interpreted in the following way: it specifies the head $h$ necessary in any point $(x, y)$ to sustain a discharge $Q$ in a well with coordinates $(x_0, y_0)$.

Equation 7.3 is known as the equilibrium or Thiem [37,38] equation. It allows for the computation of $Q$ or $K$, given one or the other, if the values of $h$ or $s$ are measured in two observation wells. For two wells, respectively at distances $r_1$ and $r_2$ from the pumped well, the measured heads are $h_1$ and $h_2$, so that

$$K = \frac{Q}{2\pi b(h_1 - h_2)} \log_e \frac{r_1}{r_2} \tag{7.6}$$

The observation wells should be close enough to the pumped well to provide for significant and easy-to-measure drawdowns. Irrespective of its hypothetical nature, Equation 7.6 has been widely used to determine the hydraulic conductivity $K$ of water-bearing strata.

UNCONFINED FLOW (Figure 7.2)

Under conditions similar to those just given, that is, complete well penetration (to the horizontal base of an isotropic, homogeneous medium),

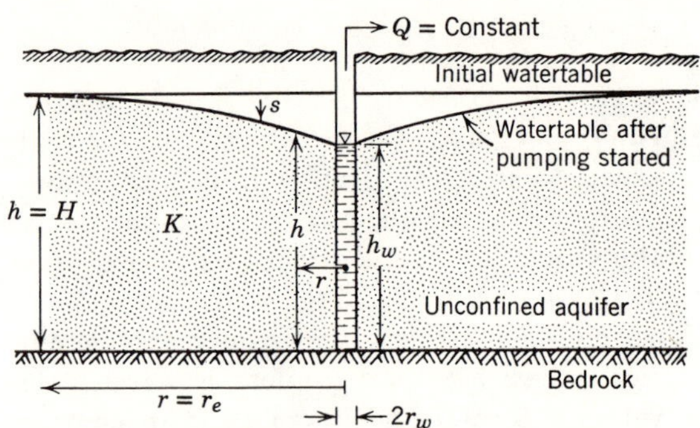


Figure 7.2 Radial flow in water table aquifer.

laminar flow and coaxial boundary of constant head, the equation for the drawdown of the watertable is derived. In this derivation $h$ measures the height of the watertable above bedrock. Under the simplifying assumptions made by Dupuit (Section 6.9), the flowrate $Q$ may be expressed as

$$Q = 2\pi rhK \frac{dh}{dr} \tag{7.7}$$

The boundary condition is, as before,

$$h = H \quad \text{for} \quad r = r_e \tag{7.8}$$

The integration of Equation 7.7 leads to

$$H^2 - h^2 = \frac{Q}{\pi K} \log_e \frac{r_e}{r} \tag{7.9}$$

or

$$h^2 = \frac{Q}{\pi K} \log_e r + \text{constant} \tag{7.10}$$

and

$$h^2(x, y) = \frac{Q}{2\pi K} \log_e [(x - x_0)^2 + (y - y_0)^2] + \text{constant} \tag{7.11}$$

Equations 7.9, 7.10, 7.11 ignore the existence of a seepage surface above the water level in the well. Because of the underlying Dupuit assumptions they fail to describe accurately the drawdown curve near the well where the strong curvature of the watertable contradicts these assumptions. $K$ or $Q$ may be determined as before and Equation 7.9 may be replaced by

$$K = \frac{Q}{\pi(h_1^2 - h_2^2)} \log_e \frac{r_1}{r_2} \tag{7.12}$$

Remarks similar to those made for Equation 7.6 also apply to Equation 7.12. Equations 7.3 and 7.9 for radial flow may be generalized and extended to the case where there is no radial symmetry if the concept of average $\bar{h}_e$ on the circle of radius $r_e$ is introduced. For confined flow,

$$\bar{h}_e - h = \frac{Q}{2\pi Kb} \log_e \frac{r_e}{r} \tag{7.13}$$

with $\bar{h}_e = \dfrac{1}{2\pi}\displaystyle\int_0^{2\pi} h_e(\theta)\, d\theta$ = average head on the circle of radius $r_e$ about the center of the well.

Similarly, for unconfined flow, one finds,

$$\bar{h}_e^2 - h^2 = \frac{Q}{\pi K} \log_e \frac{r_e}{r} \tag{7.14}$$

These formulas are derived from solutions of Laplace's equation in polar coordinates $r$, $\theta$.

## 7.2 *Steady Flow to Several Wells*

Equations 7.5 and 7.11 may be extended without difficulty to the case of any number of wells ($i = 1, 2, \ldots n$) located at $(x_i, y_i)$. This follows from the linearity of Laplace's equation and the fact that solutions may be superimposed. In the case of confined flow, for example, $h_i$ may be considered to be the head required in any point $(x, y)$ to sustain a discharge $Q_i$ of the $i$th well located in the point $(x_i, y_i)$. Hence

$$h_i = \frac{Q_i}{4\pi Kb} \log_e [(x - x_i)^2 + (y - y_i)^2] + C_i$$

To sustain simultaneous discharges $\sum_{i=1}^{n} Q_i$ of all the wells, a head $h(x, y) = \sum_{i=1}^{n} h_i$ will be needed. Therefore $h(x, y) =$

$$\frac{1}{4\pi Kb} \sum_{i=1}^{n} Q_i \log_e [(x - x_i)^2 + (y - y_i)^2] + C \tag{7.15}$$

where $C$ is an arbitrary constant determined conveniently in each problem. The corresponding formula for unconfined flow is

$$h^2(x, y) = \frac{1}{2\pi K}\sum_{i=1}^{n} Q_i \log_e [(x - x_i)^2 + (y - y_i)^2] + C \qquad (7.16)$$

NUMERICAL EXAMPLE

Two wells are drilled 150 feet apart in a stratum of sand which has a hydraulic conductivity $K = 500 \times 10^{-4}$ cm/sec and which is underlain by a horizontal impervious base. The original height $H$ of the ground-water table above the impervious base was 40 feet . A discharge $Q = 300$ gal/min is pumped from each well and it is estimated that a steady condition of flow is reached within a cylinder of radius $r_e = 2000$ feet and that the original head of 40 feet is unaffected beyond this cylinder. Both wells have a radius of 1 feet. Assume that for each individual well, the radius of the cylinder, beyond which the original head is unaffected by pumping, is also equal to 2000 feet.

*a.* Plot the drawdown curve along the vertical plane through the two wells.

*b.* How much does the existence of the second well increase the pumping head for the first well?

In general as one goes far away from a well field, the drawdown surface approaches that which would exist for a single well at the center pumping the sum of the discharges of the individual wells. (This applies for either confined steady flow [8].)

SOLUTION

The coordinate system is chosen so that its origin $x = 0, y = 0$ coincides with the middle of the distance between the wells and the vertical plane through the two wells has the equation $y = 0$. Equation 7.16 renders

$$h^2 = \frac{1}{2\pi K} [Q_1 \log_e (x + 75)^2 + Q_2 \log_e (x - 75)^2] + C$$

and after insertion of the values for $K$ and $Q_1$, $Q_2$

$$h^2 = 65 \text{ ft}^2 [\log_e (x + 75)^2 + \log_e (x - 75)^2] + C$$

$C$ is determined from the boundary condition that $h = 40$ feet for $x = 2000$ feet. Its value is $C = -380$ ft$^2$ and the final equation of the drawdown curve in the plane through the wells is

$$h^2 = 65 \text{ ft}^2 [\log_e (x + 75)^2 + \log_e (x - 75)^2] - 380 \text{ ft}^2.$$

The drawdown curve is represented in Figure 7.3. Because of the symmetry of the total drawdown curve, Well 1 is not sketched in this figure.

This well produces a drawdown curve that is the mirror image of the drawdown curve shown in Figure 7.3. The drawdown at both wells is found to be 40 ft − 16.4 feet = 23.6 ft.

If one well existed alone, Equation 7.9 for $r_w = 1$ ft would render

$$H^2 - h_w{}^2 = \frac{Q}{\pi K} \log_e r_e = 995 \text{ ft}^2$$

after insertion of known numerical values. This would lead to $h_w = 24.6$ ft and therefore the increase in pumping head would be 24.6 ft − 16.4 ft = 8.2 ft.

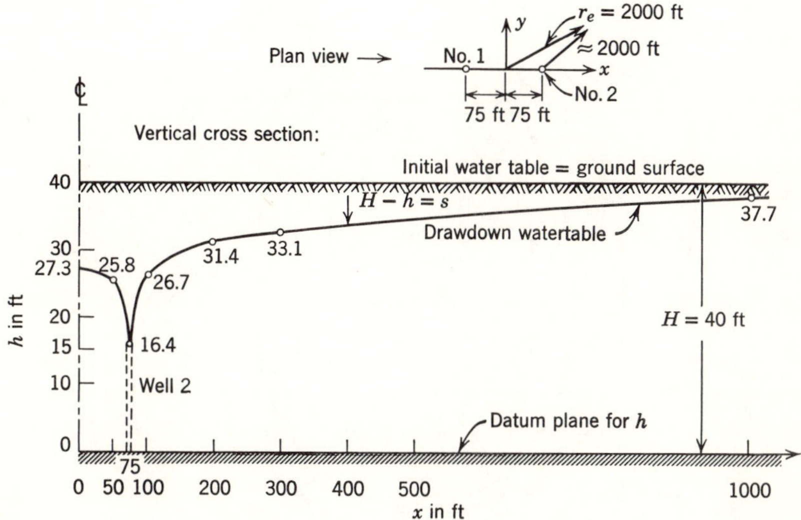


Figure 7.3 Drawdown curve in vertical plane through two wells in water table conditions. (Well 1 is to the left of the diagram and produces a drawdown that is a mirror image of the drawdown shown above.)

Finally, in Figure 7.4 for each well contour maps are drawn of $(H^2 - h^2)$ equal to 275, 300, 325, · · · 550 ft², where $h$ is the elevation of the free surface above the impervious base. By graphical superposition the contours for $H^2 - h^2$ equal to 650, 700, 750, 800, and 850 ft² for the combined effect of both wells are constructed. For each contour the value of $h$ has been labeled. The exact intersection of the composed drawdown curves of Figure 7.4 with $y = 0$ can be determined from the curve of Figure 7.3.

## 7.3 *Flow between a Well and a Recharge Well* (*Figure* 7.5*a*)

The center of the pumped well is on the $x$-axis at a distance $x_0$ from the origin, that of the recharge well at $(-x_0)$, and the flowrate of the recharge

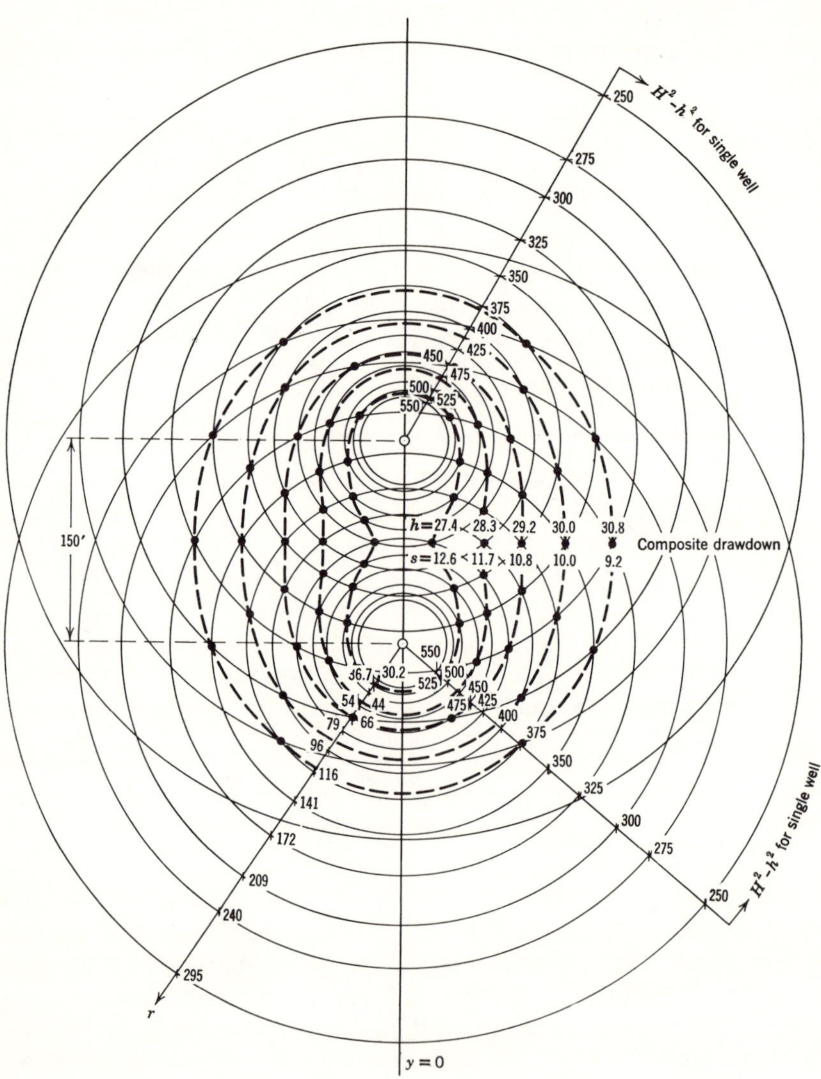


Figure 7.4 Composite drawdown for two wells by graphical superposition.

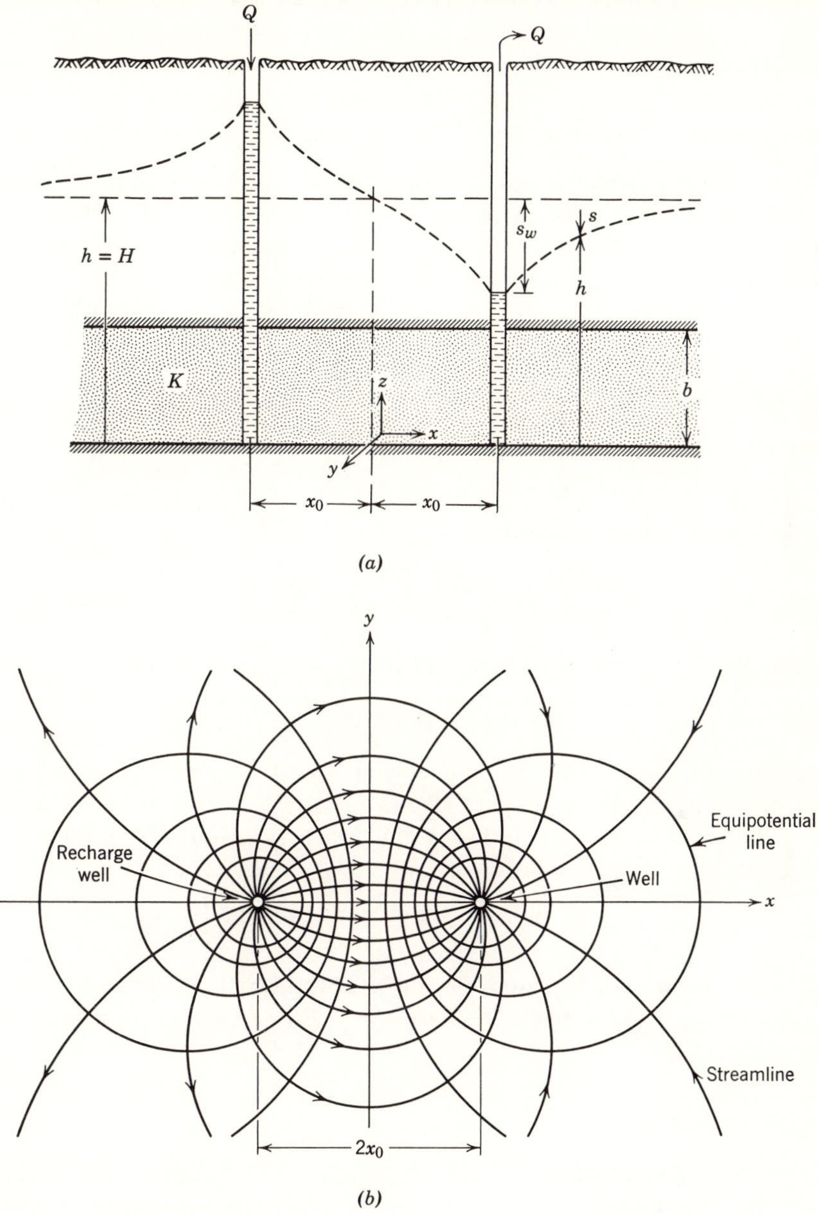


Figure 7.5 (*a*) Well and recharge well (confined aquifer). (*b*) Streamlines and equipotential lines for system of well and recharge well.

well is assumed to be $(-Q)$, where $Q$ is the discharge of the pumped well. The recharge takes place into a confined aquifer and Equation 7.15 reduces to

$$h = \frac{Q}{4\pi Kb} \log_e \frac{(x - x_0)^2 + y^2}{(x + x_0)^2 + y^2} + C \tag{7.17}$$

The constant $C$ is determined by the condition that for very large $r(r \to \infty)$, the head $H$ is not affected by the wells. Therefore $C = H$, which also implies that $h = H$ for $x = 0$, $y = 0$. The final equation for the head $h$ becomes

$$h = \frac{Q}{4\pi Kb} \log_e \frac{(x - x_0)^2 + y^2}{(x + x_0)^2 + y^2} + H \tag{7.18}$$

Equipotential lines $h$ = constant follow from Equation 7.18 as

$$e^{[4\pi Kb(h-H)]/Q} = \frac{(x - x_0)^2 + y^2}{(x + x_0)^2 + y^2} = \text{constant, say } m, \tag{7.19}$$

or, after some rearrangement,

$$x^2 + y^2 - 2xx_0\frac{1 + m}{1 - m} + x_0^{\,2} = 0 \tag{7.20}$$

Equation 7.20 represents a family of circles with center at $\left(x_0\dfrac{1 + m}{1 - m}, 0\right)$ and radii $\dfrac{2x_0\sqrt{m}}{1 - m}$. Since the flowlines have to be perpendicular to the equipotential lines, it follows immediately that they are circles with their center on the $y$-axis (see Figure 7.5*b* [11,21]).

The drawdown $s_w$ at the discharging well may be found from Equation 7.18 for $x = x_0 - r_w$, $y = 0$

$$s_w = H - h_w = \frac{Q}{4\pi Kb} \log_e \frac{(2x_0 - r_w)^2}{r_w^{\,2}} \approx \frac{Q}{2\pi Kb} \log_e \frac{2x_0}{r_w}$$

A comparison with Equation 7.3 shows that in first approximation this drawdown is equal to that at the face of a well in a circular island aquifer of radius $2x_0$.

## 7.4 *Method of Images*

The right half of Figure 7.5*a* also represents the solution of the problem of a single well near a stream. The recharge well may be considered as the negative image of the discharging well reflected in the line $x = 0$ which is coincident with the axis of the stream. The system *well stream* is replaced by the system *well-image well with negative discharge* and both systems observe the same boundary conditions along the line $x = 0$. This artificial device called method of images makes it possible to combine solutions of the type given by Equations 7.3 and 7.4 in the case of noncircular aquifers. Indeed, if Equation 7.4 is applied to both the well and the image well the result is, according to Figure 7.6*a*

$$h = \frac{Q}{2\pi Kb} \log_e r_1 + \text{constant}$$

and

$$h = \frac{-Q}{2\pi Kb} \log_e r_2 + \text{constant}$$

The head for the composite system is

$$h = \frac{Q}{2\pi Kb} \log_e \frac{r_1}{r_2} + C$$

in which $C$ again is determined by the boundary condition that far from the wells, for very large $r_1$ and $r_2$ $\left(\text{where } \log_e \frac{r_1}{r_2} \text{ approaches zero}\right)$, the head $h$ is not affected and preserves its original value $H$. Hence $C = H$ and the final result, with the distances squared for convenience, becomes

$$H - h = \frac{Q}{4\pi Kb} \log_e \frac{r_2^2}{r_1^2} \tag{7.21}$$

This equation may be identified with Equation 7.18. It is possible to prove that Equation 7.21 is correct by computing the flowrate through a cylinder

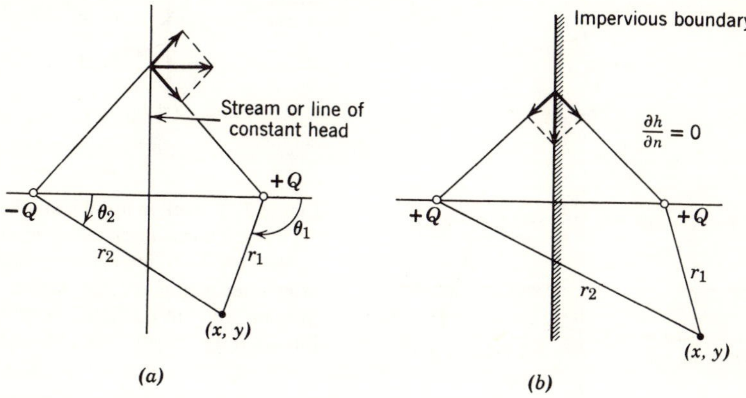


Figure 7.6 Method of images. (*a*) Well near a stream. (*b*) Well near impervious boundary.

of radius $r_2$ and height $b$ concentric with the image well

$$\text{Flowrate} = \int_0^{2\pi} bK\left(-\frac{\partial h}{\partial r_2}\right) r_2\, d\theta_2$$

The result is $Q$ if $h$ in this integral is replaced by its expression from Equation 7.21. For a well pumped in a semi-infinite aquifer near an impervious boundary, the condition that this boundary is a streamline is observed by putting a positive-image well opposite the real well (Figure 7.6*b*). By superposition,

$$H - h = \frac{Q}{2\pi Kb} \log_e \frac{r_e^{\,2}}{r_1 r_2}$$

or

$$h - h_w = \frac{Q}{2\pi Kb} \log_e \frac{r_1 r_2}{r_w^{\,2}} \tag{7.22}$$

if the pair $(r_w, h_w)$ is used instead of $(r_e, H)$.

The method may be extended to aquifers with boundaries intersecting at right angles to form quadrants or even rectangles and strips [14]. For strips and rectangles, infinite series of images are used as the insertion of one image to satisfy the conditions along one boundary disturbs the flow conditions along the other boundaries. This requires a correction which is accomplished by the insertion of new images and in this way an infinite series of image wells arise. The reader is referred to reference

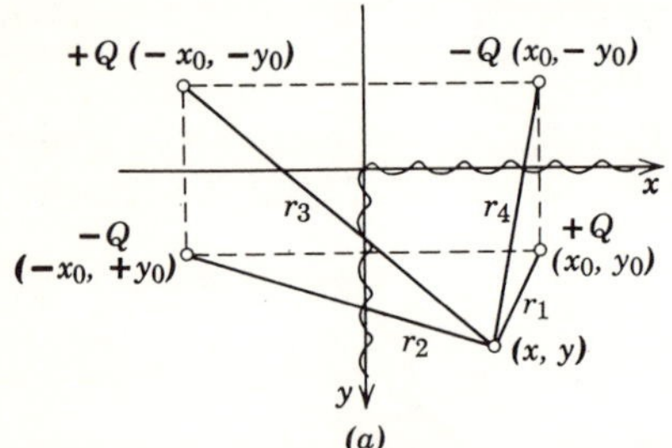


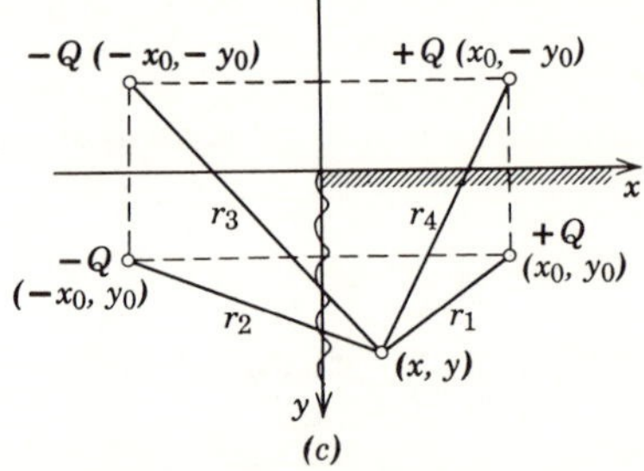


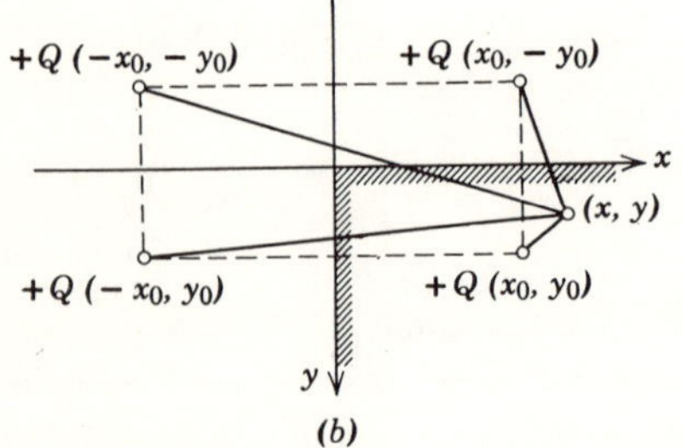


Figure 7.7 Well in infinite quadrant. (*a*) Two streams intersecting at right angle. (*b*) Two impervious boundaries intersecting at right angles. (*c*) Stream and impervious boundary intersecting.

[14] for a detailed review of this technique. Figure 7.7 illustrates systems of images with a well in an infinite quadrant:

a. a well near the intersection of two rectilinear canals,

b. a well near the intersection of two rectilinear impervious boundaries,

c. a well near the intersection of a canal and an impervious boundary.

In *a* the head $h$ may be found by superposition as

$$H - h = \frac{Q}{4\pi T}\left(\log_e \frac{r_e^{\,2}}{r_1^{\,2}} - \log_e \frac{r_e^{\,2}}{r_2^{\,2}} + \log_e \frac{r_e^{\,2}}{r_3^{\,2}} - \log_e \frac{r_e^{\,2}}{r_4^{\,2}}\right)$$

$$H - h = \frac{Q}{4\pi T} \log_e \frac{r_2^{\,2} r_4^{\,2}}{r_1^{\,2} r_3^{\,2}}$$

or

$$H - h = \frac{Q}{4\pi T} \log_e \frac{[(x + x_0)^2 + (y - y_0)^2][(x - x_0)^2 + (y + y_0)^2]}{[(x - x_0)^2 + (y - y_0)^2][(x + x_0)^2 + (y + y_0)^2]} \tag{7.23}$$

It is left as an exercise to the reader to derive similar expressions for *b* and *c*.

## 7.5 *Radial Flow to a Well in an Extensive Confined Aquifer*

This problem was investigated by Theis [36] who made an analogy with the conductive flow of heat to a sink in a plate. The solution given is that for an aquifer of infinite extent where there is no lateral inflow from surrounding waterbodies and the entire discharge must be provided by the release of stored water. According to the definition of storage coefficient (see Section 6.8), the discharge must equal the product of the storage coefficient and the rate of decline in head integrated over the area affected by pumping. According to Figure 7.8

$$dQ = -Sr\, d\theta\, dr \frac{\partial h}{\partial t}$$

and in general $$Q(t) = -S \int_{r_w}^{\infty} \int_0^{2\pi} r \frac{\partial h(r, \theta, t)}{\partial t}\, d\theta\, dr$$

For an aquifer of infinite extent tapped by a single well, the distribution of head is radially symmetric and therefore

$$Q = -2\pi S \int_{r_w}^{\infty} r \frac{\partial h(r, t)}{\partial t}\, dr \tag{7.24}$$

From Equation 7.24 it becomes evident that the rate of decline of head must decrease continuously as the area affected by pumping spreads out (as time goes on) in order to make the integral a constant. This is required if the well is pumped at a constant flow rate $Q$ and if it is assumed that $S$ remains constant as a first approximation, during pumping. Consequently, there can be no steady flow in a confined aquifer of infinite extent.

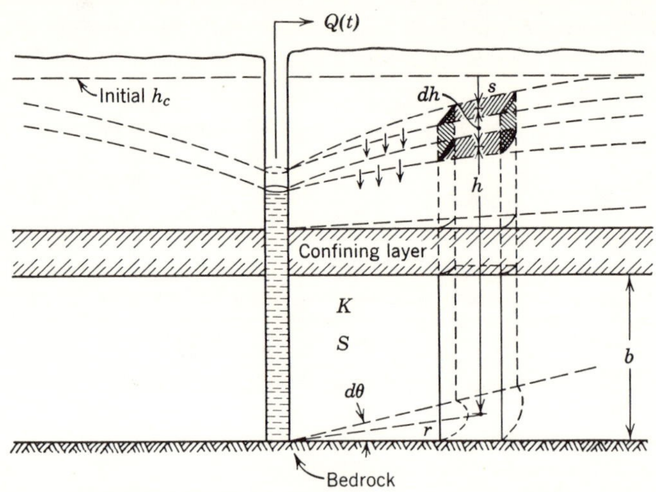


Figure 7.8 Unsteady flow in an extensive aquifer.

Theis replaced the well by a mathematical sink of constant strength and solved Equation 6.38, which may be written as

$$\frac{\partial^2 h}{\partial r^2} + \frac{1}{r}\frac{\partial h}{\partial r} = \frac{S}{T}\frac{\partial h}{\partial t} \tag{7.25}$$

for the boundary conditions

$$\left.\begin{aligned} h \to h_0 \quad &\text{as} \quad r \to \infty \\ \lim_{r\to 0}\left(r\frac{\partial h}{\partial r}\right) &= \frac{Q}{2\pi T} \end{aligned}\right\} \qquad \text{for } t > 0$$

and the initial condition

$$h(r, 0) = h_0 \qquad \text{for } t \leqslant 0$$

The latter condition means that the head in the aquifer was uniform up to the time when pumping started. The solution is

$$h = h_0 - \frac{Q}{4\pi T}\int_{r^2S/4Tt}^{\infty} \frac{e^{-x}\,dx}{x} \tag{7.26}$$

The foregoing integral is a function of the lower limit

$$u = \frac{r^2 S}{4Tt} \tag{7.27}$$

and is tabulated as the exponential integral [23], under the symbol $-Ei(-u)$ (see also [3]). It may be verified that Equation 7.26 is indeed a solution of Equation 7.25 in the foregoing conditions by substitution of the expression for $h$ in the differential equation. The exponential integral can be expanded in a convergent series so that the drawdown $s = h_0 - h$ from Equation 7.26 may be written as

$$s = h_0 - h = \frac{Q}{4\pi T}[-Ei(-u)]$$
$$= \frac{Q}{4\pi T}\left[-0.5772 - \log_e u + u - \frac{u^2}{2 \cdot 2!} + \frac{u}{3 \cdot 3!} - \frac{u^4}{4 \cdot 4!} + \cdots\right] \tag{7.28}$$

The drawdown at the well face is found from Equation 7.28 for $r = r_w$

$$s_w = \frac{Q}{4\pi T}\left[-Ei\left(-\frac{r_w{}^2 S}{4Tt}\right)\right] \tag{7.29}$$

If $Q$ is not constant, $s_w$ may be computed by summing the increments of drawdown produced by increments of $Q$.

Equation 7.26 is known as the nonequilibrium equation or Theis equation. Although its derivation has been based on several assumptions which are seldom justified in field tests, the nonequilibrium formula has been applied with reasonable success in the determination of the formation constants $T$ and $S$ of an aquifer. Among the limiting assumptions, those of complete penetration and of infinite areal extent of the aquifer should be emphasized. Other conditions require a homogeneous and isotropic medium, an infinitesimally small well diameter, and an instantaneous removal of water with decline in head.

Wenzel [43] gave a tabulation of the exponential integral written symbolically as $W(u)$, or well function of $u$, for values of $u$ from $10^{-15}$ to 9.9. A condensed version of this table is given in Table 7.1. This table is helpful in the application of a graphical method devised by Theis [36] to compute the formation constants $T$ and $S$ from pumping test data.

### THEIS METHOD

With $-Ei(-u) = W(u)$ and for U.S. practical hydrology units, Equation 7.28 may be written as

$$s = \frac{114.6Q}{T} W(u) \tag{7.30}$$

$$u = \frac{1.87S}{T}\frac{r^2}{t} \tag{7.31}$$

where $s$ = drawdown, in feet, measured in an observation well due to constant discharge of a pumped well
$Q$ = discharge of a pumped well, in gallons per minute
$T$ = transmissivity, in gallons per day and per foot
$r$ = distance, in feet, from pumped well to observation well
$S$ = coefficient of storage, dimensionless
$t$ = time in days since pumping started

From Equations 7.30 and 7.31 it follows that

$$\log s = \log W(u) + \log \frac{114.6Q}{T} \tag{7.32}$$

and

$$\log \frac{r^2}{t} = \log u + \log \frac{T}{1.87S} \tag{7.33}$$

This rearrangement is useful in order to separate the constants $\frac{114.6Q}{T}$ and $\frac{T}{1.87S}$ from the variables. The logarithm is taken for convenient plotting of the variables which have a wide range of values. Theis used the following graphical method of superposition. First a plot of $W(u)$ versus $u$ is made on logarithmic paper (Figure 7.9). This plot is known as the type curve and may be prepared by means of Table 7.1

*Table 7.1 Values of W(u) for Values of u (After Wenzel* [43])

| $u$ | 1.0 | 2.0 | 3.0 | 4.0 | 5.0 | 6.0 | 7.0 | 8.0 | 9.0 |
|---|---|---|---|---|---|---|---|---|---|
| | 0.219 | 0.049 | 0.013 | 0.0038 | 0.0011 | 0.00036 | 0.00012 | 0.000038 | 0.00 |
| $\times 10^{-1}$ | 1.82 | 1.22 | 0.91 | 0.70 | 0.56 | 0.45 | 0.37 | 0.31 | 0.26 |
| $\times 10^{-2}$ | 4.04 | 3.35 | 2.96 | 2.68 | 2.47 | 2.30 | 2.15 | 2.03 | 1.92 |
| $\times 10^{-3}$ | 6.33 | 5.64 | 5.23 | 4.95 | 4.73 | 4.54 | 4.39 | 4.26 | 4.14 |
| $\times 10^{-4}$ | 8.63 | 7.94 | 7.53 | 7.25 | 7.02 | 6.84 | 6.69 | 6.55 | 6.44 |
| $\times 10^{-5}$ | 10.94 | 10.24 | 9.84 | 9.55 | 9.33 | 9.14 | 8.99 | 8.86 | 8.74 |
| $\times 10^{-6}$ | 13.24 | 12.55 | 12.14 | 11.85 | 11.63 | 11.45 | 11.29 | 11.16 | 11.04 |
| $\times 10^{-7}$ | 15.54 | 14.85 | 14.44 | 14.15 | 13.93 | 13.75 | 13.60 | 13.46 | 13.34 |
| $\times 10^{-8}$ | 17.84 | 17.15 | 16.74 | 16.46 | 16.23 | 16.05 | 15.90 | 15.76 | 15.65 |
| $\times 10^{-9}$ | 20.15 | 19.45 | 19.05 | 18.76 | 18.54 | 18.35 | 18.20 | 18.07 | 17.95 |
| $\times 10^{-10}$ | 22.45 | 21.76 | 21.35 | 21.06 | 20.84 | 20.66 | 20.50 | 20.37 | 20.25 |
| $\times 10^{-11}$ | 24.75 | 24.06 | 23.65 | 23.36 | 23.14 | 22.96 | 22.81 | 22.67 | 22.55 |
| $\times 10^{-12}$ | 27.05 | 26.36 | 25.96 | 25.67 | 25.44 | 25.26 | 25.11 | 24.97 | 24.86 |
| $\times 10^{-13}$ | 29.36 | 28.66 | 28.26 | 27.97 | 27.75 | 27.56 | 27.41 | 27.28 | 27.16 |
| $\times 10^{-14}$ | 31.66 | 30.97 | 30.56 | 30.27 | 30.05 | 29.87 | 29.71 | 29.58 | 29.46 |
| $\times 10^{-15}$ | 33.96 | 33.27 | 32.86 | 32.58 | 32.35 | 32.17 | 32.02 | 31.88 | 31.76 |

Next, values of the drawdown $s$ are plotted against values of $r^2/t$ on logarithmic paper of the same size as that used for the type curve. If the discharge $Q$ is constant, then Equations 7.32 and 7.33 show that $W(u)$ is a function of $u$ in the same way that $s$ is a function of $r^2/t$. Therefore it will be possible to superimpose the data curve on the type curve, holding the coordinate axes of the two curves parallel, and in such a way that the data best fit the type curve. A common point, the match point, arbitrarily

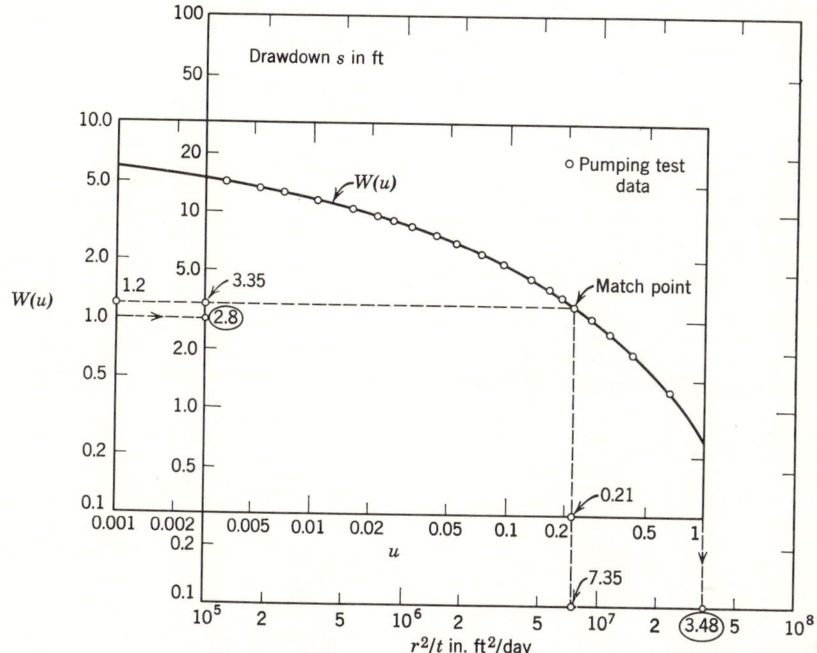


Figure 7.9 Theis' graphical method of superposition. Nonequilibrium equation.

chosen on the overlapping part of the curves, or even anywhere on the overlapping portion of the sheets, determines mutual values of $s$, $W(u)$, $\frac{r^2}{t}$ and $u$ which may be inserted in Equations 7.30 and 7.31 so that these equations may be solved for $S$ and $T$. Once the values of $S$ and $T$ are determined, it is possible to make a check on the trial by computing the values of $\frac{T}{1.87S}$ and $\frac{114.6Q}{T}$. According to Equations 7.30 and 7.31 these values must correspond to $\frac{r^2}{t}$ and $s$ respectively for $u = 1$ and $W(u) = 1$. This determines the translation of the curves uniquely as is illustrated in Figure 7.9.

NUMERICAL EXAMPLE

A completely penetrating well is pumped at a constant rate of 500 gpm. Drawdowns during the pumping period are measured in an observation well 150 feet from the pumped well, at times varying from 2 minutes to

*Table 7.2 Aquifer Test Data: Observation Well $r = 175$ ft*

| Time Elapsed since Beginning of Pumping, $t$ | | Drawdown $s$ in Observation Well | $r^2/t$ |
|---|---|---|---|
| Minutes | Days | Feet | Feet²/Day |
| 0 | 0 | 0 | $\infty$ |
| 2 | $1.39 \times 10^{-3}$ | 1.2 | $2.20 \times 10^7$ |
| 3 | $2.09 \times 10^{-3}$ | 1.9 | $1.47 \times 10^7$ |
| 4 | $2.78 \times 10^{-3}$ | 2.45 | $1.10 \times 10^7$ |
| 5 | $3.48 \times 10^{-3}$ | 2.90 | $8.80 \times 10^6$ |
| 6 | $4.17 \times 10^{-3}$ | 3.35 | $7.35 \times 10^6$ |
| 7 | $4.86 \times 10^{-3}$ | 3.65 | $6.31 \times 10^6$ |
| 8 | $5.57 \times 10^{-3}$ | 4.10 | $5.51 \times 10^6$ |
| 10 | $6.96 \times 10^{-3}$ | 4.60 | $4.40 \times 10^6$ |
| 14 | $9.72 \times 10^{-3}$ | 5.50 | $3.16 \times 10^6$ |
| 18 | $1.25 \times 10^{-2}$ | 6.15 | $2.45 \times 10^6$ |
| 24 | $1.67 \times 10^{-2}$ | 7.00 | $1.84 \times 10^6$ |
| 30 | $2.09 \times 10^{-2}$ | 7.75 | $1.47 \times 10^6$ |
| 40 | $2.78 \times 10^{-2}$ | 8.50 | $1.10 \times 10^6$ |
| 50 | $3.48 \times 10^{-2}$ | 9.00 | $8.80 \times 10^5$ |
| 60 | $4.17 \times 10^{-2}$ | 9.50 | $7.36 \times 10^5$ |
| 80 | $5.57 \times 10^{-2}$ | 10.05 | $5.50 \times 10^5$ |
| 120 | $8.33 \times 10^{-2}$ | 10.30 | $3.68 \times 10^5$ |
| 180 | $1.25 \times 10^{-1}$ | 10.50 | $2.45 \times 10^5$ |
| 240 | $1.67 \times 10^{-1}$ | 10.65 | $1.83 \times 10^5$ |
| 360 | $2.50 \times 10^{-1}$ | 10.80 | $1.22 \times 10^5$ |

6 hours. They have been recorded in Table 7.2. The type curve is first drawn, and then the data curve is superimposed on it. The match point gives values of $u = 0.21$, $r^2/t = 7.35 \times 10^6$ ft²/day, $s = 3.35$ ft, and $W(u) = 1.2$. From Equation 7.30

$$T = \frac{114.6 \times 500 \times 1.2}{3.35} = 20{,}500 \text{ gpd/ft}$$

and from Equation 7.31

$$S = \frac{0.21 \times 20{,}500}{1.87 \times 7.35 \times 10^6} = 0.000315$$

The same results would have been obtained if the choice had been the arbitrary point $s = 1$ ft, $r^2/t = 10^5$ ft²/day, for which $u = 2.9 \times 10^{-3}$ and $W(u) = 0.355$.

To check the translation of the graphs, the values of

$$\frac{T}{1.87S} = \frac{20{,}500}{1.87 \times 0.000315} = 3.48 \times 10^7 \text{ ft}^2\text{/day}$$

and

$$\frac{114.6Q}{T} = \frac{114.6 \times 500}{20{,}500} = 2.80 \text{ ft}$$

are computed. From Figure 7.9 it is evident that these respective values for $r^2/t$ and $s$ are the opposites of the values of $u = 1$ and $W(u) = 1$.

Figure 7.9 indicates that the test has been run under almost ideal conditions. Sometimes there are boundary effects, either through recharge from a nearby stream or because of the presence of impervious boundaries [27]. In the case of a recharging boundary, the drawdowns would be smaller than in the absence of such a boundary and the data would fall below the type curve, with deviations more pronounced as time goes on after pumping started. Finally, Figure 7.9 could have rendered $s$ as a function of $1/t$ because only one observation well was used.

### JACOB'S METHOD

For small values of $u$ [that is, for small $r$ and (or) large $t$] compared to $\log_e u$, series 7.28 may be terminated after the first two terms. Considering that $0.5772 = \log_e 1.78$, Equation 7.28 becomes

$$s = \frac{Q}{4\pi T}\left[\log_e \frac{1}{u} - \log_e 1.78\right],$$

or, in decimal-logarithms,

$$s = \frac{2.30Q}{4\pi T} \log \frac{2.25Tt}{r^2S} \qquad (7.34)$$

Jacob [6,21] first introduced this simplification, which should be restricted to values of $u$ less than about 0.01. The data of Table 7.2 therefore would not be very suitable for the approximation. Equation 7.34 may be applied 1° to observations in one well at different times, 2° to observations in different wells at same time, and 3° to observations in various wells at various times.

### EXAMPLES

In the first case, where observations are made in a single well, only $t$ varies in Equation 7.34 which may be rewritten as

$$s = \frac{2.30Q}{4\pi T} \log \frac{2.25T}{r^2S} + \frac{2.30Q}{4\pi T} \log t \qquad (7.35)$$

This is the equation of a straight line on semilogarithmic paper $s$ versus $\log t$. The slope of the straight line is equal to $\frac{2.30Q}{4\pi T}$ and is found as the vertical projection of the intercept of the straight line between two numbers on the time scale that have logarithms one unit apart, say 1000 and 100 on Figure 7.10. Thus $T$ may be found from the slope. $S$ may be found from any point on the straight line and Equation 7.34, with $T$ known.

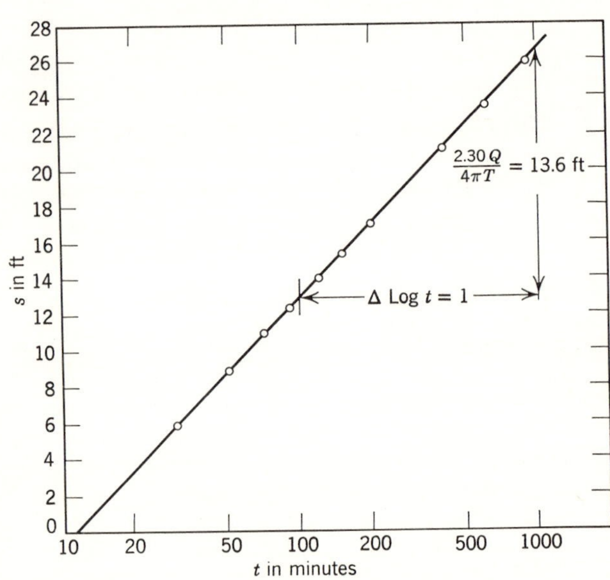


Figure 7.10 Jacob method for solution of nonequilibrium equation. One observation well.

Also, it may be determined from the intercept $t_0$ of the straight line on the $\log t$ axis. Indeed, for $s = 0$, from Equation 7.34

$$\frac{2.25Tt_0}{r^2S} = 1 \tag{7.36}$$

NUMERICAL EXAMPLE

Tabulated below are data on an observation well 50 feet from a well which is pumped for a test at 250 gpm. Find the transmissivity $T$ and storage coefficient $S$ for the aquifer.

| Time, min | 30 | 50 | 70 | 90 | 120 | 150 | 200 | 400 | 600 | 900 |
|---|---|---|---|---|---|---|---|---|---|---|
| Drawdown, ft | 6.5 | 9.0 | 11.0 | 12.4 | 14.1 | 15.4 | 17.0 | 21.2 | 23.6 | 26.0 |

From the slope of the plot on Figure 7.10

$$\frac{2.30Q}{4\pi T} = 13.6 \text{ ft}$$

or $$T = \frac{2.30}{\pi} \times \frac{250}{449}\frac{\text{ft}^3}{\text{sec}} \times \frac{1}{13.6 \text{ ft}} = 7.48 \times 10^{-3}\frac{\text{ft}^2}{\text{sec}} = 4{,}860 \text{ gpd/ft}$$

From Equation 7.36

$$S = \frac{2.25 \times 7.48 \times 10^{-3} \text{ ft}^2/\text{sec} \times 672 \text{ sec}}{2{,}500 \text{ ft}^2} = 0.00045$$

Once the values of $T$ and $S$ are computed, the assumption that $u$ is sufficiently small for all data is checked. Here it sufficies to examine the point for $t = 30$ minutes.

In the second case, only $r$ varies in Equation 7.34 which may be rewritten as

$$s = \frac{2.30Q}{4\pi T}\log\frac{2.25Tt}{S} - \frac{2.30Q}{2\pi T}\log r \tag{7.37}$$

This is the equation of a straight line on semilogarithmic paper, $s$ versus $\log r$. The slope of the straight line is equal to $\left(-\frac{2.30Q}{2\pi T}\right)$, and is found in a similar way as before (see Figure 7.11). Thus $T$ may be found from

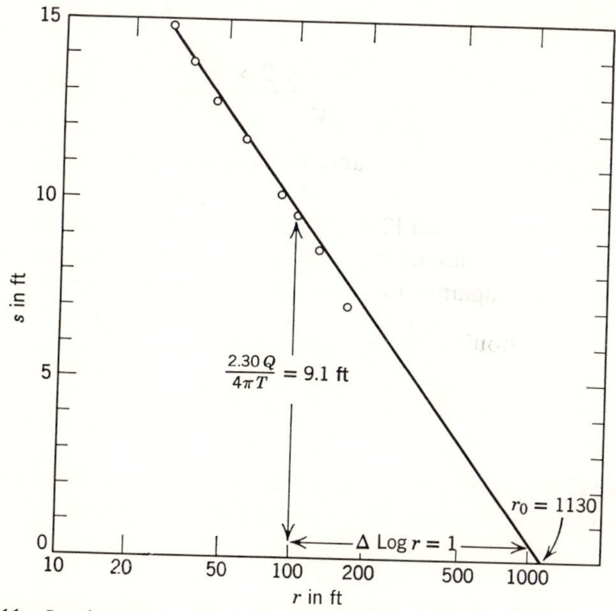


Figure 7.11 Jacob method for solution of nonequilibrium equation. Simultaneous observations.

the slope of the straight line. $S$ may be found in a similar way as before. Here, it may be determined from the intercept $r_0$ of the straight line on the log $r$ axis. Indeed, for $s = 0$, from Equation 7.34

$$\frac{2.25Tt}{r_0^2S} = 1 \tag{7.38}$$

NUMERICAL EXAMPLE

Tabulated below are data of a two hour aquifer test at 350 gpm. The drawdowns are measured in a number of nearby observation wells. Determine $T$ and $S$.

| Well | 1 | 2 | 3 | 4 | 5 | 6 | 7 | 8 |
|---|---|---|---|---|---|---|---|---|
| Distance, ft | 29 | 35 | 44 | 60 | 85 | 100 | 125 | 163 |
| Drawdown, ft | 14.9 | 13.8 | 12.7 | 11.7 | 10.1 | 9.6 | 8.6 | 7.0 |

From the slope of the plot on Figure 7.11

$$\frac{2.30Q}{2\pi T} = 9.1 \text{ ft}$$

$$\text{or } T = \frac{2.30}{2\pi} \times \frac{350}{449}\text{ ft}^3/\text{sec} \times \frac{1}{9.1 \text{ ft}} = 3.14 \times 10^{-2} \text{ ft}^2/\text{sec} = 20{,}400 \text{ gpd/ft}$$

From Equation 7.38

$$S = \frac{2.25 \times 3.14 \times 10^{-2} \text{ ft}^2/\text{sec} \times 2 \times 3600 \text{ sec}}{(1030)^2 \text{ ft}^2} = 0.00048$$

With the computed values of $S$ and $T$, the values of $u$ are sufficiently small for Wells 1 to 7.

Finally, as shown by Jacob [21], a composite drawdown graph may be obtained by having values of drawdown measured at various times in several wells plotted against the logarithms of the respective values of $\frac{t}{r^2}$. Equation 7.34 should now be rewritten as

$$s = \frac{2.30Q}{4\pi T} \log \frac{2.25T}{S} + \frac{2.30Q}{4\pi T} \log \frac{t}{r^2}$$

On semilogarithmic paper, this equation is plotted as a straight line, from which slope $T$ can be determined. $S$ is determined as before.

THEIS' RECOVERY METHOD

Theis [36] and Wenzel [43] pointed out that the nonequilibrium formula could be used to determine the formation constants from the analysis of the recovery of a shutdown well. If a well is pumped at a constant rate and then shut down, the head will recover from its lowest value $h_{T'}$ at time $T'$ when pumping stopped to attain a value $h' > h_{T'}$ at times $t'$ counted from the time of shutdown. ($t' = 0$ corresponds to $t = T'$). (See Figure 7.12.) If $H$ is the initial value of the head before pumping started, then $H - h' = s'$ is called the residual drawdown. For the computation of the residual drawdown, it is assumed that the discharge goes on at the same

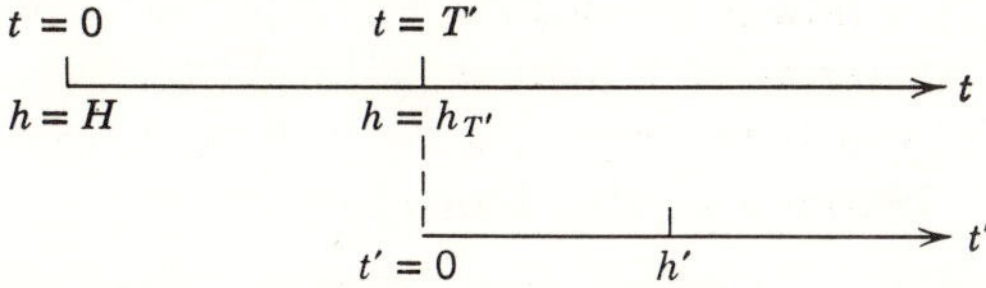


Figure 7-12 Time axes for Theis' Recovery method.

$Q$ for $t > T'$, but that at $t = T'$ a recharge well of strength $-Q$ is superimposed on the discharging well so that the net discharge is zero from $t = T'$ on. Therefore

$$s' = (H - h) + (h - h') = \frac{Q}{4\pi T}\left[\int_{r^2S/4Tt}^{\infty} \frac{e^{-x}}{x}\,dx - \int_{r^2S/4Tt'}^{\infty} \frac{e^{-x}}{x}\,dx\right] \tag{7.39}$$

For $\dfrac{r^2S}{4Tt'}$, sufficiently small, the approximation made in the derivation of Equation 7.34 is also valid here, and therefore

$$s' = \frac{2.30Q}{4\pi T}\left[\log\frac{2.25Tt}{r^2S} - \log\frac{2.25Tt'}{r^2S}\right] = \frac{2.30Q}{4\pi T}\log\frac{t}{t'} \tag{7.40}$$

This equation again is represented by a straight line on semi-logarithmic paper and its slope allows for the determination of $T$, if measured residual drawdowns are plotted against log $\left(\dfrac{t}{t'}\right)$. $S$ may be determined from the value of the drawdown at the time of shutdown.

## 7.6 *Boundary Effects on Unsteady Well Flow*

### METHOD OF IMAGES

The equations derived in Section 7.5 were all valid in the case of aquifers of infinite extent. Actually, such idealized aquifers are not the rule and it should be possible to account for the presence of nearby rivers and impervious boundaries. The principles explained in Section 7.4 are also applicable here. If, for example, nonsteady flow takes place to a well near a river, positioned as in Figure 7.6*a*, an image well with strength $-Q$ is placed across the river and the drawdown is

$$s = \frac{Q}{4\pi T}\left[-Ei\left(-\frac{r_1^2 S}{4Tt}\right)\right] - \frac{Q}{4\pi T}\left[-Ei\left(-\frac{r_2^2 S}{4Tt}\right)\right] \tag{7.41}$$

Unsteady flow towards a well is expected in the early times after pumping started and before the flow pattern was fully established. To have at least an idea of when the flow is established, a comparison is made between $u$ and $\log_e u$ in both terms of Equation 7.41. When these terms may be replaced by their logarithmic approximations, time drops out of the equation and the flow becomes steady. Indeed at that time

$$s = \frac{Q}{4\pi T}\left[\log_e \frac{2.25Tt}{r_1^2 S} - \log_e \frac{2.25Tt}{r_2^2 S}\right] = \frac{Q}{4\pi T}\log_e \frac{r_2^2}{r_1^2}$$

and this is exactly steady state Equation 7.21. This result is very significant and shows that the presence of a recharge boundary in the aquifer tends to offset the unsteady character of the flow. The method of unsteady images may further be applied to other configurations analogous to those of Section 7.4.

## 7.7 *Flow through Semipervious Strata. Leaky Aquifers*

Aquifers in nature are not always perfectly confined between completely impervious strata. This became more evident when the recharge conditions were studied for aquifers that apparently had larger yields than were available from the replenishment in the outcrop area. It was found that those aquifers were overlain by poorly pervious yet water-transmitting strata, which, over large contact areas, contributed significantly to the recharge of the aquifer. The semiconfining beds, in turn, were overlain by ponded water, as in the case of the Dutch polder areas, or by other more pervious strata which again were confined or semi-confined. Sometimes water from the main aquifer seeped through underlying semipervious beds and then again, for other head conditions in adjacent water bodies, the aquifer was recharged from below. The percolation of water into and away from the main aquifer through the semiconfining strata is called leakage. The phenomenon was subjected to mathematical analysis first by Dutch hydrologists and engineers [7,14] and later in this country by Jacob [22] and Hantush [13,14,15]. Jacob [22] pointed out that the state of semiconfinement may come about on a large scale by the presence of lenticular zones in which clays predominate, as in many coastal plain deposits, rather than by a single well-defined stratum. The effect of leakage on Equation 6.38 may be expressed mathematically by applying the principle of conservation of mass to the aquifer of Figure 7.13, in which for reasons of simplicity only leakage from above takes place. The semiconfining stratum is overlain by a sand, with capacity for lateral inflow sufficient to maintain essentially constant head in spite of the downward leakage into the main aquifer [22]. The $x$-$y$ plane coincides with the completely impervious bedrock and is also datum plane for head. Before water was withdrawn from the main aquifer, the head $h$ in the main

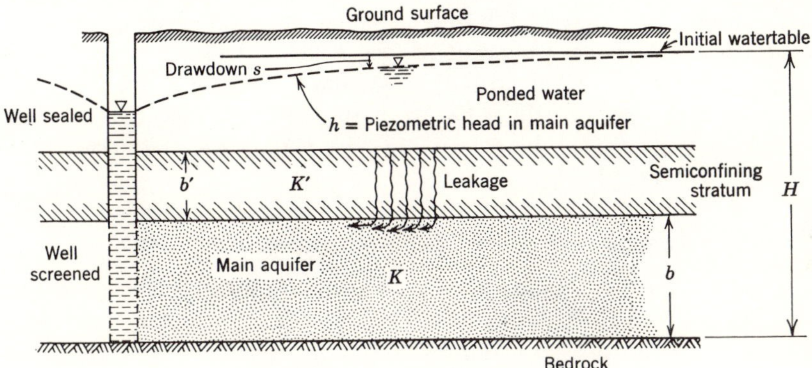


Figure 7.13 Leakage into aquifer.

aquifer was uniform and equal to the height $H$ of the watertable above bedrock. After water is withdrawn from the main aquifer, the head $h$ is represented by a curved line as indicated on Figure 7.13, and because the ponded water remains under head $H$, a head difference is established between the top and bottom of the semiconfining stratum which will induce leakage through this stratum. The hydraulic conductivity $K$ of the main aquifer is so large compared to that of the semiconfining stratum $K'$ that it is safe to assume that water seeps vertically through the semiconfining layer and is refracted over 90° to proceed horizontally in the main aquifer, according to the principles of Section 6.11. The drawdown $s = H - h$ in the main aquifer satisfies the equation

$$\nabla^2 s - \frac{s}{B^2} = \frac{S}{T}\frac{\partial s}{\partial t} \tag{7.42}$$

in which $S$ is the storage coefficient, $T$ is the transmissivity of the main aquifer, and $B$ is the leakage factor, defined as

$$B = \sqrt{\frac{Kbb'}{K'}} \tag{7.43}$$

For a detailed derivation of Equation 7.42, the reader is referred to reference [8].

The leakage factor $B$ has been introduced only because of convenience in the computations. In fact, a large leakage factor means that little leakage takes place and vice versa. To characterize the amount of leakage, values of another coefficient the leakance or leakage coefficient $K'/b'$ are usually given numerically. This coefficient may be defined as the quantity of water that flows across a unit area of the boundary between the main aquifer and its semiconfining bed, if the difference between the head in the main aquifer and that of the ponded water supplying leakage is unity. Values of the leakage coefficient for the Roswell artesian basin in New Mexico as given by Hantush [16] vary from a $4.8 \times 10^{-8}$ sec$^{-1}$ to $10^{-10}$ sec$^{-1}$, as compared to values of $3.5 \times 10^{-7}$ sec$^{-1}$ to $8 \times 10^{-9}$ sec$^{-1}$ reported by Walton [41] for glacial drift deposits in the southern half of Illinois.

## 7.8 *Determination of the Formation Constants of Leaky Aquifers*

In view of the fact that leakage may account for a significant percentage of water pumped from deep wells (up to 11 per cent for the Maquoketa formation in northeastern Illinois, or about 8,400,000 gpd [41]), the hydrogeologist should become familiar with the methods available to determine the leakance and the formation constants. A type-curve method for nonsteady state drawdown was proposed by Hantush and Jacob [15], through the solution of Equation 7.42 in the case of a uniform aquifer of infinite extent, completely penetrated by a well of infinitesimal diameter. Furthermore, the same simplifying assumptions were made as in the derivation of Equation 7.25, supplemented by the assumptions of linear leakage, constant head of the ponded water supplying the leakage, constant head of the ponded water supplying the leakage, and horizontal refraction of the leakage. The solution given by Hantush and Jacob is

$$s = \frac{Q}{4\pi T} W\left(u, \frac{r}{B}\right) \tag{7.44}$$

in which

$$W\left(u, \frac{r}{B}\right) = \int_u^\infty \frac{1}{x} e^{-x - \frac{r^2}{4B^2x}} \, dx \tag{7.45}$$

is the well function for leaky artesian aquifers and in which

$$u = \frac{r^2S}{4Tt} \tag{7.27}$$

All symbols are as defined before. For $B \to \infty$, that is, when leakage $\to 0$, Equation 7.45 reduces to Equation 7.26, implying that a graphical superposition method similar to that devised by Theis would possible, if

$W(u, r/B)$ were tabulated. Hantush [16] suggested the following method in which a number of observations are made in one well. Let $t$ be expressed in minutes instead of in days; $Q$ in gal/min; $T$ in gpd/ft $s$ in ft; $r$, distance from the pumped well to the observation well, in ft. The formulas used in the method become

$$s = \frac{114.6Q}{T} W\left(u, \frac{r}{B}\right) \tag{7.46}$$

$$u = \frac{2700r^2S}{Tt} \tag{7.47}$$

$$\frac{r}{B} = \frac{r}{\sqrt{T/(K'/b')}} \tag{7.48}$$

From Equations 7.46 and 7.47 the following equations are derived

$$\log s = \log W(u, r/B) + \log \frac{114.6Q}{T} \tag{7.49}$$

$$\log t = \log \frac{1}{u} + \log \frac{2700r^2S}{T} \tag{7.50}$$

The analogy with the Theis method becomes more obvious when Equation 7.31 or Equation 7.33 are rewritten as

$$\log t = \log \frac{1}{u} + \log \frac{1.87r^2S}{T} \tag{7.51}$$

In this instance, $\log W(u, r/B)$ is plotted against $\log \frac{1}{u}$ for a series of values of $r/B$. Figure 7.14 shows curves for $r/B$ varying from 2.5 to zero, the latter of course corresponding to the nonequilibrium-type curve. Next, values of the drawdown $s$ are plotted against values of $t$ on logarithmic paper of the same size as that used for the series of type curves. For a constant discharge $Q$, Equations 7.49 and 7.50 show that $W(u, r/B)$ is a function of $\frac{1}{u}$ in the same way $s$ is a function of $t$. Therefore, it will be possible to superimpose the data curve on the series of type curves, holding the coordinate axes of the plots parallel and in such a way that the data curve fits one of the type curves for a given $r/B$, or may be interpolated between two such curves. A match point is chosen anywhere on the overlapping portion of the sheets and its coordinates $W(u, r/B)$, $\frac{1}{u}$, $s$ and $t$ together with the specific value of $r/B$ that caused the fitting of the curves

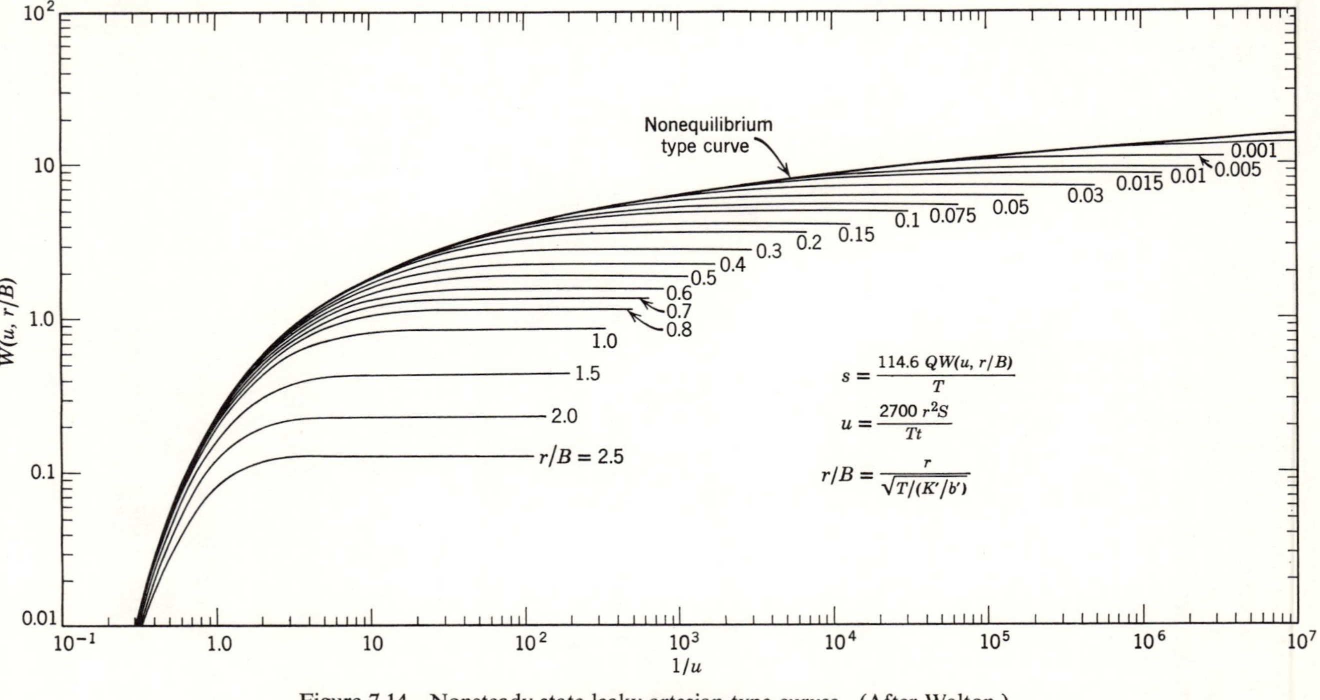


Figure 7.14 Nonsteady-state leaky artesian type curves. (After Walton.)

are inserted in Equations 7.46, 7.47, and 7.48. $S$ and $T$ follow from the first two equations and the knowledge of $r/B$ and $T$ allows for the solution of the leakance $K'/b'$ from Equation 7.48. Walton [41] has applied this method in an aquifer test near Dietrich, Illinois (Figure 7.15). This figure shows that during the first thirty minutes the effect of leakage was not measurable as the data coincide with the nonequilibrium- (nonleaky) type

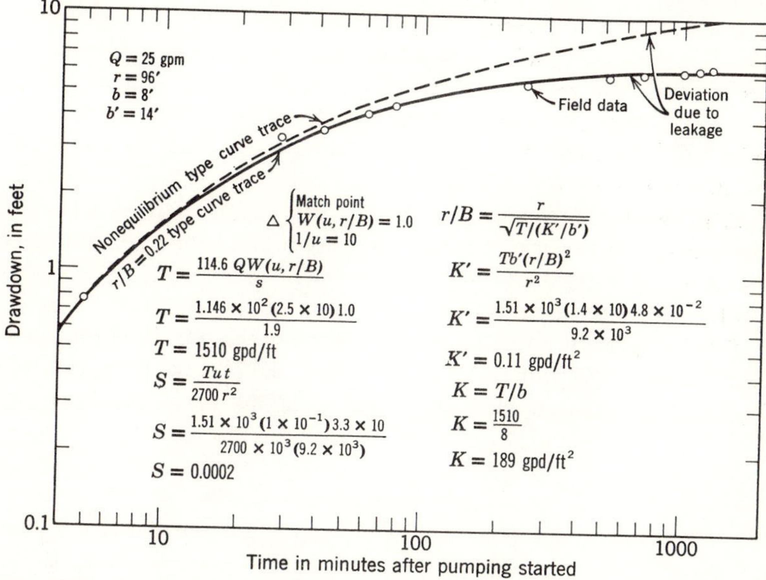

Figure 7.15 Time-drawdown graph for Well 19 near Dieterich, Ill. (After Walton.)

curve but that subsequently the field data deviated to follow the curve for $r/B = 0.22$. As time goes on, more and more well discharge is derived from leakage and ultimately, as the flow becomes steady, the entire yield of the well is derived from leakage. This was first observed by Jacob [14] who devised a graphical method of superposition for steady state drawdown [8].

# PART B

# MULTIPLE-PHASE FLOW

## 7.9 *Movement of Oil and Gas under Hydrodynamical Conditions*

The first attempt to establish the laws of oil and gas accumulation in subterranean media, the so-called anticlinal theory, conceived of a static system of gas, oil, and water with horizontal interfaces, resting on the crests of anticlines. This concept prevailed in the early days after the discovery of oil at Titusville, Pennsylvania, in 1859, and except for some sporadic thinking along fluid-dynamics lines in the 1920's and 1930's which made the flow of water an essential requirement for oil and gas movement, the anticlinal theory with its hydrostatic connotation was by and large the most accepted in the first half of the twentieth century. In 1953 Hubbert [19] put a new emphasis on the hydrodynamical aspect of the problem and at present there remains little doubt as to the basic correctness of this approach. The salient features of his paper, which are outlined in this paragraph, are the derivation of force potentials for the fluids in question and the immiscibility of the fluids. This leads to the determination of distinct fluid-fluid interfaces and the study of the surface tension effects along these interfaces.

Petroleum and gas, formed by the decomposition of organic matter deposited in sedimentary rocks, are normally found in a water-saturated environment. For this reason it is convenient to express the potential energy of these fluids in terms of that of the ambient ground water. In the study of the movement of oil and gas in porous rocks, a distinction should be made between the primary migration from the source rocks, in which the fluids are originated in a dispersed state, to the reservoir rocks, and the movement in the reservoir rocks themselves.

PRIMARY MIGRATION

The force potential per unit mass for a given liquid petroleum at a given point may be derived by extension from Equation 6.16 to include, besides the terms expressing the work against gravity and pressure, one to account for the interfacial energy between petroleum, water, and rock. Thus

$$\Phi_l^* = gz + \frac{p}{\rho_l} + \frac{p_c}{\rho_l} + \text{constant} \tag{7.52}$$

where $p$ is the pressure of the ambient water at the point, $\rho_l$ is the density of the oil, and $p_c$ is the difference of pressure across the interface between oil and water. The value of $p_c$ depends on the wettability of the rocks. In general, rocks are preferentially wet by water with respect to oil, and then $p_c$ is positive, which means that the pressure inside the oil exceeds the pressure inside the water by the amount $p_c$. If petroleum occurs in the gaseous phase, Equation 7.52 should be replaced by

$$\Phi_g^* = gz + \int_o^p \frac{dp}{\rho} + \int_p^{p+p_c} \frac{dp}{\rho} + \text{constant} \tag{7.53}$$

where the variable density $\rho$ is a function only of the pressure $p$.

To find the forces acting on the mass of fluid in a given point, the gradient $\nabla$ of Equation 7.52 must be taken as was demonstrated in the derivation of Equation 6.19. Here, in the primary migration of oil from source rocks to reservoir rocks, $p_c/\rho_l$ is the important term (Equation 7.52) to be considered in the region of the boundary. This term alone depends on the geometry of the porous medium and therefore its gradient taken across the boundary of the different rocks shows the discontinuity in permeability to be paramount in the driving force. This follows from the expression of $p_c$

$$p_c = \frac{C\sigma \cos \theta}{2r} \tag{7.54}$$

where $C$ is a dimensionless factor of proportionality, $2r$ is the mean grain diameter of the rock, $\sigma$ is the surface tension, and $\theta$ is the contact angle in the water phase which the oil-water interface makes with the rock (see Section 6.10 for $h_c = p_c/\gamma$). The force due to capillarity may be derived from Equation 7.54 as

$$-\frac{1}{\rho_l} \operatorname{grad} p_c = \frac{C\sigma \cos \theta}{2\rho_l} \frac{\operatorname{grad} r^*}{r^2} \tag{7.55}$$

Hubbert [19] assumes that grad $r = \beta r$, where $\beta$ is a proportionality constant, so that Equation 7.55 may be written as

$$-\frac{1}{\rho_l} \operatorname{grad} p_c = \frac{C'}{r} \tag{7.56}$$

where $C'$ lumps all constants together.

Equations 7.55 and 7.56 show that the force acting on the liquid petroleum is directed along the steepest rate of increase of the grain size of the rock (meaning of grad $r$) and also that this force, for a given grad $r = \beta r$, is inversely proportional to the grain size of the rock. The steepest rate of change in grain size normally occurs perpendicularly to the planes of sedimentary deposition so that oil globules have a tendency to move in that direction under influence of capillarity. Once a globule reaches the boundary between rocks of different texture, as in Figure 7.16, part of its mass is subjected to a capillary force proportional to $1/r_1$ on the medium with grain size $2r_1$ and part of its mass is subjected to a force proportional

* $\operatorname{grad} \frac{u}{v} = \frac{v \operatorname{grad} u - u \operatorname{grad} v}{v^2}$. Here $u = 1$, $v = r$, grad $1 = 0$. Rule is analogous to the rule for differentiation $d\frac{u}{v} = \frac{v\,du - u\,dv}{v^2}$.

to $1/r_2$ in the medium with grain size $2r_2$, the net resultant force being proportional to the difference $(1/r_1 - 1/r_2)$, with $r_1 < r_2$. The magnitude of the forces is indicated by the lengths of the arrows in Figure 7.16. It follows that an oil globule, in a liquid or gaseous state in a water saturated environment can only pass from fine- to coarse-textured rocks and not in the opposite direction. Once more it should be emphasized that capillary action would not be present if oil were not surrounded by water. If only

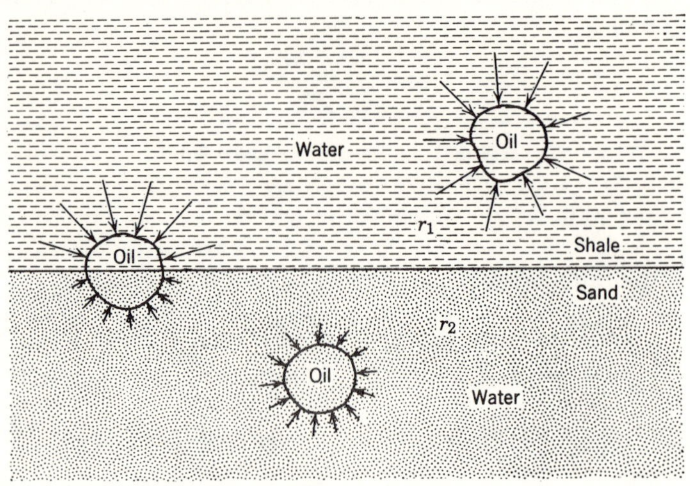


Figure 7.16 Passage of oil globules due to unbalanced capillary pressures. (After Hubbert.)

single-phase flow were to be considered, the flow would proceed equally well in both directions and obey the law of Equation 6.48.

This is the case for water that constitutes the environment in which oil is dispersed. The sand-shale boundary of Figure 7.16 may then be considered as an impermeable barrier for oil present in the sand, but not for water. In the source rock, oil is also subjected to gravity and pressure forces derived from the two remaining terms in Equation 7.1. They have to be superimposed on the capillary forces which for oil in water-saturated shales are of the order of tens of atmospheres.

### MOVEMENT IN RESERVOIR ROCKS

In the flow of oil in reservoir rocks, the role of the capillary force expressed by Equation 7.55 may be neglected because of the large value of the grain size. The capillary pressure of oil in sandstone is only of the order of tenths of an atmosphere. To find the impelling force $\mathbf{F}_l$ per unit mass of liquid petroleum, the gradient is taken of Equation 7.52 with the negative sign as explained in the derivation of Equation 6.19

$$\mathbf{F}_l = -\text{grad}\,\Phi_l^* = \mathbf{g} - \frac{1}{\rho_l}\,\text{grad}\,p \tag{7.57}$$

as compared to

$$\mathbf{F}_w = -\text{grad}\,\Phi_w^* = \mathbf{g} - \frac{1}{\rho_w}\,\text{grad}\,p \tag{7.58}$$

which is the equivalent of Equation 6.19 for the force on a unit mass of water. The pressure $p$ is the same in both equations and may therefore

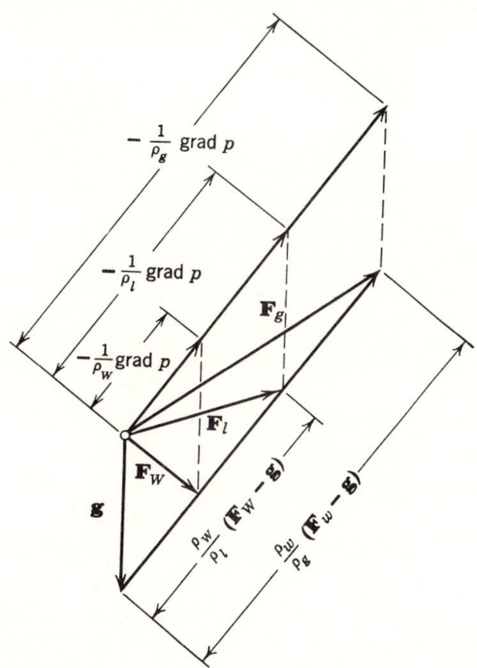


Figure 7.17 Impelling forces on water, oil, and gas in hydrodynamic environment. (After Hubbert.)

be eliminated between these equations. The results is

$$\mathbf{F}_l = \mathbf{g} + \frac{\rho_w}{\rho_l}(\mathbf{F}_w - \mathbf{g}) \tag{7.59}$$

The force $\mathbf{F}_g$ per unit mass of petroleum in the gaseous state is derived in a similar way

$$\mathbf{F}_g = \mathbf{g} + \frac{\rho_w}{\rho_g}(\mathbf{F}_w - \mathbf{g}) \tag{7.60}$$

Equations 7.57 and 7.59 may be represented by a vector diagram as shown in Figure 7.17. From this vector it is evident that for a given flow of the ambient water, the variation in density of the hydrocarbon is responsible for the direction of the total force $\mathbf{F}_l$ or $\mathbf{F}_g$ and for the separation of liquid petroleum and gas in given geological conditions. Such a case is represented in Figure 7.18 where the angle of the homoclinal dip is such that

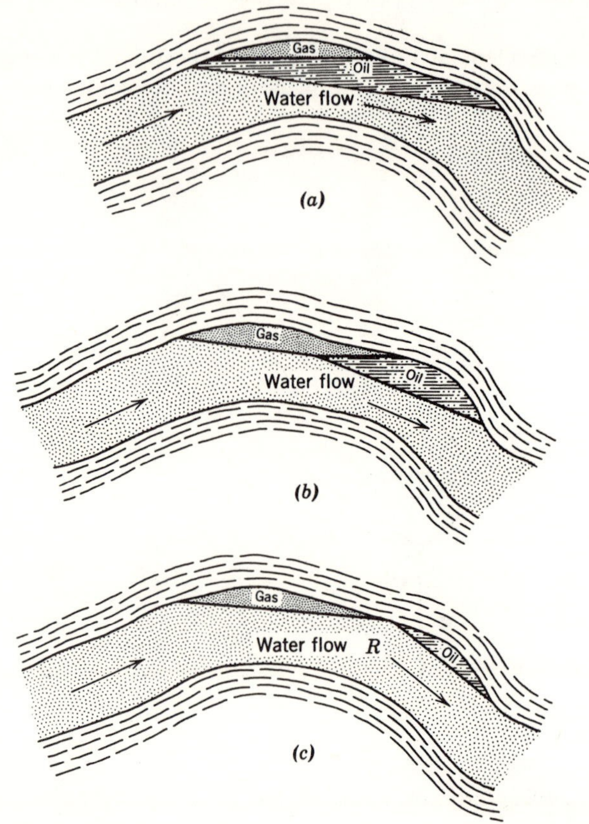


Figure 7.18 Hydrodynamic oil and gas accumulation in gently folded thick sand. (*a*) Gas entirely underlain by oil. (*b*) Gas partly underlain by oil. (*c*) Gas and oil traps separated. (After Hubbert.)

the gas would be deflected up the dip and the liquid petroleum down the dip, since the boundary sand-shale constitiutes an impermeable barrier for the hydrocarbons because of the surface tension effects. A change in the magnitude of $\mathbf{F}_w$ could change this picture so that both fluids would migrate in the same direction. From the foregoing derivations, where the force potential is used, it is clear that elements of petroleum will move from regions of higher energy to regions of lower energy and will come to rest when the energy of their surroundings is higher than their own energy or when they are trapped between regions of higher energy and impermeable barriers. Types of such traps or oil and gas accumulations are given in Figure 7.19. In Figure 7.19*a*, oil and gas form a static horizontal interface and the oil-water interface is sloping because water is flowing underneath the oil. Figure 7.18 may be considered a close-up of region *R* of Figure 7.19*c*. In all three cases, the hydrocarbons and the water

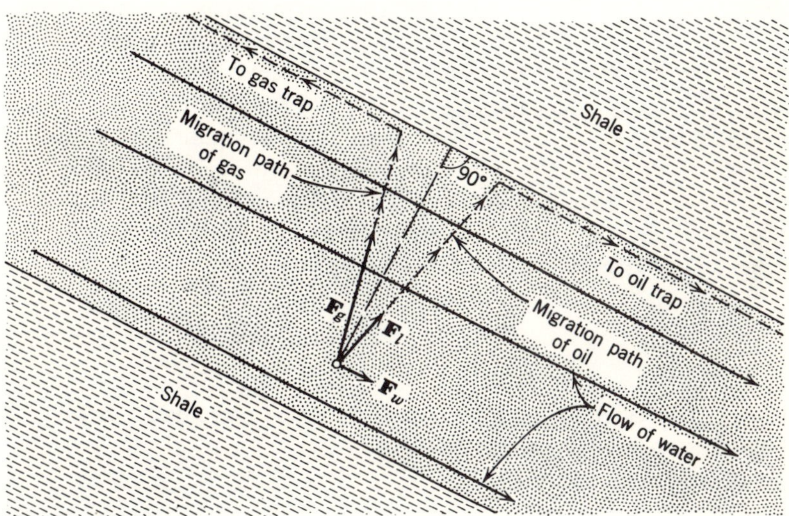


Figure 7.19 Divergent migration of oil and gas in hydrodynamic environment. (After Hubbert.)

occupy different regions which are separated by sharp interfaces because the fluids have been assumed to be immiscible. In the next paragraph, the slope of the interface between two regions will be examined.

### SLOPE OF THE HYDROCARBON-WATER INTERFACE

According to the concept of potential, pressure $p$ and elevation $z$ which are unique for any point in the two regions may be expressed in either $\Phi_w{}^*$ or $\Phi_l{}^*$ by the use of Equation 6.17

$$\Phi_w{}^* = gz + \frac{p}{\rho_w} \tag{7.61}$$

$$\Phi_l{}^* = gz + \frac{p}{\rho_l}$$

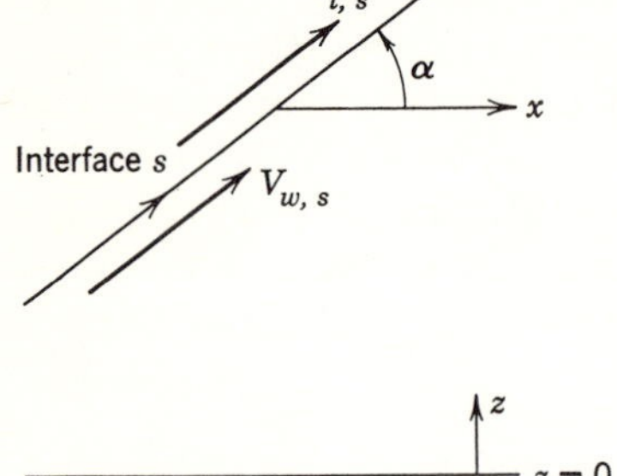


Figure 7.20 Slope $\alpha$ of the interface immissible flow.

By elimination of $p$ from Equations 7.61, the elevation $z$ of any point in terms of its potentials in the two regions is found [20]

$$z = \frac{1}{g}\left[\frac{\rho_w \Phi_w^*}{\rho_w - \rho_l} - \frac{\rho_l \Phi_l^*}{\rho_w - \rho_l}\right] \quad (7.62)$$

Similar computations may be made for a system of water and gaseous hydrocarbon. In particular, Equation 7.62 is valid for points along the interface. This equation, however, does not make it possible to compute the elevation of the interface, for the values of $\Phi_w^*$ and $\Phi_l^*$ are in general not known along the interface. Equation 7.62 becomes useful, however, if the potentials are related to the velocities in each region by means of Darcy's law. This is done by computing

$$\sin\alpha = \frac{\partial z}{\partial s} = \frac{1}{g}\left[\frac{\rho_w}{\rho_w - \rho_l}\frac{\partial \Phi_w^*}{\partial s} - \frac{\rho_l}{\rho_w - \rho_l}\frac{\partial \Phi_l^*}{\partial s}\right] \quad (7.63)$$

where $s$ is the trace of the interface in a vertical plane and $\alpha$ is the angle of the trace with the horizontal (Figure 7.20).

In view of Equation 6.20

$$\frac{\partial \Phi_w^*}{\partial s} = -\frac{g}{K_w} V_{w,s}$$

$$\frac{\partial \Phi_l^*}{\partial s} = -\frac{g}{K_l} V_{l,s}$$

where $V_{w,s}$ and $V_{l,s}$ are components of the specific discharge along the trace of the interface s so that

$$\sin\alpha = \frac{\partial z}{\partial s} = -\left[\frac{\rho_w}{\rho_w - \rho_l}\frac{1}{K_w} V_{w,s} - \frac{1}{K_l}\frac{\rho_l}{\rho_w - \rho_l} V_{l,s}\right] \quad (7.64)$$

From this equation it is easy to predict the trend in the slope of the interface in general. In particular, when one of the fluids is stagnant $\alpha$ assumes a definite sign. When the hydrocarbon is trapped as in Figure 7.19, $V_{l,s} = 0$ and $\sin\alpha < 0$ or the interface slopes downward when $\rho_w > \rho_l$ and the slope will be steeper with increasing $V_{w,s}$. On the other hand, this slope is milder when $\rho_l$ is replaced by $\rho_g$, that is, when the hydrocarbon is in the gaseous state. In the case where one fluid is stagnant it is convenient to replace Equation 7.64 by

$$\tan\alpha = \frac{dz}{dx} = -\frac{1}{K}\left[\frac{\rho_w}{\rho_w - \rho_l}\right]V_{w,x} = \frac{\rho_w}{\rho_w - \rho_l}\frac{dh_w}{dx} \qquad (7.65)$$

and the slope of the interface may be visualized by the reading of two standpipes introduced in Figure 7.21.

The influence of capillarity on the slope of the interface is zero when the medium is homogeneous, even where the fluids are immiscible [20].

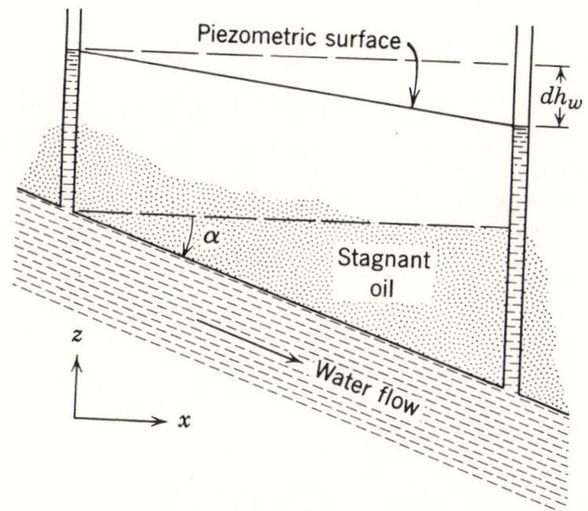


Figure 7.21 Slope $\alpha$ of interface and standpipe reading.

For further details on the different types of hydrodynamic traps and on their shifts in position with changes in the state of flow, the reader is referred to the original papers by Hubbert.

## 7.10 *Salt-Water Encroachment*

Salt-water encroachment or intrusion is the shoreward movement of water from the sea or ocean into confined or unconfined coastal aquifers and the subsequent displacement of fresh water from these aquifers. If, as a first approximation, fresh water and salt water are treated as two immiscible fluids, then they are separated by a sharp interface with a slope that may be computed by formulas similar to those of Section 7.9. Because of the slope of the interface, the front of the salt water is sometimes compared to a tongue, progressing into the land as a result of overdraft of the overlying fresh water and pushed back seaward when fresh water is replenished through precipitation. Actually, fresh water and sea water mix in a region of dispersed water and it is sometimes necessary to take account of the phenomenon of salt-water dispersion in order to obtain a more accurate picture of the nature of ground water flow in a coastal aquifer. The study of salt-water encroachment that may reach several thousand feet inland is by its very definition of interest for the water supply of islands and coastal regions which rely on ground water. Wells may become contaminated with salt water and may have to be abandoned, or fresh water may have to be injected in order to stop the inland movement of the salt water and to establish a fresh-water barrier. Furthermore, a study of the location of even the idealized interface is useful to obtain an idea of the fresh-water losses to the ocean through discharge under sea level.

### GHYBEN-HERZBERG HYDROSTATIC CONDITIONS

The hydrostatic equilibrium between immiscible fresh water and salt-water bodies in contact with each other along a certain interface was

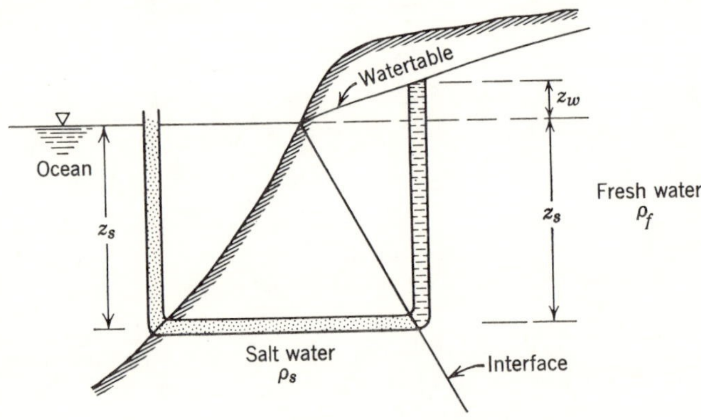


Figure 7.22 Salt water intrusion according to the Ghyben-Herzberg theory.

studied by Ghyben [1] and Herzberg [18]. The equation for the depth of the interface was in accord with measurements made in the field indicating that for every foot of fresh water above mean sea level, the thickness of the fresh-water lens resting on the salt water was about 40 feet. Near the shore the interface is sloping and hence narrows the fresh-water lens. In this elementary interpretation, as indicated in Figure 7.22, the cross sections of coastal line, interface, ocean level, and water-table meet all in one point. The depth $z_s$ of the interface below sea level may be determined from the hydrostatic law in the $U$ tube of Figure 7.22. Hence, if the datum plane for $z_s$ is taken at the bottom of the $U$ tube

$$z_s \rho_s g = g \rho_f z_s + g \rho_f z_w$$

or

$$z_s = \frac{\rho_f}{\rho_s - \rho_f} z_w \tag{7.66)*$$

where $\rho_s$ is the density of salt water, $\rho_f$ is the density of fresh water, and $z_w$ is the height of the watertable above mean sea level.

For $\rho_s = 1.026$ gram-mass/cm³ and $\rho_f = 1.000$ gram-mass/cm³,

$$z_s = 38z_w \tag{7.68}$$

The limitations of this theory are obvious: if both fluids were truly in static condition, then the watertable would have zero slope and the interface would become horizontal, with fresh water overlying salt water by mere density difference. Furthermore, fresh water is in a continuous state of motion due to changes in the watertable, for example, because of replenishment, evaporation, and discharge. Seepage surfaces above sea level have been found along many shores and have been tapped in earlier times as potable water for use on sea-going vessels. The existence of such surfaces is not considered in the Ghyben-Herzberg theory, nor is there any provision for the escape of fresh water below sea level.

The foregoing relationship, derived for watertable conditions, is also valid for confined aquifers if the height of the watertable is replaced by the piezometric head.

### HUBBERT'S HYDRODYNAMICAL APPROACH

Hubbert [20], pointing at the dynamic rather than at the hydrostatic equilibrium of the fresh-water/salt-water interface, showed the discrepancy between the actual depth to salt water and the depth as calculated by the Ghyben-Herzberg formula for flow conditions near the shore line (Figure 7.23).

The slope of the interface may be derived by direct analogy from Equation 7.63 in which the subscripts $f$ and $s$ replace the subscripts $w$ and $l$ respectively to indicate fresh and salt water. Hence

$$\sin \alpha = \frac{\partial z}{\partial s} = -\left[\frac{1}{K_f}\frac{\rho_f}{\rho_f - \rho_s} V_{f,s} - \frac{1}{K_s}\frac{\rho_s}{\rho_f - \rho_s} V_{s,s}\right] \tag{7.69}$$

If it is assumed that the salt water is stagnant and the fresh water flows over it, then $V_{s,s} = 0$ and $V_{f,s} \neq 0$. It follows that $\sin \alpha > 0$, because $\rho_f < \rho_s$, and $\alpha$ will increase with $V_{f,s}$. This explains the value of $\alpha$

* If the datum plane for $z$ is taken at mean sea level, then Equation 7.66 should be written

$$z_s = -\frac{\rho_f}{\rho_s - \rho_f} z_w \tag{7.67}$$

with all $z$'s above the datum plane measured as positive.

approaching 90° in Figure 7.23. Hubbert further shows that a relation exactly equal to Equation 7.67 holds between the points of intersection of any equipotential line of the flowing fresh water with the watertable and with the fresh-water/salt-water interface, as indicated by points $A$

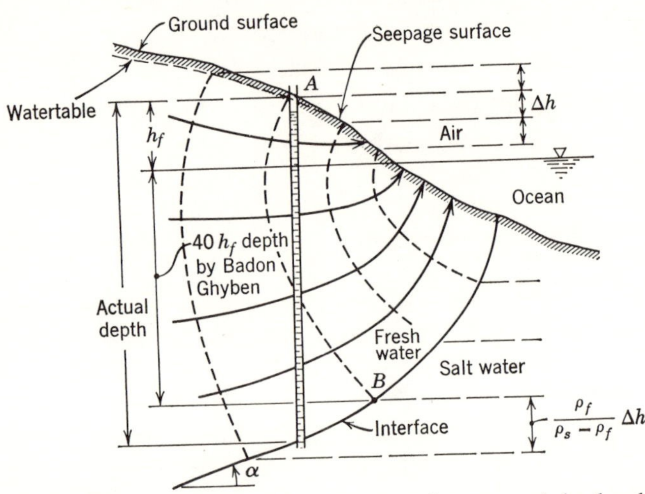


Figure 7.23 Discrepancy between actual depth to salt water and depth calculated by Ghyben-Herzberg relation. (After Hubbert.)

and $B$ of Figure 7.23. These points do not lie in a vertical line and because the equipotential line is curved the vertical distance between the watertable and the fresh-water/salt-water interface will be greater than that given by Equation 7.67.

#### LUSCZYNSKI'S POTENTIAL CONCEPTS

Lusczynski [8] made a significant contribution to the theory of salt-water encroachment by taking into account the existence of a zone of dispersion, rather than a sharp interface, between fresh water and salt water.

Lusczynski introduced the concepts of point-water head, fresh-water head and environmental-water head to compute the velocity distribution in the zone of dispersion and to determine the front of the dispersed water in the fresh water. Contingent upon the reading of water levels in some observation wells, it allows for the computation of the velocity vector field in a medium where the salt content of the water gradually varies. The value of this theory for practical purposes cannot be overemphasized because the velocity-vector field may be constructed regardless of the often complicated boundary conditions created by the geological nature of the water-bearing strata, which preclude a complete analytical solution of the problem.

Hubbert's Equation 7.62 follows from Lusczynski's theory in the limit when the thickness of the zone of dispersion is reduced to zero [8].

## PART C
## EXPERIMENTAL METHODS IN GROUND-WATER FLOW

Many problems in ground-water flow not amenable to analytical treatment may be solved in the laboratory through the use of model techniques, whereby the word model is used in the hydraulic sense as being the equivalent of an analog. Analogs are physical systems or mathematical models obeying partial differential equations with boundary conditions similar to those in the prototype, for example, the Hele-Shaw viscous flow analog, the R-C resistance capacitance electrical network analog, and so forth. In this chapter we describe some analogs which are at present most frequently used in studies of ground-water flow. For a complete description of analogs not treated here the reader is referred to the specialized text of Karplus [25].

### 7.11 *The Parallel-Plate Analog or Hele-Shaw Apparatus*

In 1899 Sir G. G. Stokes [35] presented a mathematical analysis of the flow of a viscous liquid between closely spaced parallel plates and thus proved that such flow can be derived from a potential as had been assumed by Hele-Shaw [35], who designed the first apparatus of this kind in 1897. Since then, the Hele-Shaw analog has been used extensively to simulate two-dimensional laminar flow of water through porous soil, which may be expressed by the same differential equation as the flow of the liquid in the analog. The proof presented by Stokes may be found in many textbooks [8,17,30], and valuable information about the construction and scaling of the analog has become available through numerous publications [see reference 8, Chapter 8].

The average velocities $u_{\text{ave}}$ and $w_{\text{ave}}$ of the viscous fluid between two vertical parallel plates may be expressed as

$$u_{\text{ave}} = -K_m \frac{\partial h_m}{\partial x}$$
$$w_{\text{ave}} = -K_m \frac{\partial h_m}{\partial z} \qquad (7.70)$$

in which

$$K_m = \frac{b^2}{12\mu}\gamma \qquad (7.71)$$

and

$$h_m = z + \frac{p}{\gamma} + \text{constant} \qquad (7.72)$$

in perfect analogy with Equation 6.2. For constant $\rho$ and $\mu$, the insertion of Equations 7.70 into the continuity equation

$$\frac{\partial u_{\text{ave}}}{\partial x} + \frac{\partial w_{\text{ave}}}{\partial z} = 0$$

shows that the head $h_m = z + \frac{p}{\gamma} + \textit{constant}$ in the analog satisfies Laplace's equation. In this case, the existence of a velocity potential

$$\Phi_m = K_m h_m = \frac{b^2\gamma}{12\mu}\left(z + \frac{p}{\gamma}\right) \tag{7.73}$$

in analogy with Equation 6.21 is eatablished. When the plates are horizontal, Equation 7.73, reduces to

$$\Phi_m = \frac{b^2}{12\mu}\, p = \frac{b^2\gamma}{12\mu}\frac{p}{\gamma} \tag{7.74}$$

which shows that $K_m$ has the same form for both vertical and horizontal Hele-Shaw analogs. The requirement of constant $\rho$ and $\mu$ in order to have potential flow in the analog is hard to satisfy because it is difficult in many cases to keep the temperature of the fluid in the analog constant and because the viscosity of most oils and glycerine is very much temperature dependent. Hele-Shaw analogs utilizing such fluids should be set up in temperature controlled rooms and, if the fluid is pumped, on some occasions it may be necessary to cool the fluid after it leaves the pump.

The analogy between Equations 6.21 and 7.73 shows that the average flow in the Hele-Shaw analog, far enough from obstacles placed between the plates, may simulate the flow of ground water. In the immediate vicinity of obstacles, such as the cut-off walls below the concrete weir represented in the analog, the hypothesis that the variations of the velocity components in the $x$- and $z$-directions are negligible as compared to variations of those components in the $y$-direction, is not longer valid. Both Santing [31] and Aravin [see reference 30] have proposed a limit of 1000 for the Reynolds number

$$N_R = \frac{V_{\text{ave}} b}{\nu}$$

above which the flow would cease to be laminar. This requires a very narrow spacing of the order of 1 millimeter or fractions of a millimeter when liquids of low viscosity such as light oils and water are used, and a spacing of the order of a few millimeters when heavy oils or glycerine are used.

The versatility of the analog as a tool for investigation of both steady and unsteady ground-water flow lies in the easy simulation of different values of the hydraulic conductivity of the soil through variation of the components of $K_m$ for the analog, that is, through variation of the spacing $b$ and of the fluid properties $\gamma$ and $\mu$. Regional variations of $K$ may be accounted for by the insertion of thin strips to reduce the spacing in the desired region. Figure 7.24 [31] shows a vertical analog in which rainfall is imitated by sprinkling over the length of the channel and ground-water withdrawal or artificial replenishment is reproduced by withdrawing or

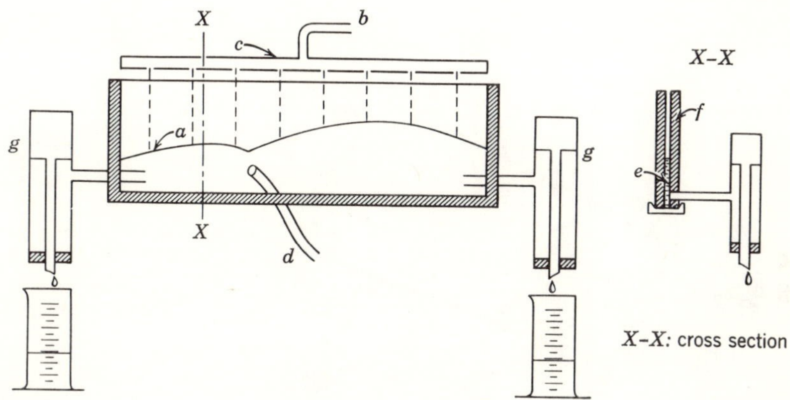


Figure 7.24 Vertical type of Hele-Shaw analog. (After Santing.) *a*, ground-water table; *b*, supply of rain; *c*, sprinkler; *d*, discharge tube; *e*, viscous liquid; *f*, transparent plates; *g*, vertically adjustable overflow.

adding liquid at the respective places in the channel. The simple boundary condition of constant (in place) and variable (in time) head may be materialized by means of vessels which are vertically adjustable and connected to the interspace; impervious boundaries in the field are represented by impervious boundaries in the analog; two fluids of slightly different density may be used for imitating fresh and salt ground water; fluids of different color but of equal density and viscosity may be used to study multiple (connected or parallel) aquifers.

### (A) HORIZONTAL ANALOGS—SCALES

Santing [31] used the horizontal analog to investigate the effects of artificial replenishment of the ground-water table in the well fields of the water works company of Zealand Flanders. To make the analog suitable for the study of unconfined aquifers, the assumption has to be made that the fluctuations in the ground-water level are negligibly small compared to the thickness of the aquifer. This is equivalent to assuming a transmissivity $T = K\tilde{h}$ for the watertable aquifer, in which $\tilde{h}$ is the average depth of flow.

Figure 7.25 represents a sketch of the horizontal analog used by Santing. A grid of vertical vessels, open at the top, is established on top of the upper plate, and each vessel is connected to the interspace of the analog as indicated in Figure 7.26. These vessels introduce storage capacity: a rise of the level of the liquid in the vessels means that liquid is stored, a drop

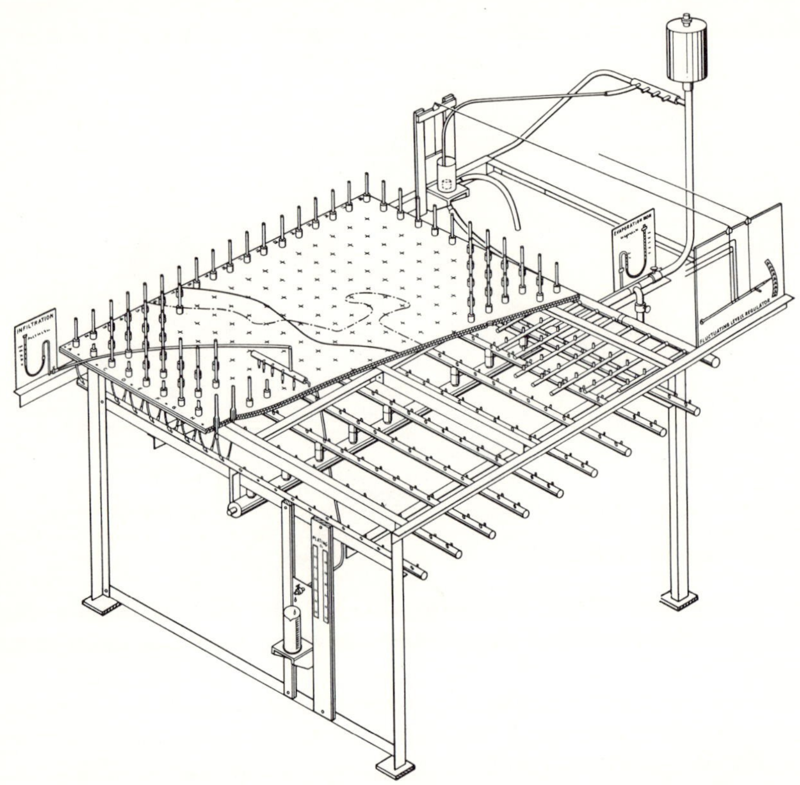


Figure 7.25 Horizontal type of Hele-Shaw analog. (After Santing.)

means a release from storage. In most cases the storage vessels will all be the same size and equally spaced on the upper plate of the analog, thus representing an area with uniform storage. In the case of two parallel overlying aquifers, however, storage of the lower aquifer may be introduced by connecting the storage vessels to the lower plate of the bottom aquifer. Variations in storage capacity may be simulated by varying either the spacing or the diameter of the storage vessels. The storage coefficient $S_m$ of the analog may be defined in complete analogy with the coefficient $S$ defined in Section 4.8. It is the amount of liquid in storage released from a column of interspace with a unit cross section under a unit decline of head. Because there is one storage vessel per area $A_m$ of the analog, it follows that

$$S_m A_m = \pi r_m^2 \tag{7.75}$$

in which the subscript $m$ refers to analog characteristics. In nature the storage capacity of the aquifer is more or less uniformly distributed; in the analog it is concentrated at the intersections of a grid, as in a finite-difference approach. Theoretically, the closer the spacing the better the

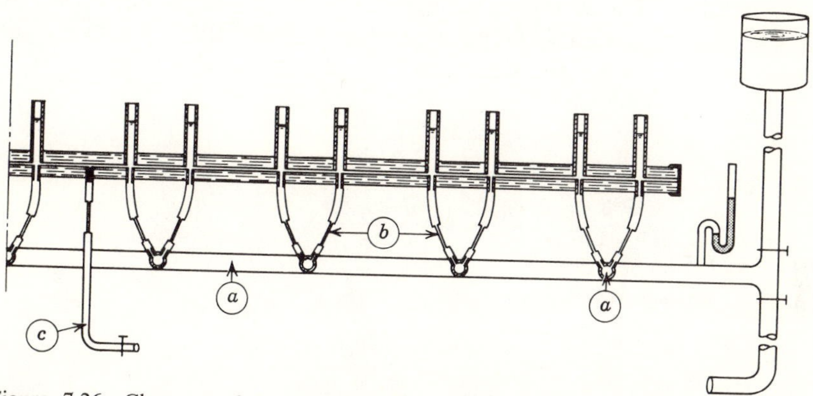


Figure 7.26 Close-up of horizontal model, with simulation of storage capacity, replenishment and evaporation. (After Santing.) *a*, replenishment and drainage system to simulate precipitation and evaporation; *b*, capillaries to increase the resistance to flow; *c*, drainage tube to simulate water withdrawal by pumpage.

approximation of the true conditions becomes. There is, however, a practical limit to the spacing as well as to the diameter of the storage vessels.

Replenishment of the watertable through percolating rainwater or evaporation from the watertable may be reproduced by means of a distribution system (Figure 7.26). Withdrawal by pumping is accomplished by draining the interspace as indicated in Figure 7.26.

SCALES: TYPICAL EXAMPLE

*Confined aquifer (S, T), unsteady flow, without radial symmetry*

The differential equation is given by Equation 6.38, or in polar coordinates

$$\frac{\partial^2 h}{\partial r^2} + \frac{1}{r}\frac{\partial h}{\partial r} + \frac{1}{r^2}\frac{\partial^2 h}{\partial \theta^2} = \frac{S}{T}\frac{\partial h}{\partial t} \tag{7.76}$$

The corresponding equation for the analog is:

$$\frac{\partial^2 h_m}{\partial r_m^{\ 2}} + \frac{1}{r_m}\frac{\partial h_m}{\partial r_m} + \frac{1}{r_m^{\ 2}}\frac{\partial^2 h_m}{\partial \theta_m^{\ 2}} = \frac{S_m}{T_m}\frac{\partial h_m}{\partial t_m} \tag{7.77}$$

in which the subscript $m$ refers to analog characteristics. The scales are taken as the ratios of the characteristics of the analog over those of the prototype. Therefore

$$\bar{u}_r = \frac{r_m}{r}, \quad \bar{u}_z = \frac{h_m}{h}, \quad \bar{u}_t = \frac{t_m}{t}, \quad \bar{u}_S = \frac{S_m}{S}, \quad \bar{u}_T = \frac{T_m}{T} \tag{7.78}$$

in which $\bar{u}$ indicates scale and the subscript refers to the quantity under consideration. The scales are determined by transforming Equation 7.76 by means of Equation 7.78 and by expressing compatibility between the transformed equation and Equation 7.77. This is done as follows:

$$\frac{\partial h}{\partial r} = \frac{\bar{u}_r}{\bar{u}_z}\frac{\partial h_m}{\partial r_m}$$

$$\frac{\partial^2 h}{\partial r^2} = \frac{\partial}{\partial r}\frac{\partial h}{\partial r} = \bar{u}_r\frac{\partial}{\partial r_m}\frac{\bar{u}_r}{\bar{u}_z}\frac{\partial h_m}{\partial r_m} = \frac{\bar{u}_r^{\,2}}{\bar{u}_z}\frac{\partial^2 h_m}{\partial r_m^{\,2}}$$

In the same way it is found that

$$\frac{\partial h}{\partial t} = \frac{\bar{u}_t}{\bar{u}_z}\frac{\partial h_m}{\partial t_m}$$

$$\frac{\partial h}{\partial \theta} = \frac{1}{\bar{u}_z}\frac{\partial h_m}{\partial \theta_m} \quad \text{and} \quad \frac{\partial^2 h}{\partial \theta^2} = \frac{1}{\bar{u}_z}\frac{\partial^2 h_m}{\partial \theta_m^{\,2}}, \qquad \text{since } \bar{u}_\theta = 1$$

The transformed Equation 7.76 becomes:

$$\frac{\bar{u}_r^{\,2}}{\bar{u}_z}\frac{\partial^2 h_m}{\partial r_m^{\,2}} + \frac{\bar{u}_r^{\,2}}{\bar{u}_z}\frac{1}{r_m}\frac{\partial h_m}{\partial r_m} + \frac{\bar{u}_r^{\,2}}{\bar{u}_z}\frac{1}{r_m^{\,2}}\frac{\partial^2 h_m}{\partial \theta_m^{\,2}} = \frac{S_m}{T_m}\frac{\bar{u}_T}{\bar{u}_S}\frac{\bar{u}_t}{\bar{u}_z}\frac{\partial h_m}{\partial t_m} \tag{7.79}$$

Equation 7.79 may be multiplied by $\bar{u}_z/\bar{u}_r^{\,2}$ and then compared with Equation 7.77. Both equations are compatible if and only if

$$\frac{\bar{u}_T}{\bar{u}_S}\frac{\bar{u}_t}{\bar{u}_r^{\,2}} = 1 \tag{7.80}$$

It should be noted that the compatibility condition does not contain $\bar{u}_z$ so that $\bar{u}_z$ may be chosen freely in any case of confined flow, as was to be expected. The scale $\bar{u}_S$ is chosen to have a reasonable spacing of the storage vessels and a vessel diameter which permits easy readings during the experiment. Also, $\bar{u}_T$ is chosen after considering practical values of $K_m$ in regard to the interspace and the fluid properties. Finally, there is one more free choice between $\bar{u}_t$ and $\bar{u}_r$, whereas Equation 7.80 determines the alternate choice of these scales. Scale computations for unconfined aquifers with replenishment and for leaky aquifers are given in reference 8.

(B) VERTICAL ANALOGS

Vertical analogs have been used more extensively than horizontal models [8]. Figure 7.27 shows the analog used by De Wiest [9] to study the unsteady behaviour of the free surface of the flow through an under-drained earth embankment.

Figure 7.27 Vertical Hele-Shaw analog. (After De Wiest.)

## 7.12 *Resistance-Capacitance Network Analogs*

The theory of the R-C network analog may be found in many papers [24], some of which have been condensed in books on analog simulation; the most recent one by Karplus [25] gives a thorough and comprehensive treatment of the subject. Skibitzke [32], Stallman [33], and Walton [42] applied the R-C analog to problems in ground-water flow, and De Wiest used the same tool to corroborate his analytical studies on leaky aquifers involving Dirac's delta function [10].

In the R-C analog network, solutions to the difference equations replacing the differential equation of ground-water flow may be visualized directly on an oscilloscope. Lumped electric-circuit elements are used to simulate the distributed properties of the subsurface strata; solutions to the problem are found only for these points of the medium which correspond to the nodes of the network that consists of an assembly of resistors and capacitors. Voltage and current sources simulating the excitations are applied to the boundaries of the network and at the interior nodes when required.

The resistor-capacitance network dissipates electrical energy in somewhat the same way that a porous medium consumes ground-water energy to let water travel through its voids. Electrical conductance and hydraulic conductivity are of the same nature; electrical charges are stored in capacitors and the storage of water in an aquifer is related to its storage coefficient, so that a simple relationship between capacitance of R-C analog network and storage coefficient of an aquifer can be established. The head $h$ in an aquifer is analogous to the electric potential $\Phi$ in the network.

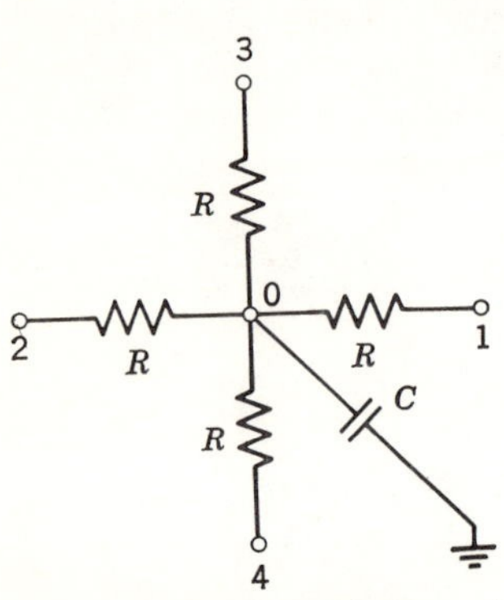

Figure 7.28 Simulation of ground-water flow with help of R-C analog. Two-dimensional flow.

The analogy becomes more apparent when the differential equations governing ground-water flow and electrical current respectively in the aquifer and in the network are compared. Therefore a finite difference approach is used in the $x$-, $y$-coordinates of Equation 6.38, in two dimensional form say

$$\nabla^2 h = \frac{S}{T}\frac{\partial h}{\partial t} \tag{6.38}$$

by covering the area of the aquifer with an equilateral grid of mesh size $\Delta x = \Delta y = a$, whereby $a^2$ is small compared with the area of the aquifer. By using this approach, Equation 6.38 may be rewritten as

$$\frac{h_1 + h_2 + h_3 + h_4 - 4h_0}{a^2} = \frac{S}{T}\frac{\partial h_0}{\partial t} \tag{7.81}$$

Consider now Figure 7.28 in which four resistors of equal magnitude $R$ and one capacitor $C$ are connected to a common terminal $O$; the capacitor is also connected to ground. Application of Kirchhoff's current law to node $O$ gives [25]:

$$\frac{\Phi_1 - \Phi_0}{R} + \frac{\Phi_2 - \Phi_0}{R} + \frac{\Phi_3 - \Phi_0}{R} + \frac{\Phi_4 - \Phi_0}{R} = C\frac{\partial \Phi_0}{\partial t} \tag{7.82}$$

in which $\Phi_0$, $\Phi_1$, $\Phi_2$, $\Phi_3$, $\Phi_4$ are the potentials at the nodes 0, 1, 2, 3, 4. Equations 7.81 and 7.82 are similar if $R$ is made proportional to the reciprocal of $T$ and if $C$ is chosen proportional to the product of $a^2$ and $S$.

### CONVERSION FACTORS

A number of conversion factors are needed to express the measured electrical quantities in equivalent ground-water terms [42]. The conversion factors for practical U.S. hydraulic units (gallons for quantity $q$; gallons per day flowrate $Q$; feet for head loss $\Delta h$) and corresponding electrical units (coulombs for charge $Q^*$; coulombs per second, or amperes, for current $I$; volts for potential loss $\Delta\Phi$) may be defined as follows:

$$q = C_1 Q^* \tag{7.83}$$

$$h = C_2 \Phi \tag{7.84}$$

$$Q = C_3 I \tag{7.85}$$

and

$$t_d = C_4 t_s \tag{7.86}$$

in which $t_d$ expresses the time in days, $t_s$ denotes the time in seconds, and $C_4$ is the fraction of a day equivalent to one second. Likewise $C_1$ refers to gallons per coulomb, $C_2$ stands for feet per volt and $C_3$ is gallons per day per ampere. The relation between the conversion factors $C_1$, $C_3$, and $C_4$ may be found as follows. By definition

$$Q = \frac{q}{t_d} \tag{7.87}$$

If $Q$, $q$, $t_d$ in Equation 7.87 are replaced by their respective values from Equations 7.83, 7.85, and 7.86, and if it is noticed that $Q^*/t_s = I$,

$$\frac{C_3 C_4}{C_1} = 1 \tag{7.88}$$

### DESIGN OF ELECTRIC CIRCUIT ELEMENTS

The value of the resistance $R$ may be determined from Ohm's law and the aforementioned conversion formulas. Ohm's law states

$$R = \frac{\Phi}{I} \tag{7.89}$$

in which $\Phi = h/C_2$ and $I = Q/C_3$, so that

$$R = \frac{hC_3}{QC_2} \tag{7.90}$$

However, by definition $T = \frac{Q}{h} = \frac{\text{gal/day}}{\text{feet}}$, and therefore Equation 7.90 may be rewritten

$$R = \frac{C_3}{C_2 T} \tag{7.91}$$

in which $R$ is expressed in ohms because $\Phi$ is expressed in volts and $I$ in amperes.

In a similar manner, the capacitance $C$ may be determined from Coulomb's law and the aforementioned conversion formulas. Coulomb's law for the electrical charge of a capacitor states

$$C = \frac{Q^*}{\Phi} \tag{7.92}$$

in which $Q^* = q/C_1$ and $\Phi = h/C_2$, so that

$$C = \frac{q}{h}\frac{C_2}{C_1} \tag{7.93}$$

However, $q/h$ has the dimensions of $L^2$, and therefore may be replaced by $a^2S$. The conversion is computed as follows:

$$\left[\frac{q}{h}\right]^{\text{gal/ft}} \times \frac{\text{gal}}{\text{ft}} = [a^2S]^{\text{ft}^2} \times \text{ft}^2$$

or

$$\left[\frac{q}{h}\right]^{\text{gal/ft}} = [a^2S]^{\text{ft}^2} \times \frac{\text{ft}^3}{\text{gal}} = 7.48[a^2S]^{\text{ft}^2}$$

Therefore

$$C = 7.48a^2S\,\frac{C_2}{C_1} \tag{7.94}$$

in which $a$ is expressed in feet, $S$ is the dimensionless storage coefficient, and $C$ is the capacitance in farads because $Q^*$ is expressed in coulombs and because $\Phi$ is in volts.

### EXTENSION TO THREE-DIMENSIONAL FLOW—ANISOTROPIC AND LEAKY ARTESIAN CONDITIONS [2]

The finite difference equation corresponding to Equation 6.34 for a grid with dimensions $\Delta x$, $\Delta y$, $\Delta z$ is

$$\frac{h_1 + h_2 - 2h_0}{(\Delta x)^2/K_x} + \frac{h_2 + h_4 - 2h_0}{(\Delta y)^2/K_y} + \frac{h_5 + h_6 - 2h_0}{(\Delta z)^2/K_z} = S_s\frac{\partial h_0}{\partial t} \tag{7.95}$$

The corresponding network equation for the potential $\Phi$ is (Figure 7.29*a*)

$$\frac{\Phi_1 + \Phi_2 - 2\Phi_0}{R_x} + \frac{\Phi_2 + \Phi_4 - 2\Phi_0}{R_y} + \frac{\Phi_5 + \Phi_6 - 2\Phi_0}{R_z} = C\frac{\partial \Phi_0}{\partial t}$$

The analogy requires

$$R_x = C_5 \frac{(\Delta x)^2}{K_x}$$

$$R_y = C_5 \frac{(\Delta y)^2}{K_y} \tag{7.96}$$

$$R_z = C_5 \frac{(\Delta z)^2}{K_z}$$

and $C$ must be proportional to $S_s$. The factor $C_5$ stands for ohm/ft day if U.S. practical units of gallons, day, and feet are used. The relationship between $C_2$, $C_3$, and $C_5$ follows from Ohm's Law, Equation 7.90, and any of Equation 7.96, written in dimensional form, in which $L$ stands for length and $T^*$ stands for time.

$$\frac{C_3}{C_2} = \frac{Q}{h} R = \frac{L^3}{T^* L} C_5 \frac{L^2}{L/T^*}$$

or

$$\frac{C_3}{C_2} = C_5 L^3 \tag{7.97}$$

The value of the capacitance $C$ is determined in a similar manner as before, from Equation 7.93, in which $q/h$ is replaced by $\Delta x \, \Delta y \, \Delta z S_s$, so that

$$C = 7.48 \, \Delta x \, \Delta y \, \Delta z S_s \frac{C_2}{C_1} \tag{7.98}$$

in which $\Delta x \, \Delta y \, \Delta z$ is expressed in cubic feet and $C$ is the capacitance in farads.

Leakage through semiconfining beds or aquitards into the main aquifer is vertical and proportional to the drawdown. It adds a term, proportional to the unknown potential, to Laplace's equation (see Equation 7.42) and therefore may easily be simulated [25,42] by the addition of resistors connected to ground and to each node of the network (see Figure 7.29*b*, resistance $R_L$). By analogy with Equations 7.91, and 7.96 Walton [42]

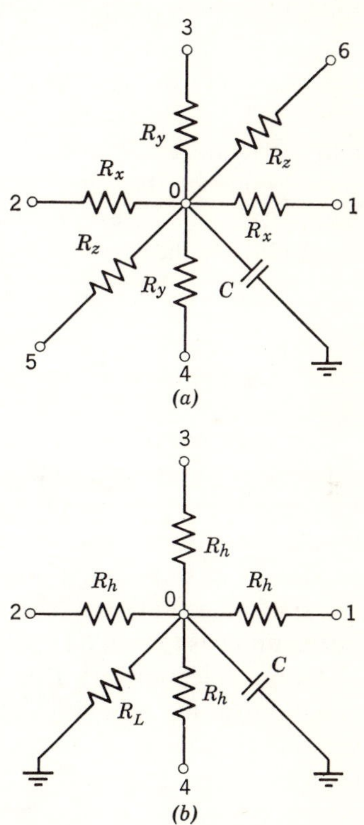


Figure 7.29 (*a*) Three-dimensional ground-water flow simulation by R-C analog. (*b*) Leaky aquifer conditions.

expresses the value of $R_L$ as

$$R_L = \frac{C_3}{C_2\left(\dfrac{K'}{b'}\right)a^2} \tag{7.99}$$

in which $K'/b'$ is the leakance as defined in Section 7.7, and in which $R_L$ is expressed in ohms.

BOUNDARY CONDITIONS [25,42]

Boundaries of constant head may be simulated by terminating the corresponding part of the network in a short circuit, while barrier boundaries, across which no flow takes place, may be reproduced by an open circuit. A radiant boundary condition, in which the head along the boundary is proportional to its normal derivative along that boundary may be simulated by connecting resistors between the nodes along the boundary and the ground. Irregular shapes of the boundary are duplicated with the help of the vector-volume technique [25] whereby resistors and capacitors in the proximity of the boundary are modified to suit the correspondence between network and aquifer parameters. Network boundaries can be extended to infinity through the use of termination strips [24]. In the case of a watertable aquifer, the free boundary condition may be observed as explained by Stallman [33].

## 7.13 *The Electrical Resistance Network*

The electrical resistance network [25] has been a favorite tool of agricultural engineers and soil physicists in studies of drainage and infiltration. In many cases the dimensions of the prototype are one order of magnitude smaller than those of interest to geological and civil engineers. Local conditions of soil nonuniformity and anisotropy become more important for smaller prototypes; for example, in some agricultural problems a distinction has to be made between fillable porosity (for rising watertables) and drainable porosity (for falling watertables [3]). To simulate these variations in the properties of the medium, use is made of a network of variable resistors or rheostats. The total voltage drop across the network is of the order of 10 volts or higher but always below the danger level for electrocution, the source being ordinary 60-cycle, 120-volt alternating current rectified to give about 10-volt direct current, or regular batteries of 6 or 12 volt. Potentials at the nodes of the network are measured with a vacuum-tube voltmeter and the total current through the network is measured with an ammeter (milliammeter).

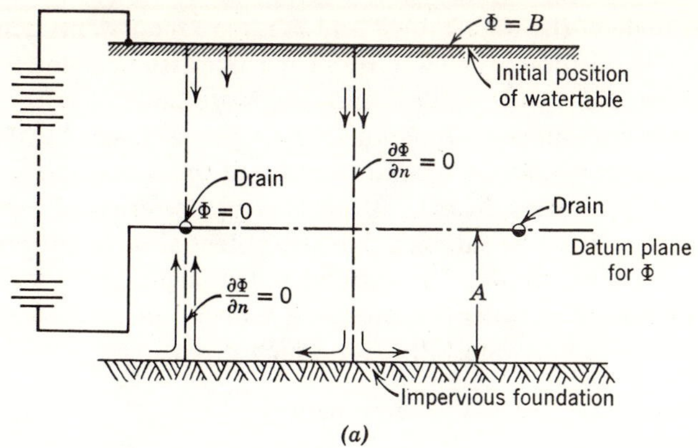


(a)

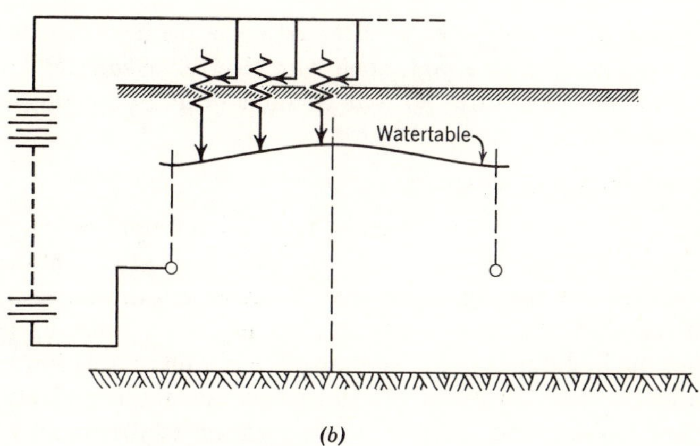


(b)

Figure 7.30 (*a*) Drainage to parallel drains, watertable at ground level. (*b*) Falling watertable by iteration. (After Brutsaert.)

The resistance network essentially yields solutions for steady-flow conditions, because it contains no electrical reactance (capitance inductance). Any transient flow must therefore be handled as a succession of steady states, for example, moving ground-water mounds [3] or watertable drawdown during tile drainage [4]. An example of computation for the falling watertable by iteration is given in Figure 7.30. The boundary condition at the watertable is [26]

$$\frac{\partial z_f}{\partial t} = \frac{K}{S_y}\left(\frac{\partial \Phi}{\partial z} - \frac{\partial \Phi}{\partial x}\tan\theta\right) \tag{7.80}$$

in which $S_y$ is the specific yield; $\Phi = z + \dfrac{p}{\gamma}$; $z_f$ is the height of the watertable above the impervious datum plane; $K$ is the hydraulic conductivity;

$\theta$ is the slope angle of the watertable; and $x$, $z$, $t$ are the independent variables. The initial position of the watertable must be assumed, and from the measured potentials at the nodes near the watertable, $\tan\,\theta$, $\Delta\Phi/\Delta z$ and $\Delta\Phi/\Delta x$ are computed as functions of $x$. For a given $\Delta t$, $\Delta z_f(x)$ is found from Equation 7.80. A second position of the watertable is drawn with the help of the values $\Delta z_f(x)$. Along this new position of the watertable the condition must be satisfied that the potential $\Phi$ be proportional to the height above the drain. The variable potential along the watertable is obtained by means of potentiometers. A convenient starting value is offered by an horizontal watertable, for which $\theta = 0$.

CONVERSION FACTORS: see R-C-analog network.

## 7.14 *Conductive-Liquid and Conductive-Solid Analogs* [24]

The voltage distributions in a sheet or solid made of a conductive material as well as that in a tank filled with electrolyte satisfies Laplace's equation and may therefore be used to simulate the potential fields of ground-water flow.

### (A) CONDUCTIVE-LIQUID ANALOGS

The electrolytic tank normally consists of a watertight container of a nonconductive material, of shallow depth, filled with a few centimeters of electrolyte. The conducting liquid should have the following properties. It should have no electrical reactance; its resistivity must be uniform throughout the liquid and linear so that there is a linear relation between voltage and current; chemical reactions between liquid and electrodes must be impossible; and the rate of evaporation of the liquid must be slow in order to prevent alteration of the resistivity in time.

Figure 7.31 [39] shows a typical conductive liquid-analog model with the electrolytic tank, the electrical equipment, and the drawing board with pantograph. As in the case of the resistance-network analog, the voltage is reduced to avoid electrocution and also to avoid heating of the electrolyte. A potential divider allows for accurate division of the voltage into a large number of subdivisions. The oscilloscope helps in the identification of the potential at the probe position: when the probe is of the desired potential, no current flows through the oscilloscope and the pencil of the pantograph is depressed to leave a mark on the drawing paper. Equipotential lines may be sketched by connecting points of the same potential.

The theory underlying the use of the electrolyte tank has been known for a long time [29]; its application to ground-water flow is based on the similarity between the differential equations which describe the flow of

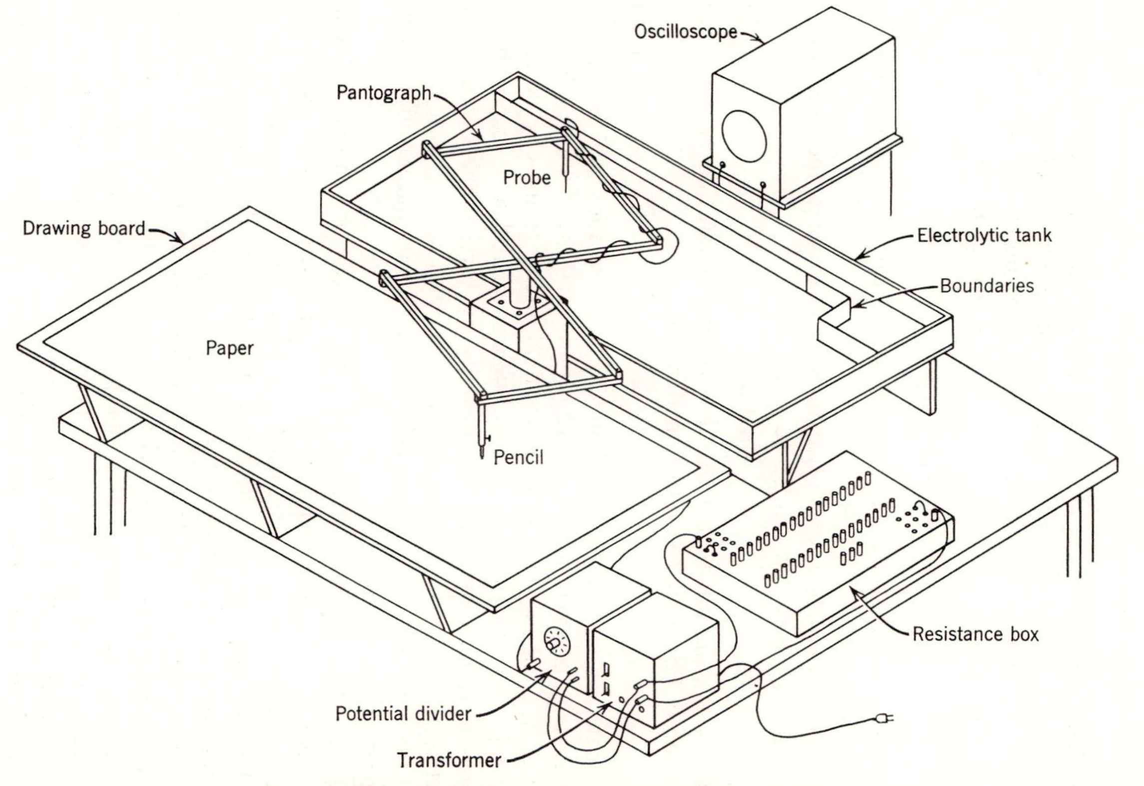


Figure 7.31 Sketch of conductive liquid analog. (After Bear.)

ground water and those which govern the flow of electric current through conducting materials. Let Ohm's law for the flow of electricity through a conducting material be written

$$\mathbf{I} = -\kappa \operatorname{grad} V \tag{7.81}$$

in which **I** is the vector of electric current per unit area; $\kappa$ is the electric conductivity of the material; and $V$ is the electric potential. Darcy's law for the laminar flow of liquid through porous media, on the other hand, may be expressed by Equation 6.13, or, using the symbol **q** to avoid confusion,

$$\mathbf{q} = -K \operatorname{grad} h \tag{7.82}$$

in which **q** is the specific discharge, $K$ is the hydraulic conductivity, and $h$ is the head.

For steady flow in a region without sources or sinks, the conservation equation for the electric charges requires

$$\operatorname{div} \mathbf{I} = 0 \tag{7.83}$$

and if $\kappa$ is constant, that is, for an isotropic-conducting material, substitution of Equation 7.81 into Equation 7.83 gives

$$\nabla^2 V = 0 \tag{7.84}$$

This equation may be compared directly with Equation 6.51 and proves the analogy between head and electric potential, hydraulic conductivity and electric conductivity, and specific fluid discharge and electric current per unit area. In the case of a two-dimensional field it is possible to represent variable-hydraulic conductivity by varying the depth of the electrolyte. The electric potential, however, must not vary with the depth and therefore it is better to use different electrolyte concentrations of the same depth in the analog.

A boundary of constant head in the prototype is reproduced by an electrode at constant potential in the analog, whereas impervious boundaries in the prototype require insulating boundaries in the analog. When the head or its normal derivative vary along the boundary the corresponding boundary of the analog is usually covered by a large number of small electrodes which are insulated from each other. In the case of variable head along the boundary, each electrode is raised to a potential $V$ corresponding to the average head of that part of the boundary reproduced by the electrode; for variable normal derivative of the head, each electrode is set to allow a certain current $I$ to flow.

Conditions of anisotropy are treated by a scale transformation similar to that obtained by Equation 6.53. Maasland [28] derived the equation

$$\frac{x_r}{z_r} = \sqrt{\frac{K_z}{K_x}} \tag{7.85}$$

in which $x_r = x_m/x_p$ is the length scale in the $x$-direction and $z_r = z_m/z_p$ is the length scale in the $z$-direction; $m$ and $p$ stand for analog and prototype; and $K_z$ and $K_x$ are the hydraulic conductivities of the prototype. It suffices to assign the value of one to $z_r$ in Equation 7.85 to obtain the previously derived Equation 6.53.

CONDUCTIVE-SHEET ANALOGS [5,25,44].

The principle of the conductive-sheet analog does not differ from the one which underlies the electrolyte tank. Sheets of electrically conducting material of sufficiently uniform, isotropic, and high resistance are suitable. They may consist of filter paper soaked in a suspension of colloidal graphite, of woven grids of metal wire and silk thread, of conductive rubber, of metallized paper, and so forth. The most commonly used sheet appears to be Teledeltos paper, manufactured by the Western Union Telegraph Company [25]. This paper is formed by adding carbon black, a conductive material, to paper pulp in the pulp-beating stage of the paper-manufacturing process. The conductive paper is then coated on one side with a lacquer, which acts as an electrical insulator, and on the other side with a layer of aluminum paint.

The advantages and disadvantages of the conductive-sheet analog are reviewed in the literature [25,40].

## **7.15** *Other Analogs*

Among the other analogs which are still in use, the sand box should first be mentioned. The effect of capillary rise, however, is greatly exaggerated in this type of analog [30]; furthermore, this analog lacks versatility and has lost ground to the aforementioned analogs as far as application is concerned. It is still in use in soil mechanics and agricultural laboratories.

Stretched membranes have the property that the vertical deflection of the membrane is approximately governed by Laplace's equation in two dimensions and have also been used successfully, especially in the case of well fields [12,44].

## REFERENCES

1. Badon Ghyben, W., 1888 and 1889, *Nota in Verband met de Voorgenomen Putboring naby Amsterdam:* The Hague, Tydschrift van het koninklyk Instituut van Ingenieurs, p. 21.
2. Bear, J., 1964, Hydraulic electric analog computers, *Journal of the Hydraulics Division*, vol. 90, no. Hy4, July 1964, pp. 321–323.

3. Bouwer, H., 1962, Analyzing ground-water mounds by resistance network; *ASCE Journal of the Irrigation and Drainage Division*, September, pp. 15–36.
4. Brutsaert, W., G. S. Taylor, and J. N. Luthin, 1961, Predicted and experimental water table drawdown during tile drainage: *Hilgardia*, v. 31, November, pp. 389–418.
5. Childs, E. C., 1943, 1945, 1956, 1947, 1951, The water table, equipotentials, and streamlines in drained land: *Soil Sci.*, v. 56, pp. 317–330 vol. 59, pp. 313–327 (1945); vol. 59, pp. 405–415 (1945); vol. 62, pp. 183–192 (1946); vol. 63, pp. 361–376 (1947); vol. 71, pp. 233–237 (1951).
6. Cooper, H. H., Jr., and C. E. Jacob, 1964, A generalized graphical method for evaluating formation constants and summarizing well-field history: *Trans. Am. Geophys. Union*, v. 27, pp. 526–534.
7. De Glee, G. J., 1930, *Over grondwaterstromingen by wateronttrekking by middel van putten:* Delft, T. Waltman, Jr., 175 pp.
8. De Wiest, R. J. M., 1965, *Geohydrology:* Chapter 6, New York, John Wiley and Sons.
9. ——, 1962, Free surface flow in homogeneous porous medium: *Trans. ASCE*, v. 127, Part 1, pp. 1045–1089.
10. ——, 1963 and 1964, Replenishment of aquifers intersected by streams: *ASCE Journal of the Hydraulics Division*, November, 1963, pp. 165–191; September 1964, pp. 161–168.
11. Ferris, J. G., D. B. Knowles, R. H. Browne, and R. W. Stallman, 1962, *Theory of aquifer tests:* p. 148, Washington, D.C., U.S. *Geol. Survey Water-Supply Paper*, 1936-E.
12. Hansen, V. E., 1952, Complicated well problems solved by the membrane analogy, *Trans. Am. Geophys. Union*, v. 33, no. 6, pp. 912–916.
13. Hantush, M. S., 1949, Plain potential flow of groundwater with linear leakage, Ph.D. dissertation: University of Utah, 86 pp.
14. Hantush, M. S., and C. E. Jacob., 1954, Plane potential flow of groundwaters with linear leakage: *Trans. Am. Geophys., Union*, v. 35, pp. 917–936.
15. Hantush, M. S., 1964, *Hydraulics of wells, advances in hydro-sciences:* v. 1, New York, Academic Press, pp. 281–432.
16. ——, 1956, Analysis of data from pumping tests in leaky aquifers: *Trans. Am. Geophys. Union*, v. 37, pp. 702–714.
17. Harr, M. E., 1962, *Ground water and seepage:* New York, McGraw-Hill Book Co., 315 pp.
18. Herzberg, A., 1961, Die Wasserversorgung einiger Nordseebader, Munich, *Jour. Gasbeleuchtung und Wasserversorgung*, v. 44, pp. 815–819, 842–844.
19. Hubbert, M. K., 1955, Entrapment of petroleum under hydrodynamic conditions: *Bull. Am. Assoc. Petroleum Geologists*, v. 37, pp. 1954–2026.
20. ——, 1940, The theory of ground-water motion: *Jour. Geol.*, v. 48, no. 8, pp. 785–944.
21. Jacob, C. E., 1950, *Flow of groundwater*, p. 346, in *Engineering hydraulics* edited by H. Rouse: New York, John Wiley and Sons.
22. ——, 1946, Radial flow in a leaky artesian aquifer, *Trans. Am. Geophys. Union*, v. 27, pp. 198–205.
23. Jahnke, E., and F. Emde, 1945, *Tables of functions:* New York, Dover, pp. 6, 7.
24. Johnson, A. I., 1963, Selected references on analog models for hydrologic studies: Appendix F, *Proceedings of the Symposium on Transient Ground Water Hydraulics*, Colorado State University, July 25–27.

25. Karplus, W. J., 1958, *Analog Simulation:* New York, McGraw-Hill Book Co., 427 pp.
26. Kirkham, D., and R. E. Gaskell, 1950, Falling water table in tile and ditch drainage: *Proc. Soil Sci. Soc. Am.*, v. 15, pp. 37–43. (Published in 1951.)
27. Lang, S. M., 1906, Interpretation of boundary effects from pumping test data: *Jour. Am. Water Works Assoc.*, v. 52, no. 3 March, pp. 356–364.
28. Maesland, M., 1957, Soil anisotropy and sand drainage; see *Drainage of agricultural sands*, T. N. Luthin (ed.): Madison, Wisconsin, Am. Soc. Agron., pp. 216–285.
29. Muskat, M., 1937 and 1946, *The flow of homogeneous fluids through porous media:* New York, McGraw-Hill Book Co., 763 pp; Second Printing, Ann Arbor, Michigan, T. W. Edwards, 1946.
30. Polubarinova-Kochina, P. Ya., 1962, *Theory of ground-water movement:* Princeton, Princeton University Press, 613 pp.
31. Santing, G., 1951, A horizontal scale model, based on the viscous flow analogy, for studying ground-water flow in an aquifer having storage: Toronto, IASH General Assembly, pp. 105–114.
32. Skibitzke, H. E., 1961, Electronic computers as an aid to the analysis of hydrologic problems: *IASH Publication* 52.
33. Stallman, R. W., 1963, Electric analog of three-dimensional flow to wells and its application to unconfined aquifers: *U.S. Geol. Survey Water-Supply Paper* 1536-H, pp. 205–242.
34. Steggewentz, J. H., and B. A. Van Nes, 1939, Calculating the yield of a well taking account of replenishment of the groundwater from above. *Water and Water Engrg.* v. 41, pp. 561–563.
35. Stokes, G. G., 1899, see article by Hele-Shaw on the Streamline motion of a viscous film. *Report of the British Association for the Advancement of Science*, 68th meeting, p. 136.
36. Theis, C. V., 1935, The relation between the lowering of the piezometric surface and the rate and duration of discharge of a well using groundwater storage: *Trans. Am. Geophys. Union*, v. 16, pp. 519–524.
37. Thiem, G., 1906, *Hydrologische Methoden:* Leipzig, Gebhardt, 56 pp.
38. Todd, D. K., 1959, *Ground water hydrology:* New York, John Wiley and Sons, 336 pp.
39. Todd, D. K., and J. Bear, 1959, River seepage investigation: *Water Resources Center Contribution* 2020, Berkeley, University of California.
40. Van Everdingen, R. O., and B. K. Bhattacharya, 1963, Data for ground-water model studies: *Geological Survey of Canada Paper* 63–12, December, 31 pp.
41. Walton, W. C., 1960, Leaky artesian aquifer conditions in Illinois: *Report of Investigation No.* 39, Illinois State Water Survey, 1960.
42. Walton, W. C., and T. A. Prickett, 1963, Hydrogeologic electric analog computers: *ASCE Journal of Hydraulics Division*, November, pp. 67–91.
43. Wenzel, L. K., 1942, Methods for determining permeability of water-bearing materials with special reference to discharging-well methods: Washington, D.C., U.S. Geol. Survey Water-Supply Paper 887, 192 pp.
44. Zee, Phong-Hung, D. F. Peterson, and R. O. Bock, 1957, Flow into a well by electric and membrane analogy: *Trans. Am. Soc. Civ. Engrs.*, v. 122, pp. 1088–1112.

# chapter 8

# EXPLORATION FOR GROUND WATER

## 8.1 *Introduction*

Most water wells are located, drilled, and tested without the aid of professional hydrogeologists. There are a number of reasons why this is so. First, the cost of scientifically based studies commonly exceeds the expected benefits. Shallow drive-point wells, for example, can be installed for less than 200 dollars. If the wells are unsuccessful, the small-diameter casing can be taken out with the aid of jacks and installed in another locality in a few hours. Professional study could more than double the cost of such wells. Second, even if hydrogeologic studies are justified economically, many regions lack specialists within easy reach. Inquiries by mail to universities and state agencies take time, and by necessity the replies are often too generalized to be of much practical value. Third, in some regions well drillers have a knowledge of the local occurrence of ground water that is adequate for many purposes.

Despite the foregoing deterrents to the use of hydrogeologists, the drilling and construction of most deep wells and even some shallow driven wells could benefit from professional assistance. Geologic information can commonly give positive guidance for the location of wells, thus avoiding the expense of unsuccessful wells. Proper well location, however, is but one of a number of problems facing the prospective well owner. Poor water quality, biological contamination, future lowering of water levels in wells, and improper completion methods are some of the many problems dealt with frequently by hydrogeologists.

The decision to employ professional help in the location and design of water wells should be based on considerations of public health as well as economics. Aquifers that are subject to easy contamination should be assessed by specialists. The horizontal and vertical distances between wells and potential sources of pollution are not always reliable measures of safety. Monetary savings can be achieved with professional help by the proper location of wells for maximum yield, the specification of total depths of wells, and the calculation of the optimum pumping rate possible for wells. The latter is often a neglected consideration. Although some well capacities are underrated, most are overrated with the result that expensive pumps and pipelines are installed at wells that fail to produce enough water to justify the investment.

The economic advantages of using scientific methods for locating ground water are generally much greater in regions in which aquifers are difficult to locate than they are in more water-favorable regions. The basic reason for this advantage can best be stated in terms of the relative size of the aquifer, or target, in relation to the search area. Random, unguided drilling has a chance of encountering a target that is equal to the ratio of the target area to the total search area [6]. Thus, if an aquifer underlies 60 acres of a 120-acre farm, half of the randomly located wells will be expected to encounter the aquifer provided that the problem is considered only in two dimensions. Unguided drilling would, therefore, require four test holes in order to obtain two water wells. The hydrogeologist, however, might locate the boundaries of the aquifer and insure that all wells will be successful. Hydrogeologic work would save the cost of two test holes. In contrast, the same area underlain by only twelve acres of aquifer would require an average of twenty randomly located test holes in order to develop two water wells. Hydrogeologic work could save the cost of eighteen test holes. Even if the hydrogeologic work could only reduce the search area to thirty-six acres, the cost of fourteen test holes might still be saved.

No description of ground-water exploration would be complete without some mention of dowsing, or water witching. There is little question of the utter uselessness of sticks, pendulums, wires, and the like used as instruments to aid in the location of water wells. The matter has been discussed at some length by Vogt and Hyman [34] and need not be repeated here. Even though reliable statistics [35] indicate that wells located by dowsers are no more successful than wells located at random, a few individual dowsers undoubtedly have a practical knowledge of hydrogeology which serves to guide, either consciously or subconsciously, the action of the rod or pendulum.

Dowsing cannot be recommended as a method of locating wells. It can,

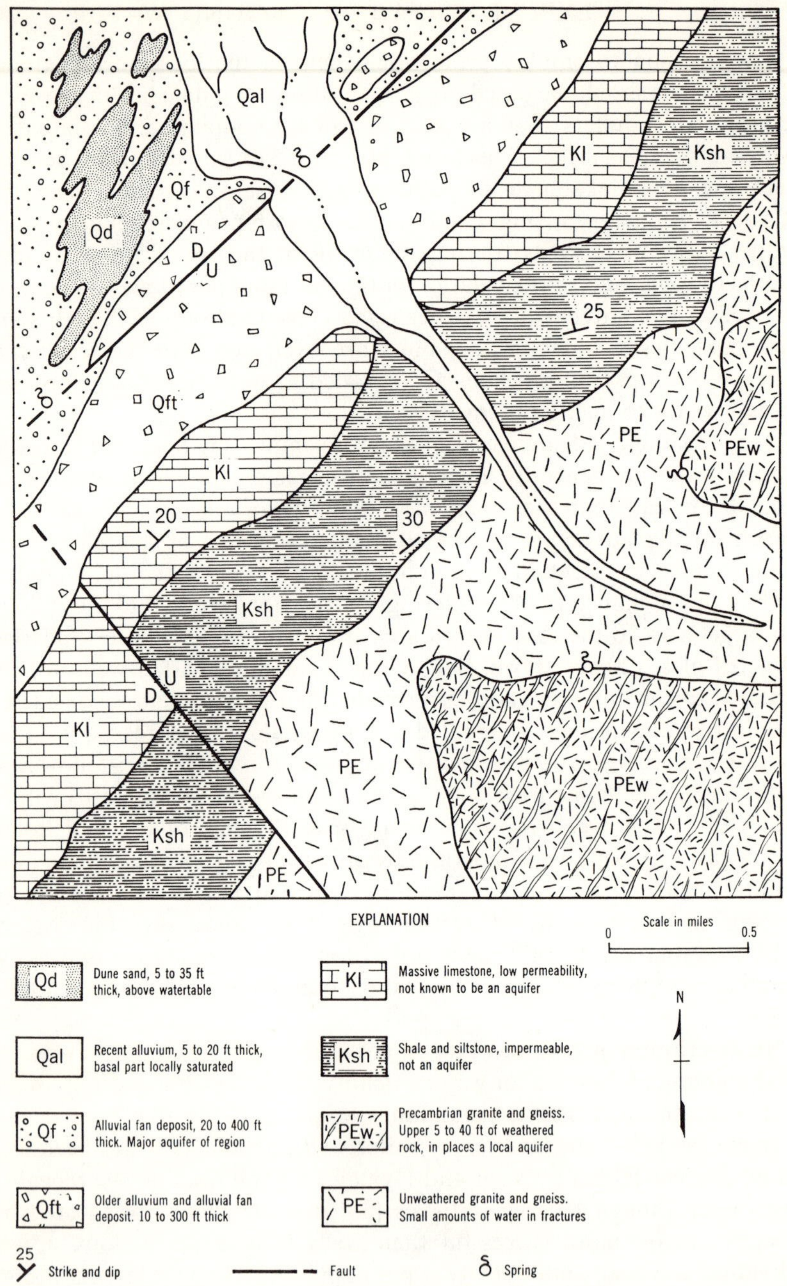


Figure 8.1 Hydrogeologic and geologic maps of the same hypothetical area showing the contrast in detail that reflects the difference in purpose between the two maps. Descriptions of the units have been abbreviated.

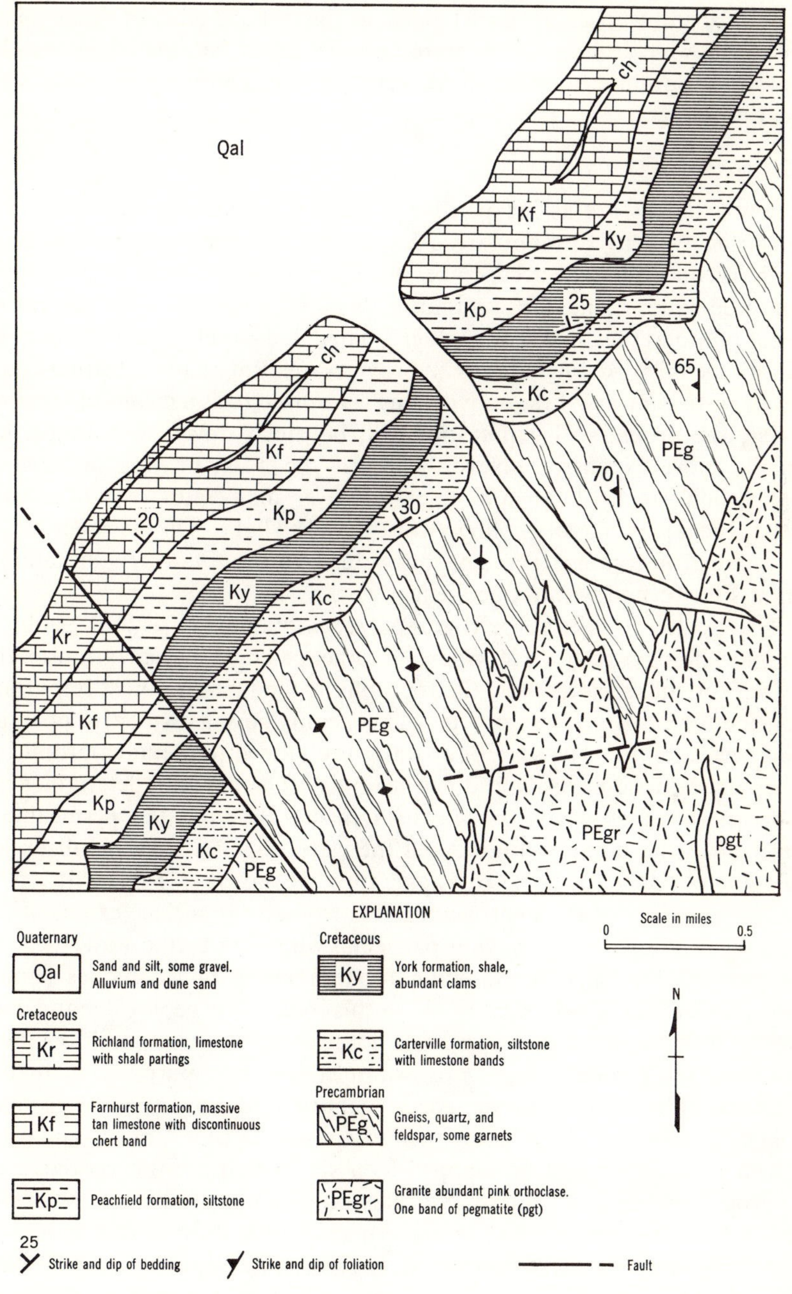
Qal
ch
Kf
Ky
Kp
25
Kc
65
PEg
70
ch
Kf
Kp
30
20
Ky
Kc
Kr
Kf
PEg
Kp
Ky
Kc
PEg
PEgr
pgt
EXPLANATION
Scale in miles
0
0.5
Quaternary
Qal
Sand and silt, some gravel. Alluvium and dune sand
Cretaceous
Kr
Richland formation, limestone with shale partings
Kf
Farnhurst formation, massive tan limestone with discontinuous chert band
Kp
Peachfield formation, siltstone
Cretaceous
Ky
York formation, shale, abundant clams
Kc
Carterville formation, siltstone with limestone bands
Precambrian
PEg
Gneiss, quartz, and feldspar, some garnets
PEgr
Granite abundant pink orthoclase. One band of pegmatite (pgt)
N
25
Strike and dip of bedding
Strike and dip of foliation
Fault

nevertheless, be useful for those who are unable to obtain scientific assistance and who do not care to assume the responsibility of the possible failure of well drilling. It is more comforting to be able to blame the dowser for failure than to have remorse concerning one's own well location.

## 8.2 *Geologic Methods*

Few activities are as inexpensive as geologic studies in the search for subsurface water. In many areas a geologist can make preliminary conclusions concerning the occurrence of subsurface water in an area of more than 100 square miles in less than a day. This is done with the aid of aerial photographs, regional geologic maps, and rapid ground reconnaissance. Detailed geologic work can progress easily at a rate of 1 square mile per day. In contrast, if random drilling is resorted to as a means of locating water, the progress of exploration will be much slower and almost infinitely more expensive. Nevertheless, if the water-bearing material is rather uniform in depth and thickness over many hundreds of square miles, the advantage of geologic work may be slight. On the other hand, in regions of complex geology the water-bearing zones may be almost impossible to find without geological work.

The geologist utilizes petrography, stratigraphy, structural geology, geomorphology, and, to a lesser extent, other geologic specializations in the search for subsurface water. Petrography is the first and most important consideration. A given rock type will generally have a distinctive porosity and permeability. The search for water is confined to the most promising zones in terms of porosity and permeability. The porosity determines the amount of water which can be held in storage, and the permeability determines the ease of withdrawing the water for use. A generalized list of some of the common petrographic types is shown in Table 8.1.

The first duty of the hydrogeologist is to map the surface extent of the various lithologic units, paying particular attention to the water-bearing properties. Students interested in modern mapping techniques should consult Compton's [8] recent book on this topic. The geologic map which results from the work, however, does not necessarily correspond to a more conventional geologic map. For example, for some purposes of hydrogeology, units such as granite, gneiss, gabbro, and diorite can be grouped together. On the other hand, many deposits of clay, gravel, and sand which normally would be mapped as a single unit are differentiated on hydrogeologic maps. Figure 8.1 illustrates some of the differences in detail between conventional geologic maps and hydrogeologic maps.

Stratigraphy is an essential tool in the search for water in areas of widespread sedimentary or volcanic rocks. The position and thickness of

*Table 8.1 Water-Bearing Properties of Common Rocks*

| Permeability | Porosity |
|---|---|
| Highest permeability | Highest porosity |
| well-sorted gravel | soft clay |
| porous basalt | silt |
| cavernous limestone | tuff |
| well-sorted sand | well-sorted sand |
| poorly sorted sand and gravel | poorly sorted sand and gravel |
| sandstone | gravel |
| fractured crystalline rock | sandstone |
| silt and tuff | porous basalt |
| clay | cavernous limestone |
| dense crystalline rock | fractured crystalline rock |
| | dense crystalline rock |
| Lowest permeability | Lowest porosity |

water-bearing horizons and the continuity of confining beds are of particular importance. The stratigraphic work may vary in complexity from the tracing of a single bed of distinctive lithology to the solution of a difficult problem through the use of paleontology and detailed measuring of geologic sections.

Structural geology is used in conjunction with stratigraphy to locate water-bearing horizons which have been displaced by earth movements. Structural studies are also used to locate fractured areas in dense but brittle rock. In unconsolidated deposits, faults may form hydrologic barriers, so their location is important in the study of the migration of subsurface water.

Geomorphology is indispensable in studying the occurrence of subsurface water in areas of late Pleistocene and recent deposits. The presence of permeable glacial sediments such as kames, kame-terraces, eskers, and outwash can commonly be mapped by studying the surface forms in a region. Stabilized sand dunes, terrace deposits, ancient beach ridges, and other permeable sediments are also clearly reflected in landforms. Geomorphology can yield lithologic, stratigraphic, and structural information of interest. Figure 8.2 gives an example of a fault located by small variations in the topography. Methods of photogeology are particularly helpful in many geomorphic studies connected with hydrogeology.

## **8.3** *Hydrologic Methods*

Hydrologic methods of prospecting for ground water include studies of total water available for recharge, ease of recharge, and the location and quantity of ground water discharged at the surface. The total quantity of water available for recharge includes both natural precipitation and surface water in large perennial streams. In general, the opportunity for finding subsurface water is related more or less directly to the recharge

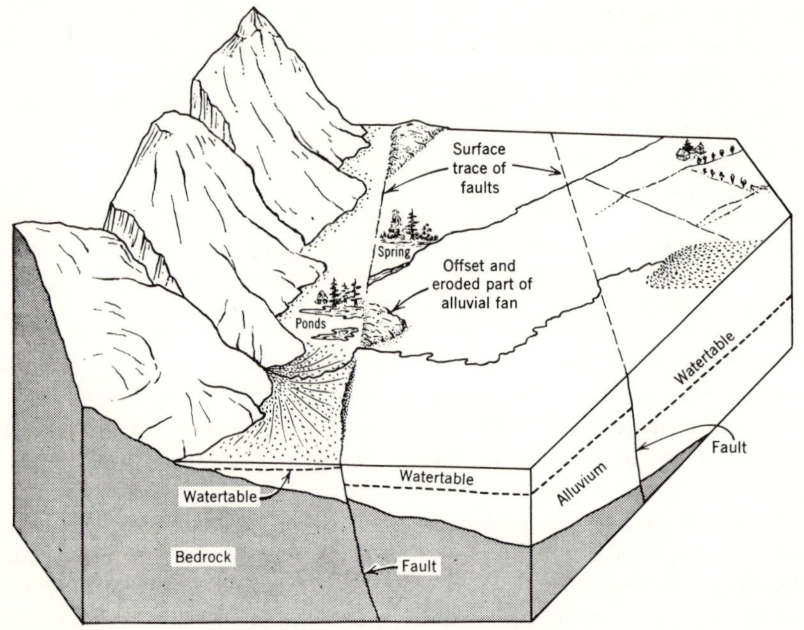


Figure 8.2 Block diagram showing topographic features common along recent faults. Offsets in stream indicate that movement along faults is largely lateral. Ponds and springs develop along the fault near the mountain front because the reduced permeability of the fault zone creates a ground-water dam.

available. Thus, in a desert region one would anticipate greater difficulty in locating subsurface water than in a humid region having a similar geologic environment.

The ease of recharge is another important hydrologic variable. Impermeable surfaces such as shale, clay, and quartzite will produce rapid runoff and prevent adequate ground-water recharge. The recharge or infiltration capacity of the surface can be estimated by various types of infiltrometers, of which the double-ring type is well adapted for rough reconnaissance work.

The location and magnitude of springs give a good indication of the general hydrologic conditions in a moist-to-humid region. Abundant small springs on valley sides and the slopes of hills generally indicate a shallow watertable with a shallow circulation of subsurface water in aquifers of poor permeability (Figure 8.3). In contrast, large springs confined to valley bottoms indicate high permeability and greater depths to the watertable. The presence of springs may also indicate the position of the watertable and thus give some indication of the depth of well which may be necessary.

Water may also be discharged at the surface by plants. If the plants are phreatophytes, the depth to the watertable may be indicated roughly by

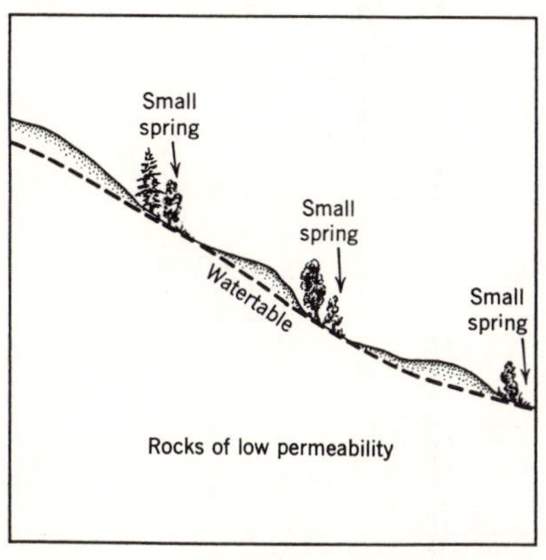


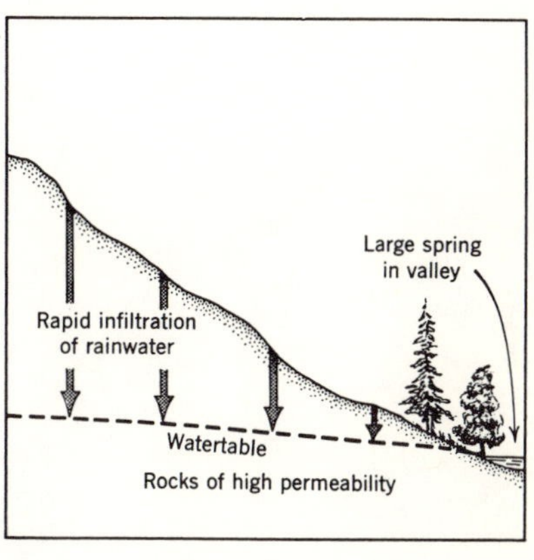


Figure 8.3 Cross sections showing the position of the watertable in regions of low and high permeabilities. (*a*) The steep hydraulic gradient needed to move water in the rocks of low permeability brings the ground water to the surface as numerous small springs. (*b*) High permeability allows the water to move easily to a small number of large springs situated in valleys.

the type of plant. The presence of phreatophytes, however, does not assure the availability of recoverable ground water, because phreatophytes can abstract water from saturated clays and fine silts which would not contribute water to a well.

In a ground-water exploration program, the quantity of water discharged by springs and evapotranspiration should be measured or estimated. The natural discharge is some indication of the maximum quantity of recoverable water in an area. If the water discharge is diffuse, however, it may be expensive to recover a very large percentage of the total amount.

Hydrologic and geologic exploration should proceed together for the best results. A region that is hydrologically favorable may not be geologically favorable for ground-water development, and the reverse is commonly true. Figure 8.4 will serve to illustrate a very simple case. The terrace material on the south side of the river is, geologically speaking, an

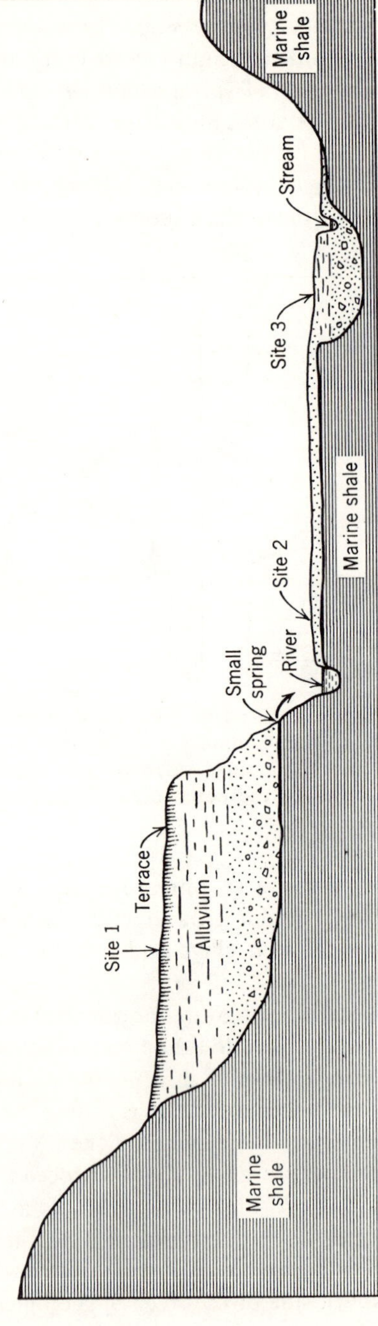


Figure 8.4 Cross section illustrating the necessity of using hydrologic information in conjunction with geologic information. Site 1 is favorably located with respect to a potentially permeable horizon but lacks a thick saturated zone. Site 2 is located to intercept a thick saturated zone, but a favorable aquifer does not exist. Site 3 is located favorably both with respect to geology and hydrology.

ideal aquifer. A hydrologic reconnaissance, however, shows only minor springs at the base of the coarse gravel. This indicates that the terrace material contains only a minor amount of water due to the rapid drainage of the gravel. The hydrologic information, on the other hand, indicates that the river is perennial and that a well on the north side of the river would be ideally suited for rapid recharge. The geologic study, nevertheless, indicates a dense impermeable shale cropping out in the river bed. The shale is undoubtedly saturated but is not pervious enough to allow significant amounts of water to seep into the well.

## **8.4** *Surface Geophysical Methods*

Most problems of well location can profit by the application of surface geophysical exploration. The geophysical work is, in general, much more expensive than geologic and hydrologic reconnaissance so that the decision whether or not to use geophysics is most often a question of economics. If the exploration project is economically important enough and if the geologic framework of the area is favorable, geophysics should, by all means, be utilized.

The success of geophysical work depends on simple but distinct variations of density, electrical conductivity, magnetic susceptibility, electrical potential, elasticity, and other measurable physical properties of the earth. If the contrasts in properties are slight, the measurements may not be precise enough to be useful. If the spatial distribution of the geologic units is too complex, the results cannot be interpreted geologically. Most failures in the application of geophysics to geohydrology arise from neglecting these two simple generalizations.

Some geophyscial methods measure directly the presence of subsurface water; others do not. The most useful applications of all geophysical techniques, however, are in the interpretation of geologic structure and stratigraphy, thus eliminating the need of an extensive drilling program [9,10,13,19,20,26].

### MAGNETIC METHODS

Measurements of variations of the earth's magnetic field are probably the most rapid to make and also the least expensive of all geophysical measurements. The magnetic intensity in vertical or horizontal planes is measured with an instrument composed of a magnetic system suspended on knife edges or more recently developed instruments that measure the interaction of the nuclear spin and the natural magnetic field. The largest natural anomalies are caused generally by the presence of magnetite, either as placer or primary deposits. Other common iron-bearing minerals such as hematite, glauconite, hornblende, and limonite may cause anomalies

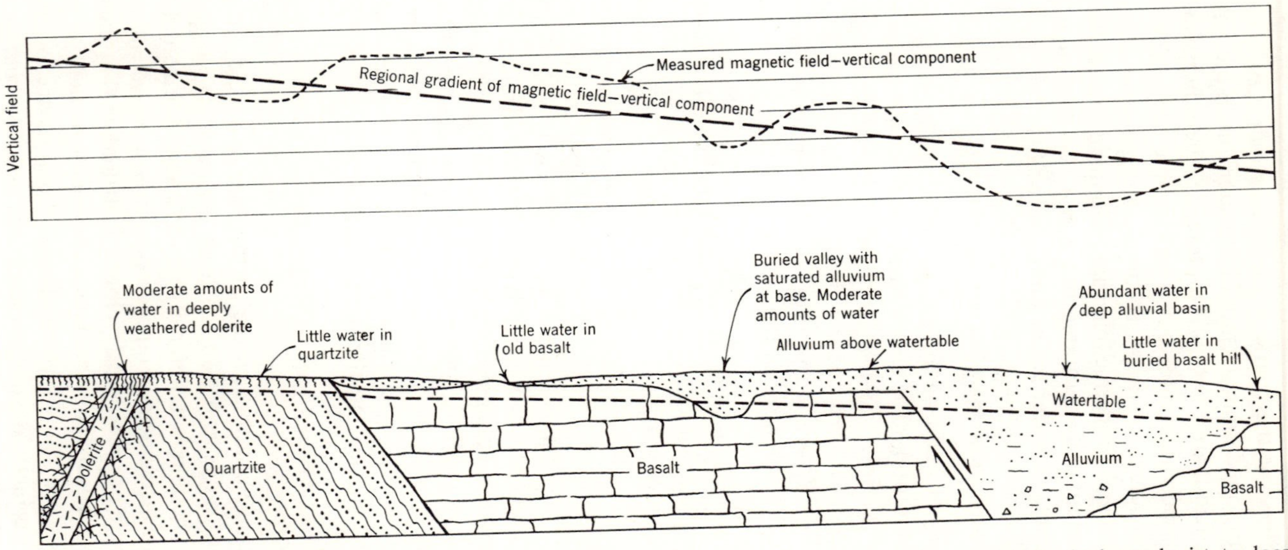

Figure 8.5 Profile illustrating how the vertical component of the magnetic field can be used to assist the hydrogeologist to locate potential drilling sites. The magnetic method, as well as other geophysical methods, depends in a large measure on a prior understanding of the local distribution of various rock types.

of considerable magnitude. The anomalies may range from about one-thousandth to one-tenth of the earth's magnetic field.

Despite the large natural contrasts in the magnetic properties of rocks, there are many difficulties in connection with the interpretation of anomalies. Railroads, buildings, bridges, automobiles, well casing, pipelines, and other cultural features will mask natural anomalies at distances of from 10 to 5000 feet. The magnetometer method is, therefore, practically useless as a key to near-surface geologic features in cities. Moreover, natural magnetic anomalies are commonly related to causes at great depth or to local variations which have no relation to hydrologic problems. For example, local concentrations of magnetite in alluvial deposits can easily produce sharp anomalies which are not related to aquifer permeabilities or to the presence of confining beds. Another difficulty in interpreting magnetic anomalies lies in the problem of calculating the spatial distribution of the material causing the anomaly. In simple cases it is possible to obtain quantitative interpretations by making magnetic surveys at various altitudes above the ground surface by means of airborne magnetometers. This, nevertheless, adds greatly to the expense of the work.

A few simple geological situations in which magnetic surveys might be useful are outlined in Figure 8.5. Magnetic surveys are also useful in locating buried pipe and abandoned well casing. The instruments most commonly used for locating metals, however, utilize electromagnetic pulses with frequencies of 500 to 2000 cycles per second. Metallic objects will distort the field of electromagnetic propagation and this distortion can be measured by a receiver placed a short distance from the center of propagation.

### GRAVITY METHODS

Studies of natural variations of the force of gravity have been used widely in connection with the exploration for oil and in theoretical studies of earth structure. In general, gravity surveys are relatively rapid and inexpensive, provided that elevations of the gravity stations to be occupied have been determined previously.

Instruments which have been used to measure variations of the force of gravity are generally of three basic types: pendulum, gravimeter, or torsion balance. Pendulums are kept at a constant length so that differences in period are related to differences in gravity. Most modern pendulums employ a rapidly vibrating rod or reed as the pendulum. This vibration is compared with vibrations of a known frequency to obtain an accuracy of more than $10^{-6}$ of the earth's gravitation field. Gravimeters measure the direct effects of the pull of gravity on a mass suspended by a delicate spring. The changes in the elongation of the spring are related directly to the vertical intensity of the gravity field. Optical or electrical methods are used to amplify movements of the spring so that accuracies of about $10^{-8}$ of the field of gravity can be obtained. Torsion balances measure the gradient rather than the direct force of gravity. Two small masses are attached to ends of a rod, which has a fixed length. The rod

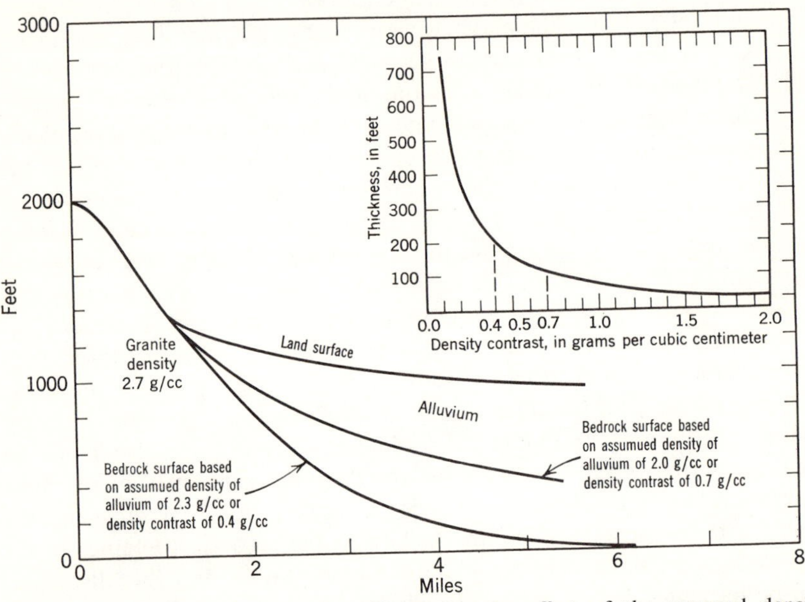


Figure 8.6 Schematic cross section illustrating the effect of the assumed density contrast on the interpretation of an alluvium-bedrock contact. Inset graph shows the relation between the density contrast and the thickness of a layer of infinite horizontal extent required to produce a 1-milligal anomaly. (From Maybey [18].) One milligal is an acceleration of $10^{-3}$ cm/sec$^2$, or, roughly, one millionth of the acceleration produced by the earth's gravitational field.

is suspended by a quartz filament which is placed under torsion when the gravitational attraction on the two masses is not balanced.

The proper interpretation of gravitational variations is dependent, first, on numerous corrections for elevation, latitude, topography, and regional geology and on geologically plausible assumptions concerning the gravitational anomalies which remain after correction. Theoretically, an infinite number of interpretations are possible. If assumptions are made concerning variations of rock density and regional geologic conditions, the presence of local features such as faults, folds and instrusions can be inferred. Under ideal circumstances, moderately accurate quantitative results, such as depths of alluvial deposits, can be obtained (Figure 8.6) [19,26,29].

The greatest weakness in the practical applications of the gravity method lies in the fact that small geologic changes are difficult to detect. The method is, therefore, of little use in most detailed problems of hydrogeologic prospecting and development, although gravity surveys have been used with some success in mapping large buried valleys [12,21].

SEISMIC METHODS

The most accurate and potentially the most useful geophysical methods are seismic methods [5,9,13,20]. Unlike magnetic and gravitational

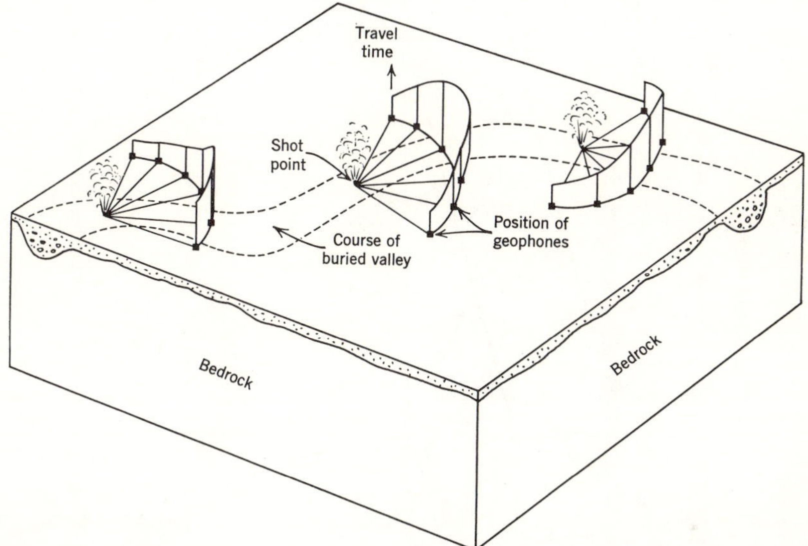


Figure 8.7 A block diagram illustrating the use of the fan-shooting method of seismic exploration. The method is well adapted to the mapping of hidden alluvial channels overlying bedrock.

methods, the seismic method does not measure a natural force field that is virtually static, but measures the reaction of geologic bodies to artificially induced vibrations. The vibrations are detected at various distances and directions from the source of energy by means of a large number of small seismometers which are commonly called geophones or detectors. The vibrations are recorded on photographic paper or on magnetic tape.

The arrangement of the detectors is dependent on the purpose of the work. An ideal arrangement for a relatively rapid reconnaissance of a buried stream channel (Figure 8.7) is a fan-like arrangement in which the detectors are spread at equal distances from the point of explosion. Loose material, such as gravel, will transmit elastic waves much more slowly than will the surrounding crystalline rock. The direction of the channel will

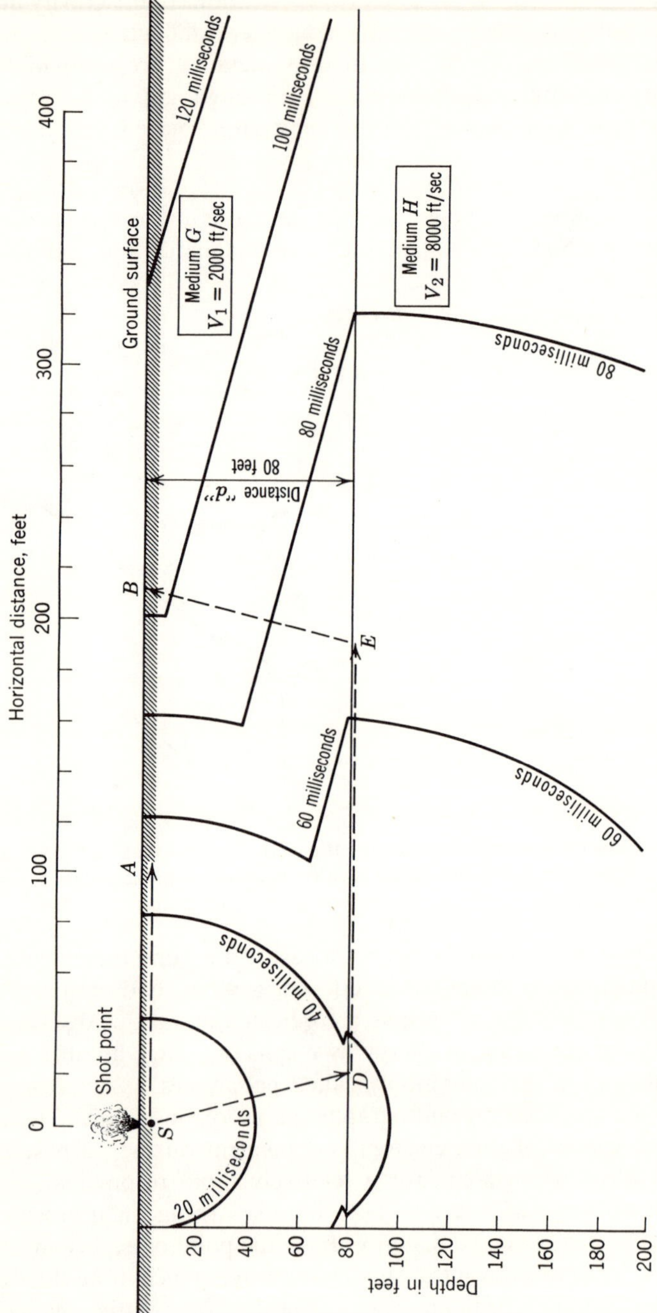


Figure 8.8 Successive positions of the initial energy front given for 20 millisecond intervals. First energy arriving at the surface beyond point $B$ is refracted through medium $H$.

be indicated by the direction of the slow travel time of the energy wave. In this method the direction and width of the channel are indicated, but not the depth of alluvium.

A more quantitative method is the refraction method in which the detectors are arranged in a single straight line from the point of explosion (Figure 8.8). The waves radiating from the point of explosion will travel in spherical fronts if the material is homogeneous and isotropic. If,

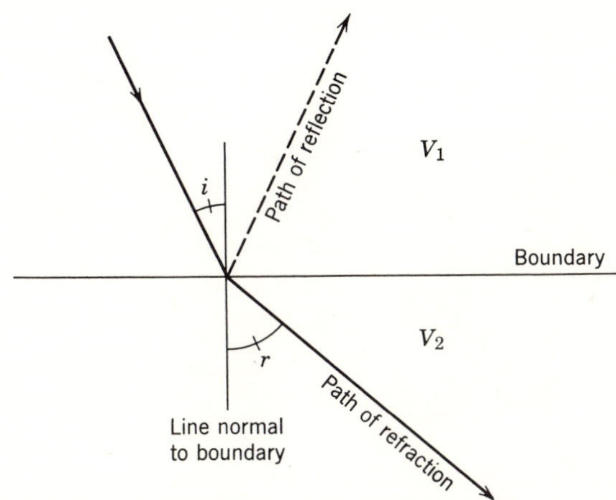


Figure 8.9 Refraction of energy across a boundary between materials having different transmission velocities.

however, variations exist in the natural material, the waves will be refracted according to the following relation:

$$\frac{\sin i}{\sin r} = \frac{V_1}{V_2} \tag{8.1}$$

in which $i$ is the angle of incidence, $r$ is the angle of refraction, $V_1$ is the velocity in the first medium, and $V_2$ is the velocity in the second medium (Figure 8.9). If the increase of velocity in the direction of propagation is large enough, or the angle of incidence is large enough, the angle of refraction may reach 90°. When this happens, total reflection of the wave takes place. When an energy wave is transmitted in stratified rocks the successive positions of the wave fronts can be indicated as in Figure 8.8. At point $A$, the first energy which arrives at the surface will travel directly through medium $G$, but beyond $B$ the first energy to arrive at the surface will travel part of the distance through medium $H$, which has a faster velocity of propagation. If a graph is made of the arrival time and distance from the source of energy (Figure 8.10) the velocities of the two media can be computed as well as the thickness of the upper layer. Inasmuch as the energy travels from *S* to *B* in a straight line, the slope of the first leg of the curve is inversely proportional to the velocity of energy propagation in medium *G*. Furthermore, since the energy reaching points beyond *B* travels an equal distance in *G*, the slope of the second leg of the curve

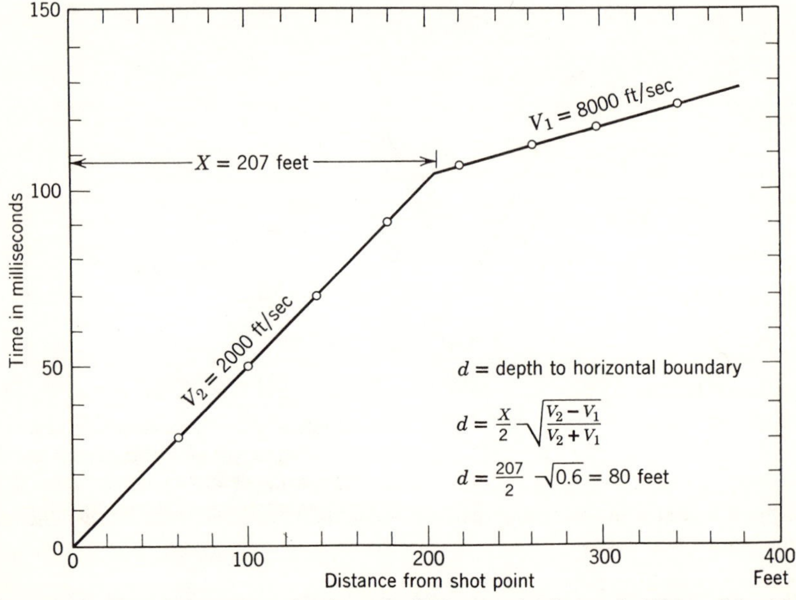


Figure 8.10 Travel-time curve for hypothetical example shown in Figure 8.8. The equation used to compute depth to boundary is valid only if land surface and subsurface boundary are parallel surfaces. Circles represent data obtained from geophones.

is proportional to the velocity of propagation to the velocity of propagation in medium *H*. The depth can be calculated with information of the two velocities by using the following relations. In Figure 8.8 the first energy arriving at *B* will travel by the path *SDEB* and by the path *SB* through medium *G*. The expression for the direct travel time, $t_1$, is:

$$t_1 = \frac{SB}{V_1} = \frac{x}{V_1}$$

The expression for indirect travel time $t_2$ is:

$$t_2 = \frac{SD}{V_1} + \frac{DE}{V_2} + \frac{EB}{V_1}$$

But since

$$SD = EB = d/\cos i$$

and

$$t_1 = t_2$$

then,

$$\frac{x}{V_1} = \frac{2d}{V_1 \cos i} + \frac{DE}{V_2}$$

but,

$$DE = x - 2d \tan i$$

then,

$$\frac{x}{V_1} = \frac{2d}{V_1 \cos i} + \frac{x}{V_2} - \frac{2d \tan i}{V_2}$$

thus,

$$d = \frac{x}{2}\sqrt{\frac{V_2 - V_1}{V_2 + V_1}} \qquad (8.2)$$

The last equation (8.2) is useful for the calculation of depths to horizontal layers which have a faster velocity than a surface layer. Similar but more involved equations have been derived for records showing three or more different inflections in the travel-time curve [13]. Irregularities of surface topography, attitude of bedding, various structural features, and certain contrasts in velocity of propagation [27] produce complications in the interpretation of travel-time curves. Measuring travel time first with energy moving in one direction and then with energy moving in another direction is a great aid in the interpretation of complicated structures. Also, it is useful to measure the velocity of the natural material by placing explosives or geophones at known depths in bore holes or by measuring travel times between relatively close points in various outcrop areas. If the velocities are known, then graphical and analytical methods can be used to solve problems of considerable complexity.

Diagrams of various types of problems related to hydrogeology which can be solved by seismic refraction are given in Figures 8.11 through 8.13. Large contrasts in the velocity of energy propagation are essential for success. If seismic refraction is contemplated two major limitations should always be kept in mind. First, it is generally impossible to gain information below a layer of dense-hard material. For example, a layer of basalt or of dense limestone will cause a deflection of the waves as indicated in Figure 8.14. No direct record of the underlying layers will be obtained. The second important limitation is in the expense of making seismic surveys. The large amount of specialized equipment and personnel normally needed makes seismic work the most expensive of all geophysical methods.

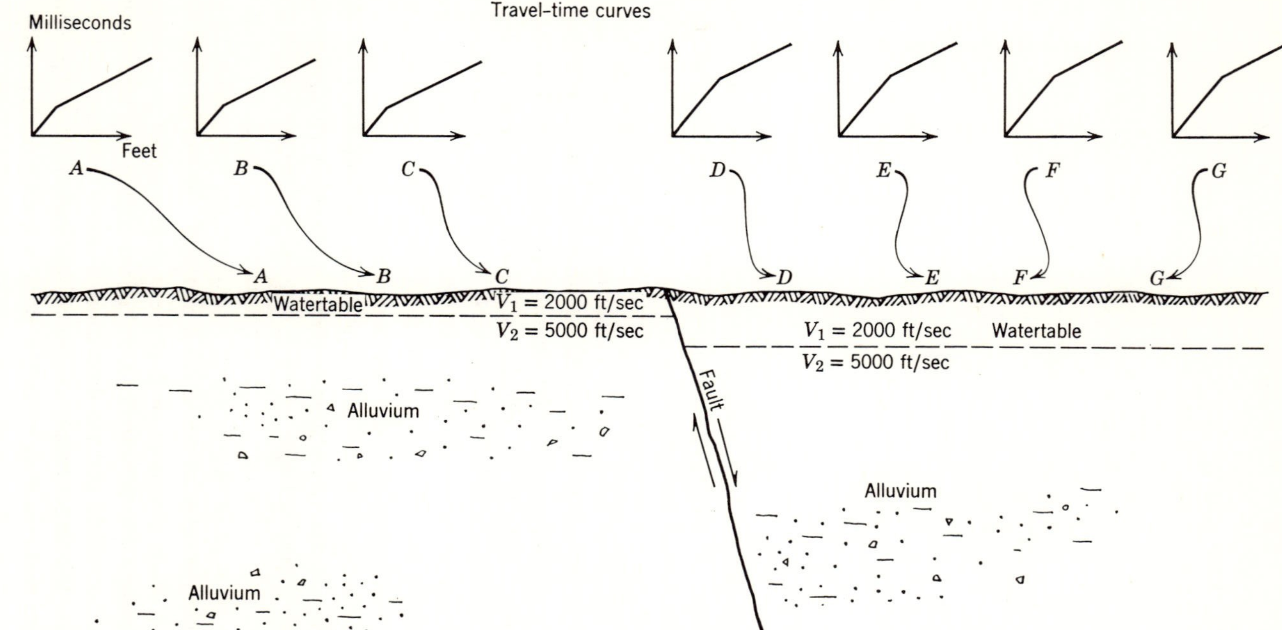


Figure 8.11 Refraction seismic studies used to locate offset watertable along fault. Alluvium is permeable and moderately homogeneous except along the fault. Horizontal scale on travel-time curves is not the same as on the geologic section; the separate seismic profiles actually overlap.

The use of reflection techniques can overcome problems of high-velocity layers at depth, although instrumentation and interpretation can be considerably more involved than simple refraction methods [9,19]. Small, compact seismic timers can be used by two people for shallow refraction surveys, thus reducing costs. If penetration of less than 50 feet is desired,

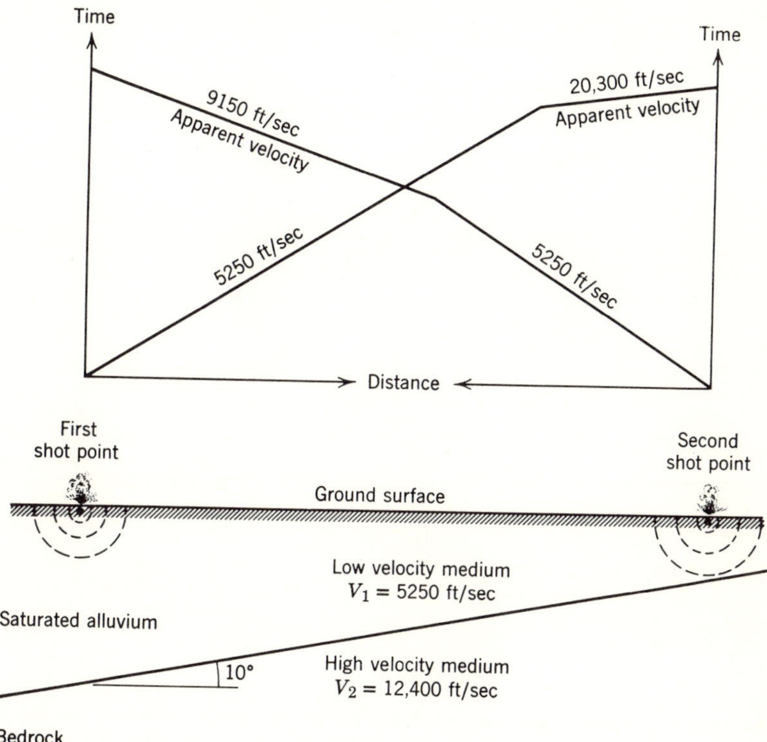


Figure 8.12 Travel-time curves illustrating the effect of a tilted boundary on the slopes of the second legs of the curves. Actual boundaries are commonly more irregular than shown in the example, consequently interpretation of travel-time curves is usually difficult and should be entrusted to experienced geophysicists.

a sledge hammer can be used to create the needed vibrations. This eliminates the use of high explosives in populated areas.

Experimental and theoretical work in the U.S.S.R. [4] has indicated that contrasts in porosity and grain size of sedimentary aquifers might be mapped by studying the absorption characteristics of refracted energy at different frequencies. It was found that the absorption characteristics were subject to rather large variations, whereas interface velocities changed only slightly. If perfected, these methods may provide a means of predicting qualitatively the porosity as well as the permeability of alluvial aquifers.

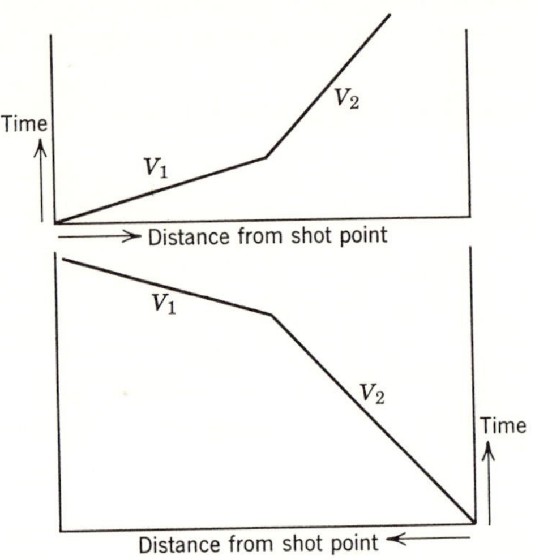


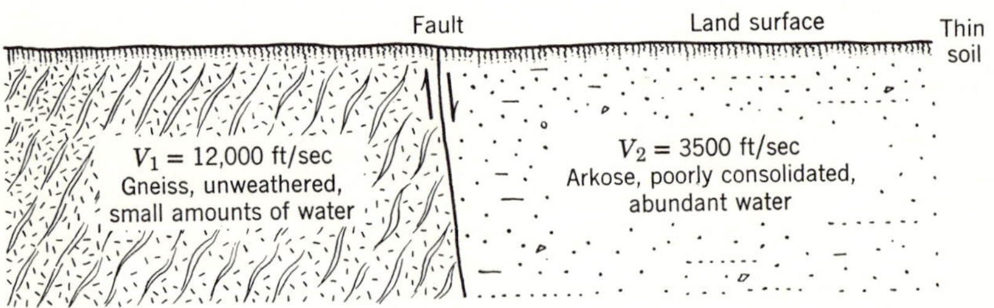


Figure 8.13 Schematic diagram showing travel-time curves obtained at the same locality but with opposite directions of energy propagation. The construction of travel-time curves from both directions is a necessary procedure for the accurate interpretation of subsurface geology. The lower curve alone could be confused with the effects of a horizontal boundary.

### ELECTRICAL METHODS

The surface geophysical method used most commonly with ground-water work is the electrical method. Instruments are relatively inexpensive, and only two or three men are required for the survey. The method is, therefore, generally economical.

Two types of electrical potentials can be measured. One is the natural electrical potential that exists between two electrodes placed in the ground. The other is an artificial potential created by passing electricity through the ground. The natural potential and resistivity are usually measured with the same instrument. The natural potential is measured first, then the resistivity.

Several electrode arrangements are possible; Wenner's and Schlumberger's are the most common (Figure 8.15). The Wenner configuration has the advantage of a more direct relation between electrode spacing and depth of current penetration. The Schlumberger configuration, on the

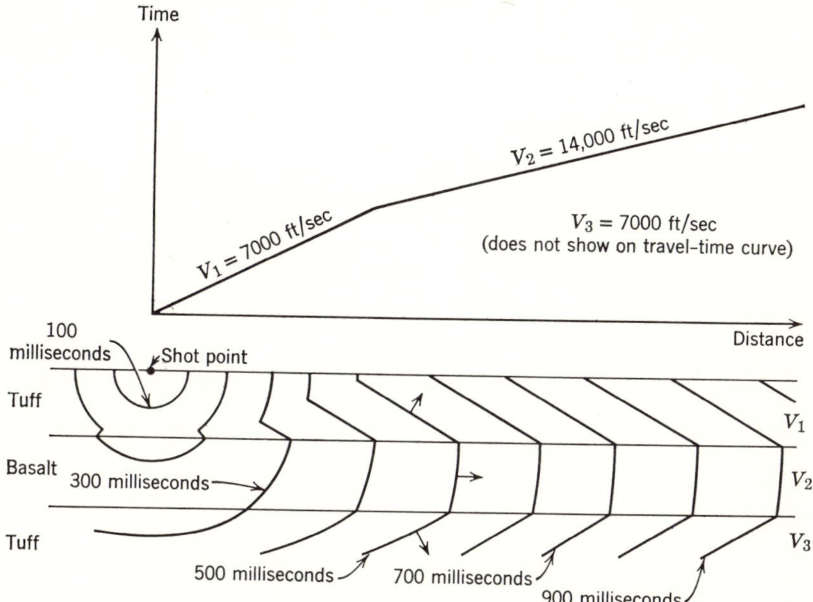


Figure 8.14 Diagram to explain why ordinary refraction techniques are not useful in detecting boundaries below a high-velocity layer. Initial wave front is refracted downward into the lower tuff so that it lags behind and can never be detected at the surface before the wave front that is refracted upward from the basalt.

other hand, allows a clearer definition of subsurface conditions for a given outer electrode spacing and uses less manpower because the central electrodes do not need to be moved every time the outer electrodes are moved [38].

Variations of natural potential have not been used widely as an exploration technique in hydrogeology. Large natural potentials are related to oxidizing sulfide minerals, corrosion of metals, waters of different chemical composition in contact with each other, and other electrochemical causes. In some localities, ore deposits, faults, and metamorphic zones can be located by measuring natural potentials. Use in hydrogeology of natural potentials has been largely limited to research work in

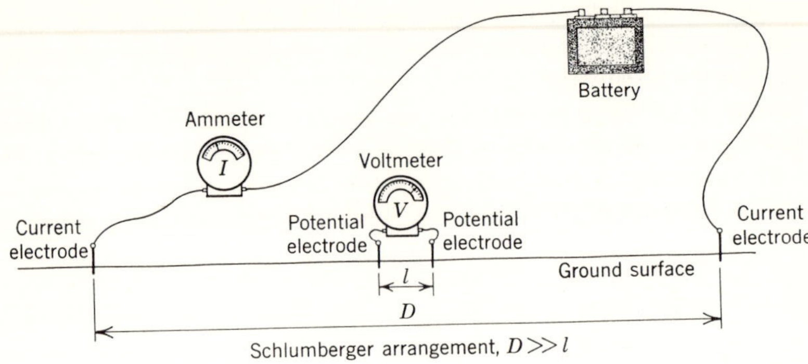


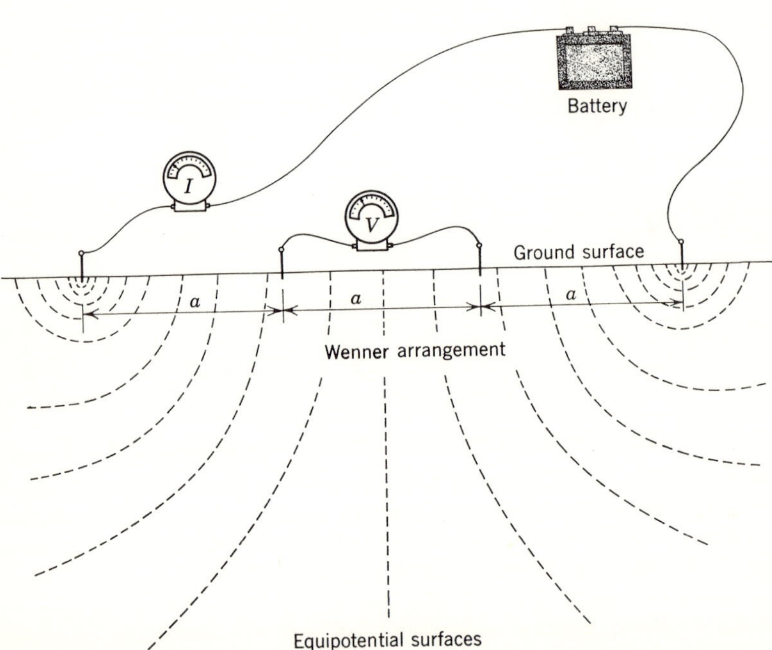


Figure 8.15 Schlumberger and Wenner electrode arrangements for the measurement of earth resistivity. Electrical circuits of ammeters, voltmeters, and power sources are shown only schematically.

areas where waters of vastly different chemical composition are in contact with each other.

The resistivity of natural materials to the flow of electricity varies widely. Dense granite may have a resistivity of $10^6$ ohm-meters, whereas clays saturated with saline water may have a resistivity of only 1.0 ohm-meter, or a natural variation of a millionfold. In general, solid minerals, such as quartz and feldspar, have very high resistivities. Brine has the lowest resistivity of abundant natural materials. Table 8.2 gives representative values for various naturally occurring materials.

Owing to the large resistivity of the common detrital minerals, quartz, feldspars, and calcite, most of the electricity passes through the fluids in the pore spaces of sedimentary rocks. The resistivity of rock is, therefore, largely a function of porosity and chemistry of the saturating fluid. The presence of clay is also important in determining the resistivity of rocks. Clays saturated with water will be surrounded by films of partially mobile ions which will migrate under a potential gradient. This migration adds to the normal migration of ions in the fluid, causing a reduction in resistivity of clay-rich rocks or sediments.

*Table 8.2 Approximate Resistivity Values for Natural Material*

| Material | Resistivity, ohm-meters |
|---|---|
| Graphite | $3 \times 10^{-4}$ |
| Pyrite | $10^{-3}$ |
| Brine | $5 \times 10^{-2}$ |
| Shale | 1.0 |
| Gypsum | 10 |
| Fresh water | 50 |
| Gravel and sand, saturated with fresh water | $10^{2}$ |
| Serpentine | $3 \times 10^{2}$ |
| Limestone | $10^{3}$ |
| Granite | $10^{6}$ |
| Quartz | $10^{11}$ |
| Calcite | $5 \times 10^{12}$ |

The potential field around two electrodes is shown in Figure 8.5. Electrodes which measure potential drop are placed far enough inside the current electrodes so that the zone of very rapidly changing potential is avoided. If the electrodes are evenly spaced, the following equation can be derived for the apparent resistivity:

$$R_a = 2\pi a \frac{V}{I} \tag{8.3}$$

in which $R_a$ is the apparent resistivity, $a$ is the electrode spacing, $V$ is the potential difference, and $I$ is the applied current. Despite this rather simple equation, no universally valid rule for relating the electrode spacing, $a$, to the effective depth of penetration can be given. The common assumption

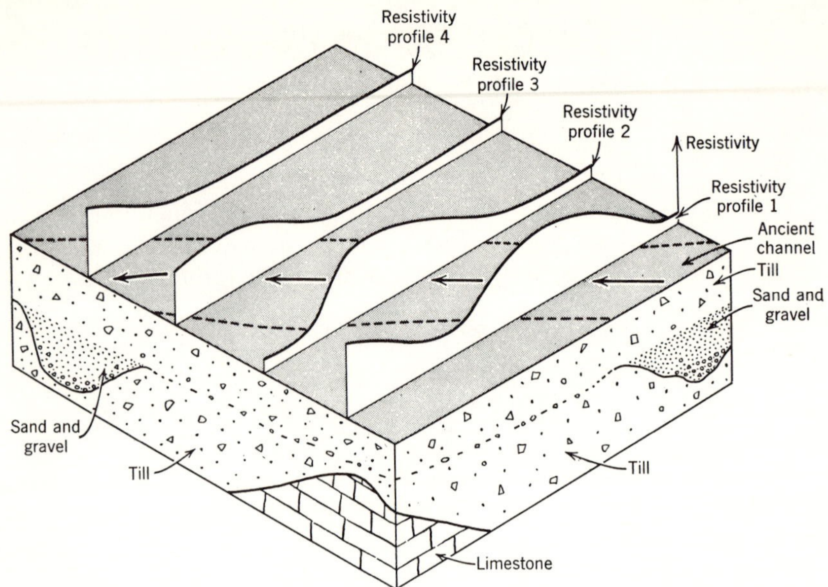


Figure 8.16 Resistivity profiles used to locate buried stream deposits sandwiched between two till sheets. After initial discovery, subsequent exploration is best made by profiles oriented perpendicular to the trend of the channel.

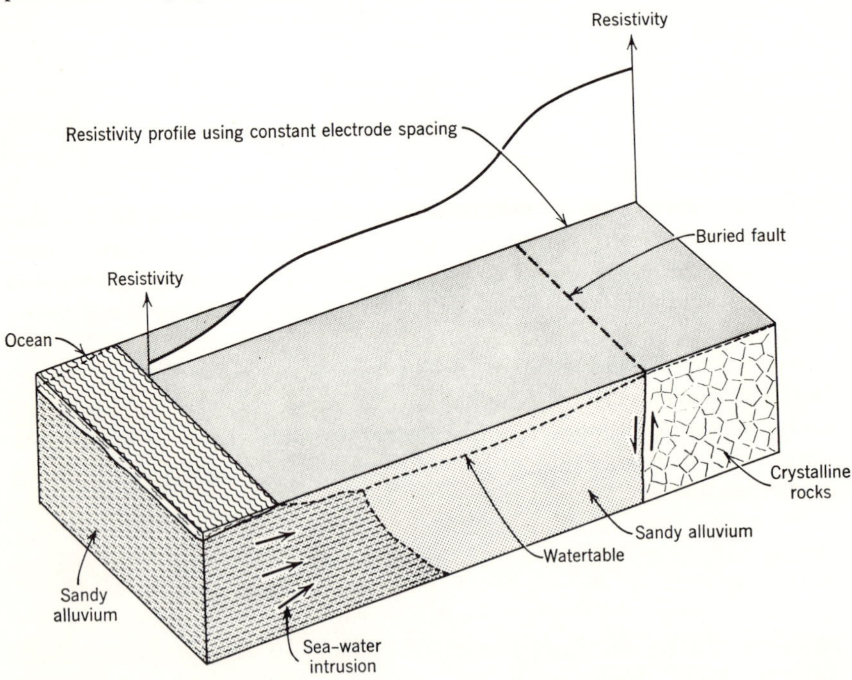


Figure 8.17 The use of a resistivity profile to locate the extent of sea-water intrusion into a homogeneous sandy aquifer. A fault is also located inland by the same profile.

that the electrode spacing in the Wenner configuration is directly related to depth of penetration can produce errors in calculated depths of several hundred per cent [23]. The determination of depths to horizontal boundaries is most conveniently made by matching field determinations with type curves that have been calculated for theoretical layers of specified contrasts in electrical resistivities.

The simplest arrangement for resistivity exploration is to use a fixed spacing, *a*, and to move the electrodes in a straight line. The resulting profile will show the apparent resistivity of rocks and sediments at a more-or-less constant depth. This method has been used with advantage to locate buried glacial gravel surrounded by till (Figure 8.16). Other potential applications of this method are shown in Figure 8.17.

Constant spacing is best adapted for vertical boundaries which are buried at a depth of less than 100 feet. For a horizontal boundary, such as the watertable or surfaces of stratified rocks, the method of vertical probing or drilling is used. The probing is done by keeping the central point between electrodes stationary and increasing the horizontal spacing by small increments. The apparent resistivity plotted against electrode spacing (Figure 8.18) gives an indication of the resistivity of various layers at successively greater depths. The apparent resistivity, however, is a measure of the effects of all the layers between the maximum depth of penetration and the surface. For this reason, the greater the number of beds, the more difficult the interpretation becomes [32]. Three or four distinct layers are about the maximum number for accurate interpretation unless other subsurface information is available.

The greatest success in the application of resistivity methods to hydrogeology has been with two-layer problems [11,14]. Salt-water fresh-water boundaries in homogeneous basalt and dune sand have been located with accuracy, particularly in areas where the depth to the boundary is less than 500 feet [31]. The method has also had some success in locating alluvium-bedrock contacts in river valleys [10,14,20,22] and in locating layers of gravel or sand below clay and silt [10,36,38]. Most attempts to locate the watertable in alluvial and glacial materials have been less successful. This is probably owing to the great variability of moisture conditions above the watertable [33].

There are three basic limitations to the resistivity method. First, metal pipes, rails, wire, and man-made structures in contact with the ground will divert the flow of electricity so that measurements of natural resistivity are difficult in urban areas. Second, only relatively simple geologic features can be interpreted without supplementary information from drilling or from other geophysical methods. Third, the depth of penetration of most portable instruments is less than 1500 feet owing to a lack of sufficient batteries for power and the geologic complexities of large volumes of earth. If these limitations are kept in mind, resistivity methods can be very useful to the hydrogeologist.

## 8.5 *Test Drilling*

If no inferences concerning geologic conditions can be gained from work on the surface, the hydrogeologist must resort to either geophysics or direct drilling to gain the desired information. Ideally, both drilling and geophysics should be used, but economic limitations often require that a choice be made. The choice is commonly dictated by two considerations. First, drilling yields information of a more positive nature. Geophysical techniques yield results which may be subject to more than one interpretation. Second, drilling costs rise rapidly with depth and with an increase in rock hardness.

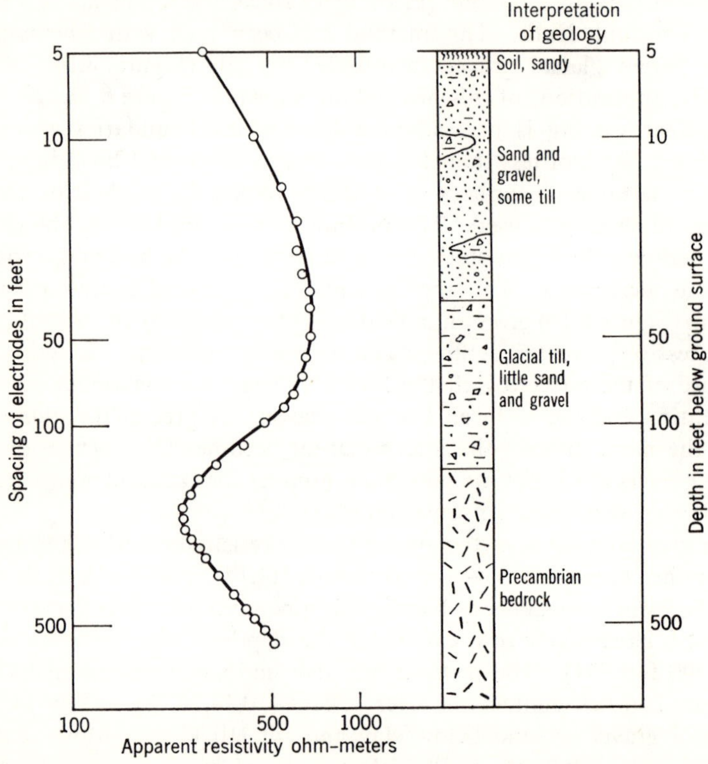


Figure 8.18 Apparent resistivity of subsurface material determined by the expanding electrode, or probing method. (From Spicer [28].)

Test drilling has the greatest advantage in flat terrain underlain by soft sediments where the aquifers are less than 300 feet below the surface. Here, holes can be drilled rapidly and usually at a very low cost.

## **8.6** *Drilling and Excavating Equipment*

The following discussion of drilling and excavation equipment is intended to be only a general orientation to a complex topic. Hydrogeologists working with exploration programs should become as familiar as possible with the practical as well as theoretical aspects of drilling and well construction. Although excellent books [3,7] cover these topics, much of the background needed can come only from firsthand experience with the great variety of equipment used in modern drilling.

### HAND TOOLS

The simplest type of well excavation, from a mechanical standpoint, is an excavation with handtools. The tools are generally confined to shovels, picks, crowbars, buckets, and hoists. If excavation is in consolidated rock, wedges, sledge hammers, hand drills, and explosives may also be used. In unconsolidated material, the sides of the hole must be supported if depths of more than five to ten feet are to be reached. Materials used for support depend on cost and availability. Boards and metal sheeting are most satisfactory, but other material such as concrete curbing, metal drums, and poles can also be used.

Hand excavation has the advantages of using highly portable equipment and requiring small investment of money. Hand excavation is limited in depth because of the danger of cave-ins and the greater work needed to remove material from the bottom of deep holes. Most hand-excavated holes are less than 100 feet deep. The volume of material which must be removed to afford working space for a man is at least ten times as great as the volume of material which must be removed to make space for pumping equipment in the final well.

Hand tools have also been developed for drilling and driving wells. The two processes are most commonly combined. A hole is first drilled 15 to 20 feet with an auger; then, a pipe with a suitable screen is driven beyond the auger hole to a position below the watertable. Auger holes can be advanced to more than 40 feet by hand, but the work is generally slow and inefficient beyond 20 feet because the auger stem must be broken and rejoined each time material is removed form the hole unless a hoist is used. Driven wells are usually limited in diameter to less than three inches in depth to less than 40 feet because friction on the pipe becomes excessive with larger depths and diameters. Through a combination of augering and driving by hand, wells up to about 60 feet can be constructed; although most such wells are less than 30 feet in depth.

Driven and augered wells are economical and rapid to construct in soft alluvium which has a watertable at a depth of less than 20 feet. A small investment of time and money makes this method one of the most efficient for obtaining small amounts of ground water. In some areas a series of driven wells can be connected to a common pump in order to yield volumes large enough for irrigation. Similar arrangements are used to lower the water level around excavations made for construction purposes.

The presence of boulders and hard strata will prevent the penetration of both auger and driven wells. In addition, saturated sand and coarse silt cannot be penetrated by the ordinary hand auger.

### JETTING RIG

The fastest type of well drilling is the simple jetting procedure which leaves the jetting pipe in the ground to serve as the well (Figure 8.19). The jet method is essentially a stream of water forced out the end of a pipe. The end of the pipe contains a special chisel-like point which helps cut the sediment when the pipe is rotated. The loose material is forced out of the hole by the rising water. The jet method is successful only in soft material free of boulders. Furthermore, the solid material which is ejected from a jetted hole is commonly a mixture of material from various levels in the hole. It is, therefore, difficult to construct an accurate log of the well. This is an important limitation if the well is to be used in connection with an exploration program.

### CABLE TOOLS

The cable-tool or percussion method is one of the oldest and still one of the most popular well-drilling techniques. The essential parts of a cable-tool machine are shown in Figure 8.20. The tools are moved up and down in the well by percussion strokes which generally vary between $\frac{1}{2}$ and 3 feet. The reciprocal action is given to the cable by the spudder. The total weight of the tools may vary from 200 pounds to more than 2 tons. The bit is worked up and down in the hole until it is filled with 3 to 6 feet of mud mixed with material loosened by the bit. The loose material, or cuttings, is removed with the bailer. If the material being drilled is loose, it is necessary to advance the casing during drilling to prevent caving of the hole. In solid rock a casing is commonly placed only in the first few feet of the hole to prevent caving of weathered and fractured rock and soil.

Rates of drilling vary with the type of formation being penetrated, depth of the hole, diameter of the hole, type of equipment, and experience of the personnel doing the drilling. Drilling will be as slow as 5 to 10 feet per

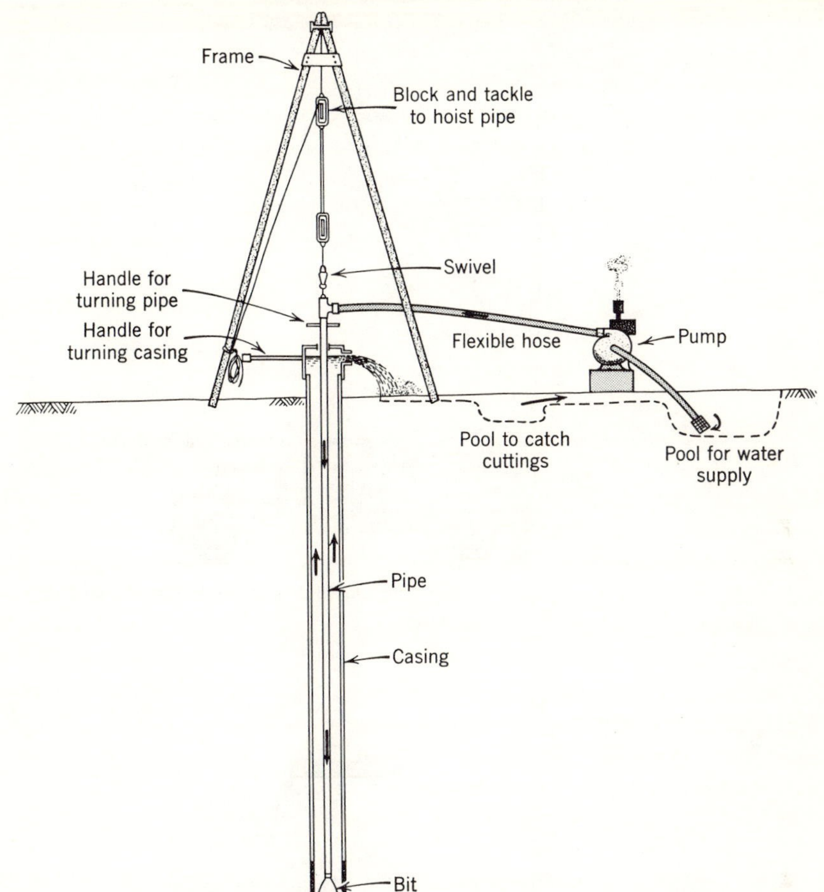


Figure 8.19 Equipment for jetting wells. Arrows indicate direction of water movement. Casing is advanced by jacks or weight dropped on casing head.

day in dense crystalline rocks and as fast as 50 to 100 feet per day in soft sandstone and sandy clay. Drilling in very hard, dense rock offers no unusual difficulties. Progress is slow because of rock hardness and the necessity of dressing bits frequently. If the rock is fractured, difficulty is commonly encountered with crooked holes which tend to follow softer zones and with bits and bailers being stuck. Unconsolidated material containing boulders is very difficult to drill. The boulders will deflect the hole, are hard to drill, and contribute to friction on the casing making the driving of the casing more difficult. Drilling through hard boulders can be even slower than drilling through solid crystaline rock. Sticky shale and clay is difficult to loosen and is commonly difficult to bail.

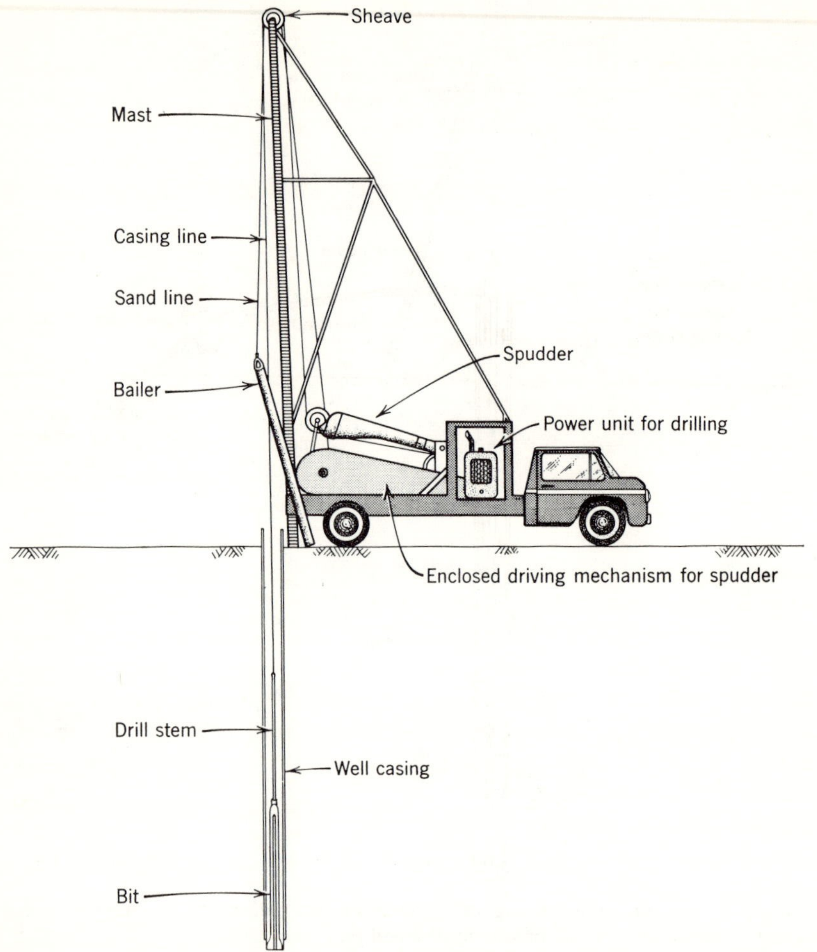


Figure 8.20 Truck-mounted percussion (cable-tool) equipment for drilling wells. Casing is commonly not used if well is being drilled in consolidated rock.

Adding sand to the hole during drilling can be helpful in reducing the stickiness of the clay and in preventing the clay from forming balls. Drilling rates in sticky shale and clay may range from 15 to 50 feet per day. Loose, fine sand is particularly hard to penetrate because it flows into the hole almost as fast as it can be bailed. Keeping the hole filled with water helps control sand inasmuch as it will reverse the flow of water from the hole into the sand, which in turn helps hold the sand in place. Drilling rates in loose, flowing sand, popularly called quicksand, may be as little as 10 to 20 feet per day.

### ROTARY RIGS

Since about 1920, rotary-drilling methods have become increasingly popular for water wells. The two main reasons for this popularity are the greater speed of drilling and the fact that well casing is rarely needed during drilling. The second advantage is very important if the hole proves to be useless for a well, because it can be abandoned with no cost beyond the initial work of drilling. In contrast, the work involved in recovering casing from cable-tool wells can be difficult and expensive if not impossible.

The basic scheme of rotary drilling is shown in Figure 8.21. Cutting of rocks and sediments is accomplished by rotating bits of various types. The power is delivered to the bit by a rotating hollow steel tube, or drill pipe. Mud is forced through the drill pipe and out at the bit. The mud then rises to the surface along the hole and removes the rock fragments, which are deposited in the settling pit. Power is delivered to the drill pipe by gears and a kelly. The kelly moves down when drilling is in progress and is pulled up temporarily when a new length of drill pipe is added. The drill pipe is prevented from falling into the hole by slips when the kelly is unfastened.

As in cable-tool drilling, rates of rotary drilling depend on a large number of factors, of which, type of geologic formations and drilling equipment are most important. In soft, unconsolidated sediments, rates of between 300 and 500 feet per day are possible. In consolidated rocks, on the other hand, rates of drilling are commonly only between 30 and 50 feet per day. Unlike cable-tool drilling, the rates of rotary drilling are not greatly affected by depth, with the exception of coring and changing bits. In which cases, the drill rods must be taken out of the hole and replaced again before drilling is resumed.

The greatest difficulties in rotary drilling are encountered in highly permeable zones such as cavernous basalt and limestone. This difficulty is caused by the loss of drilling fluid which escapes into the permeable formations. The loss of fluid, in turn, removes support in the upper part of the hole, and soft zones tend to collapse, sometimes with the loss of expensive bits and drill rod. Difficulty is also encountered in drilling through soft material which has very hard clasts and chert nodules. The bits tend to spin on top of the hard rock without penetrating them. They will also deflect the bit from the desired direction of the hole. Chert nodules in limestone and boulders or very hard rock in poorly consolidated conglomerate are particularly troublesome in this regard.

A number of variations of rotary equipment exist. The reverse rotary equipment draws the fluid upward through the drill pipe. The fluid is

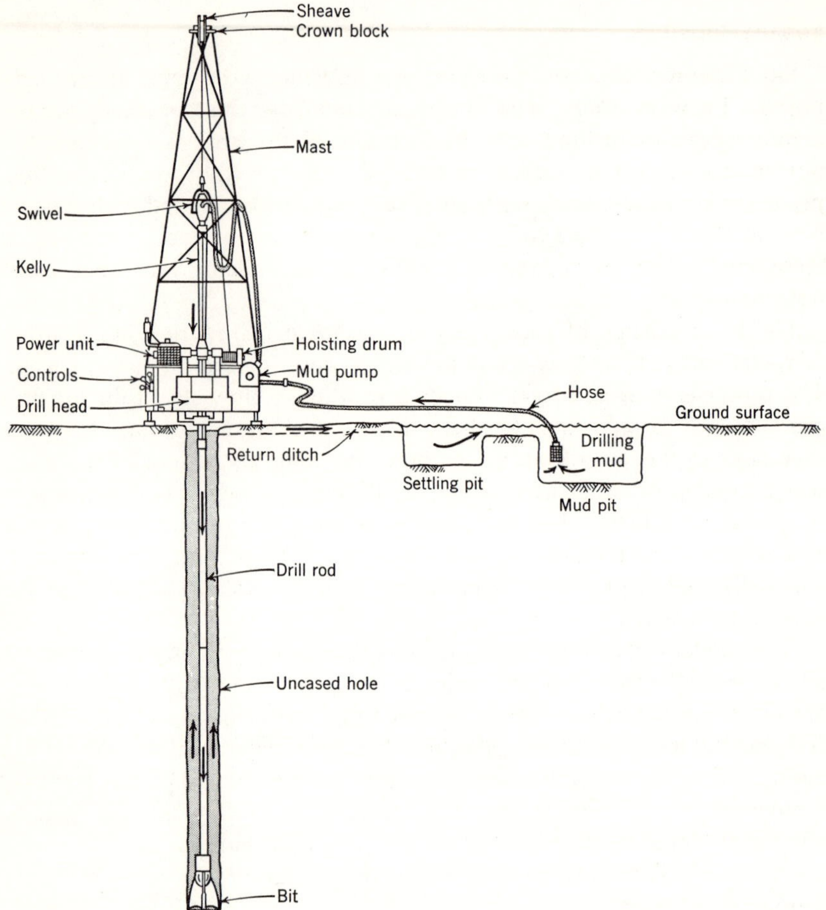


Figure 8.21 Major components of rotary equipment for drilling wells and test holes. Arrows indicate direction of mud circulation.

discharged into the mud pit through special pumps with open-blade rotors which allow large gravel to be passed. The rising fluid has a greater velocity inside the drill tube because of its smaller cross section. The higher velocity allows larger particles to be lifted and increases drilling efficiency in coarse-grained, unconsolidated deposits. If drilling is to be done in regions of little water, an air compressor can be used to circulate air instead of mud or water. Air drilling is possible if the rocks are generally impermeable and the cuttings from the bit are relatively small. Drilling equipment is also available which can be used alternately as a cable-tool rig or as rotary rig.

PNEUMATIC DRILLS

High-speed pneumatic drills have been adapted recently to water well drilling. Large volumes of air are used to lift cuttings to the surface. Drilling is generally limited to hard, consolidated rocks with small to intermediate amounts of water. Holes from 50 to 100 feet deep can be drilled in very dense granite in a single day with large pneumatic-drill rigs.

SELECTION OF EQUIPMENT

The choice of drilling equipment depends on the use to be made of the equipment as well as economic and geologic factors. If the drilling is to be for the purpose of a rapid geologic reconnaissance of a region, then rotary methods are generally better. If a detailed study of water quality is to be made during drilling, the cable-tool method is superior. Drilling equipment adapted for both rotary and cable-tool work is best for many purposes, but is generally more expensive to purchase and maintain.

## **8.7** *Geologic Logs*

The record of any phase of well drilling can be called a log. Drilling-time, casing installation, geologic formations, geophysical variations, and perforations are some of the more common items logged. Of the many types of logs, the geologic log is perhaps the most important [30]. Correlations between wells, location of water-bearing zones, and final design of well casing are dependent to a large extent on accurate geologic logs. Unfortunately, good geologic logs are quite difficult to make.

One of the sources of most difficulty is the fact that well cuttings are small and mixed with mud. The cuttings from rotary drilling should be washed on a screen which has openings smaller than 1/8 millimeter so that particles larger than fine-sized sand will be retained. [15]. In rapid rotary drilling, material finer than sand will be recirculated in the drilling mud, so that it is difficult to log zones of very fine sand, silt, and some types of coarse clay. Clay with cohesion will not be dispersed rapidly, consequently lumps of clay will be recovered on the screen. In cable-tool drilling with a casing, the entire sample including the mud will be significant. In both uncased cable-tool and rotary drilling, caving is a serious problem. Shale, silt, sand, and gravel will fall into the hole from horizons well above the zone of drilling. For this reason, percentage logs are common (Figure 8.22). Abrupt changes in rock percentages are correlated with changes in lithology even though the changes may amount to only a few per cent of the total sample.

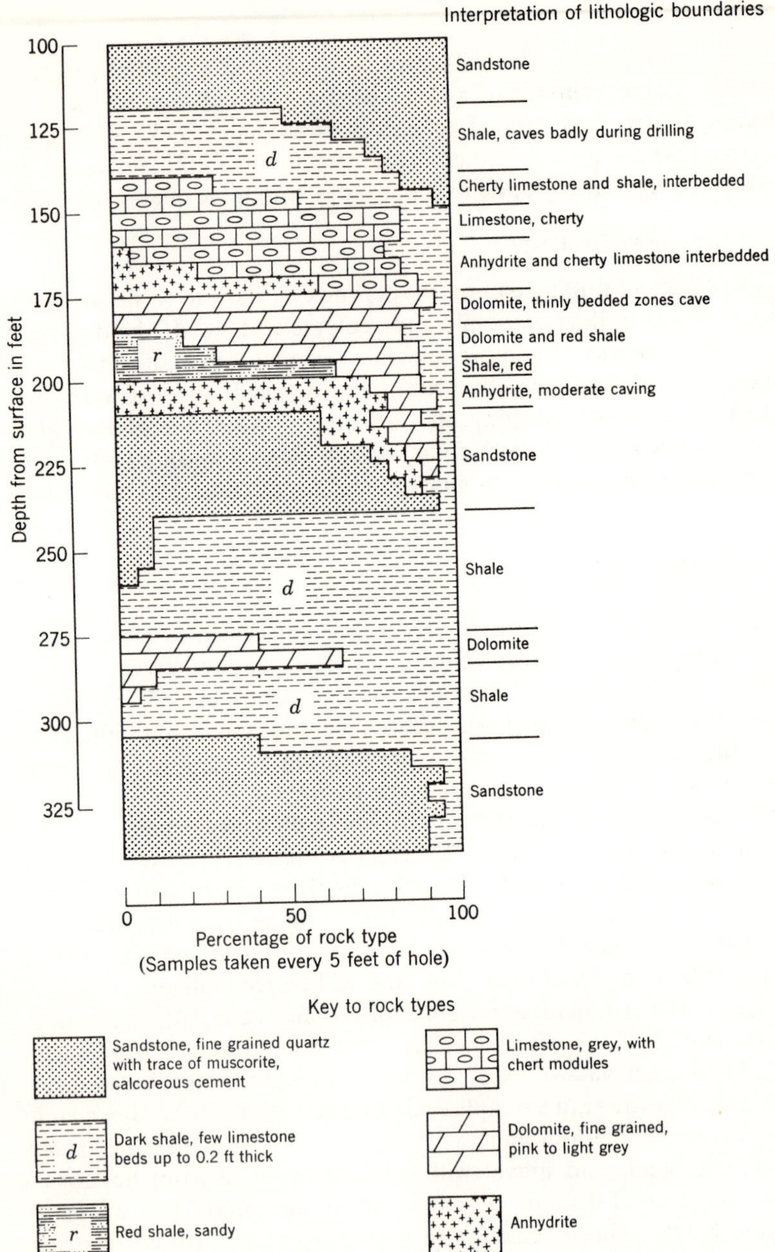


Figure 8.22 Part of a percentage log of a well penetrating sedimentary rocks.

Geologic logging should always be done in close cooperation with the driller. Changes in rock character are commonly reflected by changes in drilling rate and the reaction of the bit through rebound in cable-tool drilling or through vibration in rotary drilling. These changes can be detected by experienced drillers who should relay the information to the geologist.

Most of the standard subsurface techniques used in petroleum exploration and development are applicable to logging of water wells. Insoluble residues, calcite percentage, microfossils, megafossils, heavy minerals, porosity, permeability, and clay mineralogy are of greatest potential interest [15]. Insoluble residues can be used to study nearly homogeneous limestone and dolomite as an aid in the location of permeable zones. Calcite percentages can be used in glacial till to locate buried soil horizons which are leached of carbonates and significant in the study of glacial stratigraphy. Microfossils and megafossils can be used for stratigraphic problems and to understand past geologic environments. Heavy minerals are useful in investigating the provenance and also the weathering of sediments. Porosity and permeability are directly related to the water-bearing characteristics of the rocks and sediments. Clay minerals are responsible for many of the natural variations of permeability. The type of clay minerals present also determines the type of permeability changes which will take place if water of unusual chemical composition is injected into the well.

At the present stage of development of the water-well industry, economic considerations prevent the employment of trained geologic personnel for most work with water wells. It is common practice for the well driller himself to make the log. The well driller must also supervise other workers, handle the machinery, take charge of mechanical repairs, help place casing, and many other duties. Despite lack of formal geologic training and pressing duties during drilling, many drillers are able to produce accurate logs which are of great use to the hydrogeologist. The skillful driller learns to produce a simple log which can be written rapidly but which contains a maximum amount of information. The most important factors are size of particles, hardness, and color. In many areas a small bottle of acid is also useful to test the cuttings for effervescence, which in turn indicates a carbonate-bearing material, usually limestone, dolomite, or marble.

Inasmuch as cobbles and boulders are rarely recovered in drilling, the distinction between the two sizes is not made by many drillers. The other size distinctions should be made whenever possible. Special ambiguous names, such as "tiger clay," "chocolate sediment," and "muck" should be avoided. On the other hand, notes such as "hard pan," "sea shells," "rotten egg odor," and quicksand" are useful. Certain common rocks can be identified without much training, but attempts at difficult identifications should not be made by the driller. Misidentification can cause more difficulty than a use of less scientific sounding words. For example, granite and arkose are very similar in appearance, however their relation to regional hydrogeology may be vastly different.

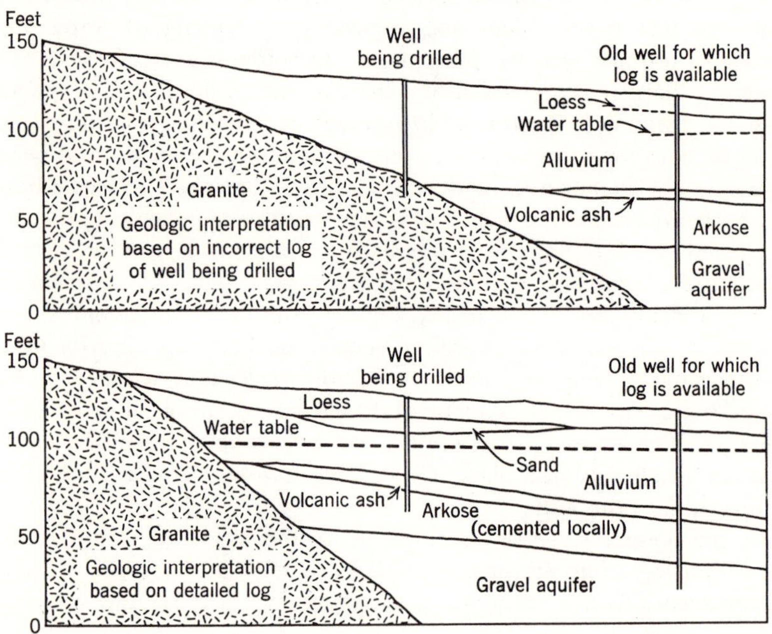


Figure 8.23 Two geologic cross sections based on the same surface information but on different geologic logs.

The importance of the correct driller's log can be seen from the two hypothetical geologic cross sections based on the interpretations of the driller's logs (Figure 8.23). The hydrogeologist knows from an inspection of outcrops along adjacent hills to expect granite and arkose in the subsurface, and also that a water-bearing gravel is commonly found between the arkose and granite. On the basis of the first log (Table 8.3) the hole would be abandoned; on the basis of the second log, the hydrogeologist would suspect the presence of arkose and would inspect all the cuttings from the well. The drilling would then be continued until the granite is reached.

Drillers rarely have the facilities for storing cuttings from wells; as a consequence, few cuttings are preserved permanently except when required by state law or by the person who has hired the driller. During

*Table 8.3 Comparison Between Types of Logs*

| Feet | Driller's Log, poor | Feet | Driller's Log, good | Feet | Geologic Log, brief type |
|---|---|---|---|---|---|
| 0–10 | Soil | 0–10 | Silt, brown, soft. | 0–2 | A and B horizons of modern soil. |
| | | | | 2–9 | Loess, light yellowish brown, friable, calcareous with gastropod remains. |
| 10–50 | Quicksand and water | 10–22 | Sand, tan, soft, some caving | 9–22 | Sand, medium to fine, well sorted, yellowish grey, friable. About $\frac{2}{3}$ quartz and $\frac{1}{3}$ feldspar. |
| | | 22–46 | Sand and silt with a little clay, tan, soft. Some water below 30 ft. Water stands in well at 29 ft. | 22–46 | Silt, sandy, poorly sorted, yellowish grey, friable. Small layers of clay less than 1 ft thick encountered every 1 to 2 ft. Clay is yellowish brown and compact. Sand, silt, and clay lack carbonates; composed of quartz, feldspar, and biotite. |
| | | 46–50 | Quicksand, white. Water rose in well, now stands at 25 ft. | 46–50 | Ash, rhyolitic, with about 10 per cent clean medium quartz, sand, light tan, friable. |
| 50–65 | Granite | 50–65 | Rock, hard, pink, looks like granite but drills easier than granite. | 50–65 | Arkose, medium grained, light reddish brown, compact. |

drilling, however, systematic samples should be kept of the materials penetrated. These, then, can be referred to as the log is written and when the perforation diameter or screen size is chosen. When possible, all the samples should be preserved permanently. In most regions, however, this is not specified as part of the driller's duty.

## 8.8 *Geophysical Logs*

Most subsurface geophysical methods have been developed during the past 40 years to meet the needs of the petroleum industry. Several of the methods also have been used in connection with the drilling of water wells. Many of the methods are restricted to use in uncased wells, although surveys such as gamma radioactivity and caliper can be used with casing in the hole. The first widely used geophysical logging methods were introduced in the early 1930's and employed two measurements, spontaneous potential and electrical resistivity. After several years of development of many other logs, the spontaneous potential and resistivity logs still remain among the most useful of all geophysical logs.

### SPONTANEOUS POTENTIAL LOGS

For water-well work, the spontaneous electrical potential is, perhaps, the most useful. The potential is caused, first by an electrochemical cell which is formed where shale, permeable material, and the bore hole join together (Figure 8.24) and, second, by the electrokinetic effect of fluids moving through permeable material. In most formations, the spontaneous potential is produced largely by the electrochemical effect. Wyllie (37) has demonstrated that the electrochemical potential formed in open drill holes filled with mud that has a much higher electrical resistivity than the formation water can be expressed by the following formula.

$$E = K \log_{10} \frac{\text{Activity of the interstitial water}}{\text{Activity of the mud filtrate}} \tag{8.4}$$

in which $E$ is the potential developed and $K$ is a factor depending mostly on temperature. Inasmuch as chemical activities are for most situations nearly proportional to electrical conductivities of solutions, the following formula can be used,

$$E = K \log_{10} \frac{R_{mf}}{R_w} \tag{8.5}$$

in which $R_{mf}$ is the resistivity of the mud filtrate and $R_w$ is the resistivity of the water which saturates the permeable formation.

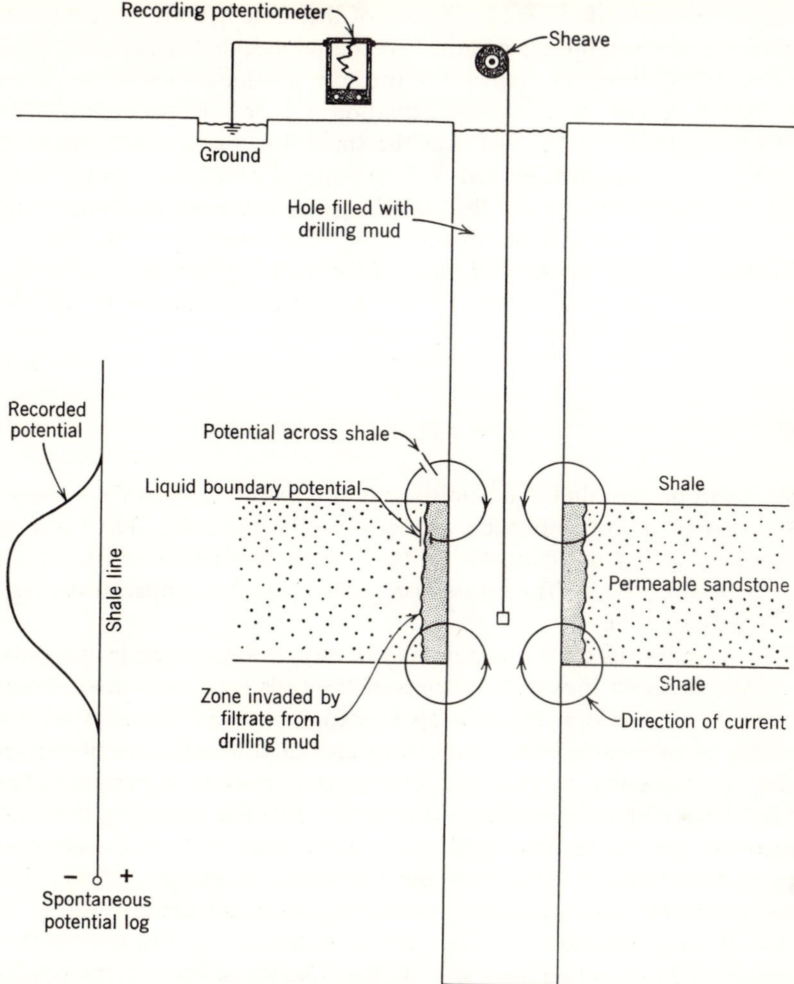


Figure 8.24 The spontaneous potential log. The direction of current flow and resulting spontaneous potential in a test hole penetrating an aquifer in which the water in the sandstone is more saline than the drilling mud. The total electrochemical potential is the algebraic sum of the liquid boundary potential and the potential developed across the shale "membrane."

The spontaneous potential in a well is measured by lowering a single electrode into an uncased hole and immersing the other electrode in a small hole at the surface filled with the drilling mud. The potential which is recorded is assumed to be zero opposite thick layers of impermeable shale or clay. A line, commonly referred to as the shale line, can be drawn on the log which connects all points of zero potential. By convention, logs are made so that deflections to the left of the shale line are negative and deflections to the right are positive. If the permeable beds are less than eight to twenty times the diameter of the hole, a value somewhat less than the total potential indicated by Equation 8.5. will be recorded. The difference is owing to the fact that the small flow of electricity between permeable and impermeable beds will be partly diverted through the drilling mud in the bore hole and through the permeable zone which has been invaded by the drilling mud (Figure 8.24). The resistivity of the shale is indicated by $R_{sh}$, of the sand $R_s$, and of the mud in the bore hole by $R_m$. If the total potential developed is $E$, the potential measured is $V$, and the current flowing is $I$, then

$$V = IR_m \tag{8.6}$$

and

$$E = I(R_m + R_{sh} + R_s) \tag{8.7}$$

Thus it can be seen that $V$ is approximately equal to $E$ when the value of $R_m$ is very large in comparison to the sum of $R_{sh}$ and $R_s$. The value of $R_m$ is large when the diameter of the hole is small or when the mud resistivity is very high. The values of $R_{sh}$ and $R_s$ will be comparatively high if the beds are thin.

The spontaneous electrical potential is used to locate permeable beds, to locate the exact tops and bottoms of the beds, and to estimate water quality. The development of a spontaneous potential depends on the presence of permeable beds. Unfortunately, an absolute value of permeability is impossible to measure, and indeed, beds with permeabilities far too low to be of interest as aquifers may give rise to large potentials. In general, the electrical potential log is better adapted to the location of tops and bottoms of beds than the conventional resistivity logs. The inflection of the curve which marks the boundary of the bed can usually be located with a precision of 1 foot on the potential log, but only with a precision of 3 to 5 feet on the resistivity log. Quality of water in the aquifer can be estimated from the potential log by means of Equation 8.5. All values in this equation except $R_w$ can be estimated; thus the equation can be solved for $R_w$. Since $R_w$ is a function of the ion concentration, a fair approximation of water quality can be obtained [37]. In general, this method is most satisfactory in the study of saline water and brine. Only qualitative information can be obtained if the formation water is fresh.

There are several reasons why the use of Equation 8.5 cannot yield exact results. In the first place, the equation is based on the assumption of an electrochemical potential arising from varying concentrations of one monovalent cation and one monovalent anion. For many connate waters this assumption does not involve a great error inasmuch as most ions are sodium and chloride, but for many waters of low ionic concentrations the monovalent ions make up less than 50 per cent of the total ionic concentration. Another source of error is in the assumption that the potential is entirely electrochemical. The electrokinetic potential under some conditions may actually exceed the electrochemical potential. The electrokinetic potential will be negative, so if the resistivity of the mudfiltrate is greater than the resistivity of the water in the aquifer, the electrokinetic potential will add to the electrochemical potential. A third source of error arises from geometric effects of bed thickness and hole diameter and has been discussed previously. If thin beds or beds with thin interstratifications of clay are avoided, the geometric errors are minor. A fourth source of error that becomes very important in fresh-water aquifers comes from the fact that $R_m$ is no longer very large in comparison to the sum of $R_{sh}$ and $R_s$ so $V$ is not approximately equal to $E$ (Equation 8.7) [24]. Despite the uncertainties involved in estimating water quality, a value of water resistivity which is within 100 per cent of the correct value can be calculated for most brackish water. An accuracy of 10 to 20 per cent is not uncommon for saline water. Greater accuracy is generally achieved with empirical correction factors derived from extended experience within small ground-water basins.

The first step in calculating water resistivity using the spontaneous potential is to measure the potential as indicated by deflections from the shale line. If the zone interest is thick enough and if several thick permeable beds are present, a sand line can be drawn which connects all maximum deflections from the shale line. The potential is then read as the difference between the sand and shale lines (Figure 8.25). The potential will be either positive or negative; this algebraic sign should be carried into Equation 8.5. Many water wells are too shallow to allow sand lines to be drawn; hence aquifers must be treated individually, and, if thin, a geometric correction factor must applied to the potential in order to estimate the true potential developed in the hole.

The second step is to estimate the temperature at the depth of interest. Most electric logs record maximum temperatures which are assumed to be bottom-hole temperatures. The mean annual air temperature of the region plus 2°C is assumed to be the temperature 60 feet below the surface. A linear change in temperature is further assumed between the bottom of the hole and a point 60 feet below the surface (Figure 8.26). In the absence of reliable bottom-hole temperatures, the regional geothermal gradient is used to estimate temperatures at a given depth.

The third step is to choose the proper value of $K$ using the estimated temperature of the formation. If the hole is shallow and in a region of temperate climate, the value of $-70$ mv can commonly be used without

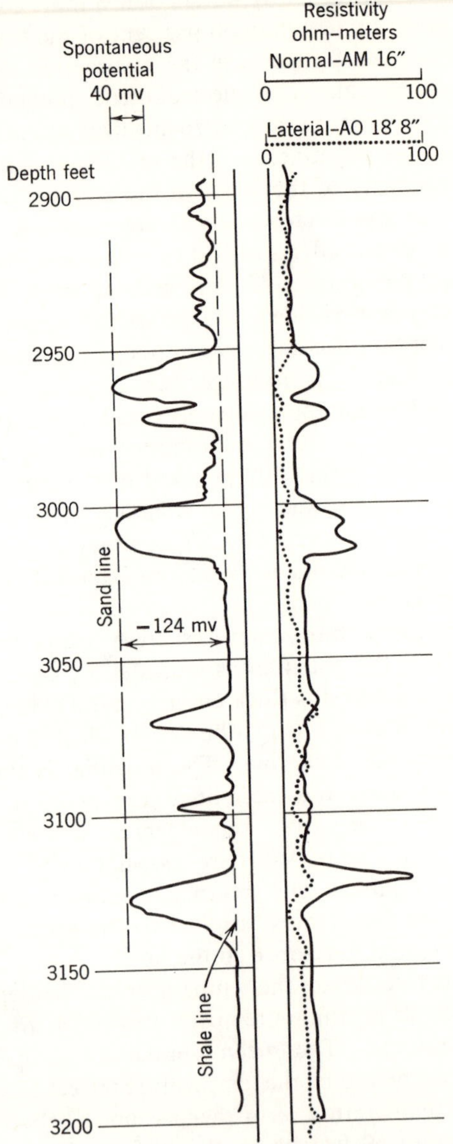


$R_m$ = 1.5 ohm-meters at 66° F
Bottom hole temperature at 4511 feet = 116° F

Figure 8.25 Electric log of an oil well that penetrates shale and sandstone saturated with brine. Ideally, sand lines should be extended to more than the three sandstone beds shown on this example.

bothering to correct for formation temperature. For water near freezing a value of $-65$ mv is better, and for water at about 150°F the value increases to $-80$ mv (Table 8.4).

The fourth step is to read the resistivity of the mud filtrate from the heading on the log. If the resistivity of the filtrate is not given, the resistivity of the unfiltered mud, which is always recorded in the log heading, can be multiplied by 0.8 to obtain a fair approximation.

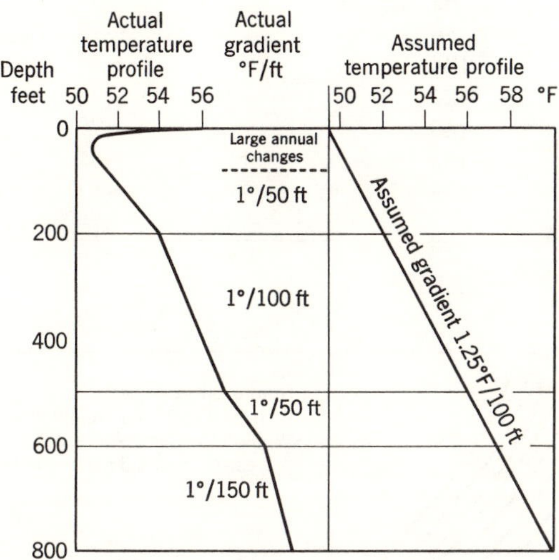


Figure 8.26 Hypothetical example of a geothermal log showing the effects of various thermal conductivities on thermal gradients at different depths. Detailed temperature profiles are difficult to obtain, so practical work commonly utilizes an assumed gradient based on bottom hole temperatures and approximate mean surface temperatures.

The fifth step is to substitute the above values for $E$, $K$, and $R_{mf}$ in Equation (8.5) and to calculate the value of the unknown, $R_w$. The value of $R_w$ can be related to ion concentration, expressed as NaCl, by use of Figure 8.27. The resistivity of the water will be in terms of a solution at the temperature of the mud filtrate. Thus in using Figure 8.27, the value of ppm NaCl is read from the resistivity curve at the temperature of the mud filtrate. This is illustrated with the numerical example shown in Figure 8.28.

## RESISTIVITY LOGS

Resistivity logs measure the effect of an artificial electric current which is produced in the logging truck and transmitted to the formation

*Table 8.4*

| Temperature °C | °F | Electrochemical Constant ($K$) Millivolts |
|---|---|---|
| 0 | 32 | −65 |
| 10 | 50 | −67 |
| 20 | 68 | −69 |
| 30 | 86 | −72 |
| 40 | 104 | −74 |
| 50 | 122 | −77 |
| 60 | 140 | −79 |
| 70 | 158 | −81 |
| 80 | 176 | −83 |

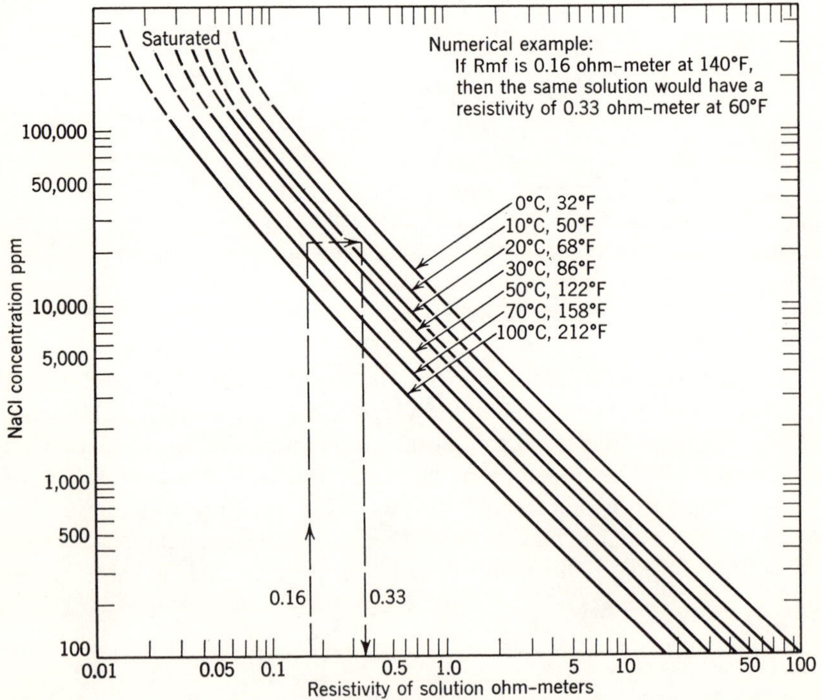


**Figure 8.27** Relation of temperature and NaCl concentration to resistivity of the solution.

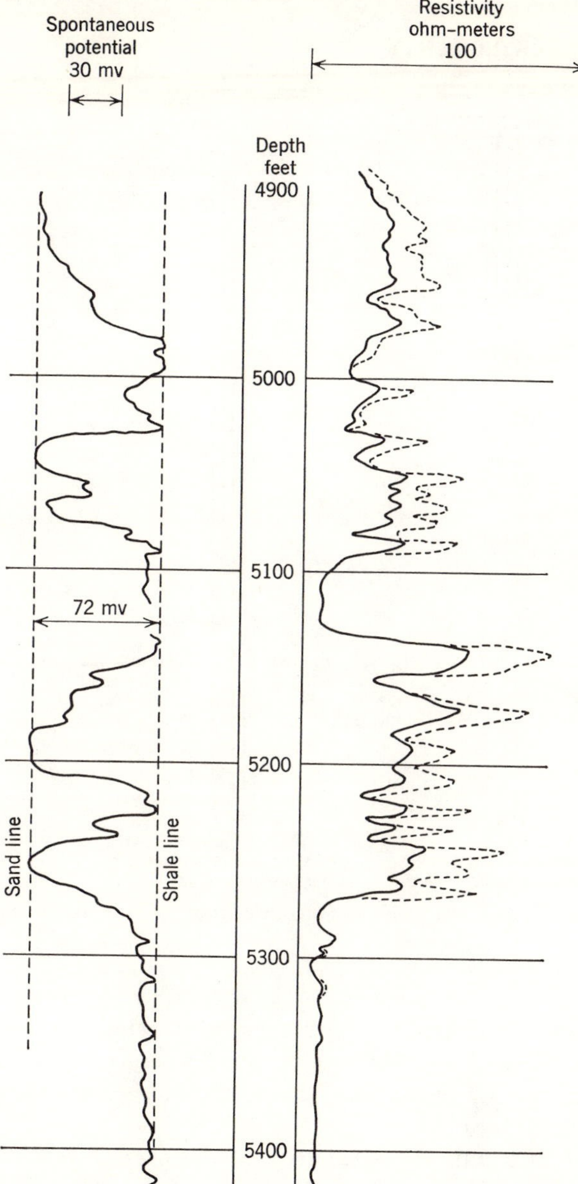


Figure 8.28 Portion of an electric log of a well penetrating sand and shale. Calculations show a method of estimating the dissolved solids content of brackish and saline water in aquifers.

*Numerical example*

From log:

$R_m = 1.36$ ohm-meters at 71°F

Spontaneous potential = −72 mv

Estimated:

Temperature at 5000 ft = 120°F

$R_{mf} = 0.8\ R_m = 1.09$ ohm-meters at 71°F

From Table 8.4:

$K = -77$ mv

Using Equation 8.5:

$$\text{S.P.} = K \log_{10} \frac{R_{mf}}{R_w}$$

Solving for $R_w$:

$R_w \approx 0.12$

From Figure 8.27:

Estimated salinity = 53,000 ppm as NaCl

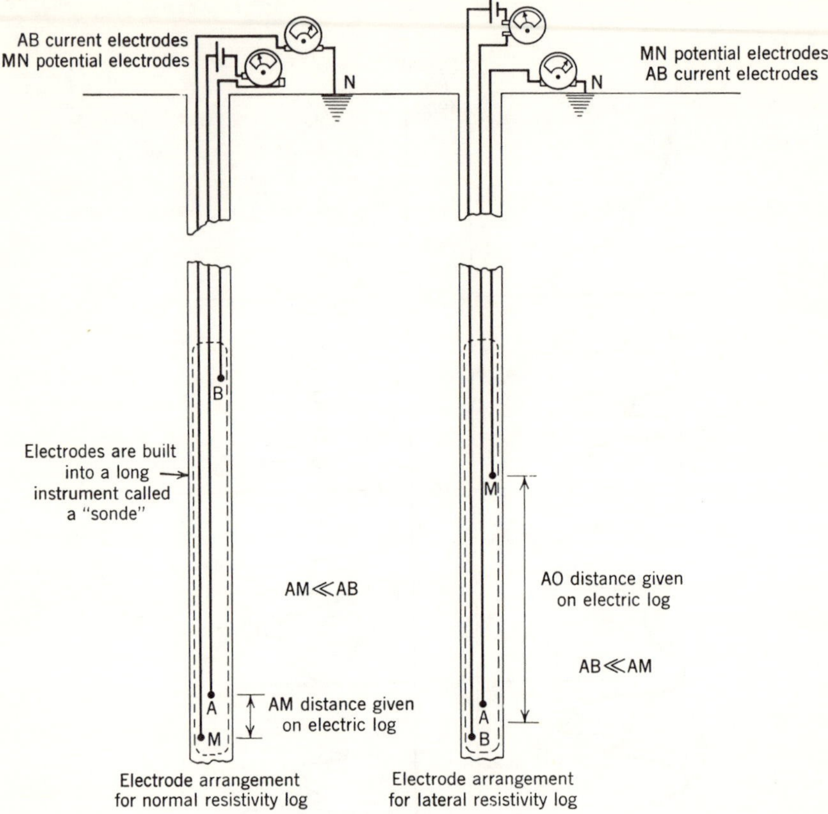


Figure 8.29 Schematic diagram showing electrode arrangements for resistivity logging.

through electrodes in the sonde. A large number of electrode spacings and arrangements are used in subsurface resistivity logging; some of these are shown in Figure 8.29. Most resistivity logs use a combination of three or four electrode arrangements. This is done to achieve varying depths of current penetration.

Resistivity of a fluid-saturated rock to the passage of electricity is a function of three primary factors. First, the salinity of pore fluid, second, the porosity of the rock, and, third, the temperature of the rock and fluid. Other factors which are of importance are the mineralogy of the solid material and the geometry of the pore spaces. Equations that express the relations among some of the major variables are

$$F = \frac{R_o}{R_w} \tag{8.8}$$

and

$$F = \frac{C}{\theta^m} \tag{8.9}$$

The value $F$ is called the formation factor, $R_w$ is the resistivity of the formation water, $R_0$ is the bulk resistivity of the formation, $\theta$ is porosity, and $C$ and $m$ are constants depending on the mineralogy and pore geometry of the rock. Values of $F$ vary from 3 to 200 for most natural materials. Most values of $C$ are close to 1.0, but values of $m$ vary from about 0.3 in unconsolidated sediments to 2.2 in consolidated rocks.

Resistivity logs are used to interpret rock types, to aid in correlation, and to estimate the chemical character of pore fluid [1,15,16,24]. As can be seen from Equation (8.9), the dense, nonporous rocks, such as limestone, will have very high resistivities. In contrast, highly porous materials saturated with saline water will have low resistivities. Clay minerals will generally reduce the resistivity because ions which cluster on the mineral surfaces will greatly increase the current-carrying capacity of the sediment. In most logs extremely low resistivities are recorded opposite shale and clay beds.

As in spontaneous potential logs, the resistivity log will commonly have distinctive configurations for certain beds or horizons. Characteristic forms of the curve may correspond to lithologic changes too subtle to detect by normal techniques of geologic logging. Thus, the resistivity log along with the potential log constitutes one of the most useful means of lithologic correlation.

Equation 8.8 can be used in some regions to estimate water quality from sand or sandstone aquifers. A rough estimate of $F$ can be obtained by assuming a reasonable porosity for the aquifer of interest. The true formation resistivity, $R_s$, can be calculated from the resistivity curves. Usually a wide electrode spacing combined with a very thick, homogeneous aquifer will give by direct inspection resistivity values close to the true resistivity. For most work, however, elaborate corrections for various factors must be made. Bore-hole diameter, electrode arrangement and spacing, mud resistivity, depth of invasion of mud filtrate, and aquifer thickness are some of the most important factors (16). Elaborate interpretation charts have been prepared by Schlumberger Corporation which take into account these factors. Several rules are given in Figure 8.30 which allow the approximation of true resistivity with sufficient accuracy for most ground-water work.

In calculating the resistivity of the water, $R_w$, it is well to remember that $R_w$ is the resistivity of the water at the temperature of the aquifer. Thus the equivalent ppm NaCl from Figure 8.27 is obtained from the curve representing the formation temperature.

An important modification of the resistivity log has been perfected with the contact log. In the conventional resistivity log the electrodes are more or less centered in the bore hole and are thus surrounded by drilling mud. Furthermore, electrode spacing varies from $1\frac{1}{2}$ to 20 feet. In contrast, contact logging employs a rubber pad which is pressed against the walls of the hole. The electrodes are placed in the pad and are spaced

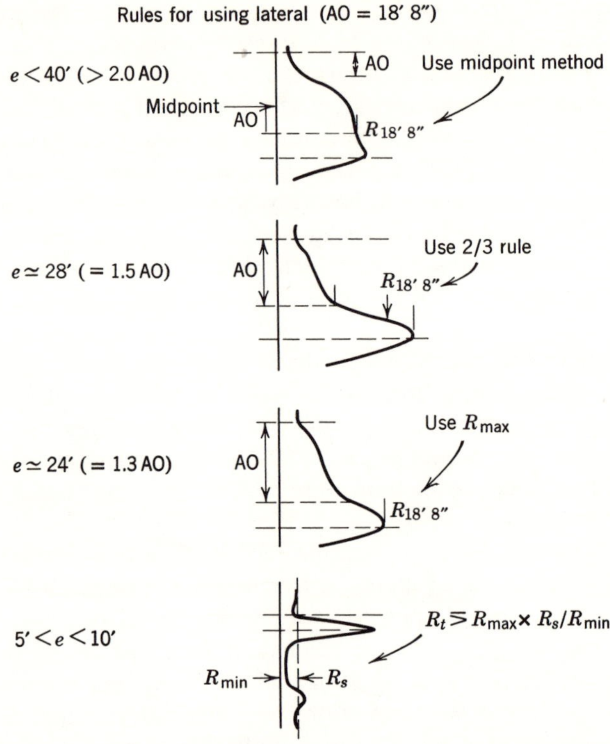


Figure 8.30 Rules for estimating the true formation resistivity, $R_t$, from resistivity logs. Key to symbols: bed thickness ($e$), maximum resistivity ($R_{max}$), minimum resistivity ($R_{min}$), and resistivity of shale ($R_s$). (Courtesy Schlumberger Corporation.)

only a few inches or less apart. The effect of the mud column is removed, and effective penetration of the resistivity measurements is limited to a few inches. The shallow penetration allows a detailed study of filter-cake development and resistivity of the invaded zone. The filter-cake development, in turn, indicates the permeability of the formation because more permeable formations will allow greater invasion of mud and therefore will develop a thicker mud cake on the walls of the drill hole. The correct resistivity of the invaded zone is a function of the formation factor. If the formation is free from clay, the following relation is found to be true:

$$F = \frac{R_0}{R_w} = \frac{R_i}{R_{mf}} \tag{8.10}$$

in which $F$, $R_0$, and $R_w$ are as defined previously, $R_i$ is the resistivity of the invaded zone, and $R_{mf}$ is the resistivity of the mud filtrate. The value of $R_{mf}$ is measured at the surface by filtering a sample of the drilling mud and measuring the resistivity of the filtrate in a Wheatstone bridge. The value of $R_{mf}$ is then corrected to formation temperature by Figure 8.27. True values of $R_0$ and $R_i$ are read from the resistivity and contact logs, respectively. Thus $R_w$ is left as the only unknown in Equation 8.10. To read correct values of $R_i$ from the contact log, corrections for mud-cake thickness, depth of invasion, and variations of $R_{mf}$ must be made [16]. Well-logging companies that perform contact-logging services also supply interpretation charts which take these factors into consideration.

A major advance in resistivity logging has been the development of induction logging. In this technique there are no electrodes in direct contact with the drilling mud; instead, the energy in the form of alternating current is supplied to a transmitter coil. The magnetic field which is produced will give rise to eddy currents flowing in the region around the bore hole. These currents in turn produce their own magnetic fields which are detected by a receiver in the instrument. The strength of the signal received by the instrument is proprtional to the conductivity or inversely proportional to the resistivity of the formation. The induction instrument is less affected by the mud column in the hole and by beds above and below the instrument than in the case of conventional resistivity logs. The induction log is most successful with high-resistivity-drilling muds such as oil-base muds.

### ACOUSTIC LOGS

The acoustic log measures the speed of sound in the rocks near a bore hole. The logging techniques were originally developed to assist with the interpretation of seismic records obtained by surface geophysical methods. Inasmuch as the travel time of the sound is strongly dependent on the porosity, the acoustic log is widely used to estimate porosity. The present techniques have been applied most widely to moderately porous, well-compacted sandstones, limestones, and dolomites. Some development work has been done on an acoustical log that will locate fractures in dense, consolidated rocks. If successful, such an instrument could have widespread use in hydrogeologic work in many regions.

### RADIOACTIVITY LOGS

Radioactivity affords another useful basis for geophysical logging. Unstable natural isotopes of elements such as thorium, uranium, radium, and actinium decay to more stable elements (see Chapter 5). In the decay process, alpha, $\alpha$; beta, $\beta$; and gamma, $\gamma$, radiation are emitted. Both alpha and beta radiation are charged particles which are stopped relatively easily by matter. Gamma radiation, in contrast, is very penetrating and is the radiation measured in subsurface logging.

Natural gamma radiation is largely owing to the presence of the unstable isotopes of uranium, thorium, and potassium, and their various decay products. The radioactivity is measured by a slowly moving instrument which houses a gamma detector, usually a scintillation counter. The gamma rays which reach the counter are detected by a small flash of light produced on a thin crystal. The light is intercepted by a photomultiplier which in turn sends an electric current to a recorder at the surface. The current recorded is proportional to the radioactivity encountered in the hole. Inasmuch as gamma emission by unstable isotopes is a randomly occurring event, the accuracy of the count is increased if the instrument remains in a given position for a greater period of time.

Logs of natural gamma radiation are used for correlation between wells and to help interpret lithology [15]. The logs are particularly useful for wells which have casings already installed, because the metal casing will have only a slight effect on the results. There are no fixed rules for interpreting lithology; however, experience in restricted areas will enable local criteria to be established. In general, organic shale will have the highest gamma activity. Rhyolitic tuff and certain sandstones may also have very high activities. Shale, shaly limestone, and shaly sandstone have moderate activities. Sandstone, limestone, and dolomite generally have low activities. Salt and coal have low activities [25].

Gamma logs are also used for depth control in well construction. Radioactive markers can be placed in the casing which will serve as depth markers in the well. If cement is used on the outside of the casing, its position can be checked by gamma logging provided proper types of radioactive material are mixed with the cement.

The natural neutron flux is very low, so artificially produced neutrons can be used in logging without the problem of correcting for natural background. The source of artificial neutrons is produced most conveniently with beryllium and an active alpha emitter such as plutonium, polonium, or radium. Neutrons are emitted when the beryllium absorbs alpha particles. Fast neutrons radiating outward from the source are slowed by the environment until they become thermal and can eventually be captured. Neutrons are slowed most effectively by hydrogen nuclei that are in water, minerals, or hydrocarbon. Capture is effected by potassium, iron, chlorine, and other nuclei in the natural media surrounding the bore hole. Gamma radiation is emitted on capture. The neutron log either measures the artificially induced gamma radiation or the slow neutrons. The detector is shielded from the source, so most of the neutrons or gamma rays arriving at the counter come through the media. Abundant hydrogen will cause slowing and capture of the neutrons near the source so only a low level of activity will be registered by the counter. A small amount of natural hydrogen will allow neutrons to penetrate to the vicinity of the counter with a resultant measure of a higher activity. Inasmuch as most natural hydrogen is in water, the activity registered on the log will be inversely proportional to the water content of the natural media near the bore hole. If the formations are fully saturated, the activity also will be inversely proportional to the porosity.

The neutron log is widely used to measure moisture changes in the unsaturated zone above the watertable. Small hand-operated probes are available for near surface work. The method has the advantages of being relatively rapid and accurate as well as being capable of making measurements without disrupting the soil beyond the initial installation of a small-diameter casing.

As in other geophysical logs, the neutron log is useful for stratigraphic correlation and aids in interpreting lithology. Both the neutron log and the gamma log are more flexible than other logs. They can be made in cased holes and holes drilled with fluids such as air and oil which cause difficulties with conventional logs.

Neutron logs must be corrected for bore-hole diameter and, if cased, details of well construction. In this regard, it is usually advisable to have a continuous log of hole diameter, or, as it is commonly called, a caliper log. Some rocks, notably bentonitic shale, cave into the hole during drilling, and caved zones with more than twice the bit diameter are common.

### OTHER TYPES OF LOGS

A host of minor geophysical, or wire-line, logging techniques exists [2,15,17]. Of these, temperature, caliper, hole-direction, fluid-resistivity, and fluid-velocity logs are most useful in hydrogeology. Temperature logs help with the interpretation of resistivity and potential logs, as well as being an important method of locating gas leaks and cemented zones (Figure 8.31). Caliper logs provide a continuous record of hole diameter which is of use in interpreting other geophysical logs and in estimating

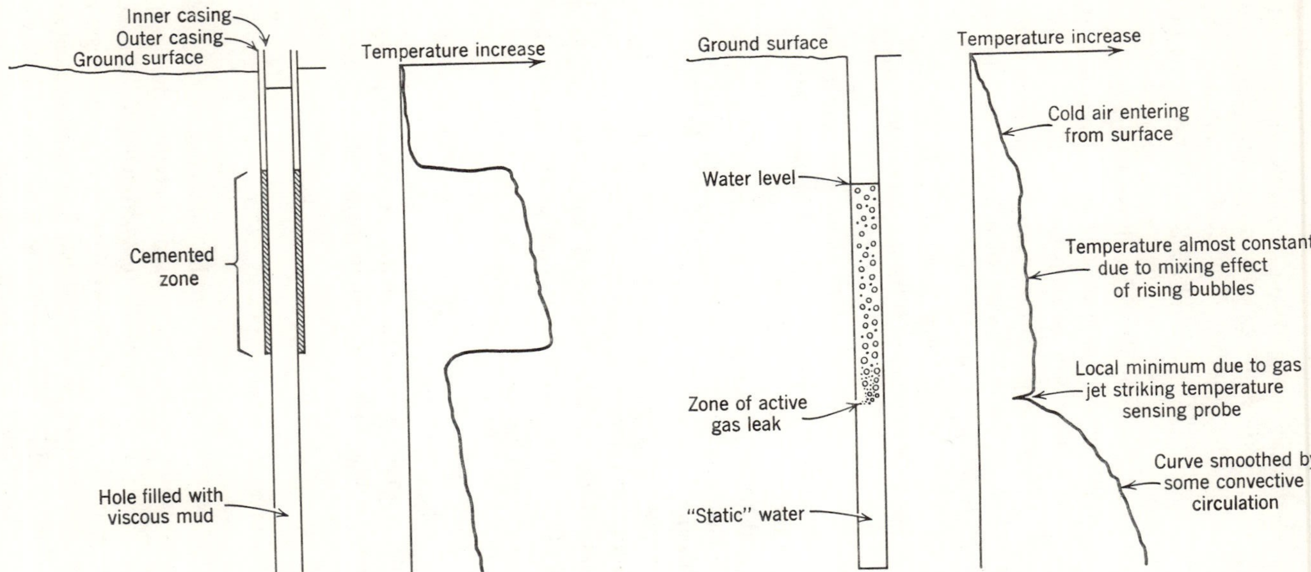


Figure 8.31 Schematic diagram showing temperature logs with anomalies caused by (*a*) heat given off by cement as it hardens and (*b*) expanding gas entering the well. Inasmuch as the well has a moderately large diameter, convection currents will be active within the well and will tend to smooth local anomalies.

cement and gravel requirements for the well. Hole direction can be used to calculate exact elevations of lithologic or hydrologic boundaries encountered by other logs. The electrical resistivity of fluids in wells is an important aid in determining zones of contamination in existing wells (Figure 8.32). Fluid-velocity surveys assist in determining the hydraulics of existing wells as well as the hydraulics of the aquifers (Figure 8.32).

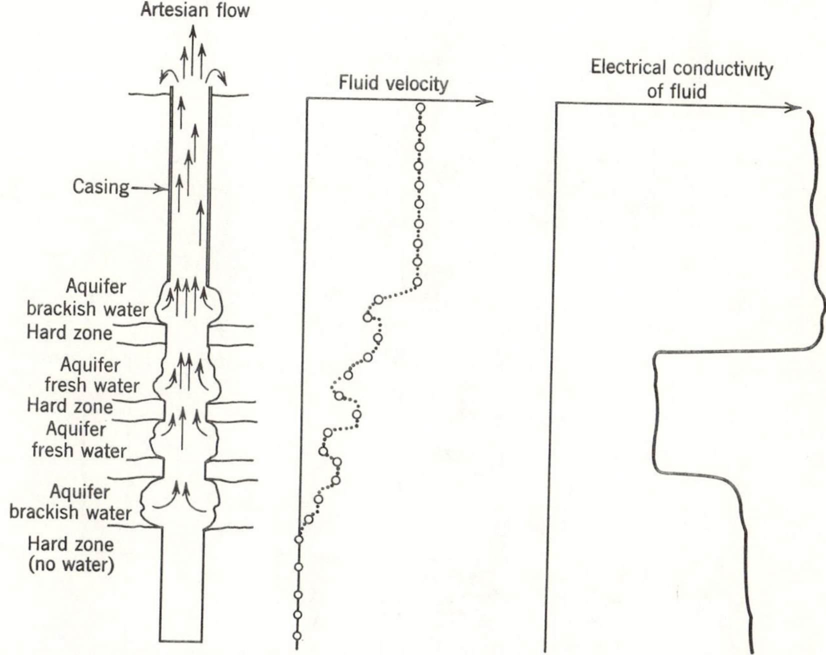


Figure 8.32 Fluid velocity and electrical conductivity logs of a hypothetical artesian well. Restrictions of the well diameter in the uncased portion cause local zones of maximum water velocity. Electrical conductivity log indicates that fresh water enters the well only through the two middle aquifers.

Information is most reliable when based on a comparison of the results of several types of surveys (Figure 8.33). Only a few water wells, nevertheless, are logged by more than three or four different devices. Added expense and delay in finishing the construction of the well tend to offset the advantages of having the additional information. An increased awareness of the usefulness of various logging techniques, however, has stimulated the development of small devices specifically made for water wells. Some of the newer devices are relatively inexpensive and convenient and provide accurate information to depths of about 2000 feet.

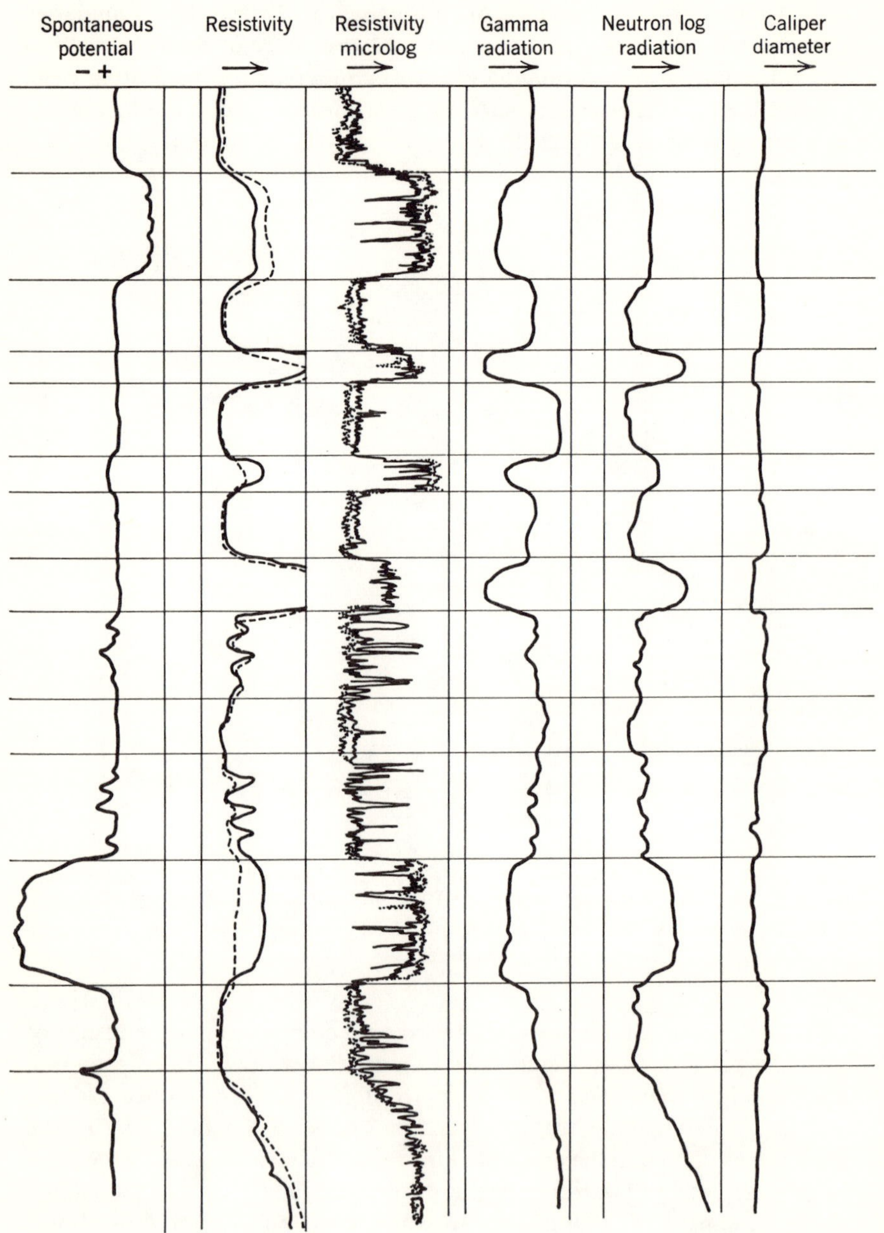

Spontaneous potential
− +
Resistivity
Resistivity microlog
Gamma radiation
Neutron log radiation
Caliper diameter

Hydrogeologic interpretation

Homogeneous clay, impermeable

Unconsolidated sand, permeable, fresh water

Homogeneous clay, impermeable

Dense rock, low porosity, impermeable, probably limestone

Shale, high gamma suggests dark shale, impermeable

Sandstone, permeable, brakish water

Shale, impermeable

Dense rock, low porosity, impermeable, probably limestone

Shale, streaks of sandstone, low permeability

Shale, homogeneous, impermeable

Shale, streaks of sandstone, low permeability

Sandstone, streaks of shale, permeable, saline water

Shale, few sand streaks, impermeable

Dense rock, weathered on upper part, high gamma suggests granite, very low permeability

Figure 8.33 A combination of six logs of a hypothetical test hole showing the hydrogeologic interpretations. The resistivity micrEOlog is also called a resistivity contact log.

## REFERENCES

1. Barnes, B. A., and P. Livingston, 1947, Value of the electrical log for estimating ground-water supplies and the quality of ground water: *Trans. Am. Geophys. Union*, v. 28, pp. 903–911.
2. Bays, C. A., and S. H. Folk, 1944, Developments in the application of geophysics to ground-water problems: *Illinois Geol. Survey Circ.* 108, 25 pp.
3. Bennison, E. W., 1947, *Ground water, its development, uses, and conservation:* St. Paul, Minnesota, Edward E. Johnson, 509 pp.
4. Berson, I. S., and others, 1959, Wave refraction by aquiferous sands: *U.S.S.R. Acad. Sci. Bull., Geophysics Ser.*, January and February. (English translation by the Am. Geophysical Union), pp. 17–29 and pp. 115–118.
5. Bonini, W. E., 1959, Seismic-refraction method in ground-water exploration: *A.I.M.E. Trans.* 1958, v. 211, pp. 485–488.
6. Brown, B. W., 1961, Stochastic variables of geologic search and decisions: *Geol. Soc. Am. Bull.*, v. 72, pp. 1675–1686.
7. Cambefort, H., 1955, *Forages et sondages, leur emploi dans les travaux publics:* Paris, Marc Eyrolles, 396 pp.
8. Compton, R. R., 1962, *Manual of field geology:* New York, John Wiley and Sons, 378 pp.
9. Dudley, W. W. Jr., and others, 1964, Geophysical studies in Nevada relating to hydrogeology: Desert Research Inst., *Univ. of Nevada, Tech. Report No.* 2, 46 pp.
10. Foster, J. W., and M. B. Buhle, 1951, An integrated geophysical investigation of aquifers in glacial drift near Champaign-Urbana, Illinois: *Econ. Geol.*, v. 46, pp. 367–397.
11. Gay, L. O., and M. Kosten, 1956, Some applications of geophysical methods to geological problems in the Gold Coast: *Gold Coast Geol. Survey Bull.* 21, 37 pp.
12. Hall, D. H., and Z. Hajual, 1962, The gravimeter in studies of buried valleys: *Geophysics*, v. 27, no. 6 Part 2, pp. 939–951.
13. Heiland, C. A., 1946, *Geophysical exploration:* New York, Prentice-Hall, p. 1013.
14. Kelly, S. F., 1962, Geophysical exploration for water by electrical resistivity: *Jour. New England Water Works Assoc.* v. 76, pp. 118–189.
15. LeRoy, L. W. (editor), 1950, *Subsurface geologic methods*, 2nd Ed.: Golden, Colorado, Colorado School of Mines, 1166 pp.
16. Lynch, E. J., 1962, *Formation evaluation:* New York, Harper and Row, 422 pp.
17. ——, and E. A. Breitenbach, 1964, Recent developments in formation evaluation: *World Oil*, v. 158, pp. 110–118.
18. Maybey, D. R., 1960, Gravity survey of the western Mojave Desert, California: *U.S. Geol. Survey Prof. Paper* 316-D, pp. 51–73.
19. McDonald, H. R., and D. Wantland, 1961, Geophysical procedures in ground water study: *Am. Soc. Civil Engineers Trans.*, v. 126, pp. 122–135.
20. McGinnis, L. D., and J. P. Kempton, 1961, Integrated seismic, resistivity, and geologic studies of glacial deposits: *Illinois Geol. Survey Circ.* 323, 23 pp.
21. McGinnis, L. D., J. P. Kempton, and P. C. Heigold, 1963, Relationship of gravity anomalies to a drift-filled bedrock valley system in northern Illinois: *Illinois State Geol. Survey Circ.* 354, 23 pp.
22. Norris, S. E., and H. C. Spicer, 1958, Geological and geophysical study of the preglacial Teays Valley in west-central Ohio: *U.S. Geol. Survey Water-Supply Paper* 1460-E, pp. 199–232.

23. Orellana, Ernesto, 1961, Criterios erroneos en la interpretación de sondeos electricos *Revista de Geofisica*, v. 20, pp. 207–227.
24. Patten, E. P., and G. D. Bennett, 1963, Application of electrical and radioactive well logging to ground-water hydrology: *U.S. Geol. Survey Water-Supply Paper* 1544-D, 60 pp.
25. Russell, W. L., 1941, Well logging by radioactivity: *Am. Assoc. Petroleum Geologists Bull.*, v. 25, pp. 1768–1788.
26. Shaw, S. H., 1963, Some aspects of geophysical surveying for ground water: *Jour. Inst. Water Engineers* (London), v. 17, pp. 175–188.
27. Soske, J. L., 1959, The blind zone problem in engineering geophysics: *Geophysics*, v. 24, no. 2, pp. 359–365.
28. Spicer, H. C., 1952, Electrical resistivity studies of subsurface conditions near Antigo, Wisconsin: *U.S. Geol. Survey Circ.* 181, 19 pp.
29. Spiegel, Z., and B. Baldwin, 1963, Geology and water resources of the Santa Fe Area, New Mexico: *U.S. Geol. Survey Water Supply Paper* 1525, 258 pp.
30. Stevens, P. R., 1963, Examination of drill cuttings and application of resulting information to solving of field problems on the Navajo Indian Reservation, New Mexico and Arizona: *U.S. Geol. Survey Water-Supply Paper* 1544-H, pp. 3–13.
31. Swartz, J. H., 1940, Geophysical investigations on Lanai in H. T. Stearns, Geology and ground-water resources of Hawaii: *Terr. Hawaii Division of Hydrography Bull.* 6, pp. 97–115.
32. Tattam, C. M., 1937, The application of electrical resistivity prospecting to ground water problems: *Colorado School of Mines Quart.*, v. 32, pp. 117–138.
33. Vacquier, V., and others, 1956, *Prospecting for ground water by induced electrical polarization:* New Mexico Institute of Mining and Technology, Research and Development Division, 41 pp.
34. Vogt, E. Z., and R. Hyman, 1959, *Water witching U.S.A.:* Chicago, University of Chicago Press, 248 pp.
35. Ward, L. K., 1946, The occurrence, composition, testing, and utilization of underground water in South Australia, and the search for further supplies: *South Australia Geol. Survey Bull. No.* 23, 281 pp.
36. Workman, L. E., and M. M. Leighton, 1937, Search for ground water by the electrical resistivity method: *Trans. Am. Geophys. Union*, v. 18, pp. 403–409.
37. Wyllie, M. J. R., 1949, Statistical study of accuracy of some connate-water resistivity determinations made from self-potential log data: *Am. Assoc. Petroleum Geologists Bull.*, v. 33, 1892–1900.
38. Zohdy, A. A. E., 1964, Earth resistivity and seismic refraction investigations in Santa Clara County, California: Unpublished Ph.D. Dissertation, Dept. of Geophysics, Stanford University, 132 pp.

# *chapter 9*

# GROUND WATER IN IGNEOUS AND METAMORPHIC ROCKS

## 9.1 *Introduction*

Few tasks in hydrogeology are more difficult than locating drilling sites for water wells in igneous and metamorphic rocks. Extreme variations of lithology and structure coupled with highly localized water-producing zones make geological and geophysical exploration difficult. Soil and vegetation commonly cover outcrops and make the necessary detailed geologic mapping impossible. Furthermore, small fractures which yield most of the water to wells in unweathered rocks are not detected by normal geophysical techniques. It is not surprising, therefore, that the percentage of unsuccessful wells in some regions may be high even though well sites have been located by experienced hydrogeologists.

Igneous and metamorphic rocks are at or near the surface in more than 20 per cent of the land surface of the world. In northern Europe and North America the bedrock is mantled with a variable thickness of glacial deposits whose aquifers yield most of the ground water used in these regions. Large parts of most other continents, however, must rely more heavily on small supplies of water from igneous and metamorphic rocks. Even in these regions ground-water development is commonly confined to alluvium within small valleys, and large interfluvial areas have remained without adequate supplies of water. The search for new supplies by governmental agencies, particularly in India, Australia, and the Union of South Africa, has been responsible for extensive drilling within the less favorable areas. This experience together with more isolated work in the United States, Sweden, Japan, and other countries has yielded data from which the more important conclusions of this chapter have been drawn.

## 9.2 *Metamorphic and Plutonic Igneous Rocks*

Common metamorphic rocks such as phyllite, slate, and schist, and common plutonic igneous rocks such as granite, diorite, and gabbro can only be classified properly by careful petrographic study. Inasmuch as important hydrologic properties can be inferred through studies of subtile petrographic variations, this work should be as thorough as possible. As an example, a single metamorphic unit in Maryland was found to yield almost twice as much water from an albite-rich facies as from an oligoclase-rich facies [6]. Also, the location of permeable zones near faults hidden under thick soil is possible through assiduous mapping of petrographic units in the vicinity of faults.

Limitations of time, money, and personnel, unfortunately, restrict the scope of petrographic studies that can be made in conjunction with most hydrogeologic studies. It is, therefore, important that field identifications of rocks be both accurate and rapid. Even though units mapped are not labeled precisely, useful information can be gained provided that the mapping itself is accurate.

### POROSITY AND PERMEABILITY

Solid pieces of fresh metamorphic and plutonic igneous rocks have porosities of less than 3 per cent and most commonly less than 1 per cent [17]. The few pores that are present are small and generally are not interconnected. As a result, permeabilities are so small that they can be regarded as zero in almost all practical problems. Appreciable porosities and permeabilities, however, are developed through fracturing and weathering of the rock. For example, laboratory determinations of the permeability of unfractured samples of metasedimentary rocks from northern Michigan yielded median values generally less than 0.01 millidarcys (Table 9.1). Aquifer tests, however, demonstrated that the same rocks as a whole were more than one thousand times as permeable as individual unfractured samples [44]. The aquifer tests further indicated that permeabilities parallel with the strike of the beds were two to three times the average permeability.

Fractures that are not associated with pronounced faults produce only a small increase in the overall porosity of rocks. Experience with grout injection in dam foundations [4,24] and observations of exposed rock in tunnels and road cuts indicate that the total volume of open space within fractures is only a small percentage of the volume of the rock. Cracks are almost always nearly closed, and widths in excess of 2 millimeters in fresh undisturbed rock are not common.

Well yields suggest that permeabilities produced by fracturing of unweathered rock within a few hundred feet of the surface generally range from 0.001 to 10.0 darcys. Microscopically, the permeability varies from nearly zero in the solid rock to as much as several hundred darcys along highly fractured zones. Owing to the single orientation of most water-bearing fractures, the permeability of the rock as a whole is strongly anisotropic.

*Table 9.1 Permeability of Metasediments from Marquette Mining District, Michigan [Stuart, Brown, and Rhodehamel* [44]]

| Rock Type | Number of Samples | Permeability of Unfractured Samples, millidarcys | | | |
|---|---|---|---|---|---|
| | | Mean | Median | Highest | Lowest |
| Iron Formation and iron ore | 36 | 1.65 | 0.006 | 38.0 | 0.00011 |
| Graywacke | 5 | 0.033 | 0.003 | 0.15 | 0.00027 |
| Slate | 9 | 0.006 | 0.0013 | 0.045 | 0.0005 |
| Chert | 1 | ... | 0.00019 | ... | ... |
| Slate with quartz seams | 1 | ... | 181.0 | ... | ... |
| Mica schist | 1 | ... | 0.0021 | ... | ... |
| Quartzite | 1 | ... | 0.0019 | ... | ... |
| Conglomerate | 1 | ... | 0.028 | ... | ... |

Effects of weathering may extend more than 300 feet into bedrock in regions of intense weathering. Depths of weathering of from 5 to 50 feet, however, are normally encountered. Hydrated minerals in weathered rock at the surface will form loose aggregates which have porosities in excess of 35 per cent (Figure 9.1). The porosity decreases with depth to zones in which the original rock-forming minerals are only partly altered. In the last few feet above fresh bedrock, the minerals are only slightly hydrated, but this is enough to produce differential expansion between mineral grains which in turn creates a porosity of 2 to 10 per cent. In general, the greatest permeability is found within the partly decomposed rock below the zone of abundant clay-size material. Well yields suggest that permeabilities in the lower part of the weathered rock are roughly an order of magnitude greater than in the unaltered rock.

Many metamorphic rocks and a small number of igneous rocks contain carbonate minerals which are subject to relatively rapid solution by circulating ground water. Even though numerous solution cavities may

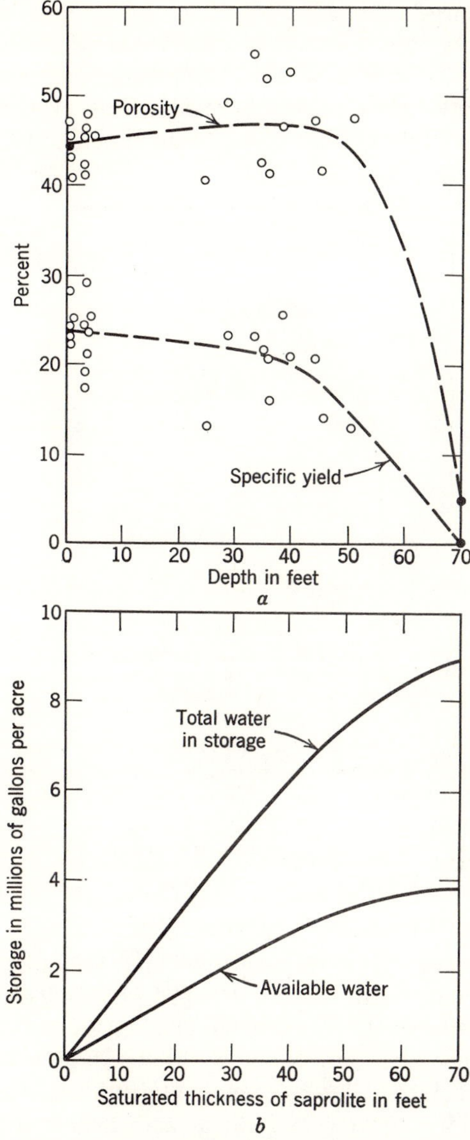


Figure 9.1 Results of laboratory tests on core samples of highly weathered metamorphic rocks from the Georgia Nuclear Laboratory area north-northeast of Atlanta, Georgia. (*a*) Porosity and specific yield. (*b*) Total water in storage and water potentially available to wells. (Diagrams from Stewart [43].)

form in rocks such as marble and the local permeability may be very large, the pore space of large volumes of rock is probably not greater than 2 to 5 per cent. This is because solution is highly localized along fractures and does not affect the rock as a whole. Depressions on granodiorite and quartzite at the land surface in certain regions were probably formed by the direct solution of the rock-forming silicate minerals [19,33]. Whether

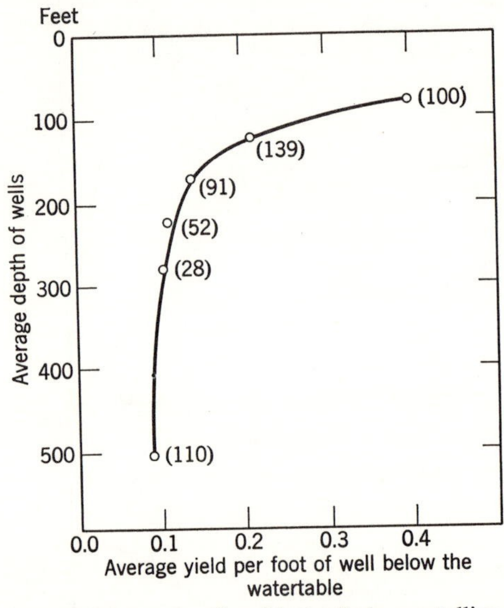


Figure 9.2 Decrease in yield of wells with depth in crystalline rocks of the Statesville area, North Carolina [20]. Numbers near points on curve indicate the number of wells used for each average shown.

or not solution is an important factor in the enlargement of narrow fractures at depth in noncarbonate rocks remains to be demonstrated.

### PERMEABILITY AS A FUNCTION OF DEPTH

The average permeability of metamorphic and plutonic igneous rocks decreases rapidly with depth (Figure 9.2) [4,6,11,23]. This decrease is a combined effect of the weight of overlying rock and the tendency of surface disturbances to penetrate only a short distance into bedrock. Joints, faults, and other fractures will tend to close at depth because of the weight of overlying material. Active rock flowage, nevertheless, is not important in most rock within a few miles of the surface. Some openings, which are of interest to hydrogeologists, can, therefore, exist at all depths. Water flow into some mines and tunnels which are hundreds, and in some cases thousands, of feet below land surface indicates that openings at great depths are large enough to supply water to wells.

Surface disturbances that produce rock permeability include landslides, rock falls, erosional unloading of underlying rock, chemical weathering, root and frost wedging, and various activities of man. Landslides and rock falls affect only the uppermost part of the bedrock and produce local deposits of rock debris which can be important zones of rapid ground-water recharge, and, if saturated, good aquifers. Erosional unloading is thought to produce sheet structure in massive granitic rocks. The narrow fractures which are more or less parallel to the surface in sheet structure are important local sources of water [18]. Distance between fractures increases rapidly with depth [14] so that they are probably rare or absent below a depth of a few hundred feet. Chemical weathering, as has been mentioned, is usually confined to the upper 300 feet of the rock. Frost and root wedging are effective only a few tens of feet below the surface.

### WELL YIELDS

In general, yields of wells are low in almost all metamorphic and plutonic igneous rocks. Average, or mean, yields for groups of wells in various regions are most commonly between 10 and 25 gpm. Median yields are considerably less. Deeply weathered rocks with substantial local recharge may have mean yields as high as 50 gpm [7,20,21]. Variations in yields within a given area are large, but there is a tendency for most yields to be small with only a limited number of high yields. Histograms of yields are, therefore, strongly skewed to the right (Figure 9.3). From 2 to 10 per cent of the wells will commonly have yields of more than 50 gpm. On the other hand, as many as 50 per cent of the wells in some areas are classed as failures; although 5 to 20 per cent failures are more common.

Differences in well yields tend to reflect differences in degree of weathering or fracturing rather than inherent differences of mineralogy or fabric within the rock. Meier and Peterson [27] cite an example in Sweden where three wells along a fault yielded between 15,000 and 25,000 liters/hour, whereas the average for the region having the same rock type was only 3000 liters/hour. Data from various regions (Table 9.2) fail to show large differences in mean yields that might be related to rock type alone. The smallest yields appear to be from phyllite and other soft foliated metamorphic rocks in which the fault and joint openings close rapidly with depth. Largest yields are from carbonate rocks in which circulating water tends to enlarge fractures by solution of calcite and dolomite.

Even though rock type alone does not always have an important effect on yield, the tectonic history and subsequent weathering of each metamorphic and igneous rock unit in a region will be distinct. Thus, in one area the average yield of a muscovite schist could be twice as large as an

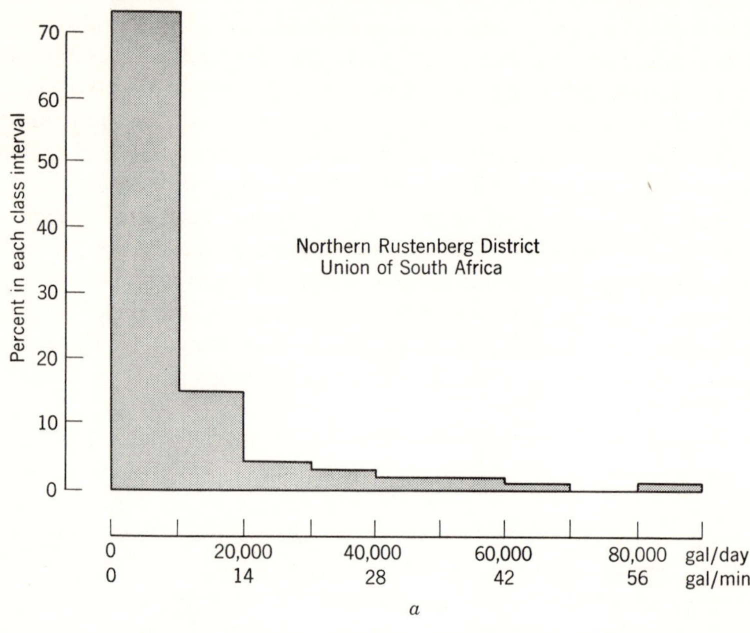

70
60
50
40
30
20
10
0
Percent in each class interval
Northern Rustenberg District
Union of South Africa
0
20,000
40,000
60,000
80,000
gal/day
0
14
28
42
56
gal/min


*a*

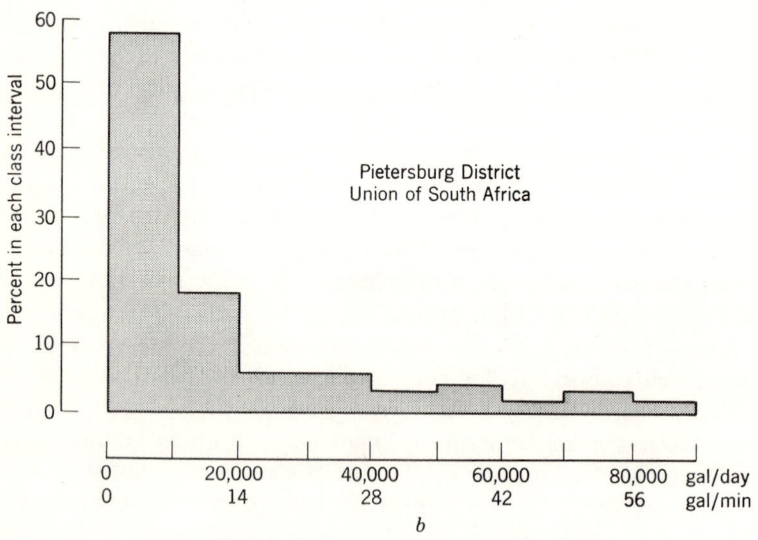

60
50
40
30
20
10
0
Percent in each class interval
Pietersburg District
Union of South Africa
0
20,000
40,000
60,000
80,000
gal/day
0
14
28
42
56
gal/min


*b*

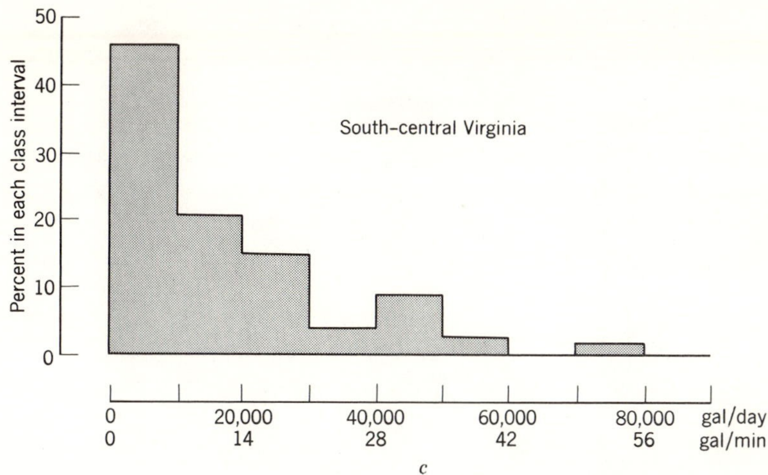


*c*

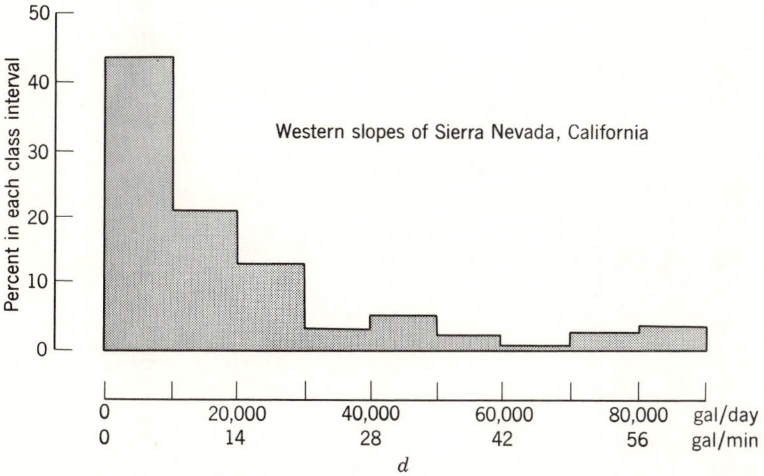


*d*

Figure 9.3 Histograms of well yields from regions of metamorphic and plutonic igneous rocks.
(*a*) Wells in the Rustenberg District are mostly in weathered granite with 16 per cent in diabase dikes. Of the 234 wells tabulated, 50 per cent were classed as failures [9].
(*b*) Wells in the Pietersburg District are in a variety of schists, gneisses, marbles, and basic dikes. Of the 269 wells tabulated, 25 per cent were classed as failures [9].
(*c*) Wells in Virginia are mostly in gneiss and schist. Of the 303 wells tabulated, 3 per cent had yields of more than 90,000 gal/day and 6 per cent had yields of less than 1000 gal/day [22].
(*d*) Wells in the Sierra Nevada are mostly in granodiorite. Of the 239 wells tabulated, 8 per cent had yields of more than 90,000 gal/day, 16 per cent had yields of 1000 gal/day or less, and 8 per cent were classified as "dry" holes [4].

*Table 9.2 Mean Yield of Wells in Various Rock Types, yield in gpm*

| | Lithology | | | | | | | |
|---|---|---|---|---|---|---|---|---|
| | Granite | Gabbro | Gneiss | Schist | Phyllite | Slate | Dolomite | Quartzite |
| Pretoria—Johannesburg Union of South Africa [9] | 10 | ... | ... | 25 | ... | ... | ... | ... |
| Northern Rhodesia [36] | ... | ... | ... | 15 | 6 | ... | 34 | ... |
| Western Rajasthan, India [45] | 6 | ... | ... | ... | ... | 8 | ... | ... |
| Sweden [27] | 9 | ... | 10 | ... | ... | ... | ... | 10 |
| United States: | | | | | | | | |
| Connecticut [11] | 13 | ... | 12 | 14 | ... | ... | ... | ... |
| Maryland [6] | ... | 11 | 12 | 24(1) | 8 | ... | ... | ... |
| Virginia [15,22] | 12 | ... | 17 | 12(2) | ... | ... | ... | ... |
| North Carolina [20] | 17 | 30(3) | 23 | 22 | ... | ... | ... | ... |
| Maine [1] | 25 | ... | ... | 20 | ... | 15(4) | ... | ... |
| Southern New England [2] | 38 | ... | 9 | 10 | ... | ... | ... | ... |

(1) Wissahickon Formation—schist, phyllite, and quartzite, "albite facies"
(2) Wissahickon Formation
(3) Also includes diorite
(4) Median yield of 91 wells is only 8 gpm

amphibole schist, whereas in another area the opposite might be true. Even the same rock type can have distinct yields in adjoining areas as might be the case with a fractured thrust sheet next to an autochthonous rock of the same composition.

Larger yields will be obtained from rocks in moist climates than in dry climates, other factors being equal [9]. This is probably owing to the fact

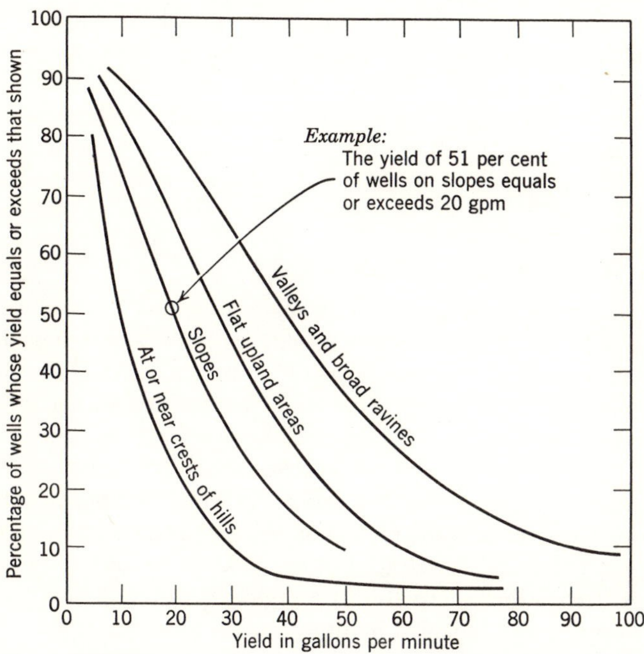


Figure 9.4 Cumulative frequency distribution of well yields in various topographic positions, Statesville area, North Carolina. (Redrawn from Le Grand [20].)

that depths to water are generally less in moist regions, and the water will saturate the more permeable rocks near the surface. Also, the greater amount of water circulating will increase permeability by accelerating near-surface weathering and increasing solution of minerals along fractures.

Topography has been found to be an important indication of well yields in certain regions (Figure 9.4) [6,20]. Wells on flat uplands and in valleys tend to yield larger amounts of water than wells on valley sides and sharp hill tops. The lack of water on or near the steeper slopes can be explained by the fact that erosion has removed much of the weathered and more permeable rock. Water levels are also further below the surface because ground water drains to points of discharge in adjacent lowlands.

Limited data suggest that the highest well yields are in or close to broad ravines [6,20]. Ravines are developed in many places along permeable fault zones, which explains the higher yield of the wells.

Despite the rather clear-cut relationships just discussed, all reported well yields are affected by testing methods and the type of well construction. This is particularly important to consider in comparing data from areas that have mostly domestic wells that are drilled only deep enough to recover a few gallons per minute with areas that have municipal and industrial wells drilled to much greater depths in order to recover a maximum amount of water. Values of average yields may, therefore, be too low unless depth, well diameter, and other details of construction are taken into account [2]. On the other hand, pumping tests are commonly terminated too soon so that the reported yields are much larger than the actual values.

### WELL LOCATION AND WATER DEVELOPMENT

The fact that a few wells in almost all regions are able to produce more than 50 gpm indicates that highly permeable zones exist and that geological and geophysical methods might be used to improve materially the chances of locating water in these zones. Few data exist to indicate the extent to which geological and geophysical methods can improve on "hit or miss" methods of locating wells which are practiced in most areas. Ward [50] documented several cases in South Australia in which geologists were able to locate water in metamorphic rocks but wells located by dowsing failed. Enslin [8] reported that five wells out of six which were located by surface electrical resistivity were successful in a region in South Africa in which only 10 per cent of wells located by other means were successful.

If rock exposures are quite numerous in the area of interest, detailed geologic mapping will be highly useful in determining the extent of jointing and the location of faults, dikes, and geologic contacts (Figure 9.5). In general, the most favorable water-bearing zones are in marble or dolomite which has been fractured by faulting and partly removed by solution. Next in importance are extensive fracture zones associated with faults. Many fault zones, however, which have been open in the geologic past, are almost completely recemented by quartz, calcite, or, less commonly, by other minerals. These faults in general should be avoided. Rocks which have been deformed considerably will tend to develop water-bearing fractures near mechanical discontinuities such as along hard, brittle, quartz veins which cut soft phyllite. If the vein or dike has the same mechanical properties of the enclosing rock, fracturing will generally be much less.

In many regions most of the faults and joints are nearly vertical. Spacing

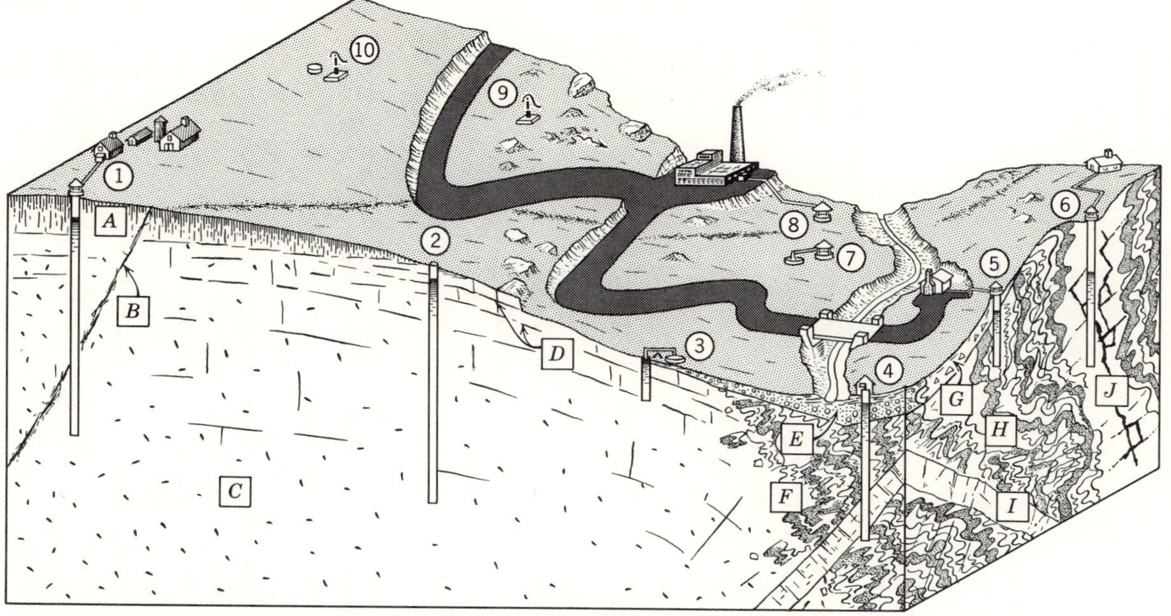


Figure 9.5. Hypothetical region showing the relationship between geologic features and expected well yields.

| Type of well | Use | Depth (feet) | Production (gpm) | Source of Water |
|---|---|---|---|---|
| 1. Drilled | Farm | 210 | 25.0 | Lower part of weathered granite and fault zone. Small amount from joints. |
| 2. Drilled | None | 200 | 0.1 | Very small amount from joints. |
| 3. Drilled | Stock | 30 | 0.5 | Small amount from joints. Water is artesian. |
| 4. Drilled | Observation | 125 | 15.0 | Lower part of alluvium and fractures and joints in and near dike. |
| 5. Drilled | Domestic | 80 | 1.5 | Lower part of colluvium and joints in schist. |
| 6. Drilled | Domestic | 130 | 45.0 | Cavernous zone in small body of marble. |
| 7. Dug | Stock | 20 | 4.5 | Alluvium. |
| 8. Drilled | Industry | 160 | 35.0 | Lower part of alluvium and same fault as in well #1. |
| 9. Dug | None | 15 | 0.2 | Small amount from joints. Well dry during droughts. |
| 10. Dug | Stock | 25 | 0.7 | Weathered granite. |

GEOLOGIC UNITS

A. Residual soil on granite.
B. Fault.
C. Granite.
D. Joints in granite.
E. Alluvium.
F. Contact between granite and schist.
G. Colluvium.
H. Schist.
I. Aplite dike.
J. Marble.

between joints will be generally from 0.5 to 10 feet, or equal to or greater than the diameter of most drilled wells. Wells that are almost vertical will, therefore, tend to intersect only one or two joint planes. For example, Gear [10] found in Uganda that after a water-bearing fracture was once encountered the chances of finding another fracture in the next 20 feet of drilling was only slightly greater than 1 per cent and the chances of finding another fracture in 140 feet was about 33 per cent. Horizontal "wells" in such a terrane will be much more successful because of the greater number of fractures which will be intercepted. The expense of drilling or excavating large vertical shafts needed to reach the depths necessary for constructing laterals within saturated rock is commonly much greater than the value of the water produced. Many horizontal "wells" have, nevertheless, been successful in deeply weathered rock where access shafts are not expensive to excavate [7] or in steep hillsides where horizontal drilling can start from the surface.

Indirect methods of locating water wells must be used in regions of heavy soil or vegetation cover. Generalizations concerning topography, well depths, and local rock types which were discussed in the preceding section serve as preliminary guides to well location. Detailed geomorphic studies may be able to locate areas of exceptionally thick weathered rock or areas in which alluvial aquifers are present. Aerial photographs commonly display slight variations in the tone of soil or vegetation, in drainage texture, and in the alignment of small ridges. These variations may indicate the presence of faults, intrusions, or bodies of carbonate rock.

Geophysical studies can be very useful in determining hidden geologic features; even though direct location of ground water itself is rarely possible. Magnetic surveys will determine the location of rocks relatively high in iron such as basic intrusives and "iron formations" in metasedimentary rocks. Mapping these rocks might assist in the location of faults. Also, basic rocks may be weathered at a greater depth than surrounding rocks, hence they may prove to be favorable locations for ground-water development. Seismic and electrical resistivity methods have been used more extensively than have magnetic methods in areas of metamorphic and plutonic igneous rocks. The prime application has been to outline the thickness of weathered rock. For example, near Lusaka, Northern Rhodesia, areas of low resistivity indicated thick mantle material filling the upper part of collapse features in dolomite [36]. Eight wells drilled into the collapse breccia had an average yield of 660 gpm as compared with an average of 34 gpm for the dolomite as a whole and about 18 gpm for all the rocks in the region. The extensive system of rather large openings in the collapse breccia was dramatically shown by the fact that small blind crustacea were pumped from one well from a depth of about 125 feet. The only other known occurrence of the species found, *Ingolfiella leleupi*, is in a cave in the Congo.

Aquifer tests in metamorphic and plutonic igneous rocks are commonly not subject to the general interpretations discussed in previous chapters. As pointed out by LeGrand [22], various combinations of fracture location, fracture width, and amount of ground water in storage near the well can give rise to almost any type of drawdown curve. A few of the possibilities are shown in Figure 9.6. A feature that is all too common in wells drawing water from fracture systems is a high or moderate initial yield that decreases rapidly with time. Usually the cause is insufficient storage of ground water near the well. It is wise, therefore, to locate wells so that they can draw water from weathered zones or from saturated alluvium or colluvium because these materials will contain from twenty to forty times the water per unit volume that is contained in the unweathered rock. If the biological purity of the water can be insured by sufficient filtration, wells may also be placed close enough to perennial streams or lakes to prevent depletion of the fracture systems.

### WATER QUALITY

The chemical quality of water from metamorphic and plutonic igneous rocks is almost always excellent. Exceptions are found in arid regions where salts may be concentrated in the recharge water by evaporation and in places where connate or ocean water has migrated into the fractures in the rocks. Water from dolomite and marble may have moderate to high hardness. Water in serpentine, dolomite, gabbro, amphibolite, and certain other rocks will have moderately high magnesium concentrations in comparison with calcium concentrations [26]. Rocks such as diorite and syenite which are low in quartz but high in content of other more soluble silicate minerals will have dissolved silica ranging from about 25 to 55 ppm in contrast with quartzite, marble, slate, and phyllite, which most commonly have less than 30 ppm silica. LeGrand [21] has shown that in North Carolina water from granite, gneiss, mica schist, and rhyolite tends to be slightly acid and has a median total dissolved solids of only 71 ppm and a median hardness of only 23 ppm. In contrast, water from gabbro, diorite, hornblende gneiss, and andesite has slightly alkaline water with a median total dissolved solids of 233 ppm and a median hardness of 145 ppm.

Biological contamination of ground water is a problem where soil is thin or absent over the water-bearing rocks. Even though fractures may be less than 1 millimeter wide, pathogenic organisms will move much more efficiently than in normal alluvial aquifers. The problem is aggravated by dug wells that penetrate only a short distance into the rock and are

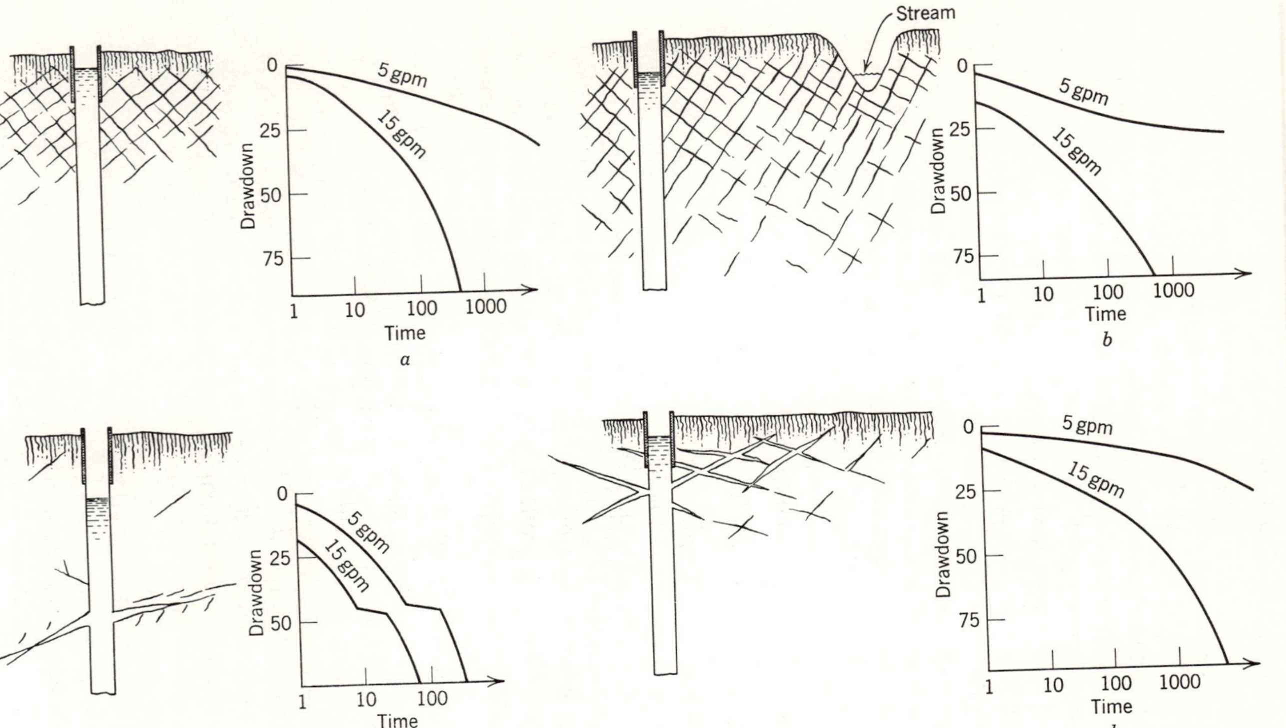


Figure 9.6 Hypothetical drawdown curves for wells in fractured crystalline rock. (*a*) Well drawing water from small fractures near the surface. A safe pumping rate for this well is less than 5 gpm. (*b*) Extensive small fractures connect this well with a nearby source of recharge. A safe pumping rate for this well is about 5 gpm. (*c*) This well is a failure except as a source for very small amounts of water. Drawdown curve reflects dewatering of well and large isolated fracture intercepted by well. (*d*) Large fractures that drain the porous weathered rock maintain a moderate yield of about 5 gpm in this well.

commonly not properly protected against direct storm runoff that enters through the well curbing into wells.

## 9.3 *Volcanic Rocks*

Volcanic rocks include materials having a wide range of hydrologic properties. Some recent basalt aquifers have close to the highest transmissivities known. This is in contrast with tuffs that generally have high porosities but very low permeabilities or dike rocks that have low permeabilities as well as low porosities. Although transmissivities of recent basalt and andesite are high, ground water may be very difficult to develop from aquifers. This is owing to the fact that ground water drains freely to points of discharge at streams or along the ocean and the depth to ground water may be excessive or water may even be locally absent. The attention of the hydrogeologist is directed, therefore, to impermeable zones which will impede the loss of water and cause the watertable to rise near the surface. In metamorphic and plutonic igneous rocks, on the other hand, attention was directed to the permeable zones.

### PERMEABILITY AND POROSITY

The porosity of unfractured volcanic rock varies from less than 1 per cent in dense basalt to more than 85 per cent in pumice [35]. Typically, rocks within dikes and sills will have less than 5 per cent porosity, dense massive flow rock will have from 1 to 10 per cent porosity, and vesicular volcanic rock will have values ranging from 10 to 50 per cent. Although porosity may be quite high, the permeability is largely a function of other primary and secondary structures within the rock. Joints caused by cooling, lava tubes [12], vessicles that intersect, tree molds [29], fractures caused by buckling of partly congealed lava, and voids left between successive flows (Figure 9.7) are some of the features that give recent andesite and basalt its high permeability. In addition to the features causing permeability, the porosity may be increased locally in the rock by weathering. Buried soils are a common feature of thick sequences of volcanic rock. Soils in most areas are less permeable than the volcanic rock and are important horizons for forming perched water. Typical zones of high porosity and permeability are shown in Figure 9.8.

If valleys are near volcanic eruptions, lava will flow down the valleys and bury any alluvium which may be present. Where the valleys contain streams from extensive drainage systems, large thicknesses of gravel may be present, which, on burial, can be important aquifers. Rivers blocked by lava will form lakes that eventually fill with silt, clay, or volcanic ash. These deposits can form important confining beds for the underlying

Figure. 9.7 Water flowing into Honolulu Board of Water Supply's water tunnel from space between successive pahoehoe flows. (Drawn from a photograph supplied by L. J. Watson, Honolulu Board of Water Supply.)

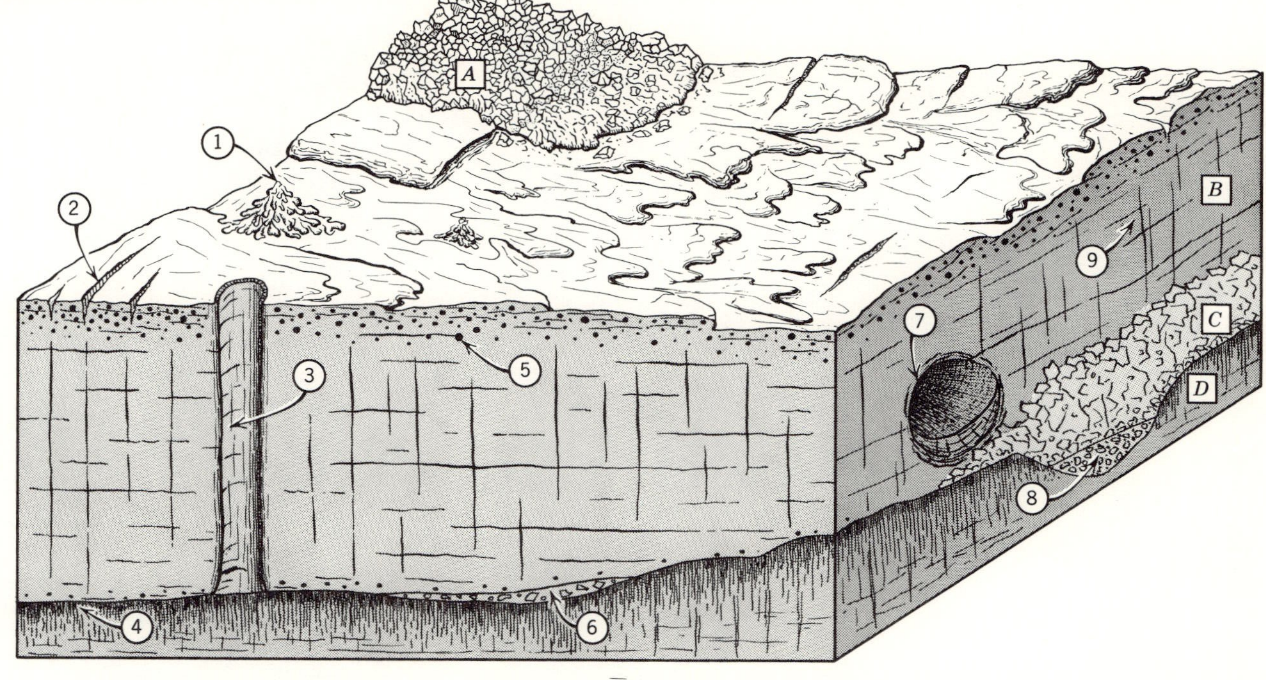


Figure 9.8 Hypothetical sequence of lava flows showing various features that produce permeability and porosity within basaltic rocks.

Features Producing Porosity

1. Orifice of spatter cone
2. Crack on small pressure ridge
3. Tree mold
4. Buried Soil
5. Vesicles
6. Small pocket of pyroclastic blocks
7. Lava tube
8. Buried stream gravel
9. Cooling joint

Sequence of Flows

A. Recent *aa* flow
B. Recent pahoehoe flow
C. Ancient buried *aa* flow
D. Very old buried pahoehoe flow

stream gravel. In areas of extensive volcanism, basins undergoing sedimentation may contain complicated sequences of alluvial, volcanic, and lacustrine material [25,48,51,52].

Volcanic rock in which interbedded sediments or pyroclastics are absent will have relatively low porosities if large volumes of rocks are considered.

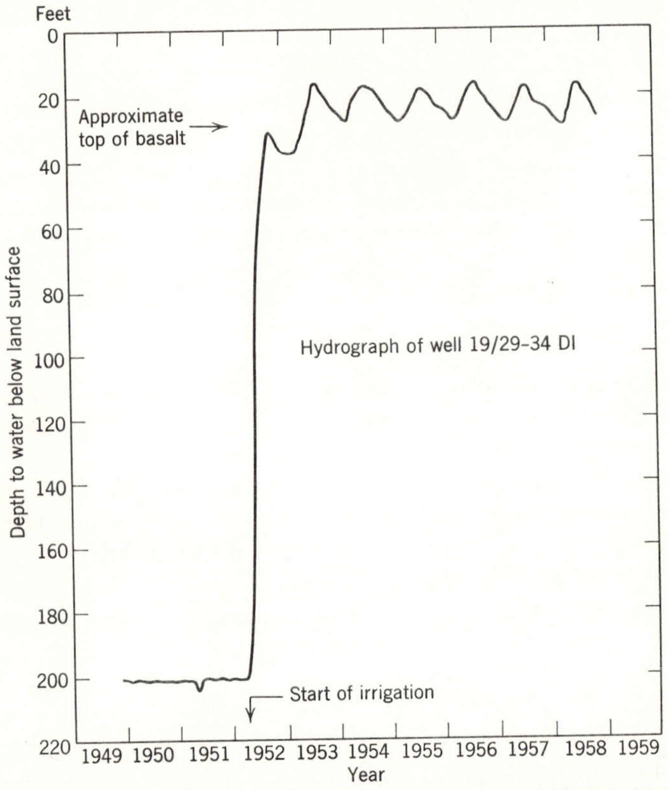


Figure 9.9 Hydrograph of a well in basalt near Moses Lake, Washington. Rapid rise of water level during 1952 was caused by the infiltration of excess irrigation water. Water-bearing zones are probably partly confined. Cyclic fluctuations after 1952 reflect recharge from seasonal irrigation [47].

Although reliable measurements of bulk porosity are not found in the literature, the rapid rise in water levels in areas recently put under irrigation [47] (Figure 9.9) indicates that the porosity can be less than 5 per cent in volcanic rocks which are good to excellent aquifers. Aquifer tests made in the Snake River Basalt of Idaho [49] yielded storage coefficients of 0.02 and 0.06, which would suggest an effective porosity of less than 10 per cent, provided that aquifers in the tests were entirely unconfined. Sediments interbedded with the lava will greatly increase the average porosity of large volumes of rocks that are predominantly volcanic. Under favorable circumstances the sediments provide storage space for the water, whereas the more permeable volcanic rock conducts the water to the wells. The amount of interbedded sediments, however, may be quite slight, as for example, near Moses Lake, Washington, where ten wells which penetrated an aggregate thickness of 8042 feet of Columbia River Basalt encountered only 320 feet of nonvolcanic material [47].

If we could quarry a sample of several hundred tons for laboratory measurement of permeability, the range of values obtained from various types of volcanic rock would be from almost zero to more than 1000 darcys. The horizontal permeability is largely owing to interflow spaces and the vertical permeability is mostly owing to fracturing of partially solidified lava in the last stages of movement together with shrinkage cracking [30,39]. Most commonly the vertical permeability is very small in comparison with the horizontal permeability. The vertical permeability is so low in many regions that separate confined aquifers are formed. Differences of head of more than 100 feet have been encountered in drilling through dense layers of lava which form the confining beds. Figure 9.10 shows typical records of water levels in a well which has penetrated a series of permeable aquifers in basaltic rock.

Both the permeability and porosity of volcanic rock tend to decrease slowly with geologic time. Some of this decrease is owing to compaction, but the filling of pores with secondary minerals is probably the most important cause of the decrease. Pre-Cenozoic volcanic rocks of southern Brazil [13,34], west-central India [5], and the eastern United States [11] are all poor aquifers, having water-bearing properties similar to metamorphic and plutonic igneous rocks.

Discussion thus far has been centered on basaltic and andesitic rocks which tend to be extruded as flows. Silica-rich lavas such as rhyolite and dacite are more viscous and will be erupted in thick, dense flows, or, more commonly, as pyroclastics. Unaltered pyroclastics, like alluvium, will have porosities and permeabilities directly related to fragment size, sorting, and degree of cementation. Poor sorting and abundance of fine material will cause most widespread pyroclastic deposits to have low permeabilities but moderate to high porosities. Welded tuffs are a special class of pyroclastic deposits formed by the fusing together of very hot fragments of volcanic rock as they settle to the ground. Welded tuffs have medium to low porosities and very low permeabilities.

Pyroclastic rocks associated with lava flows are generally porous but not very permeable. Exceptions are blocky, coarse material near volcanic vents and tuffs which have been reworked by water. A good example of an

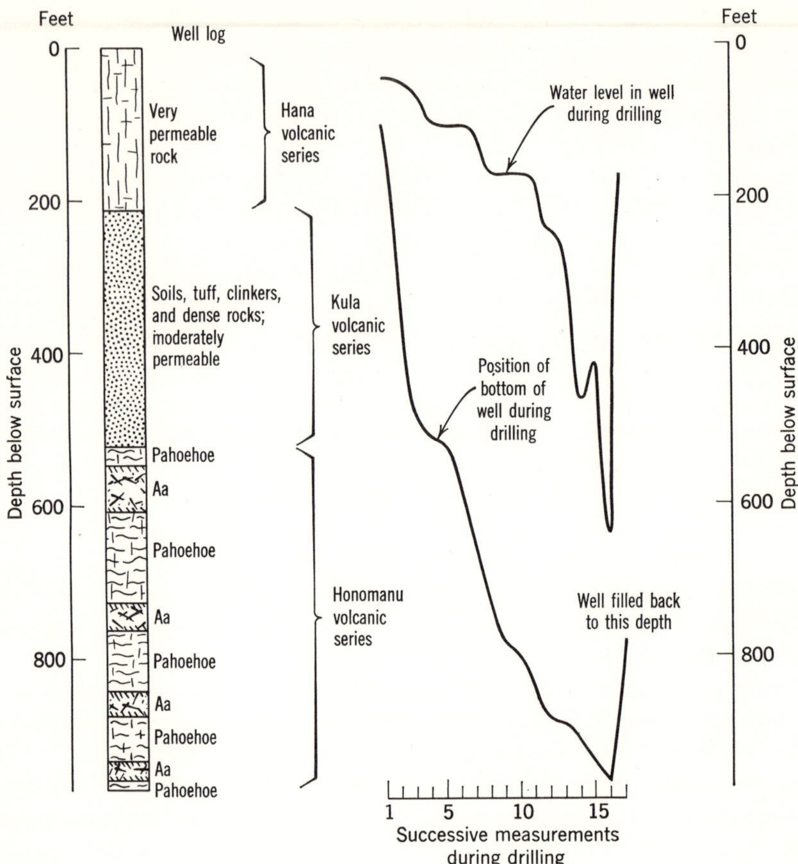


Figure 9.10 Log of well on the Island of Maui, Hawaii, showing fluctuations of water level as drilling progressed through successive basaltic aquifers. (Stearns and MacDonald [38].)

aquifer of water-sorted andesite is the Mehrten Formation of central California, which yields 500 to 1500 gpm to large irrigation wells [3,32]. Parts of the Mehrten Formation that are mud flows and altered ash act as confining layers. Rhyolitic ash and tuff in the underlying Valley Springs Formation is not known to yield significant amounts of water to wells. Similar results have been obtained from aquifer tests in rhyolitic ash and tuff of the Oak Springs Group in southern Nevada. Here, wells penetrating hundreds of feet of saturated, nonwelded tuff produce only a few gallons per minute. Results of laboratory tests on these rocks are shown in Table 9.3. Although not significantly permeable in individual samples, the welded tuff can transmit the most water because of its tendency to

*Table 9.3 Properties of the Oak Spring Group of the Nevada Test Site* [16]

| Lithology | Mean Values of Porosity, % | Mean Values of Bulk Density, g/cm³ | Mean Values of Permeability, millidarcys |
|---|---|---|---|
| Bedded tuffs, partly zeolitized | 39 | 1.50 | 0.04 |
| Bedded tuffs, pumiceous | 40 | 1.37 | 11.5 |
| Friable tuffs | 36 | 1.50 | 1.4 |
| Welded tuffs | 14 | 2.18 | 0.33 |

develop joints and other fractures which will remain open at considerable depths.

### WELL LOCATION AND WATER DEVELOPMENT

As in other geologic environments, the first step in either determining individual locations for new wells or in making systematic plans for ground-water development for entire regions in areas of volcanic rock is to conduct as thorough a hydrogeologic survey as possible. Collection of stream-flow data, water-level measurements in wells, and other aspects of collecting hydrologic data are quite similar in most regions. In volcanic terrane, however, surface streams may be virtually absent owing to the extremely high infiltration capacity of the rocks and soil. For example, a large part of eastern Maui in the Hawaiian Islands lacks surface streams even in areas having more than 200 inches of rain. If evapotranspiration is assumed to be 50 inches, then there are areas in which the annual infiltration of rain must exceed 150 inches. Even in parts of Oahu where rainfall is less, large areas may have more than 50 per cent of the rain infiltrate into the subsurface [28].

Hydrogeologic studies of volcanic rocks depart from the more conventional studies in that close attention should be paid to reconstructing the geomorphic history of an area as an aid in determining the position of ancient valleys, buried soils, and other features which exert influence on the movement of ground water (Figure 9.11) [39]. In some areas the stratigraphic sequence of lava beds can be determined and can be used to predict the subsurface position of permeable zones. In general, pahoehoe flows are much more permeable than are dense interior portions of thick *aa* flows. The upper and lower parts of the *aa* flows, however, may be brecciated and quite permeable. In addition to *aa* lava, buried soils and ash beds may form semihorizontal barriers to water movement.

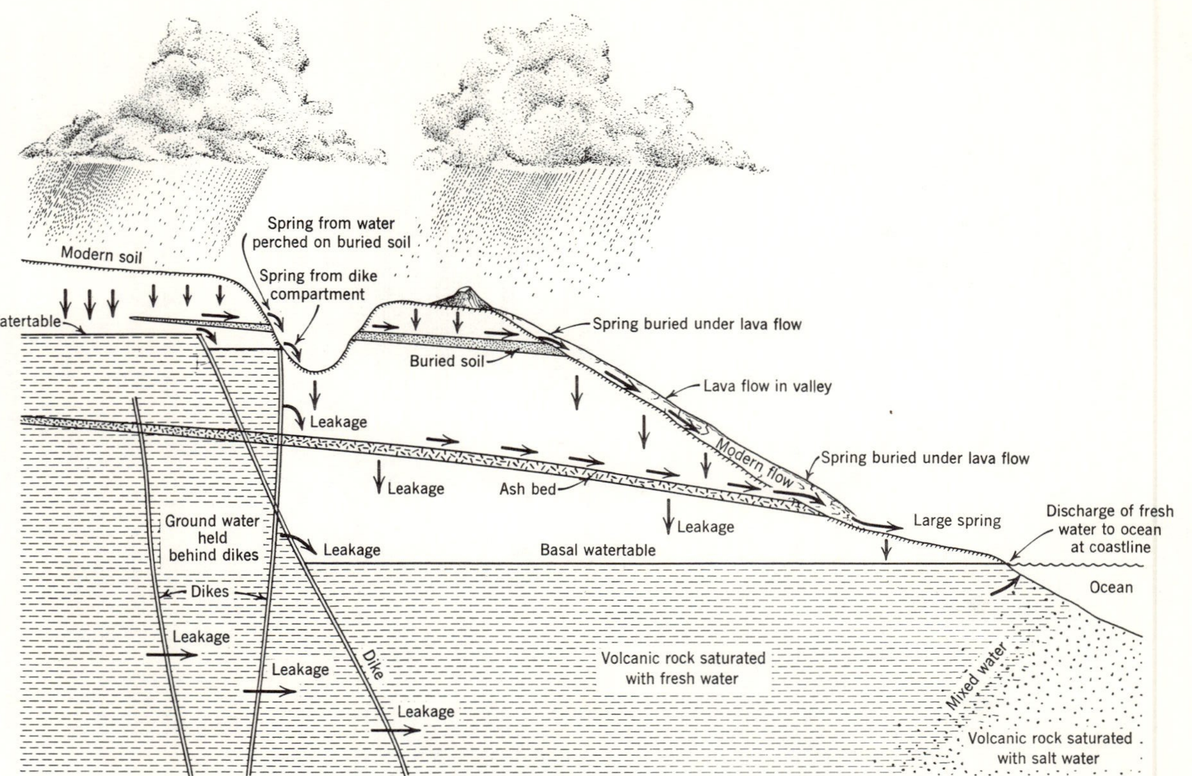


Figure. 9.11 Circulation of ground water in highly permeable basalt typical of the Island of Hawaii. Arrows indicate direction of water movement. Fresh water floating in contact with sea water has been called basal ground water in the Hawaiian Islands. (Diagram is modified from Stearns and MacDonald [39].)

Studies of aerial photographs together with a study of outcrops will determine the general area in which dikes are present. Dikes are the major vertical barriers in young volcanic rocks. Impermeable beds offset by faults against permeable beds and may also produce locally important vertical barriers [30].

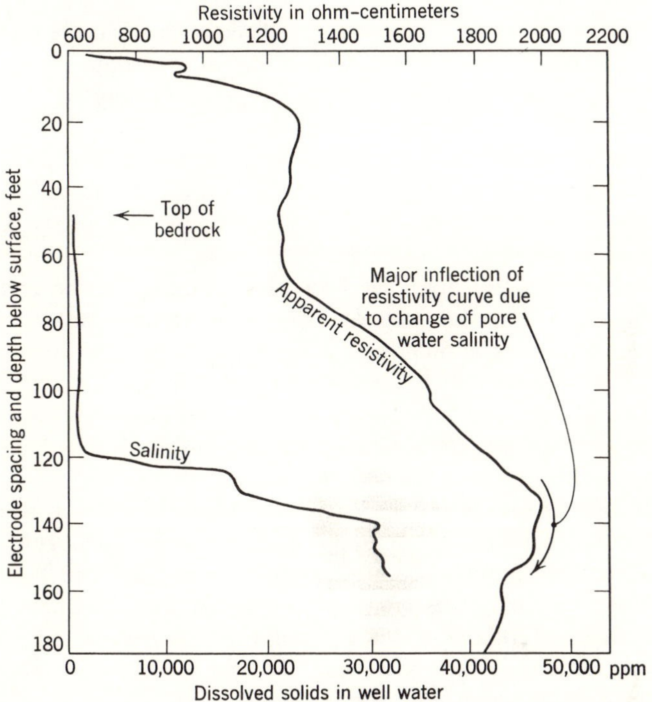


Figure 9.12 Results of a surface resistivity survey compared with ground-water salinity from a well drilled after the survey was completed. (Stearns [37].)

Geophysical techniques are generally of limited use in volcanic rocks. The main reason for this is the lack of large contrasts in magnetic susceptibilities, average densities and moduli of elasticity, and electrical conductivities. The one important exception has been in the determination of the salt-water fresh-water interface in volcanic rocks of parts of the Hawaiian Islands (Figure 9.12) [37]. Sea water saturating the pore space greatly reduces the bulk electrical resistivity of the rock so that distinct anomalies are found at the interface. If volcanic rocks are associated with alluvium or other unconsolidated deposits, many of the geophysical methods may be valuable in locating important geologic contacts. In places, however, more than two or three alternating layers of alluvium and basalt make geophysical interpretations difficult. Methods of seismic refraction will be generally useless below the first layer of basalt because it will divert the earliest signal to all the geophones. Seismic reflection has been used to determine contacts below basalt. The large quantities of explosives which are required, however, limit its use to relatively unpopulated areas.

Drilling test holes or wells in permeable lava is usually done with cable-tool machines because circulating fluid is lost in cavernous zones. Compact older lava can be drilled with any type of drilling rig which can penetrate dense rock. Loose pyroclastic material, unless excessively permeable, can best be drilled by rotary rigs circulating suitable drilling fluids.

In extremely permeable zones in basalt the capacity of the well is usually limited only by the diameter of the hole which limits, in turn, the diameter of the pump bowls which can be installed in the well. Walton [49], for example, reported 2100 gpm per foot of drawdown at the end of an eight-hour test of a sixteen-inch diameter well drawing water from the Snake River Basalt. The older and denser rocks will yield much smaller quantities. Newcomb [30] has estimated yields in the Columbia River Basalt to be about 1 gpm for every foot of well below the watertable. Wells of modern construction which penetrate the basalt may have somewhat better yields.

Tunnels have been used extensively in the Hawaiian Islands to recover water where it is perched on buried soils and tuff or ponded behind dikes. Shorter laterals or tunnels are used in some areas to skim basal fresh water without creating excessive drawdowns that would induce seawater intrusion. Wells with laterals perpendicular to the strike of dipping beds of lava generally yield the most water because a maximum number of permeable zones between flows will be intersected.

Exceptional permeabilities found in some basalts encourages overproduction of ground water. Even though excessive drawdowns are not experienced in the vicinity of individual wells, the net effect of overproduction may be the gradual lowering of the regional water levels. This was the case in the Honolulu artesian basin before careful observation and control of water withdrawals were started in the 1920's. Several new water-development plans are presently under study in the regions of volcanic rock in Washington and Idaho. In many of these areas, major difficulties are with total regional supplies and long-term trends of water levels rather than with local problems of water extraction.

### WATER QUALITY

Most water from volcanic rocks has good to excellent chemical quality. In general, it tends to be a calcium-magnesium-bicarbonate water or, in the case of acidic volcanic rocks, a sodium bicarbonate water with relatively large amounts of silica [25]. Poor chemical quality is encountered near hot springs and fumaroles where the water may be high in sodium and chloride. Even where these ions are not in excessive amounts in thermal water, the low pH or high fluoride content may not be acceptable. Water of poor chemical quality also may occur along coastal areas where sea water saturates the rock or in desert areas where evaporites or air-borne salts are dissolved in surface water as it percolates into the subsurface.

Owing to the extremely high permeability of some volcanic rocks, biological contamination is a potential danger. Fortunately, volcanic rocks tend to weather quite rapidly, so thick soils that develop will remove most pathogenic organisms. In the area of the Columbia River Basalt in the states of Washington, Oregon, and Idaho, the surface is partly covered with a discontinuous blanket of loess that is also an effective barrier to the downward migration of contaminants.

## REFERENCES

1. Clapp, F. G., and W. S. Bayley, 1909, Underground waters of southern Maine: *U.S. Geol. Survey Water-Supply Paper* 223, 268 pp.
2. Cushman, R. V., W. B. Allen, and H. L. Preen, Jr., 1953, Geologic factors affecting yield of rock wells in southern New England: *New England Water Works Assoc. Jour.*, v. 67, pp. 77–93.
3. Davis, S. N., and F. R. Hall, 1959, Water quality of eastern Stanislaus and northern Merced Counties California: *Stanford University Publ. in Geol. Sci.*, v. 6, no. 1, 112 pp.
4. Davis, S. N., and L. J. Turk, 1964, Optimum depth of wells in crystalline rocks: *Ground Water*, v. 2, pp. 6–11.
5. Deshpande, B. G., and S. N. Sen Gupta, 1956, Geology of ground water in the Deccan traps and the application of geophysical methods: *India Geol. Survey Bull.*, Series B, No. 8, 22 pp.
6. Dingman, R. J., G. Meyer, and R. O. R. Martin, 1954, The water resources of Howard and Montgomery Counties: *Maryland Dept. Geol., Mines and Water Resources Bull.* 14, 260 pp.
7. Ellis, A. J., and C. H. Lee, 1919, Geology and ground waters of the western part of San Diego County, California: *U.S. Geol. Survey Water-Supply Paper* 446, 321 pp.
8. Enslin, J. F., 1943, Basins of decomposition in igneous rocks, their importance as underground water reservoirs and their location by the electrical resistivity method: *Trans. Geol. Soc. South Africa*, v. 46, pp. 1–12.
9. Frommurze, H. F., 1937, The water-bearing properties of the more important geological formations in the Union of South Africa: *Union of South Africa Geol. Survey Memoir No.* 34, 186 pp.

10. Gear, D. J., 1951, *Underground water supplies in the granitic and gneissose rocks of eastern Uganda:* Internat. Union of Geodesy and Geophys., Assoc. of Scientific Hydrology, Brussels, 1951, pp. 252–261.
11. Gregory, H. E., and E. E. Ellis, 1909, Underground-water resources of Connecticut: *U.S. Geol. Survey Water-Supply Paper* 323, 200 pp.
12. Halliday, W. R., 1963, Caves of Washington: *Washington Dept. Conservation Information Circ. No.* 40, 132 pp.
13. Hausman, A., 1960, Estudos das possibilidades de agua subterranea no Rio Grande do Sul: *Soc. Brasileria de Geologia Boletin*, v. 9, pp. 29–41.
14. Jahns, R. H., 1943, Sheet structure in granite, its origin and use as a measure of glacial erosion in New England: *Jour: Geol.*, v. 51, pp. 71–98.
15. Johnston, P. M., 1962, Geology and ground-water resources of the Fairfax Quadrangle, Virginia: *U.S. Geol. Survey Water-Supply Paper* 1539-L, 61 pp.
16. Keller, G. V., 1960, Physical properties of tuffs of the Oak Spring Formation, Nevada: *U.S. Geol. Survey Prof. Paper* 400-B, pp. 396–400.
17. Krynine, D. P., and W. R. Judd, 1957, *Principles of engineering geology and geotechnics:* New York, McGraw-Hill Book Co., 730 pp.
18. LeGrand, H. E., 1949, Sheet structure, a major factor in the occurrence of ground water in the granites of Georgia: *Econ. Geol.*, v. 44, pp. 110–118.
19. ——, 1952, Solution depressions in diorite in North Carolina: *Am. Jour. Sci.*, v. 250, pp. 566–585.
20. ——, 1954, Geology and ground water in the Statesville area, North Carolina: North Carolina Dept. of Conservation and Development, *Div. of Mineral Resources Bull.* 68, 68 pp.
21. ——, 1958, Chemical character of water in the igneous and metamorphic rocks of North Carolina: *Econ. Geol.*, v. 53, pp. 178–189.
22. ——1960, Geology and ground-water resources of Pittsylvania and Halifax Counties: *Virginia Div. Mineral Resources Bull.* 75, 86 pp.
23. ——, 1962, Perspective on problems of hydrogeology: *Geol. Soc. Am. Bull.*, v. 73, pp. 1147–1152.
24. Linnikoff, N., 1953, *Barrages sur les terrains granitiques et cristallophylliens du Massif Central:* nineteenth Int. Geol. Congress, Sec. 8, pp. 53–58.
25. Littleton, R. T. and E. G. Crosthwaite, 1957, Ground-water geology of the Bruneau-Grand View area Owyhee County Idaho: *U.S. Geol. Survey Water-Supply Paper* 1460-D, pp. 147–198.
26. Mack, Seymour, 1960, Geology and ground-water features of Shasta Valley, Siskiyou County, California: *U.S. Geol. Survey Water-Supply Paper* 1484, 115 pp.
27. Meier, O., and S. G. Petersson, 1951, *Water supplies in the Archaean bedrocks of Sweden:* Brussels, Internat. Union of Geodesy and Geophys., Assoc. of Scientific Hydrology, 1951, pp. 252–261.
28. Mink, J. F., 1962, Rainfall and runoff in the leeward Koolav Mountains, Oahu, Hawaii: *Pacific Sci.*, v. 16, pp. 147–159.
29. Moore, J. G., and D. H. Richter, 1962, Lava tree molds of the September 1961 eruption, Kilauea Volcano, Hawaii: *Geol. Soc. Am. Bull.*, v. 73, pp. 1153–1158.
30. Newcomb, R. C., 1959, Some preliminary notes on the ground water of the Columbia River Basalt: *Northwest Sci.* v. 33, pp. 1–18.
31. Price, C. E., 1961, Artificial recharge through a well tapping basalt aquifers Walla Walla area, Washington: *U.S. Geol. Survey Water-Supply Paper* 1594-A, 33 pp.
32. Piper, A. M., and others, 1939, Geology and ground-water hydrology of the Mokelumne area, California: *U.S. Geol. Survey Water-Supply Paper* 780, 230 pp.

33. Reed, J. C., Jr., B. Bryant, and J. T. Hack, 1963, Origin of some intermittent ponds on quartzite ridges in western North Carolina: *Geol. Soc. Am. Bull.*, v. 74, pp. 1183–1188.
34. Schneider, Robert, 1963, Ground-water provinces of Brazil: *U.S. Geol. Survey Water-Supply Paper* 1663-A, 14 pp.
35. Schoeller, H., 1962, *Les eaux souterraines:* Paris, Masson, 642 pp.
36. Simpson, J. G., A. R. Dryasdall, and H. H. J. Lambert, 1963, The geology and groundwater resources of the Lusaka area: *Northern Rhodesia Geol. Survey Report No.* 16, 59 pp.
37. Stearns, H. T., 1940, Geology and ground-water resources of Lanai and Kahoolawe, Hawaii: *Terr. Hawaii Div. Hydrography Bull.* 6, 177 pp.
38. Stearns, H. T., and G. A. MacDonald, 1942, Geology and ground-water resources of the Island of Mavi, Hawaii: *Terr. Hawaii Div. Hydrography Bull.* 7, 344 pp.
39. ——, 1946, Geology and ground-water resources of the Island of Hawaii: *Terr. Hawaii Div. Hydrography Bull.* 9, 363 pp.
40. Stevens, P. R., 1960, Ground-water problems in the vicinity of Moscow, Latah County, Idaho: *U.S. Geol. Survey Water-Supply Paper* 1460-H, pp. 325–357.
41. Stewart, J. W., 1961, Tidal fluctuations of water levels in wells in crystalline rocks in north Georgia: *U.S. Geol. Survey Prof. Paper* 424-B, pp. 107–109.
42. ——, 1962, Relation of permeability and jointing in crystalline metamorphic rocks near Jonesboro, Georgia: *U.S. Geol. Survey Prof. Paper* 450-D, pp. 168–170.
43. ——, 1962, Water-yielding potential of weathered crystalline rocks at the Georgia Nuclear Laboratory: *U.S. Geol. Survey Prof. Paper* 450-B, pp. 106–107.
44. Stuart, W. T., E. A. Brown, and E. C. Rhodehamel, 1954, Ground-water investigations of the Marquette Iron-Mining District: *Michigan Geol. Survey Div. Tech. Rept.* 3, 92 pp.
45. Taylor, G. C., A. K. Roy, and D. N. Sett, 1954, *Groundwater geology of the Pali Region Jodhpur Division, Western Rajasthan, India:* Rome, Internat. Union of Geodesy and Geophys., Assoc. of Scientific Hydrology, 1954, pp. 560–573.
46. Temperley, B. N., 1960, A study of the movement of groundwater in lava-covered country: *Overseas Geol. and Mineral Resources*, v. 8, pp. 37–52.
47. Walters, K. L., and M. J. Grolier, 1960, Geology and ground water resources of the Columbia Basin Project area, Washington, Volume 1: *Washington Dept. Conservation Water-Supply Bull.* 8, 542 pp.
48. Walton, W. C., 1962, Ground-water resources of Camas Prairie, Camas and Elmore Counties, Idaho: *U.S. Geol. Survey Water-Supply Paper* 1609, 57 pp.
49. Walton, W. C., and J. W. Stewart, 1961, Aquifer tests in the Snake River Basalt: *Am. Soc. Civil Eng. Trans.*, v. 126, pp. 612–632.
50. Ward, L. K., 1946, The occurrence, composition, testing, and utilization of underground water in South Australia, and the search for further supplies: *South Australia Geol. Survey Bull.* No. 23, 281 pp.
51. Waterhouse, B. C., 1961, Note on Kawiti Basalt and hydrology of Kawakawa area, Northland: *New Zealand Jour. of Geol. and Geophys.*, v. 4, pp. 357–371.
52. Wood, P. R., 1960, Geology and ground-water features of the Butte Valley region, Siskiyou County, California: *U.S. Geol. Survey Water-Supply Paper* 1491, 150 pp.

# chapter 10

# GROUND WATER IN SEDIMENTARY ROCKS

## 10.1 *Introduction*

Popular interest in the hydrogeology of sedimentary rocks has been commonly centered on limestone caverns and deep artesian wells. Millions of people each year visit large caverns in Indiana, Kentucky, Virginia, Missouri, and New Mexico in the United States, as well as in Belgium, France, the Balkan Peninsula, and other parts of the world. The large and awesome underground chambers decorated with columnar deposits of calcite and the small tortuous passageways all show, even to the non-specialist, a common origin in the work of subsurface water. No less spectacular is the sight of hundreds of gallons of water per minute gushing from deep wells. As a consequence, the term "artesian" has an almost magical significance. Indeed, the initial pressures and flows were so large in early wells in South Dakota that the water was used to drive small flour mills and electrical generators [13]. The popularity of artesian wells here and in other regions has been responsible for an almost universal overproduction of high-yielding artesian systems.

Considerable scientific interest has also been centered on limestone caverns and deep artesian wells. The study of caverns, or speleology, has developed from a part-time hobby or sport of a few individuals to a respected branch of the earth sciences [41]. Although systematic studies of caves started more than 50 years ago, many of the hydrogeologic aspects of cave genesis are still poorly understood. The same is true of the origin of high artesian pressures. The simple analogy between an elevated storage tank with a distribution system and a natural artesian system satisfied most early students of the subject. Starting with the classical work of Meinzer on the Dakota Sandstone, however, the true complexity of artesian phenomena is rapidly becoming more evident.

In addition to the origin of caverns and artesian pressure, there are a number of other current interests in the hydrogeology of sedimentary rocks. Stratigraphic control of permeability, the origin of water of unusual chemical characteristics, and the rate of migration of water in extensive sandstone aquifers are of particular importance.

## **10.2** *Rock Types and Bed Thicknesses*

Shale, claystone, siltstone, and other fine-grained detrital rocks account for roughly 50 per cent of all sedimentary rocks. Next in abundance are sandstones, then carbonate rocks, and finally several minor types including conglomerate, gypsum, chert, tillite, salt, and diatomite. The minor types constitute less than 2 per cent of all exposed sedimentary rocks.

Bed thicknesses most commonly range between a few centimeters to a few meters. Although alternating beds of shale limestone, and sandstone are characteristic of most sedimentary sequences, individual beds may be so thick that water wells within certain regions will penetrate only one rock type even though the wells are more than 100 meters deep. Upper Cretaceous chalk of southeastern England, Mesozoic sandstones of the Colorado Plateau, and dense Mississippian limestones of Kentucky are examples of major stratigraphic units that form such outcrop areas. Even though essentially single rock types may be encountered by wells, geologic studies of structure and minor lithologic variations may afford valuable guides to the occurrence of ground water.

## **10.3** *Porosity and Permeability*

### FINE-GRAINED DETRITAL ROCKS

Most fine-grained detrital rocks have relatively high porosities but very low permeabilities. Some representative values are given in Table 10.1. Siliceous shale, some claystones, and most argillites will develop closely spaced joints if the rocks are near the surface. Also, if these rocks are involved in faulting, fractures that stay open at considerable depths may develop. The joints and fractures may yield a few gallons of water per minute to wells. Most commonly, however, the fine-grained rocks will be barriers to the movement of water. In areas of nearly horizontal strata, the fine-grained rocks serve as widespread confining beds for artesian

*Table 10.1 Porosity and Permeability of Sedimentary Rocks*

| Description and Age | Porosity, per cent | Permeability, millidarcys | Reference |
|---|---|---|---|
| Grand Saline Salt (salt dome) | less than 1.0 | 4.1 | 17 |
| Hutchinson Salt, Permian | less than 1.0 | $7.3 \times 10^{-3}$ | 17 |
| Bradford Sandstone, Devonian | 14.8 | 2.7 | 51 |
| Berea Sandstone, Mississippian | 19 | 383 | 51 |
| Oil Creek Sandstone, dolomitic, silica and carbonate cement, Ordovician | 6.7 | 4 | 50 |
| Woodbine Sandstone, medium grained, moderately cemented with silica, Cretaceous | 25.6 | 4400 | 50 |
| Repetto Sandstone, coarse grained, poorly indurated, clay cement, Pliocene | 19.1 | 36 | 50 |
| Wilcox Sandstone, shaley friable, poorly sorted, only partly cemented, Eocene | 15.3 | 0.3 | 2 |
| Limestone, Wichita Formation, compact, crystalline, no visible pores, Permian | 4.1 | less than 0.1 | 3 |
| Limestone, Wichita Formation, compact, crystalline, some visible pores, Permian | 10.1 | 7.7 | 3 |
| Limestone, chalky, Devonian | 29.5 | 37.8 | 3 |
| Limestone, oolitic | 21.6 | 339 | 3 |
| Limestone, Charles Formation, fine grained, originated as carbonate mud with enclosed pellets and fossil fragments, Mississippian | 8.4 | 0.1 | 35 |
| Dolomite, Red River Formation, sucrose, intercrystalline porosity, Ordovician | 11.9 | 16.5 | 35 |
| Dolomite, Turner Valley Formation, sucrose with fossil fragment molds, Mississippian | 27.8 | 290 | 35 |

*Table 10.1 Continued*

| Description and Age | Porosity, per cent | Permeability, millidarcys | Reference |
|---|---|---|---|
| Dolomite, Red River Formation, interlocking mosaic of subhedral crystals, Ordovician | 6.3 | 1.0 | 35 |
| Shale, Pennsylvania, depth 468 feet. (Permeability measured with 0.02 N NaCl) | . . . | $9 \times 10^{-5}$ | 18 |
| Shale, Cretaceous. (Permeability measured with 0.02 N NaCl) | . . . | $4 \times 10^{-3}$ | 18 |
| Shale, Gros Ventre Formation, Cambrian | 11.1 | . . . | 31 |
| Graneros Shale, Cretaceous | 24.9 | . . . | 31 |
| Chanute Shale, Pennsylvanian | 15.0 | . . . | 31 |
| Nonesuch Shale, siliceous, Precambrian | 1.6 | . . . | 31 |

systems. Nevertheless, it is a common mistake to assume that no significant movement of water takes place through the confining beds. For example, 100 feet of siltstone with a permeability of 0.1 millidarcy having a hydraulic head differential of 10 feet perpendicular to the bedding will transmit $1.84 \times 10^6$ gallons of water each year through each square mile of the stratum.

The large pore space in many fine-grained sedimentary rocks provides storage for vast quantities of water. Even though individual wells in most places cannot extract directly significant amounts of this water, slow drainage into aquifers can be induced by lowering the head in the aquifers. Thus, water stored in shales and similar rocks should always be considered in making ground-water inventories, particularly if differential hydraulic heads are great enough to induce drainage. Owing to the capillary effects, however, gravity drainage will be most important in the coarser sediments or in fractured zones.

Porosity of fine-grained sediments decreases with depth of burial and to some extent with age, although the relation is neither simple nor universal [34]. Newly deposited fine muds will have porosities of between 50 and 90 per cent. Compaction will force the pore water out of the fine material into adjacent permeable beds of sand so that porosities at depths of several hundred feet will be generally less than 50 per cent. At depths of several thousand feet the porosity will be less than 30 per cent and most commonly less than 25 per cent [31]. Extruded pore water will not contribute a significant volume of water to aquifers under natural conditions because of the slow rates of compaction of the fine-grained sediments. Well production may, however, increase the local rate of compaction so that measurable amounts of water can be forced from the sediments, particularly if they are poorly indurated. Pore water that is originally saline will be an important source of dissolved constituents in zones where this water moves into aquifers. If the water is channeled into near-surface aquifers along faults or stratigraphic discontinuities, localized areas of saline water are produced [14]. Where the water is forced to the surface, saline springs may result [47].

### SANDSTONE

Porosity of sandstone ranges from less than 5 per cent to a maximum of about 30 per cent. The amount of pore space in an individual sample is a function of sorting, grain shape, packing, and degree of cementation (Figure 10.1). Of these variables, cementation is the most important. Common cementing materials are clay minerals, calcite, dolomite, and quartz [42]. Original cementation may be highly localized or may be selectively removed by later leaching so that the properties of the rock as a whole can be highly variable. Sandstones that are cemented with calcite can become case hardened at the surface by redeposition of the calcite. Conclusions concerning the hydraulic characteristics of these rocks from observations of outcrops can be misleading. Silica cement in sandstone will form overgrowths on the sand grains that will tend to interlock with the overgrowths on adjacent grains. Advanced stages of silica cementation can produce orthoquartzites having extreme hardness and low porosity. Clay minerals may be present as original constituents or as products of diagenesis. Rocks cemented with clay are not usually as firm as other sandstones. The porosity of clay-cemented sandstones tends to be quite high because the clay itself has considerable porosity.

The most intensive research into the permeability of a sandstone which has been reported from the literature was a study sponsored by the Jersey Production Research Company of a homogeneous-appearing Pennsylvanian sandstone in central Oklahoma [24,25]. Here, sixty-five wells were drilled in an eight-acre plot. Cores were taken of all wells, and a total of 2000 samples were studied. Even though laboratory measurements on cores showed large variations in permeability between wells only 10 to 35 feet apart, the permeability obtained from an aquifer test agreed well with the average permeability as determined from laboratory tests. Local variations in permeability within the sandstone tested were found to be controlled chiefly by differences in the grain size of the sandstone.

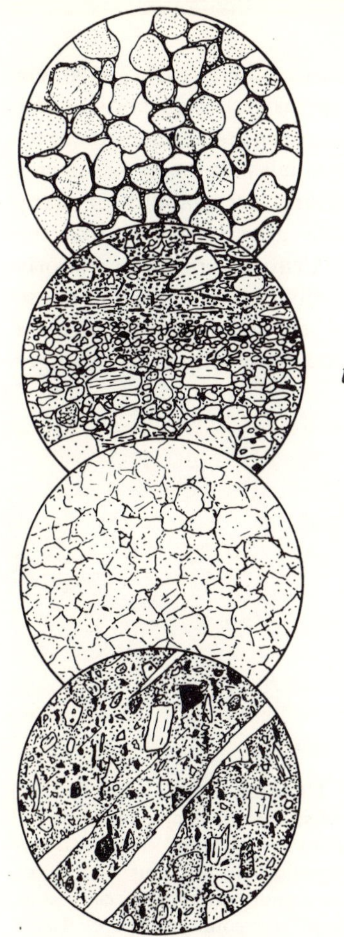


Figure 10.1 Thin sections of various types of sandstone. Drawings are composites of numerous photomicrographs from various sources. Magnification is about 10×. (*a*) Porous sandstone with a minor amount of calcite cement. Good sorting and large grain diameters would make this a good aquifer. (*b*) Porous sandstone with a minor amount of clay cement. Poor sorting would make this a very poor aquifer. (*c*) Orthoquartzite with a porosity of less than 1 per cent. This rock would not yield water to wells. (*d*) Graywacke with secondary porosity produced by fracturing. This rock would produce a moderate amount of water from the fractures.

Permeabilities of sandstones are one to three orders of magnitude lower than permeabilities of corresponding unconsolidated sediments. For example, medium-grained sand generally has a permeability between 1000 and 30,000 millidarcys, but values for the corresponding medium-grained sandstones generally range from 1 to 500 millidarcys. Some of the reduction in permeabilities between sands and corresponding sandstones is caused by a closer packing of grains in the rock, but most must be owing to the restriction of pore space by the presence of cement. There is some correlation between porosity and permeability in sandstones of similar texture and lithology [30]. The large number of variables influencing permeability, nevertheless, makes impossible the prediction of permeability on the basis of porosity alone.

CARBONATE ROCKS

Limestone and dolomite, the two common carbonate rocks, originate from a large number of different sedimentary deposits such as inorganically precipitated limey muds, shell fragments, talus deposits, calcite sand, reef masses, and accumulations of the remains of small planktonic organisms (Figure 10.2) [19,35]. The original porosity and permeability of many of these sediments are modified rather rapidly after burial, so that the original sedimentary structures of even late Cenozoic rocks are poorly preserved. In contrast, if the rocks are relatively impermeable and dense to start with and the rocks are not deformed, the sedimentary structures may persist almost indefinitely. Some of the more important changes

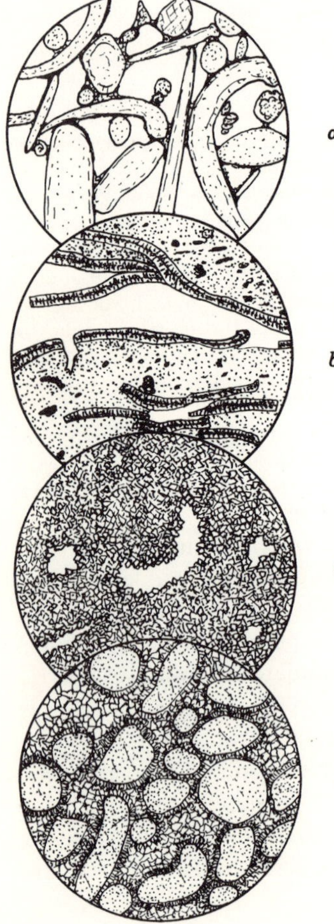


Figure 10.2 Thin sections of various types of carbonate rocks. Drawings are composites of numerous photomicrographs from various sources. Magnification is about 10×. (*a*) Coquina with minor amounts of calcareous cement. Large, interconnected pores would make this rock an excellent aquifer. (*b*) Limestone with locally high porosity formed by algal crusts in leaf-like accumulations. Pores not filled by the lithified finer carbonate mud are not sufficiently interconnected to allow rapid migration of water. (*c*) Dolomite with moderately high secondary porosity. Large pores are formed by the solution of fossil fragments. Lack of large interconnected pores would make this a very poor aquifer. (*d*) Dense limestone formed by almost complete cementation and void filling by crystalline calcite within the original calcareous sand. This rock would not yield water to wells except along solution openings and fractures.

are caused by compaction, solution of aragonite and calcite, reprecipitation of calcite cement, and formation of the mineral dolomite [20,21,35].

Original porosity is relatively high in most young limestones. Permeability, however, is generally low except in rocks such as breccias and coquina in which the large pores are not filled initially with cement.

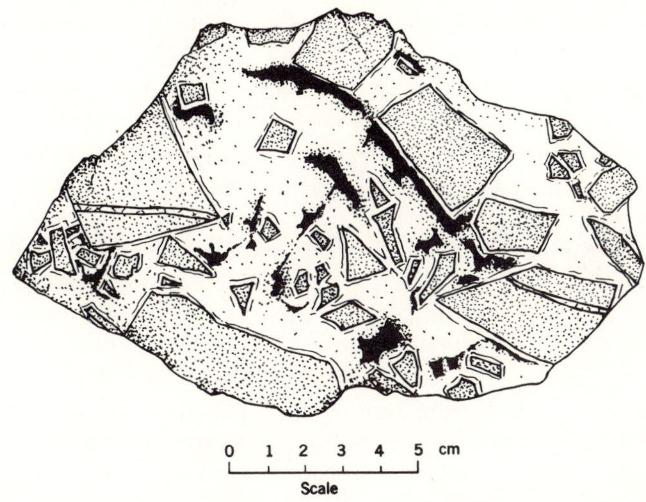


Figure 10.3 Partly cemented limestone breccia from Pyramid Lake, Nevada. Dark areas are the larger pores which have not been completely filled by the lighter colored calcareous cement.

Permeability may range from less than one millidarcy for clay-rich dense limestone to several thousand darcys for partly cemented coarse breccia (Figure 10.3). Intermediate values of 10 to 500 millidarcys are more common, however, for limestone having some original porosity. Dense crystalline limestone will generally have a permeability of less than one millidarcy (Table 10.1).

Although some original pore space may be retained in old limestone, other forms of porosity are more important from the standpoint of water production. Fractures and secondary solution openings along bedding planes and zones of primary porosity probably transmit the most water [26,43]. The postdepositional change from calcite to dolomite also creates considerable pore space [46]. If this diagenetic change takes place after the rocks are lithified, then the 13 per cent volume reduction caused by the transformation from calcite to dolomite will be left as void space. The dolomite crystals are so small, however, that the permeability generally is less than 300 millidarcys unless the pores have been enlarged by secondary solution.

Typical porosities and permeabilities of limestone and dolomite are given in Table 10.1. Extensive studies of porosities and permeabilities of petroleum-bearing horizons indicate similar values. For example, work in the Colorado Plateau area showed Paleozoic carbonate rocks with primary porosity due to algal plates had an average permeabillity of 24 millidarcys and a porosity of 9.5 per cent. Limestone with pores created by the solution of oolites had an average permeability of 3.4 millidarcys and a porosity of 11.3 per cent. Limestone which was originally a coquina had an average permeability of 5.1 millidarcys and an average porosity of 6.3 per cent. Dolomite had an average permeability of 3.0 millidarcys and an average porosity of 13 per cent. The highest permeability reported for all the rocks was 1165 millidarcys and the highest porosity was 32.4 per cent [32].

Simple assumptions concerning the hydraulics of water wells can be made which demonstrate that permeabilities common to most older carbonate rocks are not sufficient to allow even a few gallons per minute of water to flow into shallow wells. This may seem strange at first inasmuch as carbonate rocks with permeabilities of 20 to 100 millidarcys can yield large amounts of petroleum to oil wells. The apparent anomaly can be resolved by considering two important contrasts between oil and water production. First, oil wells are usually much deeper than water wells so that higher potential gradients can be developed near the oil wells. Second, the amounts produced are vastly different. Oil wells producing from 2 to 20 gallons per minute would be considered large producers in most oil fields. Water wells used for industry, municipal supplies, or irrigation would most likely be classed as failures unless they produced at least 50 gallons per minute.

The important conclusion concerning the permeabilities of Paleozoic and some younger carbonate rocks is that a search must be made for zones of secondary porosity produced through fracturing and solution. Zones of primary porosity, although permeable enough to be of interest to the petroleum industry, are not good aquifers. Primary porosity of the rock as a whole, however, is significant inasmuch as it provides storage space for ground water which is released slowly to the more permeable zones. It should be noted, however, that porosity measured in the laboratory may not have sufficient interconnections to supply ground water to wells. Carbonate rocks in Kentucky, which can be considered more or less typical Paleozoic rocks, were found to have specific yields of only 0.18 to 0.87 per cent [43], whereas estimates of specific yield based on published porosity values would be considerably larger.

Carbonate rocks with extensive solution channels or fractures primarily developed in one direction will have bulk permeabilities that will be strongly anisotropic. The direction of ground-water flow cannot be predicted, therefore, by simply drawing orthogonal lines to the ground-water contours. Arnow [4], in fact, suggested that ground-water in the outcrop area of the Edwards Limestone near San Antonio, Texas, moves almost parallel with the water-level contours. The highly cavernous nature of this limestone makes this suggestion seem quite reasonable. Figure 10.4 illustrates a less extreme example of anisotropy in two dimensional flow

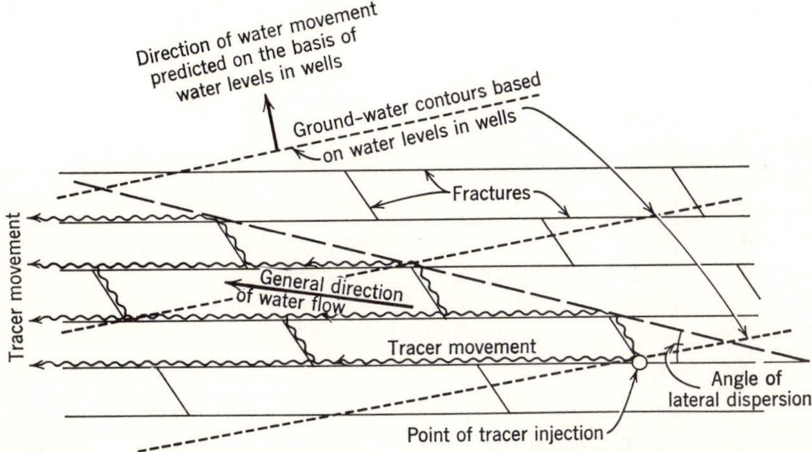


Figure 10.4 Two-dimensional fracture pattern showing the relation between true direction of ground-water flow and the direction inferred by drawing orthogonal lines to the regional water-level contours.

in which the head drop in all channels is directly proportional to the length of the channel.

Sedimentary rocks all have a certain amount of stratification that produces some anisotropy in the vertical direction as compared with the horizontal direction [36]. Extensive measurements of sandstone cores from Illinois [37] indicated that the median ratio of the horizontal permeability to the vertical permeability was 1.5, and about 12 per cent had a ratio of more than 3.0. Slightly less than 6 per cent had vertical permeabilities greater than the horizontal permeabilities and hence had a ratio of less than 1.0.

## 10.4 *Well Yields*

Most wells in moderately indurated sedimentary rocks have yields of between 1 and 500 gpm. Fine-grained rocks have yields generally less than 5 gpm, sandstones have yields between 5 and 200 gpm, and limestones

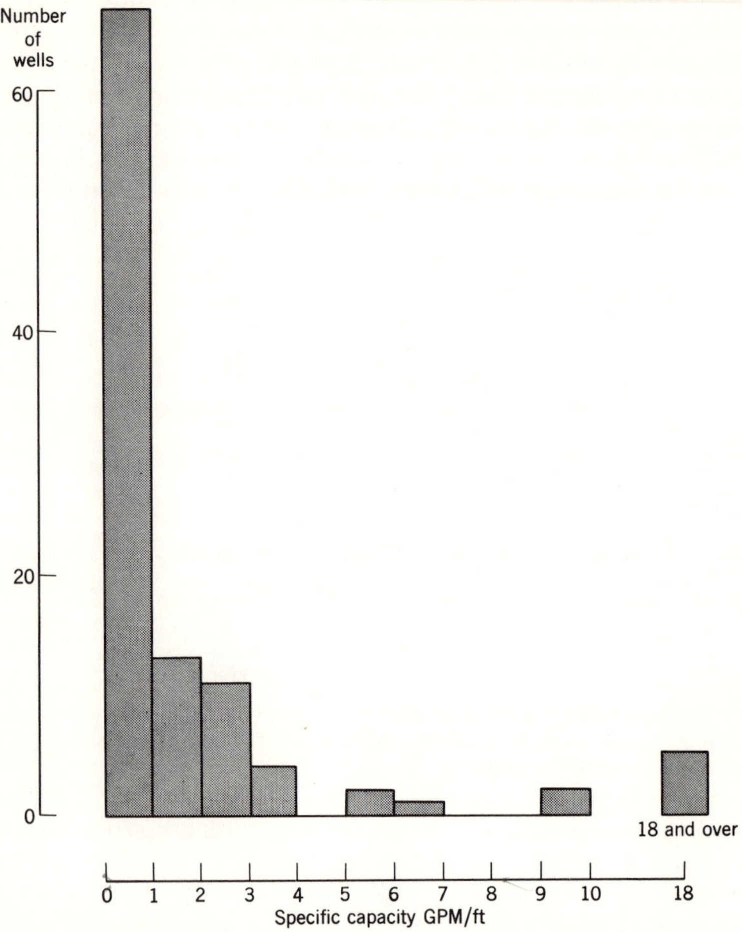


Figure 10.5 Frequency distribution of specific capacities of 105 wells in Ordovician rocks of Missouri. Most of the aquifers are limestone. (Data are from Robertson [39].)

have yields that may exceed 2000 gpm, but more commonly are between 5 and 20 gpm.

Although reported yields give some indication of water-bearing properties of aquifers, a great variety of well construction methods and testing procedures make these values subject to differences that cannot be related to aquifer characteristics. Specific capacities, expressed in gallons per minute the well produces for each foot of drawdown, will give more useful information. If aquifer thicknesses are known, the specific capacity can be divided by this thickness to obtain a specific-capacity index that can be used to predict well yields in parts of an aquifer for which well data are lacking but for which the thickness can be estimated [45].

Frequency distributions of specific capacities of wells in most moderately indurated sedimentary rocks show a strong right, or positive, skewed distribution (Figure 10.5). This skewed relation is most pronounced

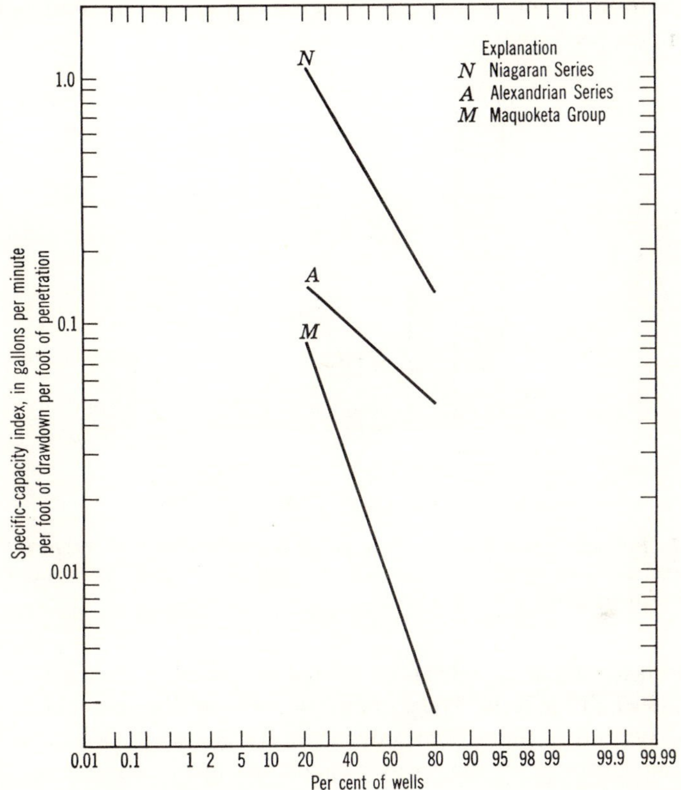


Figure 10.6 Specific capacity index graphs for Paleozoic dolomite aquifers of northern Illinois. (From Walton [44].)

in limestone and fractured orthoquartzites. Walton [44] found it useful to plot specific-capacity indices on logarithmic-probability paper to obtain approximate relations for three different stratigraphic units (Figure 10.6). These values can then be used to determine the likelihood of producing wells of a given yield for a given thickness of aquifer.

Original well yields can be increased by surging the well, by using acid to enlarge openings in carbonate rocks, by fracturing the rock with explosives or with fluid pumped into the well under high pressure, and by various combinations of these methods. Koenig [28] has shown that the success of well stimulation is greatest in wells with initially low yields. The specific capacities of these wells can commonly be improved 20 to 800 per cent. Roughly 2 to 10 per cent of all attempts by stimulation methods to improve yields fail. Nevertheless, owing to the fact that the cost of stimulation may be as much as 40 per cent of the original cost of the well, stimulation attempts may be economic failures even though well production has been improved.

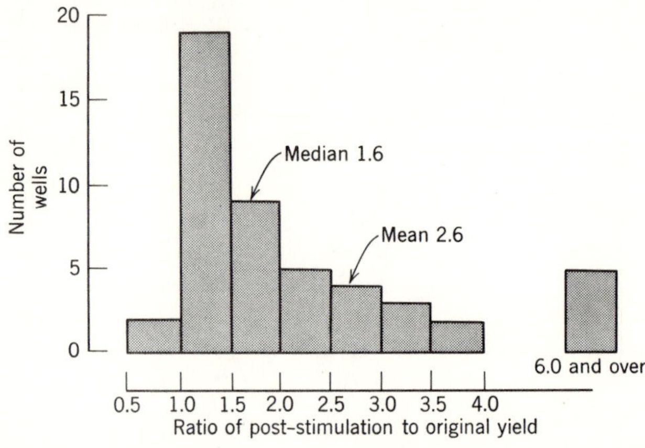


Figure 10.7 Frequency distribution of ratios of post-stimulation to original yields of sedimentary aquifers of Missouri. Most stimulation was by acid treatment. (Data are from Robertson [39].)

Actual improvement of yields is attributed to a number of modifications of the aquifer and the well. Deep sandstone wells in Illinois are shot with 100 to 600 pounds of nitroglycerine set opposite the water-bearing zones. Average improvement of the various sandstone units varies from 22 to 38 per cent. Increase in the yield of these wells has been attributed to the enlargement of the hole and the fact that fine materials deposited during drilling are removed from the face of the well [44]. Explosives also may be effective in opening fractures in crystalline and other firmly indurated rocks. In Missouri and other states, acid treatment of wells in carbonate rocks is commonly practiced. The acid probably cleans the face of the well, but also may enter into small fractures in the vicinity of the well and increases the local permeability by solution. Effects of acidizing are generally less predictable than shooting. In certain regions mean improvements due to acidizing are greater than 250 per cent (Figure 10.7).

## 10.5 *Exploration for Ground Water*

Many sequences of marine sedimentary rocks have amazingly continuous beds of limestone and dark shale. Such rocks are, perhaps, best developed in the Upper Paleozoic rocks of central United States. One such marker bed, the Leavenworth limestone, a member of the Oread Formation, is between 1 and 2 feet thick and extends from central Iowa to Oklahoma without any significant lithologic changes. The importance of these beds to hydrogeologic problems lies in the fact that they can be used to predict the position of less regular beds which may be of interest as aquifers or confining beds. Predictions are particularly successful in areas with nearly flat-lying beds. Under such circumstances it is not unusual to be able to study local stratigraphic exposures and predict the depths to major lithologic units with an accuracy of better than 10 per cent. This is particularly true of limestone beds. Ancient stream channel deposits, beaches, and bars will form irregular sandstone bodies whose positions may be hard to predict on the basis of surface exposures alone (Figure 10.8).

Various geophysical methods, particularly seismic methods, have been applied widely to the exploration for petroleum. Indurated sedimentary layers commonly can be traced for long distances by refraction techniques and detailed information of the depths to the various beds can be determined. In addition, it is possible to map faults and folds within the sedimentary sequences. Aside from geophysical logging, little geophysical work has been done to assist with the location of water wells in sedimentary rocks. This is a matter of economics. If the wells are to be drilled to great depths, the cost of an adequate geophysical survey could be greater than the cost of a single water well. If the well is shallow, information from surface outcrops can commonly be used to predict geological conditions. Ground-water surveys do, nevertheless, make extensive use of geophysical as well as stratigraphic information gathered from the petroleum industry. Water chemistry, depth to aquifers, and relative permeabilities of various units can be determined for large regions provided that enough surface and subsurface geological and geophysical work has been done for the petroleum industry.

### SANDSTONES

Firmly cemented sandstones with low porosities and permeabilities will yield water to wells along fractures. The same general guiding principles apply to the location of water in these rocks as apply to the location of ground water in crystalline rocks of plutonic origin. Most favorable areas for development of ground water are along fault zones and within

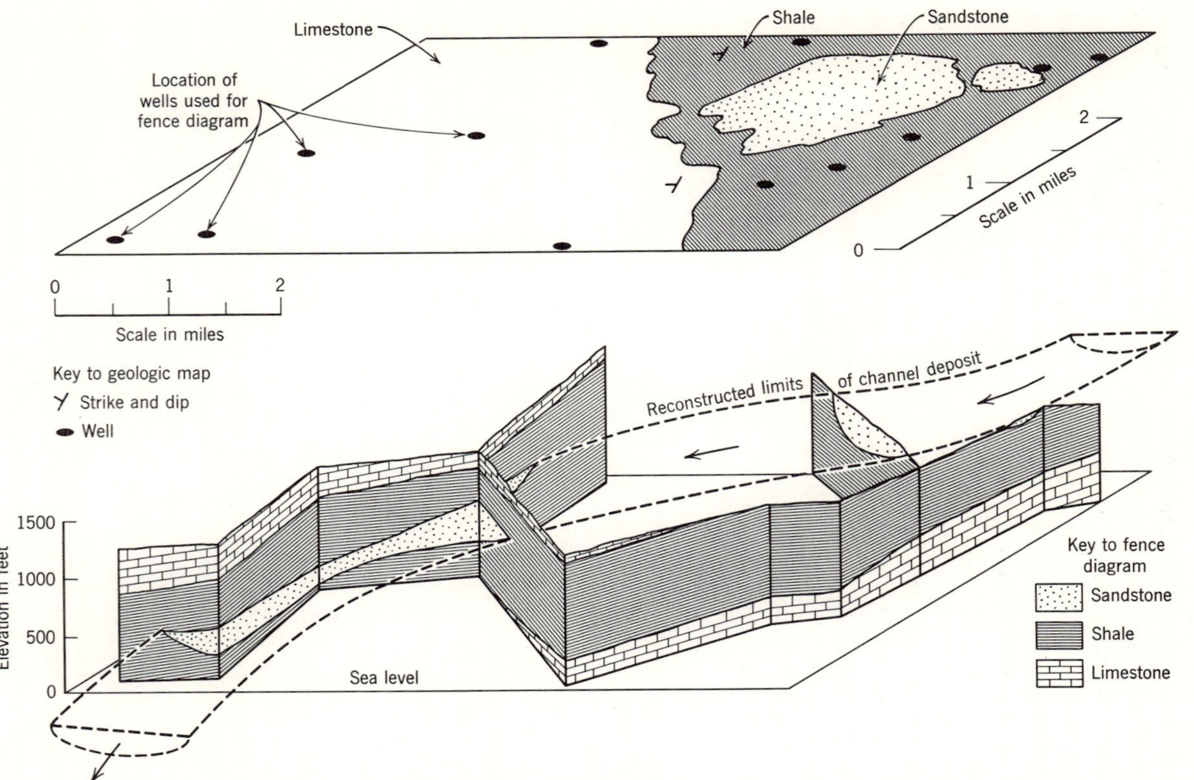


Figure 10.8 Fence diagram constructed from surface geologic map and well logs. Subsurface position of channel sandstone would have been impossible to predict on the basis of surface exposures alone.

thoroughly jointed zones. Better wells will be found in broad valleys and on flat upland areas than on hill crests and valley slopes. The permeability of the rocks in general decreases with depth.

Care should be taken not to confuse case-hardened sandstone with sandstone which has been cemented throughout its entire thickness. Case-hardened rock usually gives way to unconsolidated sands or poorly indurated sandstones at depths of less than 20 feet. Thus, sandstones which appear hard and impermeable on the surface may be fair aquifers at depth.

### CARBONATE ROCKS

Despite the fact that individual limestone beds can be located through structural and stratigraphic studies, the yields of the beds are particularly hard to predict. Minor jointing, differences of composition, faults, small layers of shale or chert, and initial permeability may all control the localization of solution openings in dense limestone and dolomite. In some places certain formations are known from empirical evidence to be aquifers. In only a few places, however, can a casual surface geologic inspection indicate whether or not a limestone bed will be an aquifer. Suggestive evidence from an outcrop would be the presence of solution openings, closely spaced jointing, or the presence of a fault in the zone of interest. Evidence indicating that a dolomite or limestone unit might not be an aquifer at depth would be the presence of numerous shale partings as well as the absence of solution openings.

Observations in quarries and other excavations in nearly horizontal carbonate rocks have shown that solution openings along vertical joints are spaced rather widely and that openings along bedding planes are more important from the standpoint of water production [26,43]. In addition, the vertical joints that are widened by solution near the surface tend to be filled by clay from the overlying soil. Horizontal openings, on the other hand, tend to remain open. Owing to the very small chance of striking vertical fractures with wells, most wells obtain water from horizontal solution openings along the bedding planes (Figures 10.9 and 10.10). The horizontal openings, nevertheless, tend to be better developed near faults, so that traces of vertical faults at the surface should be used to locate the most favorable drilling sites for water wells.

In areas of thick limestone or dolomite, wells located in valley bottoms are somewhat better than those on valley slopes. Water storage in adjacent alluvium together with a watertable closer to the surface account for some of this advantage. Wells drilled on broad uplands are also more successful than those drilled on hill slopes. Fractures and solution openings are more abundant along crests of anticlines and within syclinal troughs than

Ground Water

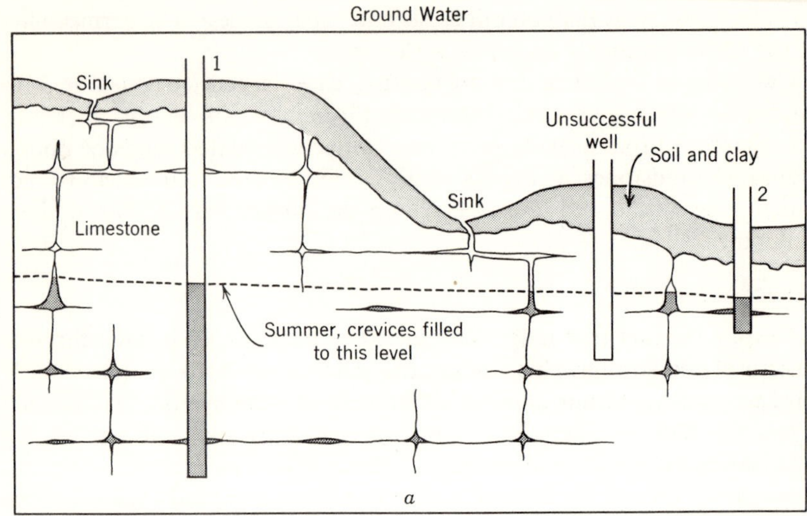


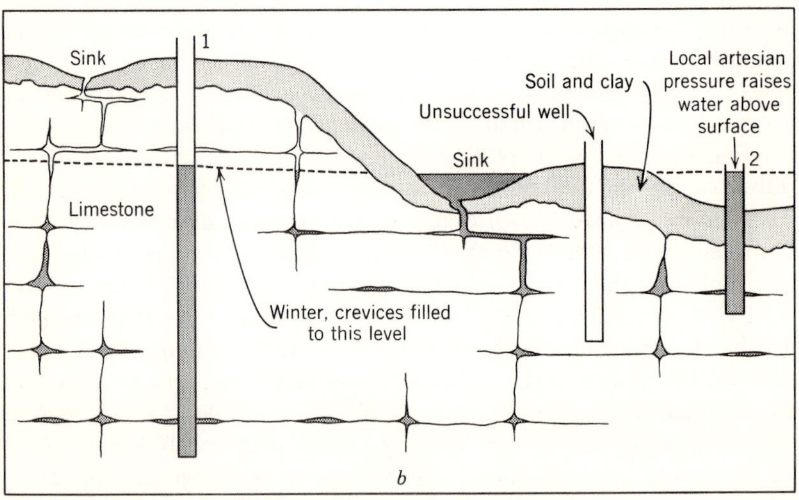


Figure 10.9 Diagrams showing crevice system and occurrence of water under water table and artesian conditions in the Hopkinsville Quadrangle, Kentucky. (Diagram by Walker [43].)

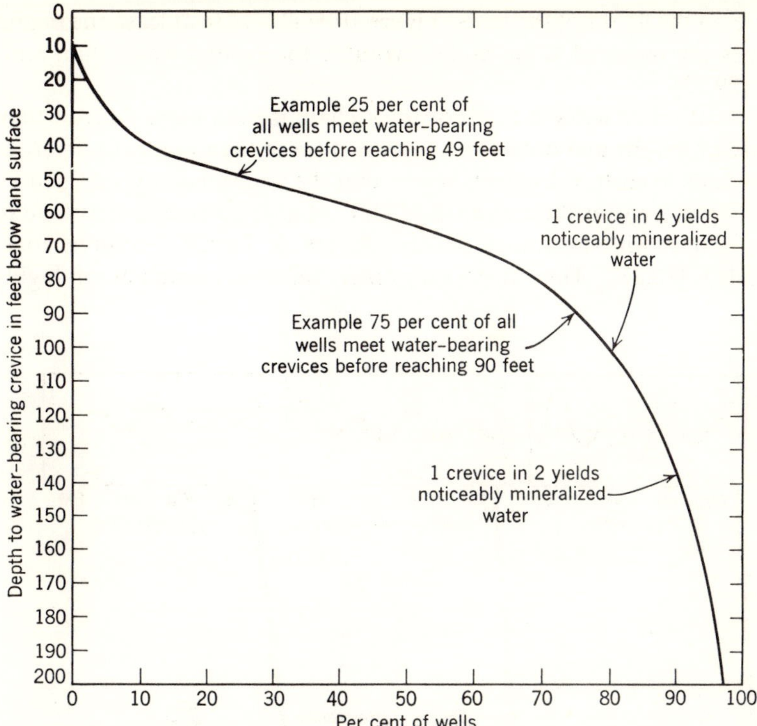


Figure 10.10 Curve showing relation between depth and per cent of wells that have encountered water-bearing crevices, Hopkinsville Quadrangle, Kentucky. (Diagram by Walker [43].)

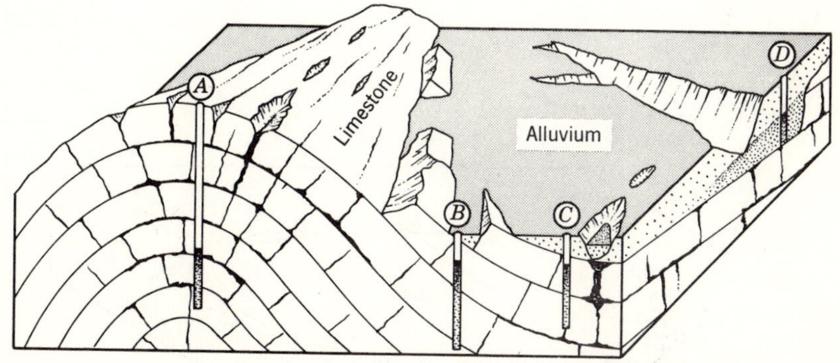


Figure 10.11 Water wells in a region of folded carbonate rocks. Dark areas indicate zones with solution openings. Well A is successful but pumping lift is excessive. Well B has a low yield. Well C is successful and has only a small pumping lift. Well D has been developed in alluvium. (Diagram adapted from Kiersch and Hughes [27].)

they are on the flanks of the folds (Figure 10.11) [27]. In at least one region synclines are reported to be more favorable for ground water than anticlines [29,33].

The location of wells in areas of thick chalk is even more difficult than in dense limestone and dolomite. Compilation of the results of numerous aquifer tests in eastern England shows that the transmissivity varies in an irregular pattern over the area studied [23]. Higher transmissivities tend to be associated with anticlinal flexures, zones of harder fractured rock, and valleys [10,23]. The valleys may reflect selective erosion along highly

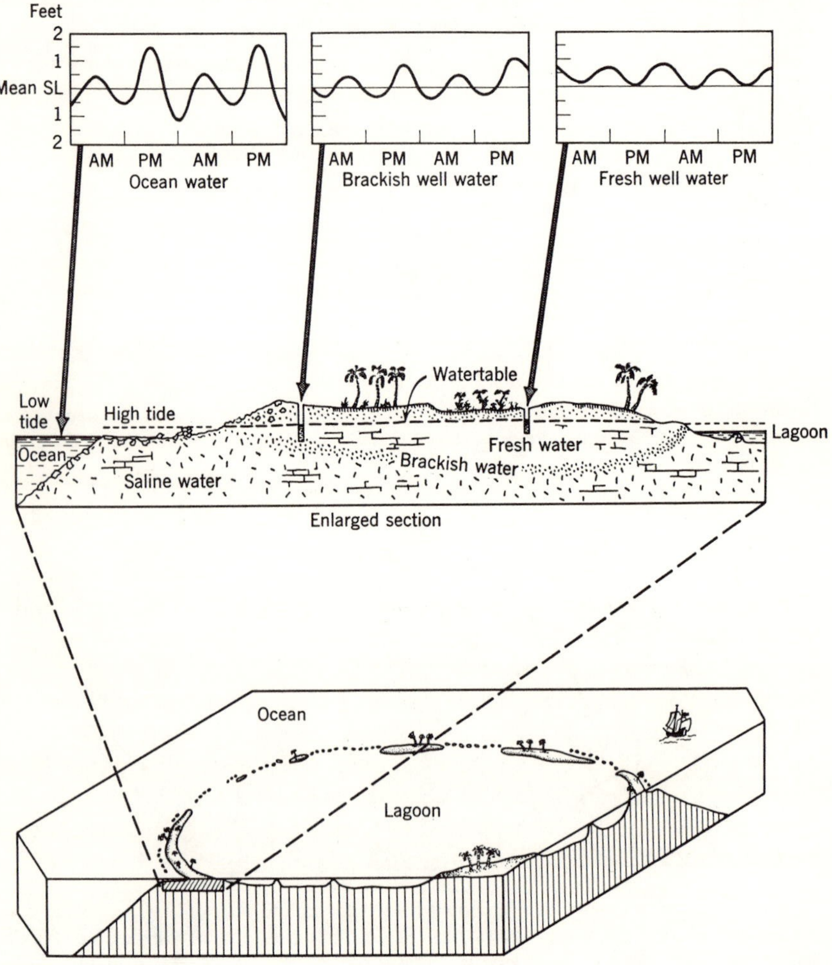


Figure 10.12 Fresh-water lense and water-level fluctuations in an island of an atoll.

fractured areas, so the higher transmissivities may be owing to the fracturing rather than to the topographic position.

Young limestones generally retain some original porosity, and if the limestone is made of coarse detrital material, permeable zones are not hard to locate. In fact, in the youngest limestones the entire rock mass may be permeable. A special case of young limestones is found on atolls.

Most of the atoll that is above sea level is composed of fragmental material that has piled on top of a denser and more consolidated reef platform. The fragmental material varies from a very coarse, blocky rubble on the ocean side of the islands to a fine, calcareous sand on the lagoon sides (Figure 10.12). The coarser rubble is commonly cemented but retains a high permeability [12].

Ground water on the islands originates from sea water that moves in and out of the sediments in response to tidal fluctuations and from the direct infiltration of rainwater. Potable water is absent on small islands, particularly in regions where the rainfall is low. On larger islands, rainwater will infiltrate and tend to float on the underlying sea water. Owing to the variation of permeability across the islands, the fresh-water lense is asymmetrical, being thickest near the lagoonal side where the permeability is the lowest and the water does not flow outward as rapidly as in the coarse rubble (Figure 10.12) [12].

Water for drinking in most of the islands is gathered from the runoff of roof tops and is collected in storage tanks. In the larger islands, ground water is used for washing and to a lesser extent for drinking [48]. The small thickness of the fresh-water lense prevents extensive pumping without the upward intrusion of saline water.

## **10.6** *Limestone Caverns*

A fascinating aspect of hydrogeology that has a close relation to the occurrence of water in limestone terrane is the study of limestone caverns. Before 1930 most geologists assumed that the caverns originated above the watertable by the combined effects of solution of vadose water and the downward cutting by cave streams. W. M. Davis [15], however, pointed out that such features as tubular shaped caverns, large blind branches to caverns, zones of delicate lacework, and reversed cavern gradients could best be explained by assuming that the caverns were excavated initially below the watertable. The part of cave formation at or above the watertable is evidenced by surface streams being connected in continuous systems with cave streams, by stream gravel within caverns, and by entrenched stream meanders on the floors of some caverns. Most caverns above the watertable show some evidence of partial filling by dripstone accumulations (stalagtite, stalagmite, and other forms), by rubble from roof collapse filling parts of the caverns, and by clay filling along the floors [8,15].

Most modern students of the subject will recognize the fact that some cavern formation takes place below the watertable and at the watertable, as well as above the watertable. The relative importance, however, of the various sites has been the subject of lively discussion in the literature for

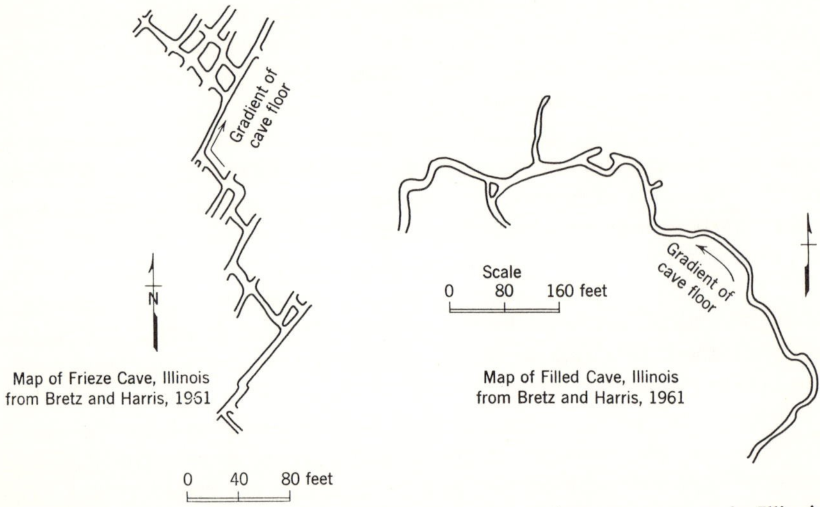


Figure 10.13 Rectilinear and sinuous patterns of two limestone caverns in Illinois. (From Bretz and Harris [9].)

many years and a number of extensive reviews of the problem have been published [8,22,38,40]. Despite the divergent views, most researchers point out that there is a strong structural control for the localizations of the larger part of the caverns that have been studied. The structural control, which can be either jointing or faulting, is seen in the strikingly rectilinear patterns as shown on maps of caves (Figure 10.13).

One of the original arguments of Davis for the formation of limestone caverns at some depth below the watertable was the fact that the equal-potential surfaces in homogeneous and isotropic media near a line sink would require the deep circulation of some of the water as shown in Figure 10.14. The possibility of deep solution of limestone is supported by recent geochemical studies that indicate the presence of water under-saturated with respect to $CaCO_3$ at depths of at least 800 feet below the surface [5]. Inspection of the flow net of Figure 10.14 indicates a marked crowding of the net near the river and some crowding near the watertable;

this indicates, in turn, zones of greater ground-water velocity. Inasmuch as the solution rate will be some direct function of amount of water per unit time coming in contact with the rock, the areas of greatest ground-water velocity should also be zones of maximum solution. Thus the homogeneity will be destroyed and an even greater amount of ground water will be diverted to near the watertable. Normal bedding of the limestone, if nearly horizontal, will add an initial anisotropy which in

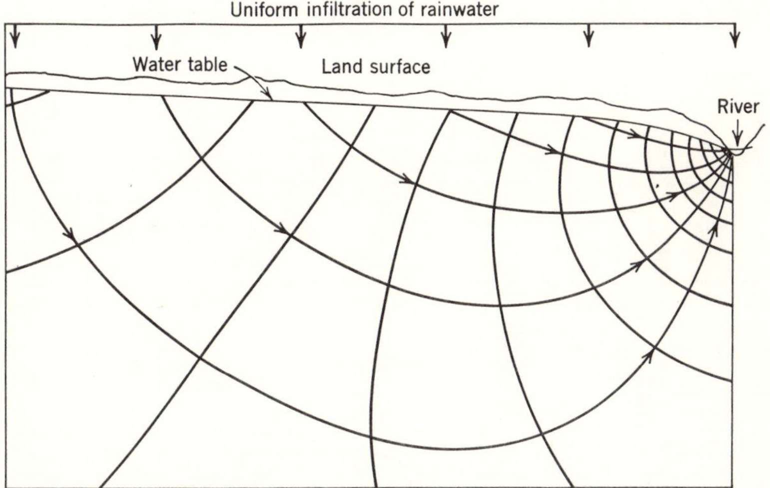


Figure 10.14 Hypothetical circulation of ground water in a homogeneous and isotropic medium with recharge.

nature should accentuate the solution effects near the watertable. Thus, based on general reasoning, in moderately homogeneous rocks the largest number of caverns which are being formed at the present should be near the watertable and near the points of natural ground-water discharge.

Records of water-well drillers generally show that the chances of encountering large water-filled caverns are quite small even in regions noted for their caverns. Even if the general trend of such a cavern could be determined, maps of caverns presently above the watertable suggest that only a few wells would encounter the cavern because of the irregular course of the cavern in detail (Figure 10.13). Large caverns, nevertheless, are surely present below the watertable, as evidenced by a few exploratory wells, studies of dam foundations, and observations of very large springs. Some of the springs have flows of more than one cubic meter per second. Such flows can only be transmitted through very large openings. The largest limestone spring described in the literature is near the town of Ras-el-Ain, Syria, which is on a tributary of the Euphrates River near the Turkish

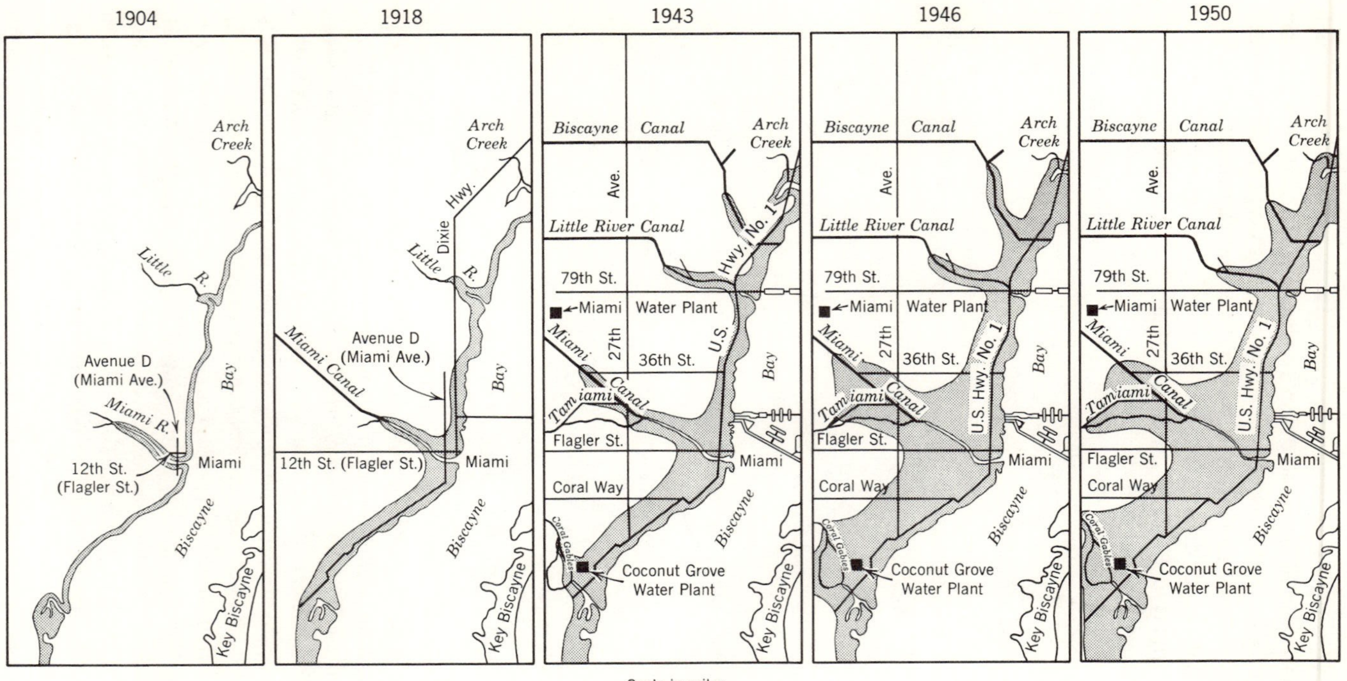


Figure 10.15 Progressive salt-water encroachment in the Miami area from 1904 through 1950. Control structures placed in tidal canals in 1944 have reduced the rate of encroachment and in some places have greatly improved water quality. (After work of G. D. Parker as presented by Black, Brown, and Pearce [6].)

border [11]. The discharge of this spring generally fluctuates between 35 and 40 cubic meters per second and averages 38.7 cubic meters per second (1370 cfs). A number of other springs in the same geologic province have discharges of more than one cubic meter per second [1].

## 10.7 *Ground-Water Management*

The moderately low specific yields of many limestones favor rapid overproduction during times of little recharge. Where cavernous interconnections exist the effects of overproduction may spread rapidly to adjacent areas. In one area around Hershey, Pennsylvania, a general decline of water levels caused by dewatering operations at a mine not only caused a detrimental lowering of water levels but triggered the formation of a large number of new limestone sinks [16].

In addition to the relatively rapid decline of water levels, human activities also cause difficulties with pollution and contamination of various types. Large-solution openings favor the rapid entrance of organic as well as inorganic contaminants. Shallow wells, abandoned quarries, and pits are sometimes used for the disposal of storm water, garbage, and other unwanted material. Commonly the holes intercept permeable zones and allow the direct feeding of contaminants into aquifers. Overdraft of coastal limestone aquifers has caused salt-water encroachment in Florida. Also, construction of drainage canals has allowed sea water to migrate inland and feed into aquifers from the canals in Miami, Florida (Figure 10.15).

## 10.8 *Water Quality*

Ground water recovered from sedimentary rocks can vary in chemical quality from saturated brines in deeply buried marine rocks to water with less than 100 ppm total dissolved solids in near-surface sandstones. If the ground water in all sedimentary rocks is considered, saline water and brine are far more abundant than fresh water. Fortunately, however, the fresh water is commonly at or near the surface and the nonpotable water is found at greater depths. In general, the salinity increases with depth, but exceptions are found in many regions. For example, in southeastern Kansas the most saline water is found in the Cherokee Formation of Pennsylvanian age, whereas potable water is recovered from the underlying formations, particularly the sandy dolomite of the Roubidoux Formation of Ordovician age [49].

The problem of the origin of brines in many of the older sedimentary rocks remains one of the fascinating problems of the geochemistry of ground water [30]. Normally in Cenozoic sediments, connate water does not exceed the salinity of sea water unless there is an obvious connection with modern or ancient evaporite deposits. On the other hand, connate water in Paleozoic rocks commonly will have total dissolved solids between 100,000 and 350,000 ppm. Most of these brines are in sedimentary beds that show no indication of ever having evaporite deposits associated with them. Extensive evaporation of the sea water during the time of deposition or the postdepositional solution of salt beds appear to be ruled out as explanations.

The preferential expulsion of the loosely attached ions during the advanced stages of the compaction of shales is one possible source of the high salinity. A similar, and more popular explanation, involves the preferential passing of water through shale beds, allowing the dissolved solids to build up in the lower beds as circulating water replenished the deeply buried zones through infiltration from the outcrop or from connate water expelled from underlying shale beds [7].

Regardless of the exact mechanism responsible for the original concentration of the brines, the subsequent migration of the brine during the exposure of the brine-bearing beds by erosion is of direct practical concern to the hydrogeologist. Saline or brackish springs throughout regions of marine sedimentary rocks attest to the slow flushing of the connate water. In regions where thick sedimentary aquifers exist, the flushing must be quite efficient because widespread fresh-water zones exist such as in the early Paleozoic aquifers in northern Illinois. The removal of saline water probably takes place in stages in many aquifers that are intersected laterally by unconformities. During each erosional cycle, some of the brine is removed. Nevertheless, if the selective passage of water through shale is an important factor, the salinity may build up between times of vigorous ground-water circulation.

In Tertiary rocks of the Coast Ranges in California, in Pennsylvanian rocks of central United States, and in many regions underlain by shale-rich sequences, saline water is encountered at depths of only a few tens of feet below the bottoms of adjacent stream valleys. Permeability is generally low enough to prevent the active circulation of meteoric water even over periods of many hundreds of thousands of years.

Contrasts in the chemical quality of water in the three common groups of sedimentary rocks, sandstone, shale, and carbonates, is usually quite marked. Shale most commonly contains higher amounts of iron, fluoride, and a low pH, typically between 5.5 and 7.0. Limestone will have a lower $SiO_2$ content, a somewhat greater amount of calcium and magnesium, and pH values generally above 7.0. Water quality of sandstones is somewhat more variable, depending on adjacent rock types, mineral composition of the sands, and depth of the aquifers. Many wells that recover potable water from deep sandstone aquifers along the Atlantic and Gulf Coastal Plain have very soft water with a combined calcium and magnesium content of less than 20 ppm. The sodium and bicarbonate contents of these waters, however, are typically quite high.

## REFERENCES


1. Abd-El-Al, Ibrahim, 1953, Statics and dynamics of water in the Syro-Lebanese limestone massifs; in *Ankara symposium on arid zone hydrology:* United Nations Educational, Scientific, and Cultural Organization, pp. 60–76.
2. Archie, G. E., 1950, Introduction to petrophysics of reservoir rocks: *Am. Assoc. Petroleum Geologists Bull.*, v. 34, pp. 943–961.
3. ——1952, Classification of carbonate reservoir rocks and petrophysical considerations: *Am. Assoc. Petroleum Geologists Bull.*, v. 36, pp. 278–298.
4. Arnow, Ted., 1963, Ground-water geology of Bexar County, Texas: *U.S. Geol. Survey Water-Supply Paper* 1588, 36 pp.
5. Back, W., 1963, Preliminary results of a study of calcium carbonate saturation of ground water in central Florida: International Assoc. Sci. Hydrology, 8th year, No. 3, pp. 43–51.
6. Black, A. P., E. Brown, and J. M. Pearce, 1953, Salt water intrusion in Florida—1953: *Florida Water Survey and Research Paper No.* 9, 38 pp.
7. Bredehoeft, J. D., and others, 1963, Possible mechanism for concentration of brines in subsurface formations: *Am. Assoc. Petroleum Geologists Bull.*, v. 47, pp. 257–269.
8. Bretz, J H., 1942, Vadose and phreatic features of limestone caves: *Jour. Geol.*, v. 50, pp. 675–811.
9. Bretz, J H., and S. E. Harris, Jr., 1961, Caves of Illinois: *Illinois Geol. Survey, Rept. Investigations* 215, 87 pp.
10. Buchan, S., 1963, Geology in relation to ground water: *Jour. Inst. Water Engineers*, v. 17, pp. 153–164.
11. Burdon, D. J., and C. Safadi, 1963, Ras-el-Ain: The great karst spring of Mesopotania, an hydrogeologic study: *Jour. Hydrology*, v. 1, pp. 58–95.
12. Cox, D. C., 1951, The hydrology of Arno Atoll, Marshall Islands: National Acad. Sci., *Atoll Research Bull. No.* 8, 29 pp.
13. Darton, N. H., 1897, Preliminary report on artesian waters of a portion of the Dakotas: *U.S. Geol. Survey 17th Ann. Rept.*, Part 2, pp. 1–92.
14. Davis, S. N., and F. R. Hall, 1959, Water quality of eastern Stanislaus and northern Merced counties, California: *Stanford University Pub. Geol. Sci.*, v. 6, No. 1, 112 pp.
15. Davis, W. M., 1930, Origin of limestone caverns: *Geol. Soc. Am. Bull.*, v. 41, pp. 475–628.
16. Foose, R. M., 1953, Ground-water behavior in the Hershey Valley, Pennsylvania: *Geol. Soc. Am. Bull.*, v. 64, pp. 623–646.
17. Gloyna, E. F., and T. D. Reynolds, 1961, Permeability measurements of rock salt: *Jour. Geophys. Research*, v. 66, pp. 3913–3921.
18. Gondouin, M., and C. Scala, 1958, Streaming potential and the SP log: *Am. Inst. Mining Metal. Eng. Petroleum Trans. Tech. Paper* 8023, 9 pp.

19. Harbaugh, J. W., 1960, Petrology of marine bank limestones of Lansing Group (Pennsylvanian), southeastern Kansas: *Kansas Geol. Survey, Bull.* 142, Part 5, pp. 189–234.
20. Harbaugh, J. W., 1961, Relative ages of visibly crystalline calcite in late Paleozoic limestones: *Kansas Geol. Survey Bull.* 152, Part 4, pp. 91–126.
21. Hohlt, R. B., 1948, The nature and origin of limestone porosity: *Colorado School of Mines Quart.*, v. 43, no. 4, 51 pp.
22. Howard, A. D., 1963, The development of karst features: *National Speleological Soc. Bull.*, v. 25, pp. 45–65.
23. Ineson, J., 1963, Applications and limitations of pumping tests: Hydrogeological significance: *Jour. Inst. Water Engineers*, v. 17, pp. 200–215.
24. Johnson, C. R., and R. A. Greenkorn, 1960, Comparison of core analysis and drawdown test results from a water-bearing Upper Pennsylvanian sandstone of central Oklahoma: *Geol. Soc. America Bull.*, v. 71, pp. 1898.
25. Johnson, C. R., and R. A. Greenkorn, 1963, Correlation of lithology and permeability for a Virgilian sandstone reservoir in central Oklahoma: *Geol. Soc. America Proceedings*, No. 73, pp. 180–181.
26. Johnston, R. H., 1962, Water-bearing characteristics of the Lockport Dolomite near Niagara Falls: New York, *U.S. Geol. Survey Prof. Paper* 450-C, pp. 123–125.
27. Kiersch, G. A., and P. W. Hughes, 1952, Structural localization of ground water in limestones—"Big Bend District," Texas-Mexico: *Econ. Geol.*, v. 47, pp. 794–806.
28. Koenig, L., 1960, Survey and analysis of well stimulation performance: *Am. Water Works Assoc. Jour.*, v. 52, pp. 333–350.
29. La Moreaux, P. E., and W. J. Powell, 1960, Stratigraphic and structural guides to the development of water wells and well fields in a limestone terrane: *Internat. Assoc. Sci. Hydrology Pub. No.* 52, pp. 363–375.
30. Levorsen, A. I., 1954, *Geology of petroleum:* San Francisco, W. H. Freeman and Co., 703 pp. p. 550
31. Manger, G. E., 1963, Porosity and bulk density of sedimentary rocks: *U.S. Geol. Survey. Bull.*, 1144-E, 55 pp.
32. McComas, M. R., 1963, Productive core analysis characteristics of carbonate rocks in the Four Corners area; in R. O. Bass (Editor), *Shelf carbonates of the Paradox Basin, a symposium:* Four Corners Geol. Soc., Fourth Field Conference, pp. 149–156.
33. McMaster, W. M., 1963, Geology and ground-water resources of the Athens area, Alabama: *Geol. Survey of Alabama Bull.* 71, 45 pp.
34. Meade, R. H., 1962, Factors influencing the pore volume of fine-grained sediments under low-to-moderate overburden loads: *Sedimentology*, v. 2, pp. 235–242.
35. Murray, R. C., 1960, Origin of porosity in carbonate rocks: *Jour. Sed. Petrology*, v. 30, pp. 59–84.
36. Muskat, M., 1937, *The flow of homogeneous fluids through porous media:* New York, McGraw-Hill Book Co., 763 pp.
37. Piersol, R. J., L. E. Workman, and M. C. Watson, 1940, Porosity, total liquid saturation, and permeability of Illinois oil sands: *Ill. Geol. Survey Report Invest. No.* 67, 72 pp.
38. Radzitzky, d'Ostrowick, I., 1953, *L'hydrogéologie des roches calcareuses:* Dinant, L. Bourdeaux-Capelle, 199 pp.
39. Robertson, C. E., 1963, *Well data for water well yield map:* Missouri Geol. Survey and Water Resources, 23 pp.
40. Thornbury, W. D., 1954, *Principles of geomorphology:* New York, John Wiley and Sons, 618 pp.

41. Trombe, F., 1952, *Traité de spéléologie:* Paris, Piyot, 376 pp.
42. Waldschmidt, W. A., 1941, Cementing materials in sandstones and their probable influence on migration and accumulation of oil and gas: *Am. Assoc. Petroleum Geologists Bull.*, v. 25, pp. 1839–1879.
43. Walker, E. H., 1956, Ground-water resources of the Hopkinsville Quadrangle, Kentucky: *U.S. Geol. Survey Water-Supply Paper* 1328, 98 pp.
44. Walton, W. C., 1962, Selected analytical methods for well and aquifer evaluation: *Illinois State Water Survey Bull.*, 49, 81 pp.
45. Walton, W. C., and J. C. Neill, 1963, Statistical analysis of specific-capacity data for a dolomite aquifer: *Jour. Geophys. Research*, v. 68, pp. 2251–2262.
46. Weyl, P. K., 1960, Porosity through dolomitization: conservation-of-mass requirements: *Jour. Sed. Petrology*, v. 30, pp. 85–90.
47. White, D. E., 1957b, Magmatic, connate, and metamorphic waters: *Geol. Soc. Am. Bull.*, v. 68, pp. 1659–1682.
48. Wiens, H. J., 1962, *Atoll environment and ecology:* New Haven, Yale University Press, 532 pp.
49. Williams, C. C., 1948, Contamination of deep water wells in southeastern Kansas: *Kansas Geol. Survey Bull.* 76, Part 2, pp. 13–28.
50. Winsauer, W. O., H. H. Shearin, Jr., P. H. Masson, and M. Williams, 1952, Resistivity of brine-saturated sands in relation to pore geometry: *Am. Assoc. Petroleum Geologists Bull.*, v. 36, pp. 253–277.
51. Wyllie, M. R. J., and M. B. Spangler, 1952, Application of electrical resistivity measurements to problem of fluid flow in porous media: *Bull. Am. Assoc. Petroleum Geologists*, v. 36, pp. 359–403.

# *chapter 11*

# GROUND WATER IN NONINDURATED SEDIMENTS

## **11.1** *Introduction*

The search for ground water most commonly starts with an investigation of nonindurated sediments. There are sound reasons for this preference. First, the deposits are easy to drill or dig so that exploration is rapid and inexpensive. Second, the deposits are most likely to be found in valleys where ground-water levels are close to the surface and where, as a consequence, pumping lifts are small. Third, the deposits are commonly in a favorable location with respect to recharge from lakes and rivers. Fourth, nonindurated sediments have generally higher specific yields than other material. Fifth, and perhaps most important, permeabilities are much higher than other natural materials with the exception of some recent volcanic rocks and cavernous limestones.

Nonindurated sediments can be subdivided on the basis of origin into a large number of categories, the most important of which are alluvium, till, ice-contact deposits, loess, dune sand, marine sands and clays, colluvial deposits, and lacustrine clays and sands. Residual soils, although not sedimentary deposits in the usual sense of the term, have many hydrogeologic characteristics in common with alluvium and colluvium.

## **11.2** *Porosity, Permeability, and Specific Yield*

Porosities of nonindurated sediments range from a minimum of about 20 per cent in coarse, poorly sorted alluvium to about 90 per cent in soft

*Table 11.1 Porosity and Permeability of Selected Nonindurated Sediments*

| Sample Number | Orientation of Sample | Type of Sediment | Dominant Size | Permeability, darcys | Porosity, per cent | Specific Yield | References |
|---|---|---|---|---|---|---|---|
| 1 | vertical | alluvium | fine sand | 26.4 | 51.1 | 45.5 | 26 |
| 2 | horizontal | alluvium | fine sand | 25.3 | 51.5 | 45.8 | 26 |
| 3 | vertical | alluvium | fine sand | 16.5 | 47.0 | 39.9 | 26 |
| 4 | horizontal | alluvium | fine sand | 13.2 | 45.7 | 39.0 | 26 |
| 5 | vertical | loess | silt | 0.33 | 49.3 | 33.1 | 26 |
| 6 | horizontal | loess | silt | 0.22 | 50.7 | 34.7 | 26 |
| 7 | horizontal | marine | clay | 0.000016 | 48.5 | 3.6 | 26 |
| 8 | vertical | marine | medium sand | 38.5 | 41.7 | 38.3 | 26 |
| 9 | horizontal | marine | medium sand | 55.0 | 40.2 | 37.6 | 26 |
| 10 | ... | alluvium | fine sand | 5.5 | 52.2 | ... | 42 |
| 11 | ... | alluvium | coarse sand | 189 | 33.3 | ... | 42 |
| 12 | ... | alluvium | gravel | 1130 | 25.1 | ... | 42 |
| 13 | ... | artificial | gravel (larger than 38 mm) | 43500 | 38.0 | ... | 2 |
| 14 | ... | ... | clay (kaolinite) | 0.0015 | 50.0 | ... | 2 |
| 15 | ... | ... | clay (montmorillonite) | 0.000015 | 66.6 | ... | 2 |
| 16 | ... | dune sand | medium sand | 28.0 | 35.8 | 34.5 | 8 |

*Note:* Samples 1 and 2, 3 and 4, 5 and 6, and 8 and 9 are paired samples from identical locations.

muds and dry organic material. Porosities of between 25 and 65 per cent, however, are more common (Table 11.1). Specific yield values range from almost zero to about 50 per cent. Values typical of fine silts and clays are less than 10 per cent. Gravel and coarse sands have values typically greater than 20 per cent. In contrast to the rather uniform porosity values, permeabilities of various nonindurated sediments vary through extremely wide values. Highest permeabilities measured are more than $10^9$ times larger than the lowest permeabilities. Despite the wide range of values, some approximations of permeability can be made if the geologic origin of the sediments are known. This matter is discussed at greater length in the later sections of this chapter.

Porosity, specific yield, and permeability are all dependent on the shape, packing, size distribution, and incipient cementation of the constituent particles in nonindurated sediments. Highly angular particles tend to be held apart by irregular, sharp corners thus producing large values of porosity, permeability, and specific yield for a given grain diameter. Slight rounding of the particle corners will facilitate interdigitation so that subangular particles will have a maximum compactness [38]. Rounded particles will not be able to interdigitate and will, therefore, be less compact and will have values of porosity, permeability, and specific yield that are similar to those of highly angular particles. In addition to the angularity of the particles, the overall shape of the particles has a large effect on packing. Tabular particles tend to form box-like openings, particularly in fine-grained material where slight adhesive forces can resist the small weight of the particle that tends to rotate into a closer packing (Figure 11.1). Theoretically, perfect spheres can be packed with porosities ranging from 25.95 to 47.64 per cent [17]. Plate-shaped particles, on the other hand, can be packed with porosities ranging from almost zero to almost 100 per cent.

The mode of deposition will have some effect on the packing of the particles. For example, till deposited beneath glacial ice will have a lower porosity than will mudflow deposits of a similar composition. Sand deposited on the lee side of a dune will be more porous than the same size sand deposited on a beach. Depth of burial will also affect packing through natural increase of intergranular pressure (Figure 11.2). The average decrease of porosity within the upper 1000 feet must amount to several per cent, although studies of core samples have shown a wide scatter of data and even local increases of porosity with depth which are caused by variations of particle size and shape. Pore reduction due to burial is most important in clays [28] and shales but is present to a minor extent in sands and gravels. Most of the pore reduction is owing to inelastic effects of intergranular movement. Some elastic deformation of grains does take place so that core expansion should be taken into account in samples taken for laboratory study from depths of more than 100 or 200 feet.

Porosity, specific yield, and permeability are all sensitive to particle size and size distribution, or sorting. The size will determine the relative

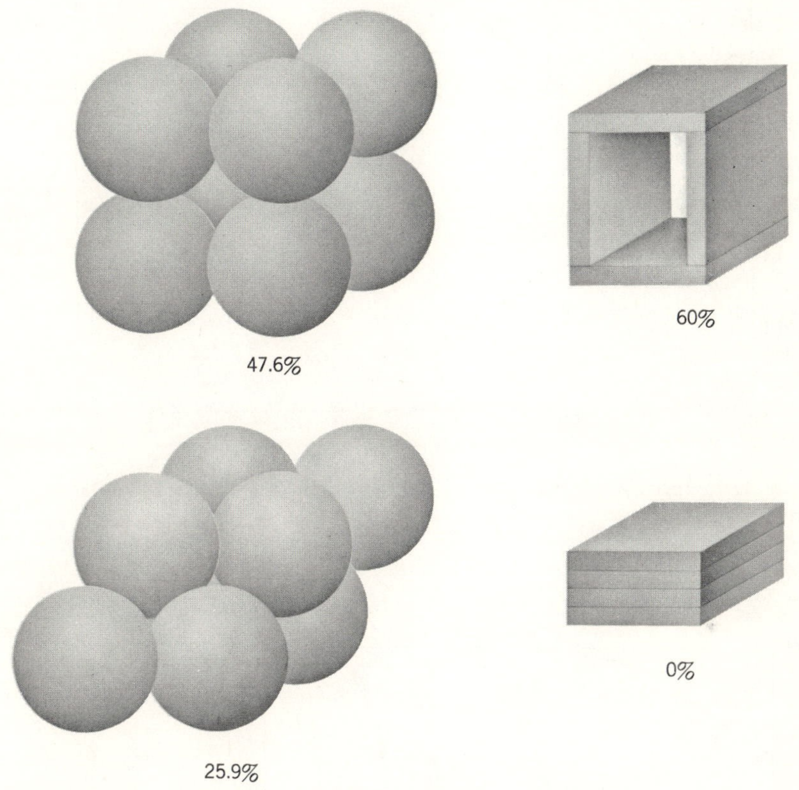


Figure 11.1 Open and close packing of spherical and tabular particles. Figures indicate maxima and minima porosities for stable configurations of the particles.

importance of surface tension and smaller molecular forces in retaining water within pores. The size will also help determine the pore diameters which have a dominant effect on permeability (Figure 11.3) and specific yield. Sorting will determine the extent to which smaller grains can occupy space within larger pores (Figure 11.4). Other things being equal, poorly sorted sediments will have lower values of porosity, permeability, and specific yield. Krumbein and Monk [23] studied the effect of both particle size and sorting in artificially mixed sands and expressed their

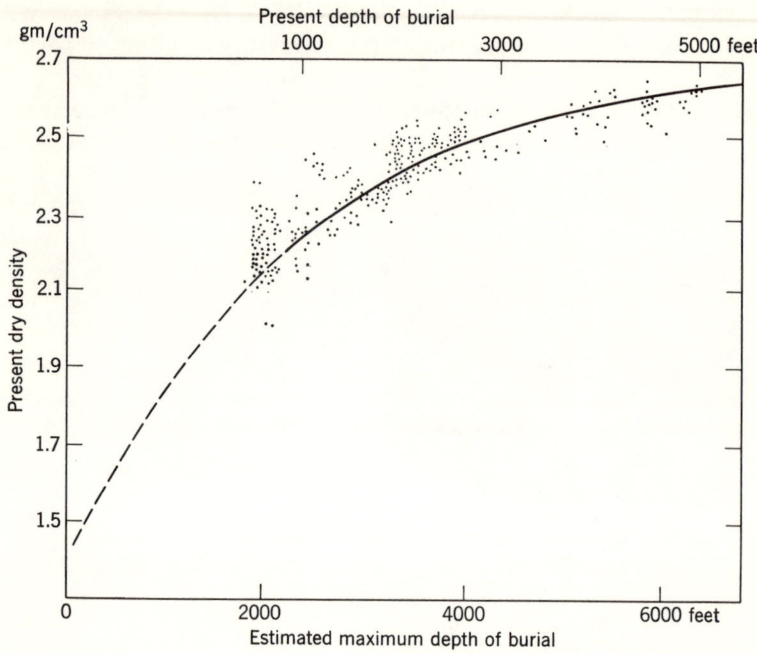


Figure 11.2 Increase of shale density with depth in rock cores from Oklahoma. (From Athy [1].)

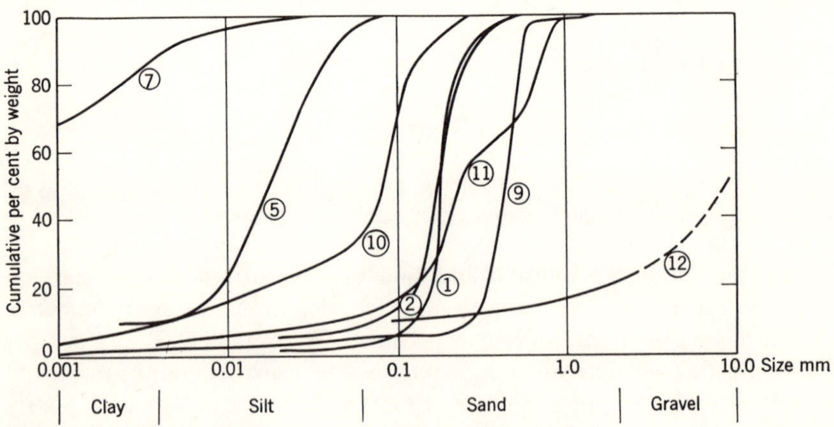


Figure 11.3 Size distribution of several samples listed in Table 11.1. Importance of size in determining permeability can be seen by comparing median sizes with permeability. Sample numbers on curves correspond with sample numbers in Table 11.1.

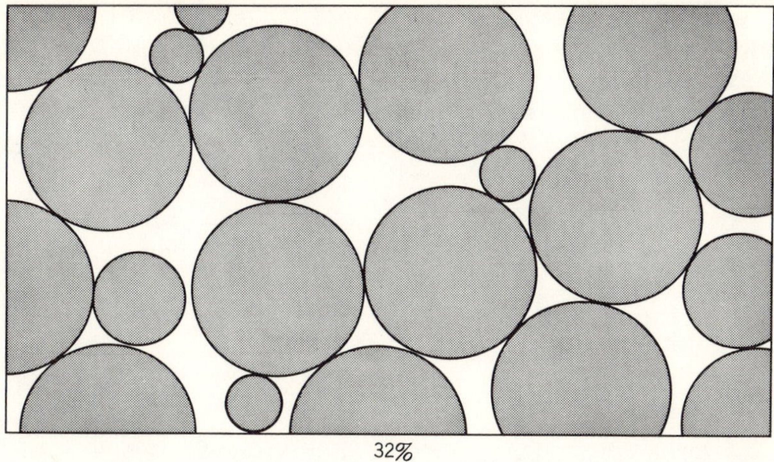

32%

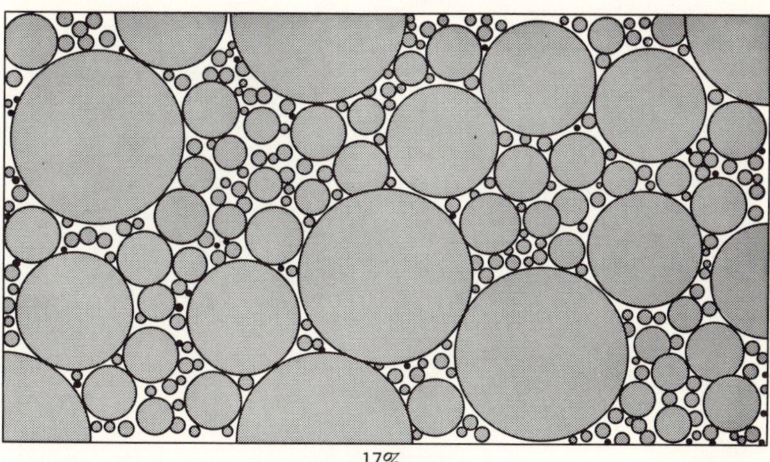

17%

Figure 11.4 Contrast in porosity caused by difference in sorting. Numbers under diagrams indicate percent porosity in each diagram.

results in the following semiempirical equation:

$$k = 760\, d_*^{\,2} e^{-1.31\sigma} \tag{11.1}$$

in which $k$ is permeability in darcys,

$d_*$ is the geometric mean diameter given in millimeters,

$e$ is the dimensionless constant 2.718,

$\sigma$ is the log standard deviation of size distribution, which is dimensionless,

and 760 is a constant for the conversion of permeability units to darcys

The foregoing equation was developed for only a restricted type of material. It may, however, reflect the general form of equations applicable to other sediments.

Almost all sediments have small amounts of cementing material even though they may give the appearance of being completely nonindurated.

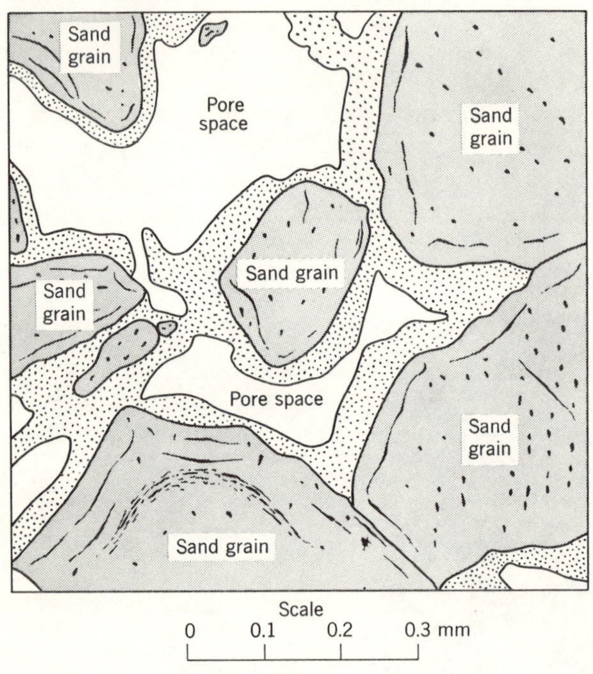


Figure 11.5 Clay skins or coatings (stippled) on sand grains in a sandy soil as seen in a thin section of soil. (U.S. Soil Conservation Service [40].)

Clay and colloidal material which are present in all but the purest sediments tend to form coatings on larger particles and open networks next to smaller particles. Clay coatings are best developed within the B horizon of soils (Figure 11.5) where permeabilities may be reduced to only a small fraction of their original values as evidenced by permeabilities of parent material. Inasmuch as clay will expand with hydration, the permeability of clay-cemented material will be strongly influenced by the amount of moisture present. Silica, calcite, limonite, and other cementing materials common in sedimentary rocks are relatively unimportant in their nonindurated counterparts.

## 11.3 *River Valleys*

### GENERAL CHARACTERISTICS

The distribution of clay, silt, sand, and gravel within river valleys is exceedingly complex in detail. General patterns are, nevertheless, quite regular and predictable. In wide valleys in which the river channel occupies a restricted part of the valley the deposits can be classified on the basis of their topographic forms and position in relation to the channel (Figure 11.6). Coarsest deposits that provide the best aquifers are sands and gravels that make up the traction load of the stream and are deposited directly within the channel. Point bar deposits are special channel deposits formed during high water at river bends. As the river migrates, point bars with intervening swales are left behind. The swales fill slowly with silt and clay deposited by flood water. The result is a series of arcuate strips of fine-grained sediments overlying a somewhat thicker layer of sand. Continued flood deposition will build up a thicker layer of fine sediment that grades from coarse silt at the natural levee into fine silt and clay within the backswamp deposits (Figure 11.6). The fine-grained flood deposits reach their maximum thickness where they have filled former cutoff meanders of the river channel. These fillings are called "clay plugs."

Not all river valleys display the entire sequence of floodplain deposits characteristic of the larger rivers of southern United States. Many rivers carrying large bed loads of coarse sediment will develop broad flood plains laced with numerous interconnected channels. These braided channels do not develop well-defined levees, swale fillings, and other deposits related to streams with straight or meandering channels.

Changes of stream regimen can cause abrupt downcutting of rivers leaving former floodplain materials as alluvial terrace deposits. These deposits generally exhibit the same types of sediments as the adjacent valley deposits except where drainage changes in the watershed have caused changes in sediment provenance and carrying capacities of the rivers.

Despite the great lateral variability of river-valley sediments, most of the valley deposits have a simple vertical succession from coarse sands or gravels near the bottom of the channels to silts and clays at the top. The relative thickness of the coarse and fine units depends on the type of sediments carried by the river and the geologic history of the river at the point of interest. Coarse glacial outwash filling a mountain valley commonly has only the slightest amount of fine-grained material at the top. Some of these valleys have been filled by more than 200 feet of coarse

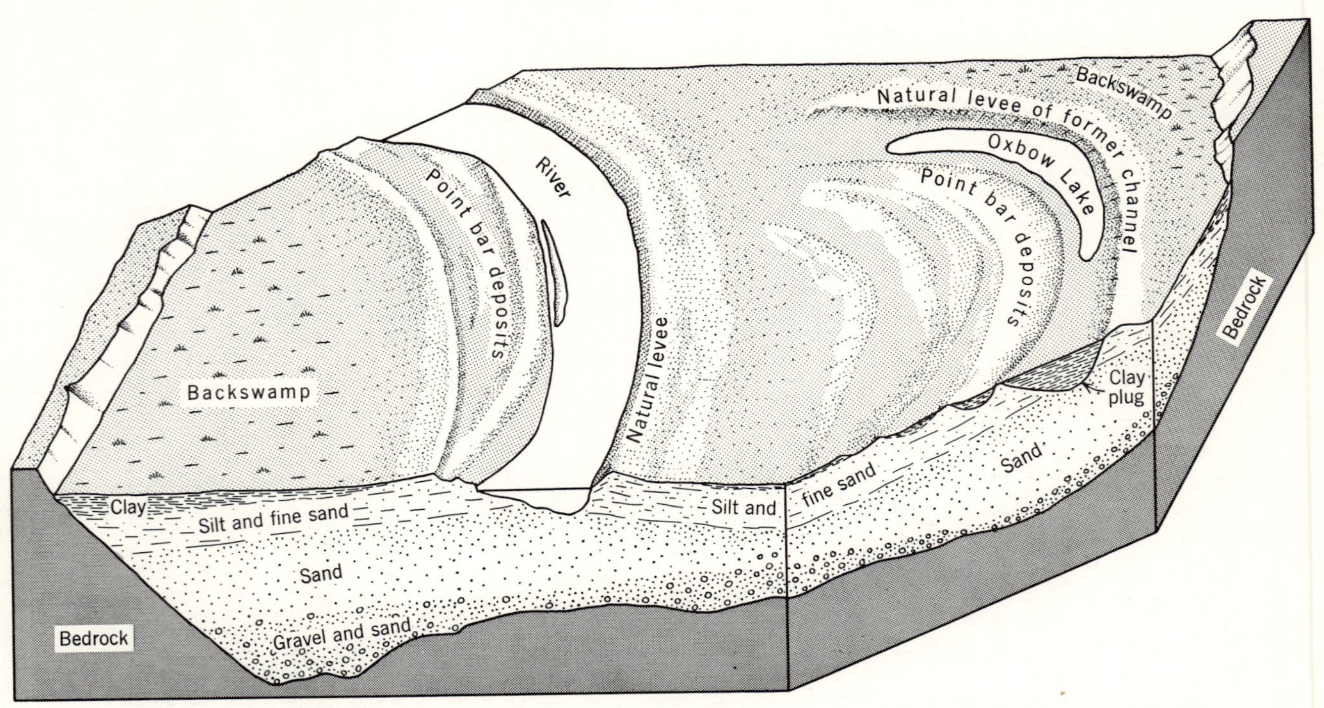


Figure 11.6 Topographic forms and deposits typical of broad flood plains of large rivers.

bed-load material. On the other hand, rivers that are suddenly blocked near their mouths may fill their valleys with silt and clay and have only small amounts of bed-load material at the bottoms of the sedimentary fills. Typically, however, deposits of modern or Late Pleistocene rivers will be from 20 to 150 feet thick and have at least five and, more commonly, several tens of feet of coarse sands and gravels near their bases. Deposits of most rivers thicken at the coasts because they fill former valleys that had been deepened in response to much lower sea levels during times of Pleistocene glaciation.

#### PERMEABILITY

A large number of studies have been made of the water-bearing properties of alluvium from normal stream valleys. Fine-grained sediments are rarely impermeable if judged by standards used in the petroleum industry. Silts and loosely compacted clays will have permeabilities of at least a few millidarcys and in many sediments the permeabilities will be several hundred millidarcys. Some of the permeability can be explained on the basis of an open structure of the original clay and silt aggregates. Most of the permeability, however, is probably owing to secondary structures such as root holes, worm burrows, and desiccation cracks. Most tests of water-bearing zones within alluvial sediments indicate permeabilities of between 10 and 100 darcys, although maximum values of more than 500 darcys are not rare. For many purposes, useful estimates can be made of permeability if the size of the sedimentary particles are known. Correlations can best be made of data from single river valleys such as the Arkansas River in southern United States (Figure 11.7).

#### WELL YIELDS

Moderate well yields of 10 to 50 gpm can be obtained from almost all river deposits that originate from large perennial streams. Much larger yields of 100 to 2000 gpm are also common where the permeable zones total at least 10 feet and the saturated zone in the alluvium is at least 40 feet thick [4,13,42].

Exact well yields cannot be calculated before drilling, but useful estimates can be made by assuming reasonable permeability values [16] and taking into account the total expected thickness of the aquifers, the position of water levels in nearby wells, and the nearness of hydrologic boundaries. As a hypothetical example, a test hole near the Arkansas River indicates 15 feet of coarse sand having a diameter of about 0.8 millimeter and 5 feet of fine sand having a diameter of 0.1 millimeter. The remainder of the hole penetrated silt and clay. Using the conductivity shown in Figure 11.7 and multiplying by the appropriate thickness, the

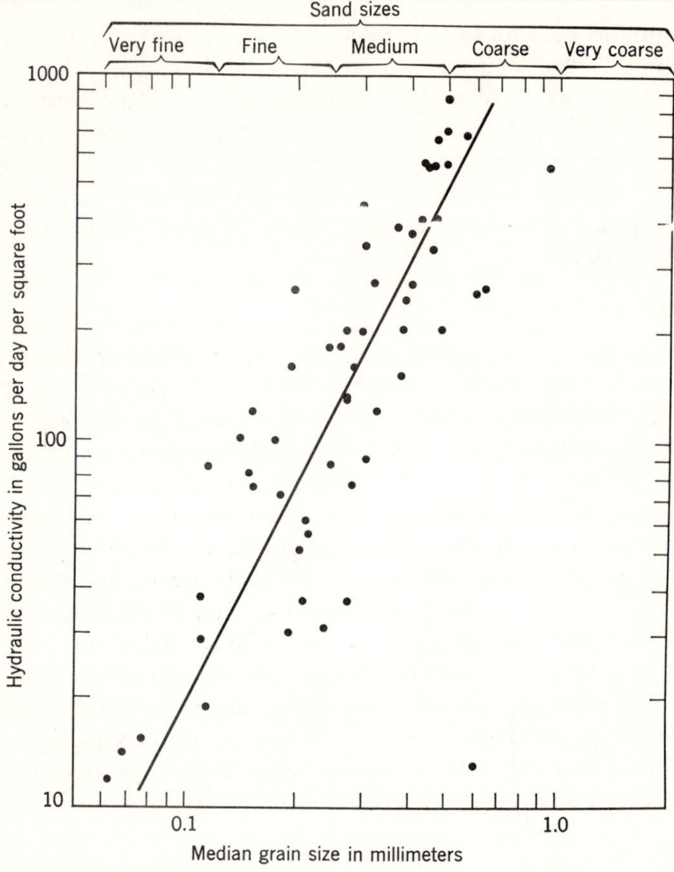


Figure 11.7 Relation between median grain size and permeability of sand from the Arkansas River valley, Arkansas. (Bedinger [3].)

transmissivity is estimated to be 16,000 gallons/day-foot. If the aquifers penetrated are confined and are some distance from hydrologic boundaries, the following approximate expression derived by Logan [25] can be used.

$$T = \frac{1.22Q}{s} \tag{11.2}$$

in which $T$ is the transmissivity,
$Q$ is the discharge of the well,
and $s$ is the drawdown of the pumped well

The value of $s$ must be estimated by taking the difference in elevation between the initial ground-water level and some conservative position of

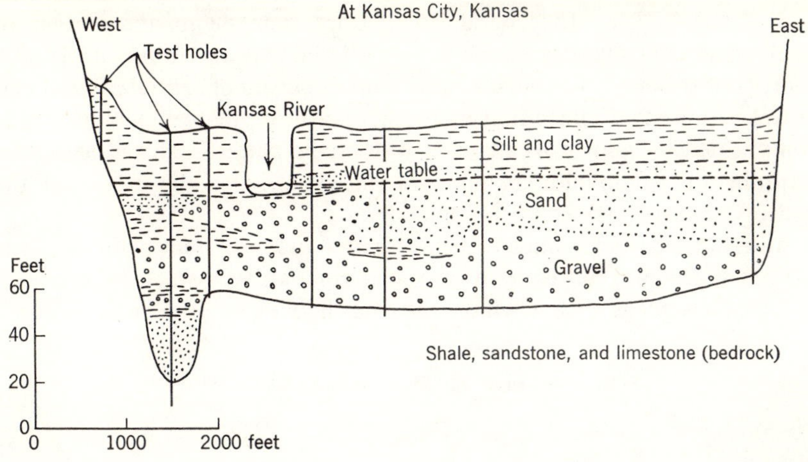


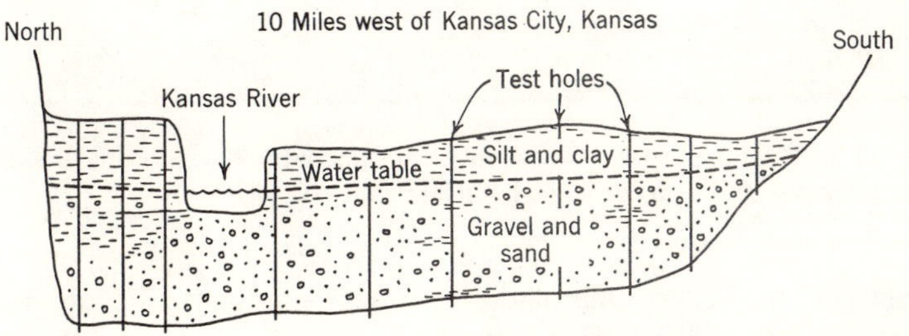


Figure 11.8 Buried inner valley beneath alluvium of the Kansas River near Kansas City, Kansas. The inner valley is well defined at Kansas City (upper cross section) but is represented by only a local thickening of the alluvium 10 miles to the west (lower cross section). (From Fishel [13].)

the future pumping level, say the top of the main aquifer. If $s$ is assumed to be 20 feet, then the expected yield of the well will be $16{,}000 \times 20/1.22 \times 1440 = 182$ gpm. This method of estimating well yields is approximate at best and should not be used if other methods are available. It can, nevertheless, be of great help in ordering test pumps, recommending the type of construction for the finished well, and deciding the need for further test drilling.

### EXPLORATION FOR GROUND WATER

Little skill is needed to obtain wells of moderate production within the alluvial valleys of large perennial streams. On the other hand, the location of wells of high production may require the skills of geophysicists as well as hydrogeologists. In general, an attempt is made to avoid areas of slack-water deposits that will be largely silt and clay and to find areas near sources of recharge that have a maximum thickness of saturated sand and gravel. In areas of glaciation or recent lava flows the geologic history is complicated by numerous changes in drainage patterns as well as by the introduction of nonalluvial material into the river valleys so that the hydrogeologists' task is indeed difficult.

Alluvium-bedrock contacts below many modern floodplain deposits are not horizontal but contain restricted inner valleys which were eroded in Late Pleistocene time when the river was flowing on or near the bedrock

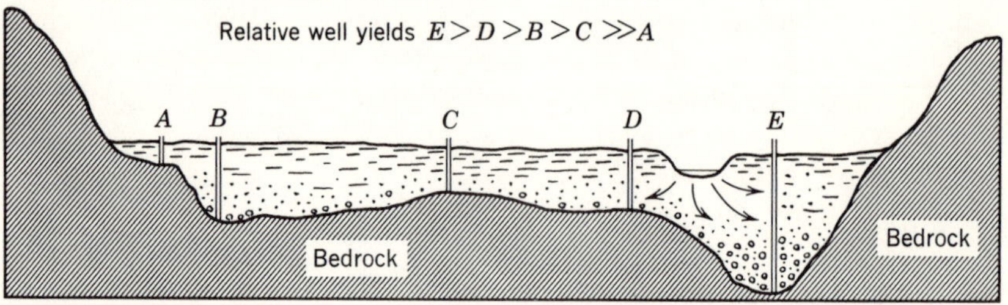


Figure 11.9 Hypothetical cross section of an alluvial valley showing the effects of aquifer thickness and hydrogeologic boundaries on well yields.

contact (Figure 11.8). One of the first tasks of the hydrogeologists is to map the course of the buried valley so that wells can be located in order to intercept the maximum thickness of sand and gravel. Once this course has been mapped, it is a relatively simple matter to locate wells as close as possible to sources of recharge (Figure 11.9).

Few geologic clues exist at the surface as to the position of the buried inner valleys. If alluviation has been quite recent or if the exposed valley sides are resistant to erosion, steep bluffs that are presently a great distance from the stream may suggest that the former river channel impinged against the valley side. Unfortunately, the channels of most modern streams have shifted so rapidly that it is impossible to relate the topography of the valley side to the position of the deepest part of the alluvium. Another clue might be the presence of small side streams that carry large amounts of debris into the master stream. These side streams would tend to force the channel of the master stream to the opposite side of the valley.

Test drilling and geophysical methods can be used to obtain reliable geologic information of the thickness and types of alluvial deposits in river valleys. Possible difficulties in interpreting drilling information come from poor logging or incorrect lithologic identifications. Clayey till can be mistaken for shale and large boulders might be misinterpreted as bedrock. Seismic data are almost always reliable provided that they are interpreted by experienced geophysicists. Although they are less expensive, electrical techniques are not universally as reliable as are seismic methods. Sand overlying sandstone, clay in contact with shale, and many other combinations of alluvium and bedrock may fail to have the necessary contrast of electrical resistivities to be located by resistivity methods. On the other hand, excellent results can usually be obtained in valley deposits within areas of metamorphic and igneous bedrock where contrasts of resistivity may be tenfold or more between the alluvium and the underlying rocks.

Careful reconstruction of the regional geologic history is particularly useful in areas of former drainage changes. Semicircular valleys presently without rivers may indicate the position of abandoned river valleys underlain by thick alluvium (Figure 11.10). Terraces are also present along most large rivers. These may be strath terraces with little or no water-bearing material or they may be fill terraces ideal for ground-water development (Figure 11.11).

### WATER QUALITY

The chemical quality of water in most river valleys is good. Except for desert regions and areas of intensive development, ground water is derived from local recharge on the valley floor and from lateral inflow of streams and aquifers of tributary valleys. The chemistry of the water is controlled, therefore, by the vegetation, culture, and rock types in the side streams and on the valley floor. In central United States, for example, where carbonate rocks are abundant most of the ground water is high in calcium, magnesium, and bicarbonate. Alluvium along many streams in western United States carries water high in calcium and sulfate derived from nearby rocks rich in gypsum and anhydrite.

Where ground-water development is intense and the aquifers are subject to direct recharge from rivers, the water quality is controlled in part by the quality of the river water. In many regions the quality of the river water is almost identical with the native ground water. In other regions the difference is distinctive (Figure 11.12). One of the troublesome aspects of river recharge is the temperature fluctuations introduced into the ground water (Figure 11.13) [22]. High summer temperatures of river water can reduce the usefulness of the water for heat-exchange purposes in office buildings and in industrial plants.

Water wells completed near bedrock may draw water of poor quality from the base of the alluvium. In most places this water is connate water that is seeping slowly into the overlying sediments.

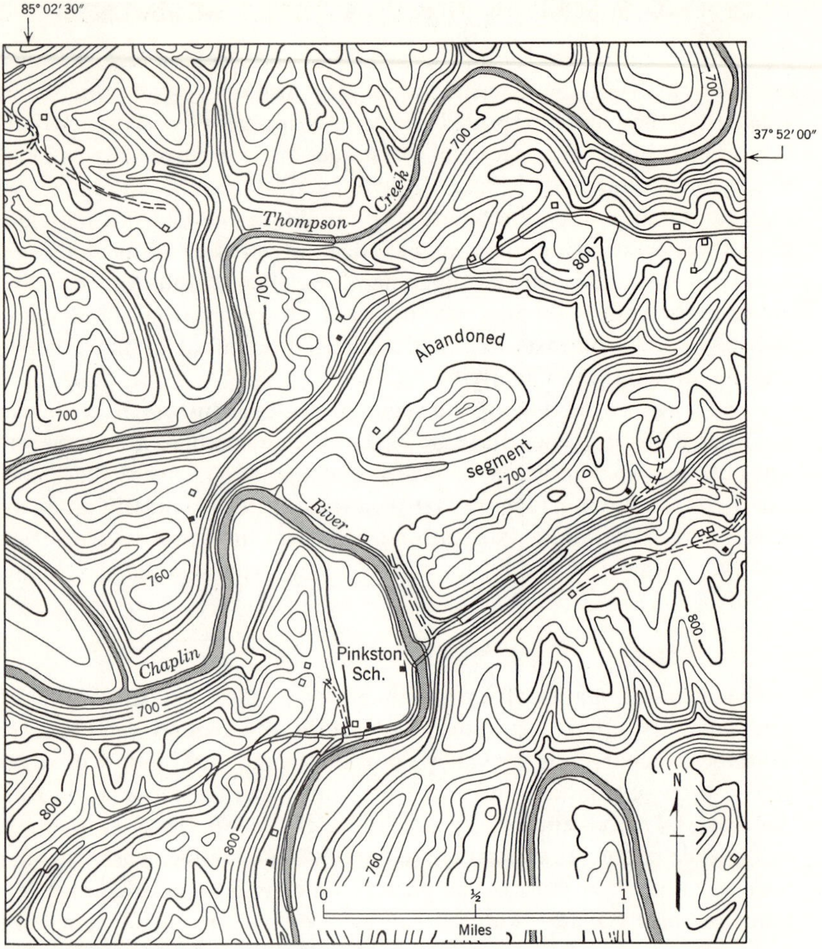

Figure 11.10 Abandoned valley segment of Chaplin River, Kentucky (Cardwell Quadrangle, *U.S. Geol. Survey 7.5 Minute Series*).

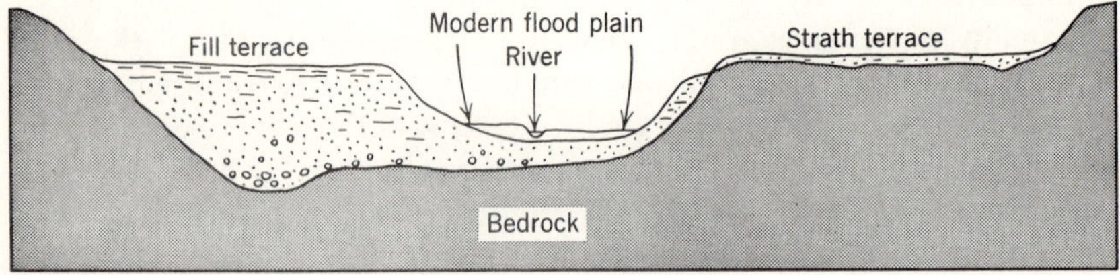


Figure 11.11 Hypothetical cross section showing a terrace made by alluvial filling on one side of the river and a strath terrace on the other. Strath terraces are formed by rivers cutting laterally with little aggradation of alluvial material. In arid regions, lower parts of pediments may be mistaken for parts of strath terraces.

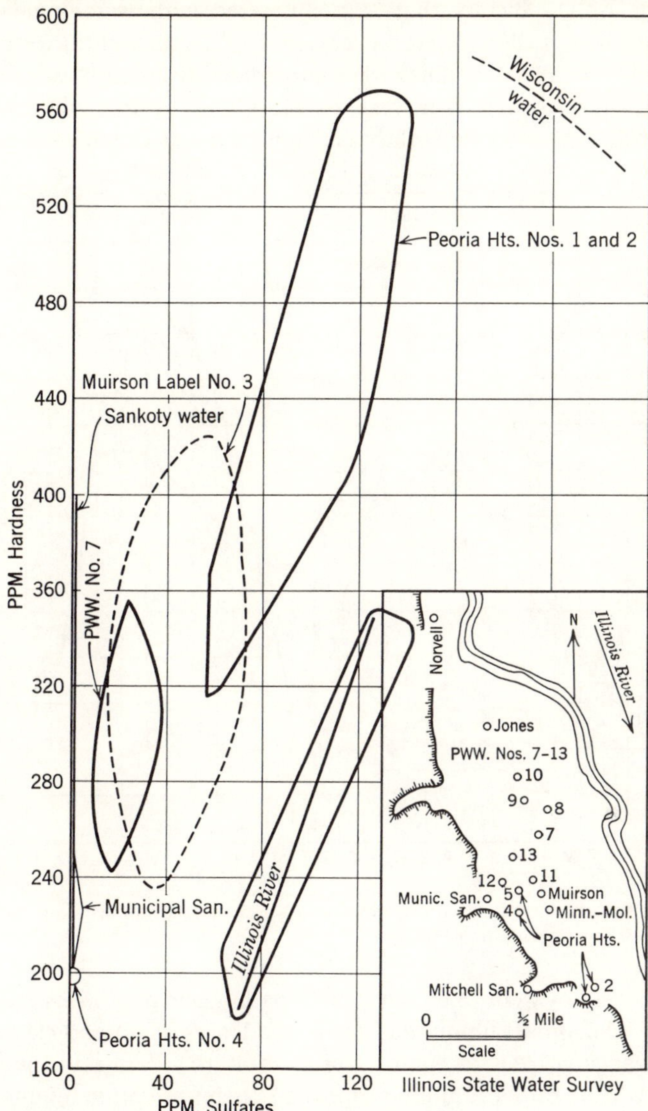


Figure 11.12 Relation between hardness and sulfate in water of the Sankoty well field in Peoria, Illinois. Ground-water levels are below those of the nearby Illinois River, so the water quality of the wells is influenced by recharge from the river.

"Sankoty water" as shown on the diagram is water from thick sandy aquifers in a buried valley that crosses the area. Quality of water is determined by relative mixtures from three sources, (1) Illinois River water, (2) shallow upland tills of Wisconsin age, and (3) water from deeper buried sands of the Sankoty Formation. (Horberg and others [20].)

About the only difficulty that is encountered with water quality in almost all regions is the scattered occurrence of water high in iron and, less commonly, manganese. The rather unpredictable distribution of these constituents suggests that they may be mobilized by low pH and Eh conditions near scattered subsurface accumulations of organic debris.

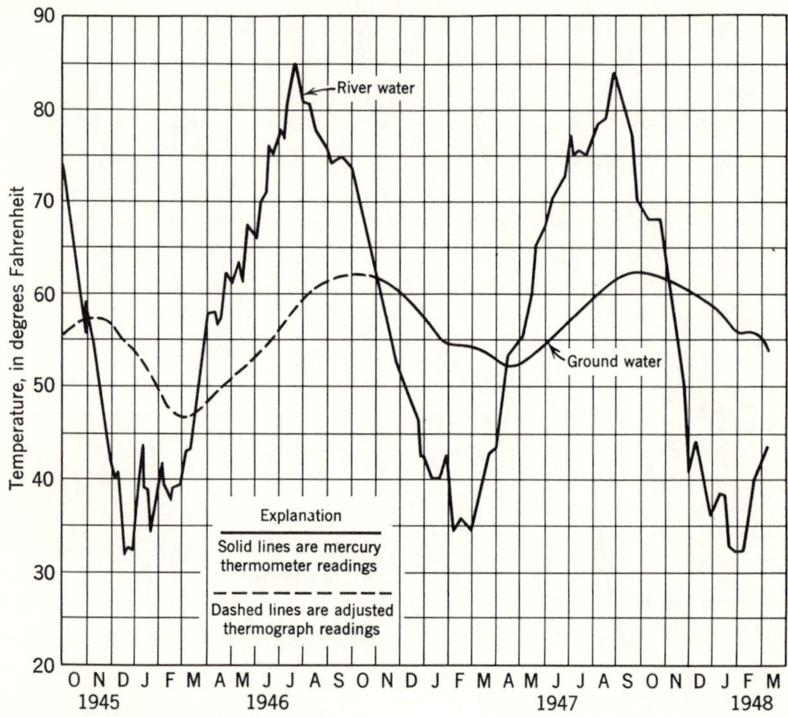


Figure 11.13 Temperature fluctuations in ground water derived in part from recharge from the Ohio River (Rorabaugh [36].)

Contamination of wells in alluvium with pathogenic organisms is a rare occurrence. The almost ubiquitous layers and lenses of silt and clay serve to filter organisms even in many gravel aquifers. Permeable gravels, however, favor the rather rapid movement of chemical contaminants that may be introduced accidentally into alluvial aquifers.

## 11.4 *Valleys of Tectonic Origin*

### TYPES OF DEPOSITS

Large valleys generally owe their origin to tectonic movements rather than to erosion by streams. Some of these valleys are bounded by long fault systems. The deep, narrow valleys of the Jordan River in the Near East and the Owens River in California are good examples. Broad alluvial plains as those of the Po River in Italy and the Ganges River in India are not so spectacular as the deeper valleys, but because of their size they contain much larger amounts of ground water. The broad valleys are localized by regional downwarping of the earth's crust. Many so-called valleys are divided by low ridges and hills. The Central Valley of Chile is an example of a large complex valley.

Large valleys are characteristically filled by rivers depositing vast amounts of debris eroded from surrounding mountains. Other types of deposits may, however, be locally more important. Dependent on the geologic history, lacustrine, pyroclastic, eolian, and even glacial material may be more important than alluvial deposits. The total thickness of such deposits commonly exceeds 2000 feet and may in some valleys be more than 5000 feet. The amount of ground water stored in such valleys is measured in hundreds of millions of acre feet.

The Central Valley of California (also known as the Great Valley) exhibits many features common to large valleys of tectonic origin. The valley is bordered on most sides by high mountains that have shed sediments into the valley for at least 50 million years and for possibly as long as 100 million years. Although the geography has changed considerably during this time, the region of the present valley has accumulated locally more than 25,000 feet of sediments. The vast majority of these sediments are of marine origin and are saturated with saline water. From 500 to 3000 feet of nonmarine sediments, however, directly underlie the surface and contain one of the most important ground-water reserves in the United States.

Fresh-water aquifers in the Central Valley range in age from Miocene pyroclastic materials that have been reworked by streams to recent alluvial channel deposits along the major streams. Although more than 90 per cent of these sediments are deposited by streams, several lacustrine clays are present and sand dunes at the surface suggest that some of the sands in the subsurface are also of eolian origin. The most important lacustrine unit is the Corcoran Clay that divides the water-bearing zones in the southern and western part of the Great Valley (Figure 11.14).

Unlike some deposits of large valleys, the aquifers of the Central Valley are mostly sands with only minor amounts of gravel. The small amount of gravel can probably be ascribed to the fact that weathering of most of the rocks in adjacent mountains has produced sediments that are predominantly fine to medium grained. Individual aquifers are mostly buried channel deposits, so it is very difficult to predict their location. The vast thickness of the entire water-yielding sequence, nevertheless, means

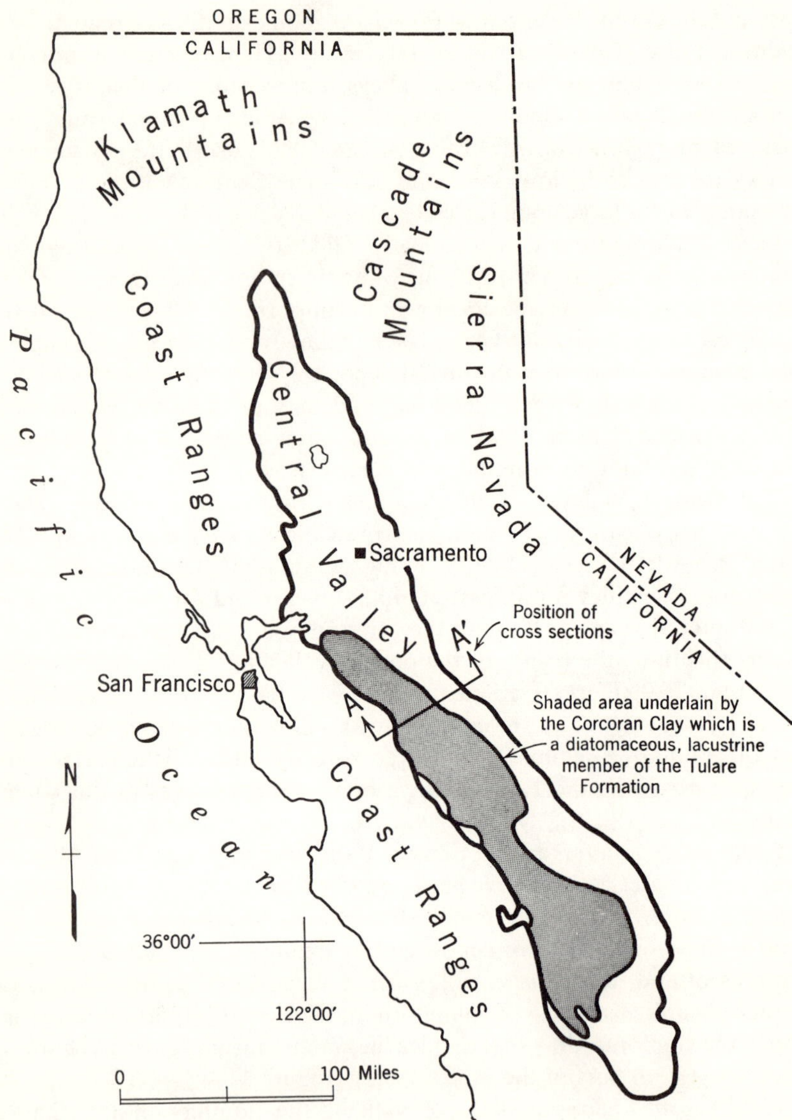


Figure 11.14 Distribution of the Corcoran Clay member of the Tulare Formation. This clay forms an important confining layer within the southern part of the Central Valley. (Modified from G. Davis [11].)

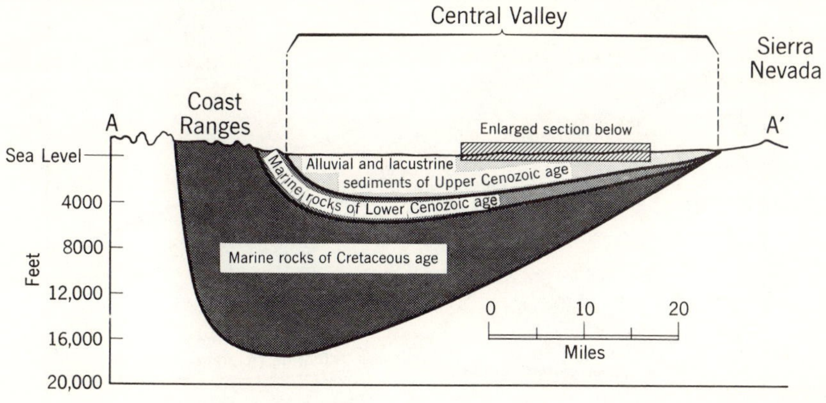


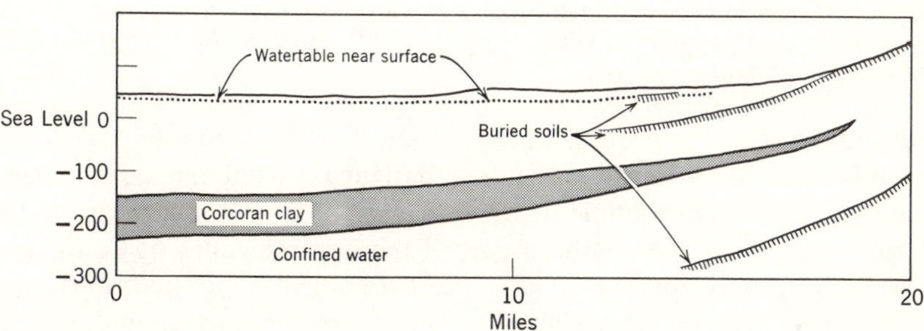


that wells are able to encounter suitable aquifers within a reasonable depth even though their exact position cannot be predicted.

## POROSITY, PERMEABILITY, AND SPECIFIC YIELD

Water-bearing characteristics of sediments from large valleys are quite similar to those of narrower river valleys with the exception that finer sized materials are more abundant in the larger valleys. Also, the finer materials that have been buried at some depth tend to have a lower porosity due to compaction by the weight of overlying sediments. In general, the water-bearing zones will have permeabilities of 10 to 100 darcys and the clay and silt zones will have permeabilities of less than 0.1 darcy. Porosities of subsurface materials will range from about 25 per cent to 60 per cent, and specific yields, from a few per cent to as much as 40 per cent (Figure 11.15).

Although data of specific yield show considerable variation, a number of studies have used "average" specific-yield values to attempt to estimate

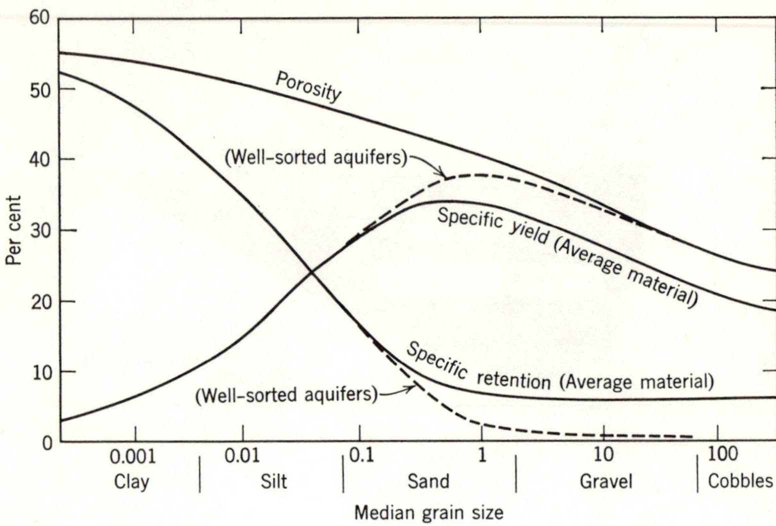


Figure 11.15 Relation between median grain size and water-storage properties of alluvium from large valleys. (From Conkling [10], modified by data from other sources [9,11,21,31].)

the total quantity of ground water present in valley deposits. Well logs are inspected and certain values are assigned to different size ranges. Table 11.2 gives an example of the assumptions used for a study of the storage capacity of the southern part of the Central Valley of California. Inasmuch as both logging practices and sedimentation conditions vary widely from one region to another, these assumptions should not be applied in other regions without considerable study.

*Table 11.2 Specific Yield Values Assumed for the Southern Part of the Central Valley, California (From G. Davis and others* [11])

| Driller's Description | Assumed Specific Yield, per cent |
|---|---|
| Gravel, sand and gravel, and similar materials | 25 |
| Fine sand, tight sand, tight gravel, and similar materials | 10 |
| Clay and gravel, sandy clay, and similar materials | 5 |
| Clay and related materials | 3 |
| Crystalline rock (fresh) | 0 |

WELL YIELDS

Wells in deep valley deposits typically have yields between 300 and 3000 gpm. Specific capacities of eighty-six wells penetrating valley deposits are summarized in Figure 11.16.

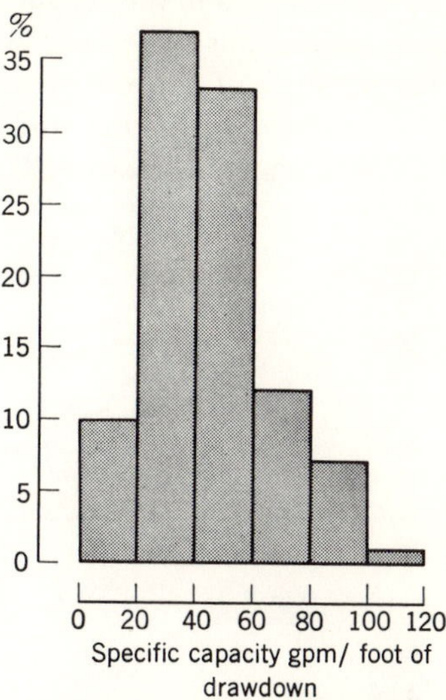


Figure 11.16 Histogram of specific capacities of 86 municipal, irrigation, and drainage wells in the eastcentral part of the Central Valley, California. Most wells are between 50 and 250 feet deep. The watertable is 5 to 20 feet below the surface. (Unpublished data, Turlock Irrigation District and other sources.)

LOCATION OF WELLS

The subsurface position of aquifers in valley deposits is exceedingly hard to predict on the basis of surface geology alone. Individual aquifers are most commonly ancient stream channel deposits that retain the sinuous and interconnected shapes of their counterparts in modern streams. It is not at all unusual to have completely different logs of wells only a few hundred feet apart. Corresponding well yields can be also vastly different.

Surface geologic methods are most successful in outlining broad hydrogeologic patterns rather than in pinpointing drilling locations. The bedrock limits of the valley can be mapped and the bedrock contact can be projected a short distance into the subsurface. Units can also be differentiated within the valley deposits. Older alluvial deposits can be differentiated by greater weathering and erosion of the surface exposures. If the older deposits form a rather straight contact with the younger deposits, then a steeply dipping subsurface contact is inferred. If, on the other hand, the older deposits have a highly irregular contact and a number of inliers are present, then a shallow dip is inferred (Figure 11.17). Areas of maximum potential recharge can be located by careful mapping of sand and gravel outcrops at the surface. Studies of drainage basins tributary to the major valleys can suggest the types of sediments to be expected in the major valley. Small drainage basins will tend to disgorge unsorted debris into the large valley as a result of flash floods, whereas larger drainage basins may have perennial streams that sort the sediment and deliver clean sands and gravels to the major valley. One of the most important surface geologic features are those associated with faulting. Old faults projected into the subsurface will give some suggestion of the thickness of sediments to be encountered in the subsurface. Active or recently active faults will also give indications of relative thickening of units in the subsurface as well as the presence of hydrologic barriers in the subsurface.

A number of factors undoubtedly account for the fact that faults in unconsolidated deposits tend to form barriers. First, fault action will tend

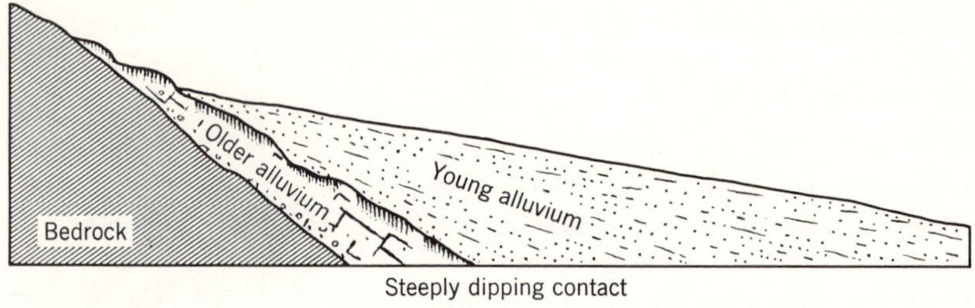


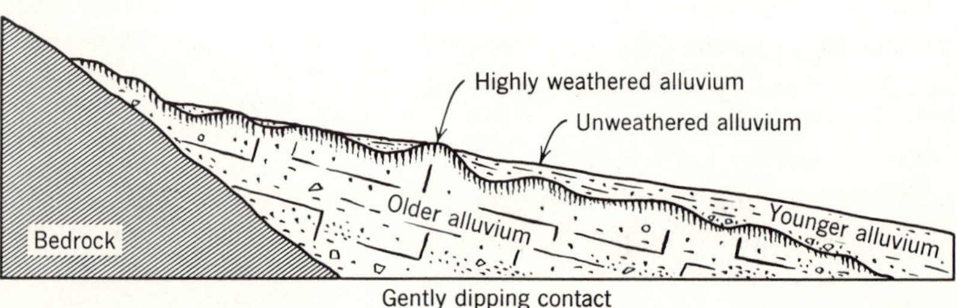


Figure 11.17 Contact relations between older and younger alluvium. In the absence of well-defined bedding, the irregularity of the contact can be used to infer the general attitude of the old alluvium.

to pulverize rocks and minerals along the fault plane. This will be particularly effective at greater depths where confining pressure will increase the interparticle friction. Second, impermeable beds may be offset along the fault to block permeable beds. This effect will be most important where the number of permeable zones are limited. Third, elongated and flat clasts will tend to be rotated parallel with the fault surface and will reduce permeability perpendicular to the fault. Fourth, deposition of minerals along the fault surface will also reduce permeability.

The overall depositional pattern in desert basins of intermediate size has been found to be significant. Apex areas of alluvial fans are composed of older material or are dominated by mudflows and other poorly sorted material from flash floods so that the average permeability is considerably lower than deposits farther down on the alluvial fan. The deposits in an intermediate position tend to be reworked by streamflow and have higher permeabilities despite smaller grain sizes. The distal parts of the alluvial fans tend to interfinger with playa deposits that are predominantly fine-grained material. Wells in the distal parts of the fans tend to be much less productive than those of the central part.

Geological and geophysical logs together with surface studies may be used to piece together a useful picture of subsurface geological conditions. The success of the work, however, will be controlled in part by the geographic density and the accuracy of the logs and other information that might be available. Although individual permeable beds are hard to trace, other useful stratigraphic information can be obtained. The continuity of impermeable or semipermeable strata is of particular interest. Distinctive colors or lithologies are useful keys. Some of the best markers are lake clays, ash beds, lava flows, buried soils, and extensive peat horizons (Figure 11.18). The marker beds will allow the determination of the dip of other beds and will help determine whether or not the beds are continuous. Also, distinctive configurations of geophysical logs may assist greatly in the determination of the continuity of clay beds (Figure 11.18). Once the extent of confining beds has been established, the interpretation of ground-water migration as related to problems of extraction becomes much easier.

Surface geophysical methods generally do not help pinpoint deeper aquifers within large valleys. The methods are most useful in determining the extent and attitude of basalt, marl, compact sand, and other horizons of contrasting resistivities or seismic properties. If the aquifers are near the surface, they can be located by electrical resistivity methods so that wells can be drilled or the permeable zones used for recharge.

### PROBLEMS OF AQUIFER COMPACTION

The reduction of artesian pressure in thick valley deposits will induce compaction of aquifers and adjacent silts and clays. Laboratory tests of sand and gravel suggest that compaction of these materials should be relatively minor and that most of the volume change is related to compaction of the fine-grained material. On the other hand, the fact that volume changes in the subsurface take place very rapidly in response to changes in pressure would suggest that only the sediments rather close to the aquifers are affected immediately after pressure reduction. A permanent reduction of artesian pressure will, however, cause eventual compaction of fine-grained sediments considerably above or below the aquifer.

The first well-documented account of land subsidence in valley sediments is contained in Tolman's classic book on ground water [39]. He pointed out the close relationship between the extent of confined aquifers and the

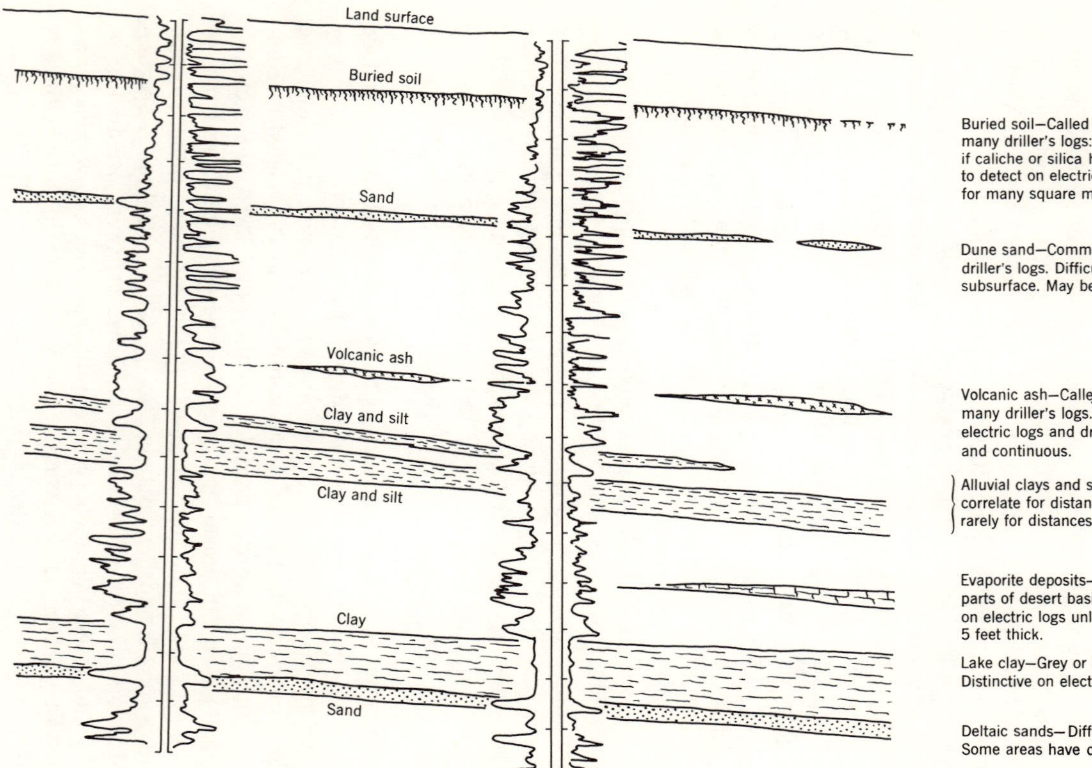


Figure 11.18 Useful stratigraphic markers within large alluvial deposits. Electric logs indicate a change from fresh water near the surface to brackish water within the lower two-thirds of the diagram.

area of subsidence in the Santa Clara Valley, California. That the subsidence was caused by compaction of aquifers and associated sediments and not by tectonic subsidence was demonstrated by well casings that protruded from the ground as the ground surface subsided in relation to the bottom of the well. Subsequent studies have shown that the rate of subsidence is also closely related to the rate of lowering of the artesian pressure (Figure 11.19)

Many other areas of subsidence due to ground-water pumping are known in the United States as well as in Japan, Mexico, and England. The most

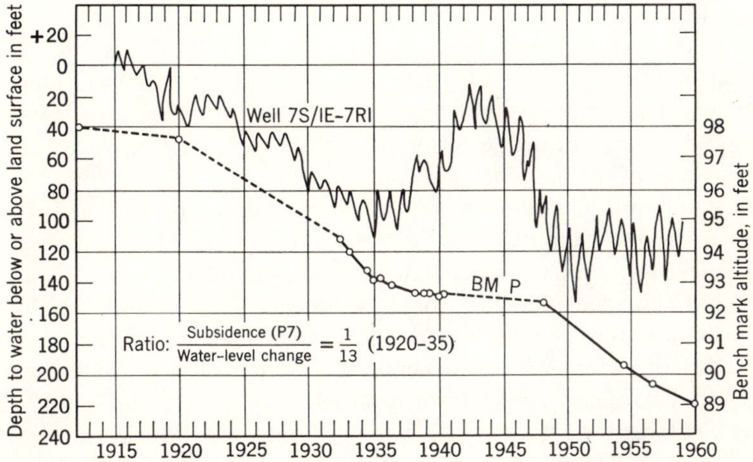


Figure 11.19. Change in altitude at bench mark P7 in San Jose, California, related to the change in water level of an adjacent well. (Poland and Green [34].)

spectacular subsidence areas, those of the southern Great Valley of California [33] and the Mexico City area, are in large valleys of tectonic origin. In both areas the maximum subsidence has probably exceeded 15 feet and most certainly has exceeded 10 feet over relatively large areas.

Extensive problems can be caused by subsidence. Ground-water storage capacity is permanently destroyed [32]. Elevations determined by expensive surveys must be reestablished periodically. Topographic maps must be redrawn. Water flow in canals and rivers may become sluggish because of the reduction of the already low gradients by subsidence. If wells extend below major zones of subsidence, the bottoms of casings will remain stationary while the overlying material settles downward and exerts a drag along the casing. The resulting stress transmitted to the casing will commonly collapse the casing. Piles for buildings and other structures may also extend into or below zones of subsidence. The structures then tend to remain stationary as the ground surface subsides. Any connections with the ground such as ramps and stairs will fail if they are not flexible. This effect has been particularly damaging in Mexico City where subsidence is caused in part by pumping from aquifers within about 150 feet or less of the surface. New buildings on piles that have been joined to older buildings resting on the land surface have almost always caused extensive damage to the other buildings. The problems caused by subsidence in Mexico City have become so serious that ground-water pumping is prohibited in the central part of the city [43].

### WATER QUALITY

Ground water in large valleys of tectonic origin is highly variable in its chemical character. Saline water originating from ancient sea water or from the solution of evaporites may be present in the deeper aquifers or near the surface in the central parts of large closed basins. Valleys that have good exterior drainage throughout their geologic history are likely to have water of the best chemical quality. Exterior drainage will prevent the accumulation of saline water through partial evaporation and it will also favor the rapid circulation of ground water that will be most effective in removing any connate water that may be migrating to the surface.

In some places the mineralogy of the sedimentary material will be almost as important as the drainage history of the valley in determining the chemical quality of water. Most sediments that are derived from sedimentary rocks will have large amounts of calcium, magnesium, carbonate, and sulfate available for solution within the ground water. Such waters have typically 50 to 200 ppm calcium, 10 to 50 ppm magnesium, 100 to 250 ppm bicarbonate, and 50 to 300 ppm sulfate. Chloride in these waters is highly variable, depending on the leaching history of the sediments. In contrast, if the sediments are derived from metamorphic and plutonic igneous rocks, the total concentration of dissolved solids is generally much lower. Sulfates and chlorides are particularly low, being generally less than 50 ppm each. Silica will be the only constituent that is more abundant than in water saturating sediments derived from sedimentary rocks. The presence of volcanic material will increase the amount of silica even more. In general, the total concentration of dissolved solids in water from sediments derived from volcanic rocks will be intermediate between the two previous types.

Biological contamination is rarely a problem in wells within large valleys. The greatest danger of organic pollution is in areas of extremely permeable aquifers that are near the ground surface. Faulty well construction or improper placement of waste-disposal facilities can, as in all other regions, be responsible for the introduction of pathogenic organisms more or less directly into well water.

## 11.5 *Coastal Plains*

### INTRODUCTION

Coastal plains vary in size from small isolated valley deposits that grade inland into normal stream deposits to vast, almost featureless plains that fringe hundreds of miles of the coasts bordering the Arctic and Atlantic Oceans. Sediments of the coastal plains generally represent both alluvial and marine environments. The marine sediments, in turn, are represented by beach, lagoonal, estuarine, as well as by deeper marine environments.

Marine and lagoonal sediments are by no means the only types of coastal plain deposits. Indeed, in some areas alluvial and deltaic types are most abundant.

### STRATIGRAPHY

Many stratigraphic units along coastal plains grade oceanward from partly alluvial deposits into entirely marine units. This gradation is accompanied by a tendency for the sediments to become progressively finer grained. The bulk of the sediments are clays and silts, although some areas, notably the southern Atlantic Coastal Plain and the eastern Gulf Coastal Plain, contain abundant marls and limestones.

### WATER-YIELDING PROPERTIES

The water-yielding properties of sediments underlying coastal plains are in general quite similar to those of sediments in large valleys. Permeabilities of water-bearing zones generally range between 1 and 100 darcys (Figure 11.20) and those of silt and clay zones, less than 0.01 darcy. Tests concerning specific yields of a large number of coastal plain sediments have not been published, but the general similarity of alluvial sediments with many coastal units would suggest that coarser materials have specific yields of 15 to 35 per cent and the finer materials less than 10 per cent.

Yields of larger irrigation, municipal, and industrial wells are between 200 and 3000 gpm, or roughly the same as wells in large valleys. Lower yields are most common in older sedimentary units and in coastal areas receiving sedimentation in protected environments.

Within the United States, the coastal plain deposits, particularly along the Gulf of Mexico, contain the most important ground-water reserves of any geologic environment [24,27]. High well yields together with abundant water for recharge suggests that expanded ground-water development for most of the Gulf Coast will be possible without rapid depletion of reserves.

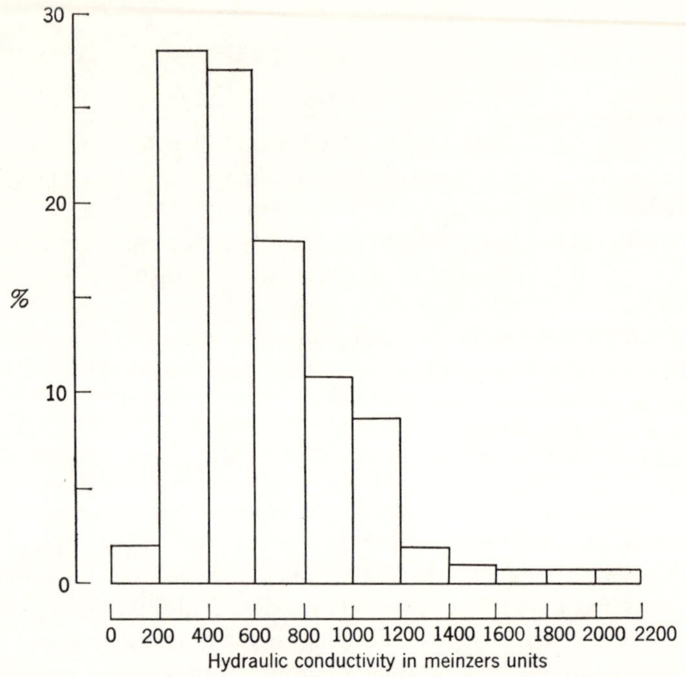


Figure 11.20 Histogram of the hydraulic conductivity of 150 samples of water-bearing sands from the coastal plain sediments of southern Mississippi. Samples tested were disturbed during recovery. (Data from Brown and others [5,6, and 7].)

DEVELOPMENT OF GROUND WATER

Stratigraphic techniques developed for the petroleum industry have also been of great service in hydrogeologic studies of coastal plain sediments. Electric logs have been of particular importance to aid in correlation, locate permeable horizons, and estimate water quality (Figure 11.21). Some of the widespread aquifers such as the Carrizo Sand of Texas [37] extend laterally for hundreds of miles. This particular unit yields potable water to a 5355-foot water well that is one of the deepest in North America. Other units change rapidly in a lateral direction and can only be correlated by paleontological methods or by study of very closely spaced wells.

Seismic techniques are adapted for the tracing of aquifers in the subsurface through known relations to reflecting or refracting horizons. To date, nevertheless, less expensive geologic methods have been relied on for purposes of subsurface correlation. Records of water wells together with abundant information from oil and gas wells generally supply the geologist working in coastal plains with basic data for hydrogeologic interpretations.

WATER QUALITY

Four types of water are most common within coastal aquifers (Table 11.3). Water of low dissolved solids content with sodium, calcium, and bicarbonate as the most abundant ions is generally found near the surface in the outcrop area of the aquifers. At depths of several hundred to more than a thousand feet the water is mostly a sodium bicarbonate water with

*Table 11.3 Typical Analyses of Ground Water from Coastal Plain Aquifers, All Values given in parts per million*

| | Fresh Water from Shallow Well near Mobile, Alabama [35] | Sodium Bicarbonate Water from "700-Foot" Sand near New Orleans Louisiana [12] | Brackish Sodium Chloride Water from a 749-Foot Well near Mobile, Alabama [35] | Water from a 60-Foot Well Contaminated with Sea Water in Mobile, Alabama [35] |
|---|---|---|---|---|
| $SiO_2$ | 9.4 | 36 | 16 | 5.5 |
| Fe | 0.1 | 0.25 | 0.4 | 1.9 |
| Ca | 0.6 | 2.0 | 35 | 159 |
| Mg | 0.2 | 1.7 | 19 | 439 |
| Na | 2.5 | 220 | 1680 | 3840 |
| K | 0.5 | 2.2 | 11 | 136 |
| $HCO_3$ | 6.0 | 385 | 408 | 427 |
| $SO_4$ | 2.5 | 0.1 | 6 | 706 |
| Cl | 2.8 | 131 | 2630 | 6760 |
| F | 0.0 | 0.4 | 0.8 | 1.0 |
| Total solids | 27 | 590 | 4780 | 13,000 |

very low amounts of sulfate, calcium, and magnesium. The sulfate is probably reduced by bacteria in the aquifers whereas the calcium and magnesium are exchanged for sodium from the clay minerals. Still deeper, connate water is encountered that mixes with fresh water from shallower zones. This water is typically a sodium chloride water. The fourth type of water is sea water that has recently invaded coastal aquifers. This water generally has a lower silica content but a higher sulfate content than normal connate water, although these criteria are not applicable in many areas where the normal connate aquifers have water relatively high in sulfates but low in silica.

Some of the purest ground water from the standpoint of dissolved solids is found in terrace deposits of the Gulf Coastal Plain. The aquifers evidently have high recharge rates with associated rapid circulation of

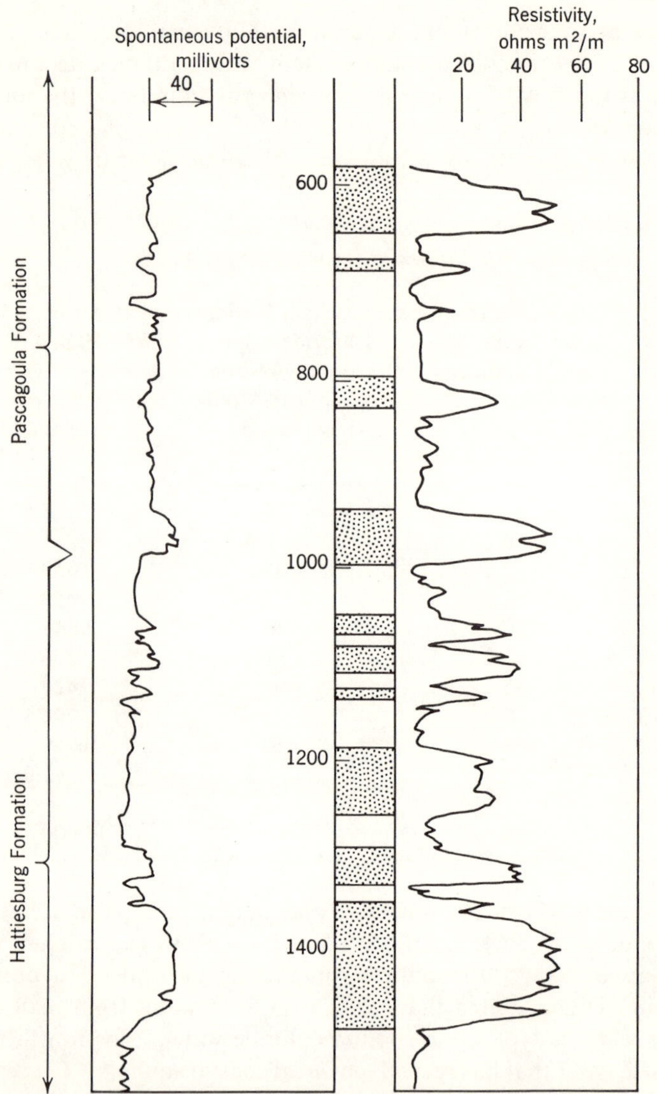


Figure 11.21 Portions of an electric log that show alternating sands, silts, and clays of coastal plain sediments in southern Mississippi. Stippled pattern indicates predominantly sandy zones that are potential aquifers. Water is fresh above 1500 feet as suggested by the positive spontaneous potential. Below 2600 feet the water is saline. Note the change of the resistivity scale between the two diagrams. (Brown and Guyton [6].)

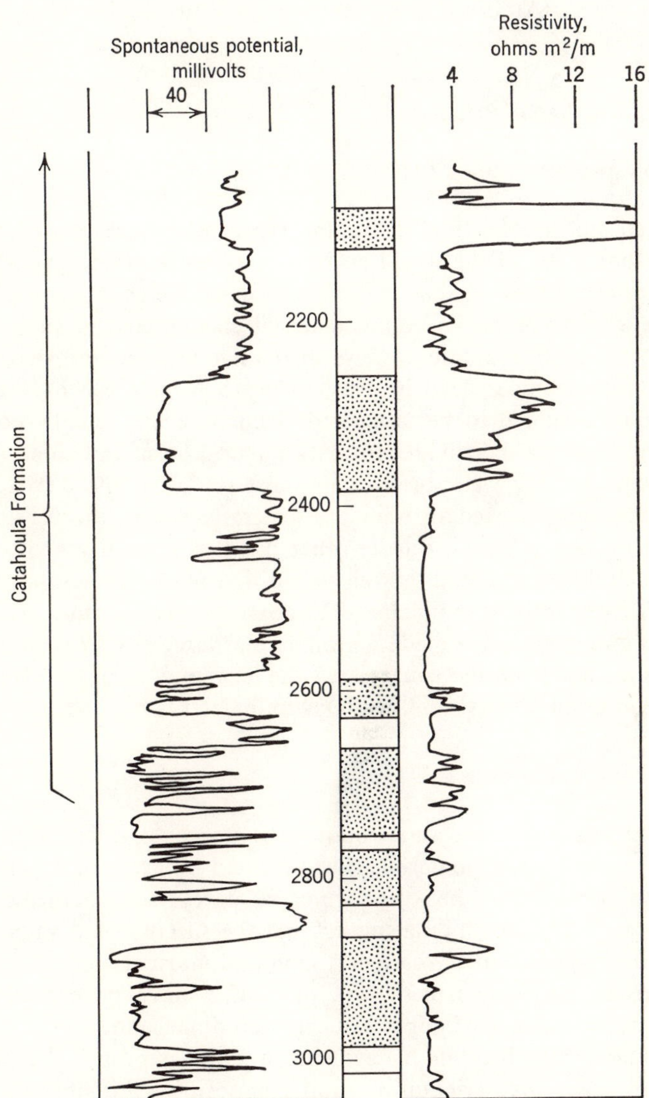
Spontaneous potential,
millivolts
40
Resistivity,
ohms m²/m
4
8
12
16
2200
2400
2600
2800
3000
Catahoula Formation

water. Over periods of many hundreds of thousands of years most of the more suitable constituents have been leached out so that water percolates through the aquifers with only slight chemical modification from the original rainwater. An example of this type of water is shown in Table 11.3.

## **11.6** *Regions of Eolian Deposits*

### DISTRIBUTION AND NATURE OF DEPOSITS

Eolian deposits are far less abundant than either stream or glacial deposits. Many parts of the world such as southern Ukraine, Nebraska, and southeastern Saudi Arabia, however, are blanketed by eolian deposits. The actual volume of eolian material encountered in subsurface exploration is hard to evaluate because distinctive textural and structural features are seldom recognized in drill cuttings. Almost certainly many silts and sands assumed to be alluvial or lacustrine are actually eolian.

Eolian deposits can be divided into two types, loess and dune sand. The dune sand is very well sorted with most grains in the 0.05 to 0.5 millimeter size range. Median sizes are generally within the 0.1 to 0.3 millimeter size range. Loess, on the other hand, is generally a silt-sized material which, like dune sand, is well sorted, but unlike dune sand, loess will vary widely in median grain size. The coarsest loess is found near the source areas and may have a median grain size as large as 0.07 millimeter. Many miles distant from the source area the loess may be partly clay size with a median grain size as small as 0.009 millimeter.

### WATER-YIELDING PROPERTIES

Dune sand has about the most uniform hydrogeologic properties of any type of water-bearing material. Porosity will be between 35 and 40 per cent and permeability between 5 and 50 darcys with a median of perhaps 25 darcys [8]. Scant data suggest that the specific yield should be between 30 and 38 per cent. Weathered sand and unusual dune materials such as gypsum and clay pellets will, of course, have properties that diverge widely from those of normal quartz-rich dune sand.

Even though cross bedding is evident in dune deposits, the slight differences in grain size, orientation, and composition probably do not cause highly anisotropic conditions to develop. Permeability measurements of oriented samples of dune sand are lacking.

Aquifers of dune sand are not widely utilized because wells that prevent entrance of the loose sand are difficult to construct by standard practices, because permeable dune sand may drain rapidly and saturated zones are not present in many dunes, and, finally, because active dune areas are not favorable for habitation. Despite these drawbacks, dune areas are favorable for water development because of high recharge rates, good water quality, and moderately high permeabilities.

Loess has porosities of from 40 to 55 per cent. The high porosities are possible because individual silt grains tend to be held in rather open networks by the cementing action of small amounts of clay always present in the loess. Specific yields have been measured in only a few samples of loess. Values of 35 per cent should be characteristic of coarse loess, and values of less than 15 per cent characteristic of fine-grained loess. Permeability of loess varies through an even greater range of values, from a minimum of about $10^{-4}$ darcy to a maximum of about 1.0 darcy. The median value of one hundred ninety-three tests from fourteen localities is slightly more than $10^{-2}$ darcy [15].

Vertical joints, root holes, and animal burrows give loess an anisotropic character with regard to permeability. Within some loess the overall vertical permeability must be much greater than the horizontal permeability. Exact determinations of the extent of this anisotropism have not been made. In many regions buried soils will separate loess sheets of different ages. The clay content of the soil horizons will be considerably greater than the surrounding loess so that the permeability will be correspondingly less. Buried soils can be inpermeable enough in some areas to form overlying bodies of perched water during wet seasons.

Loess is not commonly an aquifer because of its low permeability and because where its permeability is the highest it is usually in high topographic positions where subsurface drainage is efficient. If loess overlies impermeable soils or consolidated rocks, stock or farm wells can be developed that will give good service except during long periods of dry weather.

### WATER QUALITY

Active dunes are most commonly composed of clean, relatively inert minerals that will cause only a minor modification in the chemical character of infiltrating water. As a consequence, most ground water in dune sand is of good quality except in places where water of poor quality migrates laterally into the dunes. Water in loess should be somewhat more mineralized owing to changes of water chemistry in the soil horizons as well as to greater abundance of clays with a high exchange capacity, abundant freshly fractured minerals, and large quantities of relatively mobile calcium carbonate within the loess below the zone of weathering.

## 11.7 *Glaciated Terrain*

### DISTRIBUTION AND TYPES OF GLACIAL DEPOSITS

About 30 per cent of the land surface of the earth has been covered by glacial ice during the past million years. At present, 10 per cent of the land surface is covered. The hydrogeology of land that has been uncovered was profoundly influenced by the once extensive ice sheets. Repeated glacial advances have removed a large part of the unconsolidated deposits that once must have been present as residual soils and terrace and flood-plain sediments. Much of the debris removed by glaciers has been dumped

*Table 11.4 Glacial Deposits* (*Terminology from Flint* [14])

| Deposit | Associated Topographic Forms |
|---|---|
| Unsorted | |
| Till | Till plains |
| Ablation till, Lodgment till | Terminal, laterial, and inter-lobate moraines. Drumlins |
| Sorted | |
| Ice contact | Eskers, kames, kame terraces, and kettles |
| Outwash | Outwash terraces and outwash plains |

near the margins of the glaciated areas or deposited as outwash along the drainage channels beyond the glacier's margin. Much of the fine material ultimately came to rest on the ocean floors or in loess deposits that originated from broad outwash plains. Some of the fine-grained material was incorporated within glacial till.

Glacial deposits, or drift, can be subdivided into poorly sorted material deposited directly from glacial ice, or till, and water-sorted deposits. The water-sorted material is further divided into deposits which came to rest in close proximity to glacial ice, or ice-contact deposits, and material that was carried away from the vicinity of the ice by meltwater streams. The latter deposits are called outwash and are essentially a variety of alluvium. The more common types of glacial deposits are summarized in Table 11.4.

Glacial till is deposited directly from glacial ice. Till that is deposited from slowly melting ice at the base of glaciers is called lodgment till. It tends to be denser and more fissile than ablation till that is let down by the melting ice. All till is characterized by poor sorting and scattered boulders. Till from continental glaciers is much finer grained than till deposited by mountain glaciers.

Ice-contact deposits represent a vast variety of textural types ranging from silts and clays deposited in ponds and lakes on or adjacent to the ice to well-sorted coarse boulders and cobbles deposited by streams carrying debris from melting ice. A large variety of topographic forms are also produced by accumulations of ice-contact deposits. Kame terraces are formed by the accumulation of glacial debris along the margins of stagnant glacial ice that remains in valley areas. When the ice melts the debris is left as a terrace along the sides of the valley. The part of the terrace that was once in contact with the ice is strongly affected by collapse and slumping, thus forming irregular borders facing the centers of the valleys. Small patches of ice-contact material may be let down when the glacial ice melts. Small hills formed in this way are known as kames. Some kames also may be simply erosional remnants of larger masses of ice-contact deposits. Eskers are the most distinctive of the various land forms composed of ice-contact deposits. They are typically long sinuous ridges composed of coarse sand and gravel. Eskers are the bed-load deposits of former streams that occupied subglacial ice tunnels or, less commonly, streams on the ice surface. Most eskers were probably formed during the stagnant or near-stagnant phase of glaciation. Kettle holes are formed by the collapse of till and ice-contact sediments as isolated masses of residual ice melt. Kettle holes may be found on kame terraces, wide parts of eskers, till plains, and terminal moraines.

### WATER-BEARING PROPERTIES

Very few data are published on the porosity of till and no data are available on the specific yield of till. The unsorted character of till and the compacting weight of the glacial ice combine to make till rather dense and impermeable. Most porosities should fall in the range of 25 to 45 per cent. The higher porosities should be expected in the clay-rich tills.

Norris [30] has tabulated available data on the permeability of clay-rich tills deposited by continental glaciers. The permeability values of thirty-seven determinations ranged from about $1.7 \times 10^{-5}$ to $5 \times 10^{-2}$ darcy with a median of $2.2 \times 10^{-3}$ darcy. Although not reported in the literature, permeabilities of till in mountainous areas should be much larger with maximum permeabilities of more than 1 darcy.

Glacial till covers several million square miles of the earth's surface, yet relatively few water wells draw their supplies directly from the till. Of the wells drawing water from clayey till, most of them probably obtain their water from joints or small sand lenses within the till. A number of wells in mountain valleys probably draw water from coarse-grained till.

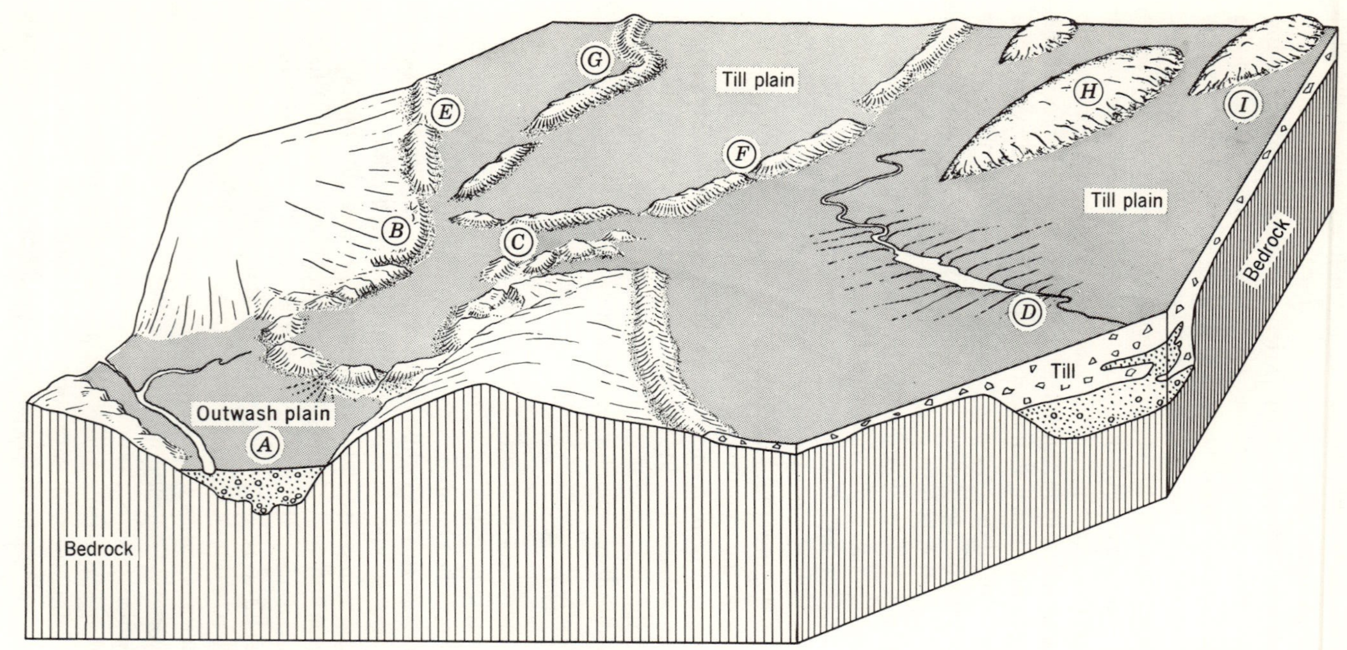


Figure 11.22 Topographic features of an area glaciated in Late Pleistocene time. *A*. Outwash plain having excellent aquifers. *B*. Kame terrace, largely underlain by permeable material. High topographic position suggests that most of the ground water has drained out of the ice-contact deposits. *C*. Kames, permeable material of limited extent, probably minor amounts of saturated sediments in basal parts of hills. *D*. Swale in till plain suggests that a large thickness of underlying sediments has been compacted. This is a likely place to explore for a buried valley. *E*. Lateral moraine, proximity to bedrock upland indicates till is underlain at shallow depth by bedrock. *F*. Recessional moraine, at this point glacial material is probably thicker. Possibility exists of minor amounts of water-sorted material that may be saturated. *G*. Esker, permeable material with thin saturated zone near base of deposit. *H*. Drumlin, mostly till, possibly with rock core, little chance of water. *I*. Till plain, little likelihood of aquifers within till.

Water-yielding properties of ice-contact deposits vary from those of clay to those of the coarsest gravel. Wells drawing water from ice-contact material are relatively common where the deposits are low enough topographically so they are not subject to rapid drainage. The limited aerial extent of many aquifers of ice-contact material, however, restricts long-term production.

Some of the most productive wells known in the United States are in outwash near the cities of Tacoma [41] and Spokane in the state of Washington. Here, specific capacities of more than 1000 gpm/ft drawdown have been measured. Yields of 200 to 2000 gpm with specific capacities of from 10 to 100 gpm/ft drawdown, nevertheless, are more typical of outwash deposits. The hydrogeologic characteristics of outwash are similar to normal valley alluvial deposits. The lack of large thicknesses of fine floodplain deposits, the prevalence of prominent forset-type cross bedding, and anomalously large boulders serve to distinguish the outwash deposits from normal alluvium.

### EXPLORATION FOR GROUND WATER

Location of areas favorable for ground-water development in glaciated regions can be exceedingly difficult. Glacial deposits that are no more than 100,000 years old retain much of their original topographic expression. Where these younger deposits are present, therefore, many of the more distinctive topographic forms can be mapped by conventional geologic methods with the aid of aerial photographs and topographic maps (Figure 11.22). Even with most of the surface features correctly identified, aquifers may be hard to locate. Eskers, kames, and kame terraces will be commonly in topographic positions well above adjacent streams so that drainage is very effective and few saturated zones within the ice-contact deposits can be found. In these regions, aquifers will most likely be within post-glacial alluvium, outwash, and buried glacial and preglacial channel deposits, or within underlying bedrock. Numerous pockets of water-sorted material may also be present within terminal moraines. If the moraines are broad enough and the recharge large enough, many of these more-or-less isolated pockets of permeable material can be important local aquifers. Although not as common as in the areas of terminal moraines, ice-contact deposits are present as isolated pockets throughout almost all thick accumulations of glacial till.

Till may also contain gravel-filled channels that were formed during retreats of the glacial margin with subsequent advances of the ice leaving the deposits sandwiched between extensive till sheets [29]. Owing to their greater extent, these channel deposits within till are much better aquifers than the isolated pockets of ice-contact material. Several outcrops, geophysical profiles, or test holes, however, are needed to properly identify these channel deposits within the till.

Valleys filled with outwash and/or preglacial alluvium are the most important type of aquifers within glaciated regions (Figure 11.23) [18,19]. In some places these valleys are not completely filled so that modern valleys mark the position of the much deeper ancient valleys. In other places the

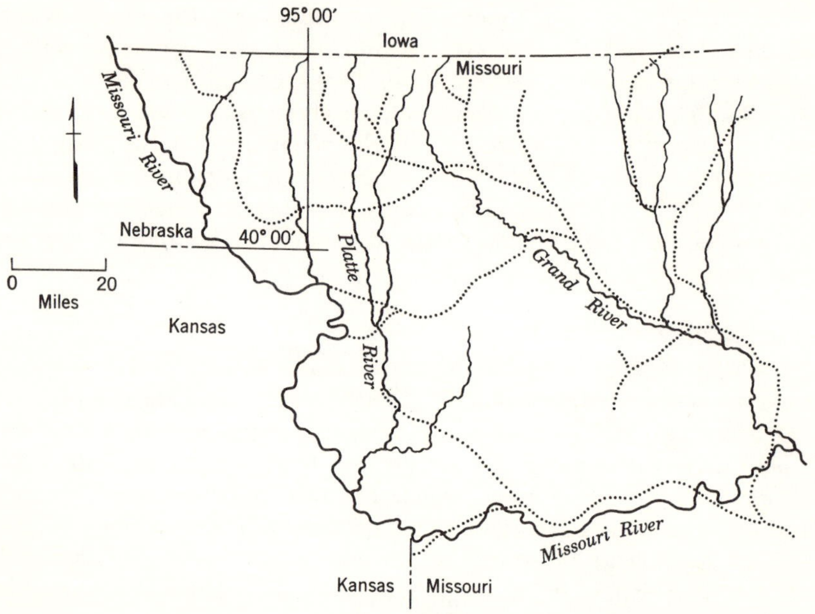


Figure 11.23 Position of buried drainage pattern (dotted lines) contrasted with the modern pattern (solid lines) in northwestern Missouri. Buried valleys filled with glacial sediments form the most important aquifers in this region. (Howe and Heim [18].)

most recent till plain completely covers buried valleys leaving little or no surface indications of their presence. Many of these valleys are filled with glacial material having aggregate thicknesses of more than 50 feet of permeable sand and gravel along with less permeable till.

Large streams flowing counter to the direction of glacier advance will be blocked and in many areas will be turned into lakes. Predominantly fine-grained sediments will be deposited in the ponded reaches of the river. One of the best known of the preglacial valleys, the Teays River valley, flowed northward into Ohio from West Virginia (Figure 11.24). In southern Ohio the valley floor is $1\frac{1}{2}$ miles wide and 200 to 300 feet below adjacent bedrock uplands. Clays and silts deposited in the northward trending segment of this valley reach a maximum thickness of 264 feet.

In this part of Ohio the sediments are not permeable enough to form good aquifers.

Exploration for ground water in glaciated regions is best accomplished by a combination of surface geologic mapping, geophysical exploration, and test drilling. Geologic mapping alone may not only lack enough detail but it may also be misleading. For example, some terminal moraines

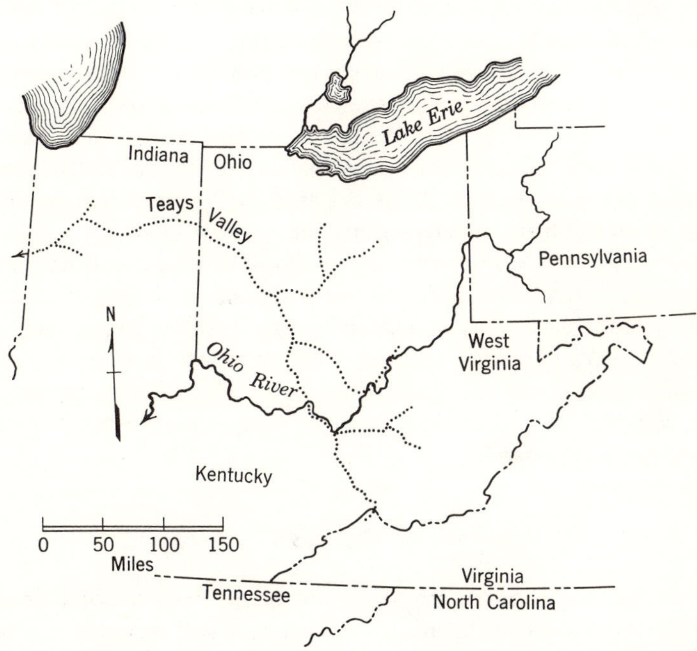


Figure 11.24 Course of the Teays drainage system (dotted line) in Ohio and adjacent states. The position of the present Ohio River is shown as a solid line. (Modified from Flint [14].)

are localized by preglacial bedrock hills. Glacial material may be a relatively thin veneer so attempts based on surface geology to locate aquifers may be futile. On the other hand, seismic exploration may locate buried valleys, but information concerning the geologic history of the area would be needed to decide whether or not the valley might be filled with fine sand and silt which would be unsuitable, or with medium sand which would be a suitable aquifer. Test drilling is, of course, the only sure way to prove the presence of aquifers, but, as has already been pointed out, the cost of drilling is many times that of most other exploration methods.

Figure 11.22 illustrates some of the favorable areas for ground water in a glaciated area. To the hydrogeologist, there is hardly another geologic environment that will tax his ability more but at the same time give such positive results when the work is completed.

WATER QUALITY

The chemical quality of water from glacial deposits that are subject to active circulation is generally good. In central North America much of the glacial material is derived from marine rocks rich in calcite, dolomite, and to a lesser extent anhydrite and gypsum. Here, ground water in glacial deposits commonly is very hard and may contain objectionable amounts of calcium and sulfate even though there is free circulation of the ground water (Figure 11.12). Where the ground water is almost stagnant, as below a thick till sheet, the water may contain so much dissolved material that it is not potable. This is particularly true where ground water seeping upward from underlying marine rocks is saline.

Well water from glacial deposits is almost always free from pathogenic organisms. Great care should be exercised, nevertheless, to protect wells in glaciated areas. This is especially important in areas such as New England where numerous shallow wells penetrate into sandy zones that may also receive effluent from septic tanks. Excessively permeable outwash deposits lack filtration capacity and could transmit pathogenic organisms long distances.

## **11.8** *Regions of Miscellaneous Deposits*

Nonindurated deposits may range in origin from artificial road fill to solifluction deposits in the arctic. No attempt will be made to discuss all possible types, but brief mention will be made of the more abundant types not previously discussed. The identity of most of these types is difficult once their mode of origin is obscured by burial beneath other deposits. When encountered in drilling they are simply classified according to size or shape of clasts.

Landslide debris can form locally important springs that issue from fractured material. If the landslide debris is initially unconsolidated, open fractures may be absent at depth and the mass as a whole is quite impermeable even though coarse material may make up part of the landslide debris.

Rockfalls produce deposits that are very permeable. These may blanket areas of several hundred acres. Talus slopes built by numerous smaller rockfalls cover large portions of mountain valleys and represent important recharge areas for the local ground water. The amount of water needed to initiate infiltration is quite small. Preliminary experiments at Stanford

University suggest that only a thickness of about 0.2 millimeter of water is needed to wet the average irregular surface of freshly broken rock. Additional water will percolate downward to the underlying rocks. The water will, moreover, concentrate in rivulets and will not wet the entire surfaces of the underlying rocks (Figure 11.25).

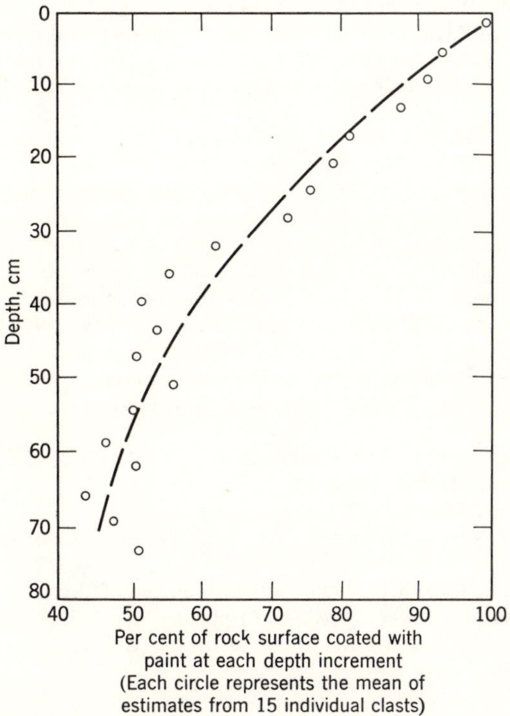


Figure 11.25 Percentage of rock surface coated with paint at different depths. Thin paint mixture was spread on the surface of a pile of rock until excess paint appeared at base of pile. The paint was then allowed to dry. The diagram illustrates the tendency of fluids to concentrate in rivulets when flowing through very coarse (40–80 mm diameter) rock rubble. (Based on results of unpublished experiments at Stanford University.)

Beach ridges formed by the ocean or by lakes are most commonly underlain by very permeable material. Where the ridges are wide enough and where the recharge is sufficient the beach deposits will yield enough water for domestic or stock wells. Beach ridges that are well above present lake or ocean levels are extensively developed around the Baltic Sea in northern Europe, the Great Lakes area in Canada and the United States, and along the many coastal plains of the world.

## REFERENCES

1. Athy, L. F., 1930, Density, porosity, and compaction of sedimentary rocks: *Am. Assoc. Petroleum Geologists Bull.*, v. 14, pp. 1–24.
2. Barber, E. S. (Chairman), 1955, Symposium on permeability of soils: *Am. Soc. Testing Materials, Special Pub. No.* 163, 136 pp.
3. Bedinger, M. S., 1961, Relation between median grain size and permeability in the Arkansas River Valley, Arkansas: *U.S. Geol. Survey Prof. Paper* 424-C, pp. 31–32.
4. Bredehoeft, J. D., 1963, Hydrogeology of the lower Humboldt River basin, Nevada: Desert Research Inst., *University of Nevada, Tech. Rept.* 3, 50 pp.
5. Brown, G. F., 1944, Geology and ground-water resources of the Camp Shelby area: *Miss. Geol. Survey Bull.* 58, 72 pp.
6. Brown, G. F. and W. F. Guyton, 1943, Geology and ground-water supply at Camp Van Dorn: *Miss. Geol. Survey Bull.* 56, 68 pp.
7. Brown, G. F., and others, 1944, Geology and ground-water resources of the coastal area in Mississippi: *Miss. Geol. Survey Bull.*, 60, 232 pp.
8. Brown, S. G., and R. C. Newcomb, 1963, Ground-water resources of the coastal sand-dune area north of Coos Bay, Oregon: *U.S. Geol. Survey Water-Supply Paper* 1619-D, 32 pp.
9. Cohen, P., 1963, Specific-yield and particle-size relations of quaternary alluvium Humboldt River valley, Nevada: *U.S. Geol. Survey Water-Supply Paper* 1669-M, 24 pp.
10. Conkling, H., and others, 1934, Ventura County investigation: *California Division of Water Resources Bull.* 6, 244 pp.
11. Davis, G. H., and others, 1959, Ground-water conditions and storage capacity in the San Joaquin Valley, California: *U.S. Geol. Survey Water-Supply Paper* 1469, 287 pp.
12. Eddards, M. L., and others, 1956, Water resources of the New Orleans area, Louisiana: *U.S. Geol. Survey Circ.* 374, 41 pp.
13. Fishel, V. C., 1948, Ground-water resources of the Kansas City, Kansas, area: *Kansas Geological Survey Bull.* 71, 109 pp.
14. Flint, R. F., 1957, *Glacial and Pleistocene geology:* New York, John Wiley and Sons, 553 pp.
15. Gibbs, H. J., and W. Y. Holland, 1960, Petrographic and engineering properties of loess: *U.S. Bureau of Reclamation Engineering Monographs No.* 28, 37 pp.
16. Golder, H. Q., and A. A. Gass, 1963, Field tests for determining permeability of soil strata; *in Field testing of soils: Am. Soc. Testing Materials Special Tech. Pub. No.* 322, pp. 29–45.
17. Graton, L. C., and H. J. Fraser, 1935, Systematic packing of spheres with particular relation to porosity and permeability: *Jour: Geol.*, v. 43, pp. 785–909.
18. Heim, G. E. Jr., and W. B. Howe, 1963, Pleistocene drainage and depositional history in northwestern Missouri: *Kansas Acad. Sci. Trans.*, v. 66, pp. 378–392.
19. Horberg, L., 1950, Bedrock topography of Illinois: *Illinois Geol. Survey Bull.* 73, 111 pp.
20. Horberg, L., and others, 1950, Groundwater in the Peoria region: *Ill. Geol. Survey Bull.* 75, 128 pp.
21. Johnson, A. I., R. C. Prill, and D. A. Morris, 1963, Specific yield—column drainage and centrifuge moisture content: *U.S. Geol. Survey Water-Supply Paper* 1662-A, 60 pp.

22. Kazmann, R. G., 1948, River infiltration as a source of ground water supply: *Am. Soc. Civil. Engineers Trans.*, v. 113, pp. 404–424.
23. Krumbein, W. C., and C. D. Monk, 1943, Permeability as a function of the size parameters of unconsolidated sand: *Am. Inst. Mining and Metall. Eng. Trans. Petroleum Div.*, v. 151, pp. 153–163.
24. LaMoreaux, P. E., 1960, Ground-water resources of the South—A frontier of the Nation's water supply: *U.S. Geol. Survey Circ.* 441, 9 pp.
25. Logan, J., 1964, Estimating transmissibility from routine production tests of water wells: *Ground Water*, v. 2, No. 2, pp. 35–37.
26. MacGary, L. M. and T. W. Lambert, 1962, Reconnaissance of ground-water resources of the Jackson Purchase region, Kentucky: *U.S. Geol. Survey Hydrologic Atlas* 13, 9 pp.
27. McGuinness, C. L., 1963, The role of ground water in the national water situation: *U.S. Geol. Survey Water-Supply Paper* 1800, 1121 pp.
28. Meade, R. H., 1964, Removal of water and rearrangement of particles during compaction of clayey sediments-review: *U.S. Geol. Survey Prof. Paper* 497-B, 23 pp.
29. Norris, S. E., 1961, Hydrogeology of a spring in a glacial terrane near Ashland, Ohio: *U.S. Geol. Survey Water-Supply Paper* 1619-A, 17 pp.
30. ——, 1963, Permeability of glacial till: *U.S. Geol. Survey Prof. Paper* 450-E, pp. 150–151.
31. Piper, A. M., and others, 1939, Geology and ground-water hydrology of the Mokelumne area, California: *U.S. Geol. Survey Water-Supply Paper* 780, 230 pp.
32. Poland, J. F., 1960, Land subsidence in the San Joaquin Valley, California, and its effect on estimates of ground-water resources: *Internat. Assoc. Sci. Hydrogeology Pub.* 52, pp. 324–335.
33. Poland, J. F., and G. H. Davis, 1956, Subsidence of the land surface in the Tulare-Wasco (Delano) and Los Banos-Kettleman City area, San Joaquin Valley, California: *Am. Geophys. Union Trans.*, v. 37, pp. 287–295.
34. Poland, J. F., and J. H. Green, 1962, Subsidence in the Santa Clara Valley, California, a progress report: *U.S. Geol. Survey Water-Supply Paper* 1619-C, 16 pp.
35. Robinson, W. H., and others, 1956, Water resources of the Mobile area, Alabama: *U.S. Geol. Survey Circ.* 373, 45 pp.
36. Rorabaugh, M. I., 1956, Ground water in northeastern Louisville, Kentucky: *U.S. Geol. Survey Water-Supply Paper* 1360-B, pp. 101–169.
37. Scalapino, R. A., 1963, Ground-water conditions in the Carrizo Sand in Texas: *Ground Water*, v. 1, no. 4, pp. 26–32.
38. Tickell, F. G., and W. N. Hiatt, 1938, Effect of angularity of grain on porosity and permeability of unconsolidated sands: *Am. Assoc. Petroleum Geologists Bull.*, v. 22, pp. 1272–1279.
39. Tolman, C. F., 1937, *Ground water:* New York, McGraw-Hill Book Co., pp. 341–346.
40. U.S. Soil Conservation Service, 1960, *Soil classification, a comprehensive system:* U.S. Dept. Agriculture, Soil Conservation Service, 265 pp.
41. Walters, K. L., 1963, Highly productive aquifers in the Tacoma area, Washington; *U.S. Geol. Survey Prof. Paper* 450-E, pp. 157–158.
42. Wenzel, L. K., and V. C. Fishel, 1942, Methods for determining permeability of water-bearing materials with special reference to discharging-well methods: *U.S. Geol. Survey Water-Supply Paper* 887, 192 pp.
43. Zeevaert, Leonardo, 1963, Foundation problems related to ground surface subsidence in Mexico City, in *Field testing of soils: Am. Soc. for Testing and Materials Special Pub. No.* 322, pp. 57–66.

# *chapter 12*

# GROUND WATER IN REGIONS OF CLIMATIC EXTREMES

## 12.1 *Introduction*

Climatic extremes found on the earth's surface have yet to be fully evaluated. Precipitation is so infrequent in some desert areas that even 100 years of exact measurements at a given locality will not determine the mean annual precipitation with any degree of precision. Many of these regions are known to have less than 1.0 inch of annual precipitation and some may have less than 0.1 inch. At the other extreme, the maximum mean annual precipitation in some areas is probably more than 450 inches. Several areas in the world have been considered by various individuals as being the wettest. Those mentioned most often are Mount Waileale, Hawaii; Cherrapunji, India; and southern Chocó province, Columbia. Temperature variations are just as great as those of precipitation. Recorded values range from about −125°F in Antarctica to 136°F in North Africa. Mean annual temperatures range from less than −10°F in parts of Antarctica to roughly 80°F in several tropical areas.

Extremes of temperature and precipitation do not necessitate the use of new laws of hydrogeology but they do require considerable modification of customary notions of the amount and distribution of recharge, magnitude of ground-water gradients, the continuity of aquifers, and the distribution of water of poor chemical quality.

## 12.2 *Regions of Exceptionally High Precipitation*

Regions of very high precipitation are relatively small in extent compared with areas of low temperature or low precipitation that are discussed in later sections. Best known areas that have a total precipitation in excess of 150 inches per year are in the mountains of Borneo, western Cameroun, northeastern Malagasy, eastern India, northeastern New Guinea, southern Chile, eastern Columbia, and southern Burma. Numerous other smaller areas exist, most of which are in mountainous regions. Precipitation is rarely divided evenly throughout the year, but most of the above areas do not experience exceptionally dry periods. Water is thus available for ground-water recharge throughout the year. Large areas of only moderately high precipitation, on the other hand, have total precipitation in excess of 50 inches per year but have marked wet and dry seasons. The dry seasons are so severe in such places as southern India, northeastern South America, and east central Africa that domestic and agricultural shortages of water are experienced each year.

Significant hydrogeologic effects of very high precipitation should be present, although studies are virtually lacking of even the most basic aspects of ground-water occurrence. Most of these areas are sparsely populated so that development of water supplies has not advanced beyond diversion of small amounts of water from springs and streams and the collection of rainfall from roofs. These rudimentary supplies are generally adequate for most needs inasmuch as almost constant replenishment exists.

Owing to an almost complete lack of information, we can only speculate concerning the hydrogeology of regions having very high precipitation. The presence of abundant recharge water would suggest that all but the most permeable aquifers should be filled to capacity, or, stated in another way, most of the rocks in the subsurface will be fully saturated except for the permeable zones that are free to drain. As a corollary, ground-water levels should be near the surface and hydraulic gradients should be much steeper than in similar aquifers in other climates (Figure 12.1).

Abundant water also suggests that chemical leaching of soluble and partly soluble material from aquifers should proceed rapidly. This means, in turn, that the chemical quality of the water should be good for most purposes. Abundant decaying organic matter that must be present in heavily forested areas would contribute carbon dioxide and organic acids to infiltrating water which would tend to decrease the pH and increase the manganese and iron content of the water. Also, the rapid circulation of ground water would favor a rapid flushing of saline and brackish connate waters from aquifers that are recently exposed by erosion.

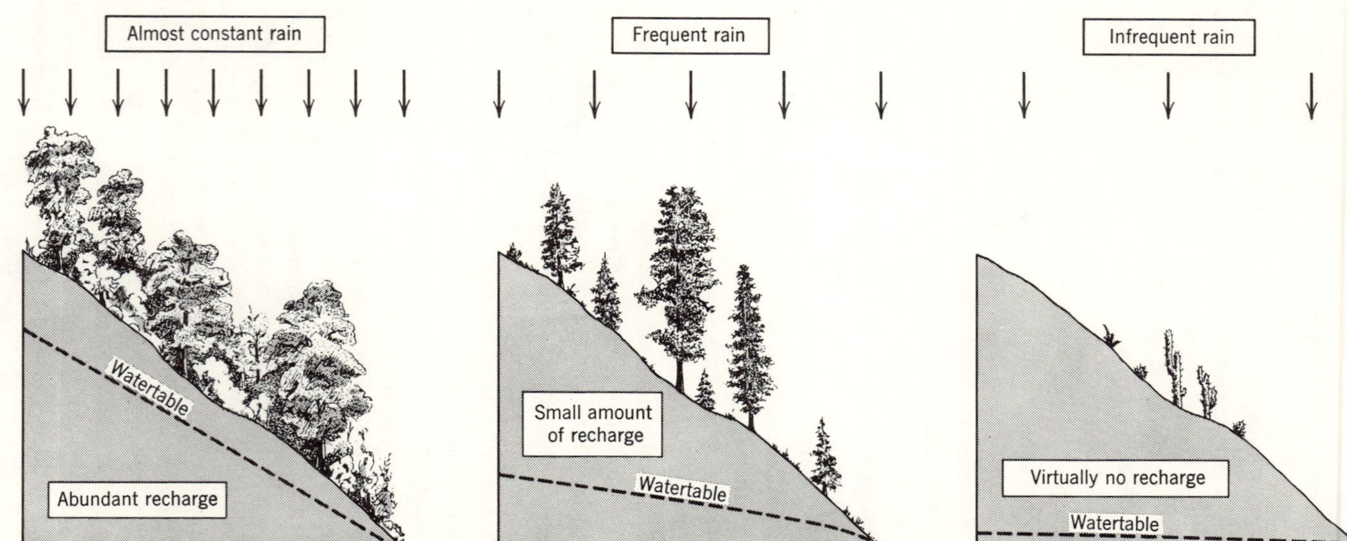


Figure 12.1 Diagrams showing the effect of varying amounts of recharge on the position of the watertable. Permeability of the rocks in the three diagrams are identical.

The economic importance of ground water in areas of very high rainfall is at present quite minor. Problems of dewatering for mining and construction projects have been encountered. In the future, however, biological purity and lack of turbidity in most ground water may justify its development in these regions having an overabundance of surface water.

## **12.3** *Ground Water in Arid Regions*

### INTRODUCTION

Few activities of the hydrogeologist are as satisfying as the development of large quantities of ground water in an otherwise parched desert. Although it is true that even large ground-water basins in such regions are destined for ultimate depletion with all but the most conservative developments, the fact that agriculture and industry can flourish for several decades is of great economic and social importance.

Arid and semiarid regions are found at all latitudes. Dry regions in the arctic and antarctic are not normally classified as deserts. Even so, parts of northern Greenland and the McMurdo Sound area of Antarctica have closed basins and saline lakes, and, except for temperature, would be hard to distinguish from deserts at lower latitudes. Most of the true deserts are between latitudes 15° and 50° either north or south of the equator. The largest single desert has an area larger than the United States and extends from the Atlantic Ocean across northern Africa and the Arabian Peninsula to the Persian Gulf. Other major deserts are in central Asia, southwestern North America, and central Australia. In all, about one tenth of the land area of the world is arid, and an equal or slightly greater area is semiarid. Ground water and imported surface water already supply agricultural and domestic needs in some of the more favorably situated areas within arid and semiarid regions. Nevertheless, vast areas remain to be developed throughout the world. This activity will occupy hydrogeologists for many years to come.

### CHARACTERISTICS OF AQUIFERS

Aquifers that antedate the formation of deserts are not greatly affected by increasing aridity. The general absence of organic material and the higher temperatures suggest that the small amount of infiltrating water that may be present will not dissolve very much carbon dioxide in comparison with water in humid temperate regions. This, together with the general lack of moisture, means that chemical weathering after deserts develop will be slow and the formation of solution openings in carbonate and other rocks will be retarded. Extensive carbonate aquifers found in some deserts may, therefore, be vestiges of climates considerably more moist than the one at present. Others, such as those of the south flanks of the Atlas Mountains, probably owe their origin to the action of large amounts of water infiltrating into intake areas in well-watered uplands adjacent to the deserts.

Desert conditions profoundly affect the type of sedimentation, which, in turn, controls the types of aquifers that are found in the nonindurated Cenozoic deposits. Streams are rarely vigorous enough to maintain their courses over volcanic or tectonic obstructions. Resulting closed basins fill with fine lacustrine deposits mixed with saline residues in their central parts. Fresh water and suitable aquifers are only found along the margins of such basins or in deeper aquifers not affected by present arid conditions. Stream-channel deposits form the most important aquifers deposited under desert conditions. Unlike streams in humid regions, desert channels carry extremely large amounts of material in suspension and traction which may reach such high concentrations that the flows are viscous mixtures of mud and debris. Resulting deposits are characterized by being poorly sorted and generally of low permeability. The permeable zones develop where streamflow persists long enough to sort the coarser debris in the stream channel. This usually takes place during the declining stages of streamflow when the stream is confined to rather narrow parts of the channel bottom. As a consequence, permeable zones constitute only a small part of the alluvial deposit and are commonly difficult to locate by drilling.

Nonindurated Cenozoic deposits that fill valleys formed by volcanism and block faulting constitute most of the important desert aquifers in southwestern United States, northwestern Mexico, northern Chile, and certain parts of central Asia. On the other hand, erosion has been the dominant process throughout most other desert regions with the result that bedrock is near the surface and nonindurated aquifers are confined to shallow deposits along ephemeral streams, basal parts of very large sand-dune tracts, and fillings of subsidence and deflation basins. Fortunately, some of the desert basins that lack thick alluvial deposits have wide-spread consolidated and semiconsolidated aquifers. The Mesozoic sandstones of the Great Artesian Basin of Australia and the Nubian sandstone of Egypt and parts of adjoining countries (Figure 12.2) are two well-known examples. Both aquifer systems contain water that ranges from potable in portions within a few hundred miles of recharge areas to brackish in the deeper buried parts of the aquifers. Upward movement of water from the Nubian aquifers has sustained the oases of Dakhla, Farafra, and Bahariya for countless centuries. Modern wells have greatly increased water discharge causing some reduction of the hydraulic head. Despite this, the total amount of water available from the Nubian aquifers is so vast that a

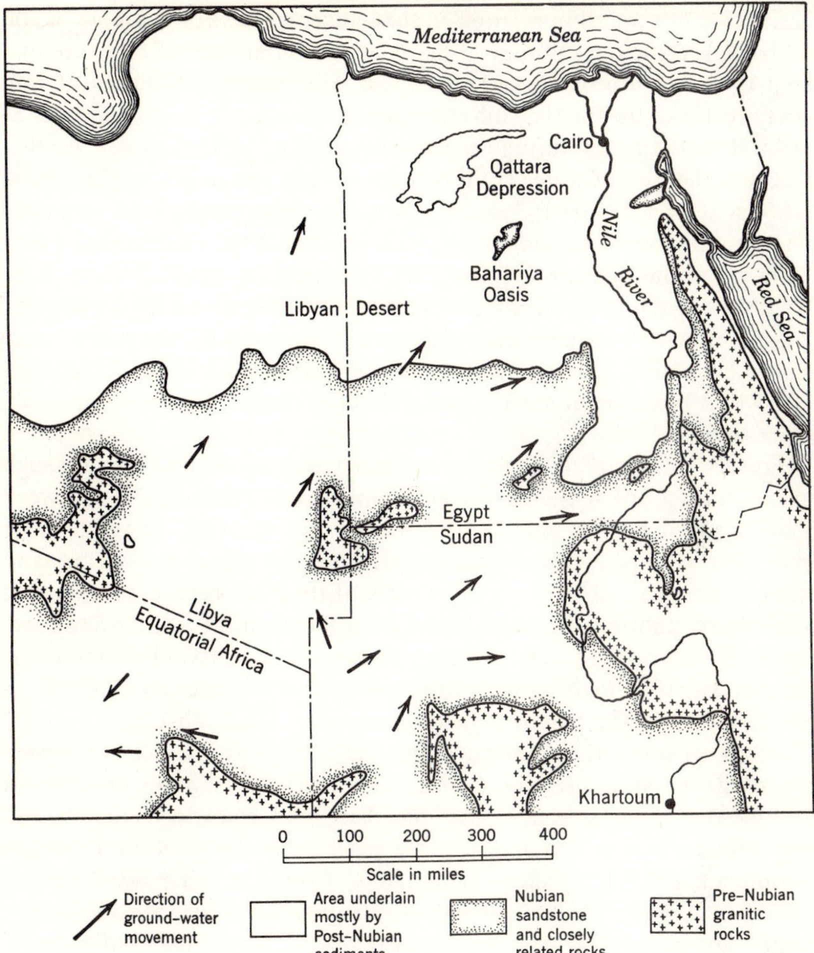


Figure 12.2 Outcrop area and general direction of ground-water movement in Nubian and related sandstones of northeastern Africa. (Unpublished map by William C. Ellis.)

considerable increase of the number of wells is probably justified provided that the water can be utilized profitably.

### GROUND-WATER RECHARGE

Most of the distinctive hydrogeologic features of arid regions are related to the quantity and quality of ground water available. All except the most permeable soils will generally cease to allow the passage of recharge water when rainfall is less than 5 to 10 inches. The exact amount depends on the permeability of the soil, the specific retention of the soil, and the distribution of rainfall in relation to the temperature. For example, a soil that has a specific retention of 15 per cent and is depleted of moisture to a depth of 2 feet during the summer heat will require 3.6 inches of rain merely to make up for the soil-moisture deficiency. If the rain occurs at several different times during the year, intervening periods of dry weather will cause the loss of water from the soil so that amounts much in excess of 3.6 inches will be needed to start recharge. In contrast, sand in a sand dune should have a specific retention of less than 5 per cent so that a rain of 2 inches should penetrate more than 3 feet into the sand. This is probably deep enough to penetrate beyond the zone of seasonal drying in parts of dunes that are experiencing deposition. Lowy [19] has made more elaborate calculations taking into account other factors and has concluded that recharge in sand dunes should be expected even in relatively dry deserts.

The fact that rainfall does not infiltrate as a saturated front through soils in at least one semiarid region is conclusively proved by recent work in the western part of the Central Valley of California. A number of small alluvial fans in this region are composed of clay-rich alluvium and mudflow deposits. The original porosity of these deposits is so high that the sediment fabric is unstable under even the slightest overburden load. Compaction, however, does not take place until the strength of the intergranular clay binding is reduced by wetting [6]. Construction of highways and irrigation works has been made difficult by local subsidence induced by irrigation water that infiltrates into the subsurface. Extensive experiments with flooded test plots as well as studies of the moisture content of numerous cores have demonstrated that the porous deposits have not been saturated since burial. In some places this dry condition extends to more than 100 feet below the surface. Even if a fairly rapid rate of deposition is postulated, the length of time needed to deposit these sediments must be measured in thousands of years. Thus, there has been no general infiltration of rain into the subsurface on these small alluvial fans for this length of time. The present mean annual rainfall in this region is about 7 inches, or greater than in a true desert. It seems rather safe to conclude, therefore, that extensive infiltration does not occur through desert soils which contain an appreciable amount of clay.

Barren rock or hard clay surfaces that are common in many desert regions will shed rains of only a few tenths of an inch. In rare cases the runoff may be relatively free of suspended material. More commonly, the runoff will erode large amounts of soil and surface debris. The resulting mixture will range from very turbid water to a viscous mudflow. If the water is not too turbid and if the channel bottom is permeable enough, water will flow into the gravel and sand of the channel bottom [17]. A large part of this water will return again to the atmosphere by evaporation. Where infiltration is most vigorous, underlying aquifers may be recharged. Most of the recharge will probably take place where the channel is somewhat constricted rather than in portions of the channel where the flow spreads out over large sections of the desert floor. This is because of the greater concentration of permeable material at the constrictions and the fact that the narrow constrictions are most likely to receive the last part of the flood flow which is made up of water draining from the earlier deposited saturated flood debris and is relatively free from suspended sediments.

Students of desert hydrogeology generally consider that most of the recharge takes place through the channel bottoms [2,25]. Extensive data to support this idea are lacking. Most of the arguments are based on generalizations such as have just been given. In a few places detailed water-level records have been able to give further support to this position. One of the most instructive studies was made of the effects of severe floods in the Tucson area in Arizona [30]. Despite the fact that a very large area was inundated and a much larger area was subject to unusually heavy rains, water-level records of numerous wells show that significant recharge was confined to a relatively small part of the flooded area (Figure 12.3).

The relatively small amounts of recharge in desert regions is reflected in the great depths to water in upland areas and the exceptionally flat hydraulic gradients that are commonly encountered. Under virgin conditions the gradients may be so small that important hydrogeologic barriers cannot be distinguished. Ground-water development increases the amount of water moving through the subsurface and steepens all gradients. This serves to accentuate the zones of lower permeability (Figure 12.4).

### GROUND-WATER CIRCULATION

In western United States many of the desert basins are filled to the ground surface with water, and a few support permanent lakes. Water percolating into the upper parts of alluvial fans will be discharged through vegetation and direct evaporation from the soil in the central parts of the basins (Figure 12.5). An interesting exception to this general condition is found in the Ash Meadows circulation system of southern Nevada. Here, despite precipitation that exceeds 10 inches on the higher mountains, the basins have not filled, and the ground-water levels in wells are more than 1000 feet below some of the valley floors. Fractured Paleozoic rocks of complex structure are permeable enough to serve as an underdrain to the region and to transmit the ground water to several points of discharge in Ash Meadows (Figure 12.6) [29,31].

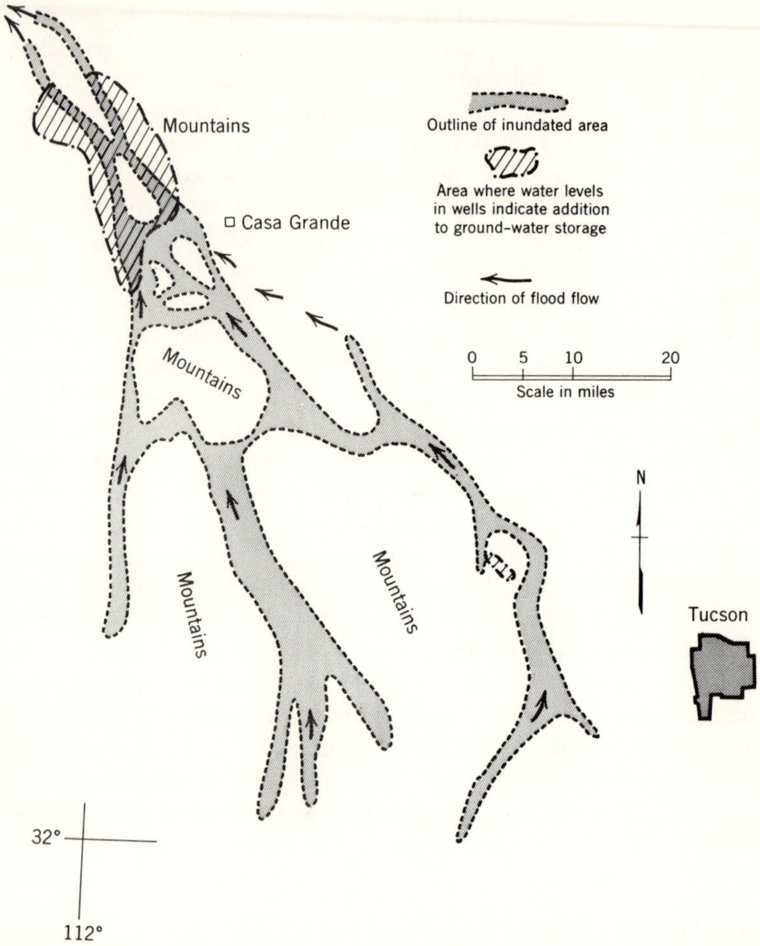

Mountains
Outline of inundated area
Area where water levels in wells indicate addition to ground-water storage
Casa Grande
Direction of flood flow
Mountains
0 5 10 20
Scale in miles
N
Mountains
Mountains
Tucson
32°
112°

Figure 12.3 Map of part of Santa Cruz Valley, Arizona, showing recharge produced by flooding in an arid region. (Map from White and others [30].)

Ground surface
Watertable before pumping
Watertable after pumping
Sandstone
K = 10
Alluvium
K = 100
Alluvium
K = 100
Fault zone
K = 1

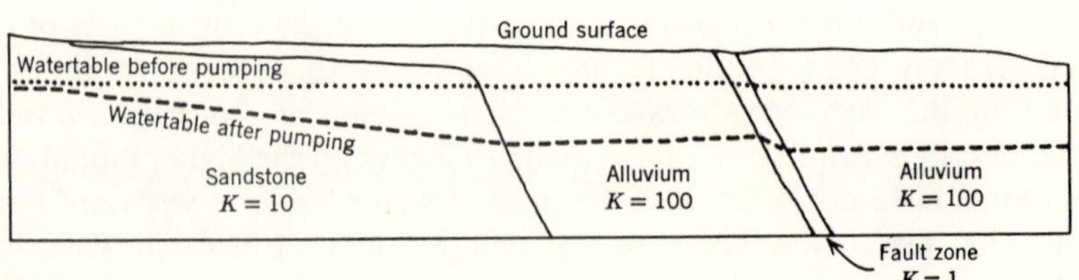

Figure 12.4 Diagram illustrating how the development of ground water will increase the regional hydraulic gradients which in turn accentuates variations in hydraulic conductivities.

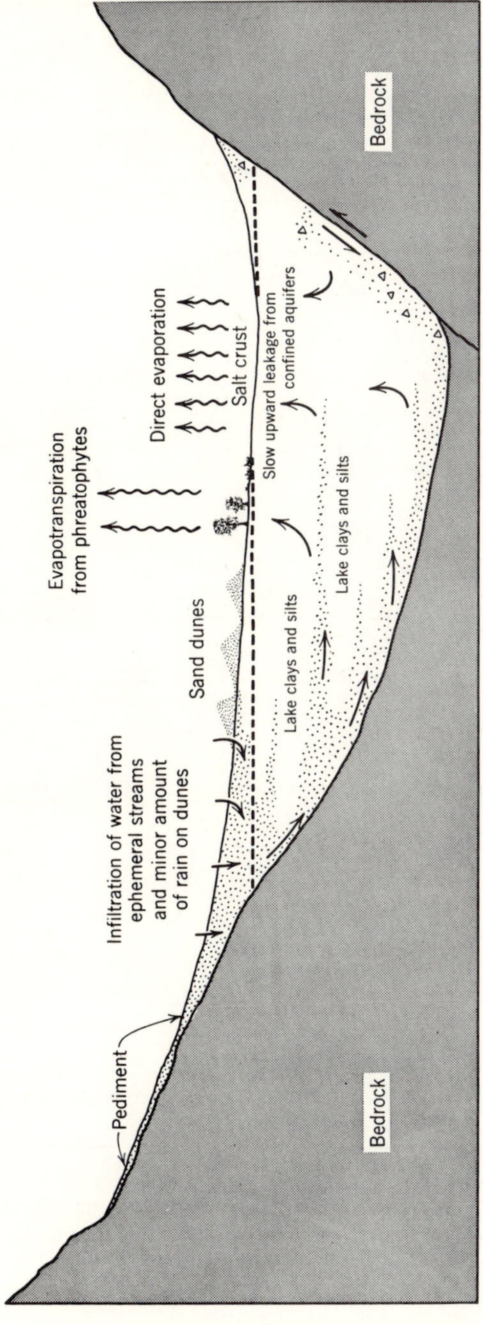


Figure 12.5 Ground-water circulation in a simple closed basin in a desert. Heavy dotted line indicates the elevation that water would stand in wells if they were drilled.

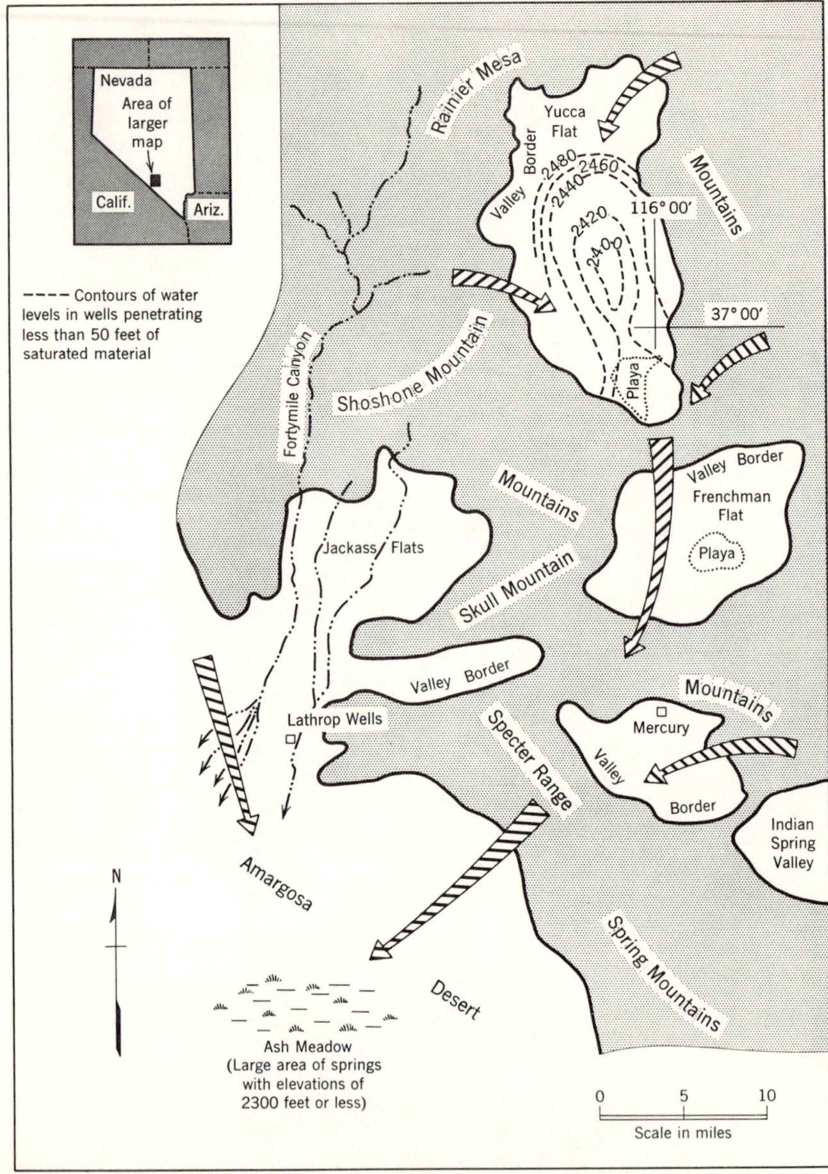


Figure 12.6 Ground-water circulation within fractured Paleozoic rocks of southern Nevada. Closed contours of water levels within overlying tuff and alluvium are caused by the movement of water partly in a downward direction into the underlying Paleozoic aquifers. Both Frenchman and Yucca Flats are closed topographic basins with ground-water levels generally deeper than 1000 feet below the surface. (Map is modified from Winograd [31].)

WATER QUALITY

It is a generally observed fact that the average chemical quality of ground water is much poorer in desert regions than in regions of normal precipitation (Figure 12.7). Representative chemical analyses are given in Table 12.1. Although water is not all brackish or saline, the following factors operate to increase the dissolved solids. First, salts contained in rain and snow will be concentrated at the surface by the high-evaporation rates. Occasional heavy rains and floods will have, therefore, a ready source of dissolved solids as the water percolates downward to eventually recharge the ground water. Second, buried saline deposits that will be dissolved by ground water are common in the near-surface deposits of desert regions. Third, the slow circulation of ground water will slow the flushing of connate water and will allow sufficient time for the dissolution of slightly soluble materials in contact with the ground water. Fourth, and last, fine dust composed in part of soluble salts blown from playas forms a ubiquitous dry fallout that accumulates until it is washed into the soil by occasional rains. The soluble components of the dry fallout form an important, but not fully evaluated, source of dissolved solids in desert ground water.

Potable water is by no means rare in many deserts [2,4,5,24,29], although brackish water is all that can be found in many large regions where inhabitants have been known to survive quite well using water with dissolved solids of more than 3000 ppm. Water with the lowest dissolved solids are generally found in small quantities in fractured metamorphic and plutonic rocks and in sediments derived from these rocks. Other favorable aquifers are basal zones in very large sand-dune tracts and in limestone and basalt aquifers that are permeable in outcrop areas and are recharged rapidly during rains.

The most reliable water from the standpoint of quality is water that is brought into desert regions by natural or artificial means and moves only a short distance in the subsurface. Good examples of this type of ground water are found in many of the valleys of western United States, along the Nile River, along numerous rivers in the coastal deserts of western South America, along the lower Indus River, and along major rivers south and east of the Aral Sea in the U.S.S.R. Where irrigation is practiced and where subsurface drainage is poor the loss of ground water by evapotranspiration may be so large that the salts originally in the water may become concentrated in the soil. Periodic irrigation will redissolve most of these salts so that water infiltrating into the upper part of the shallow ground water may be quite saline. This effect has become one of the major problems facing the immense irrigation projects of West Pakistan

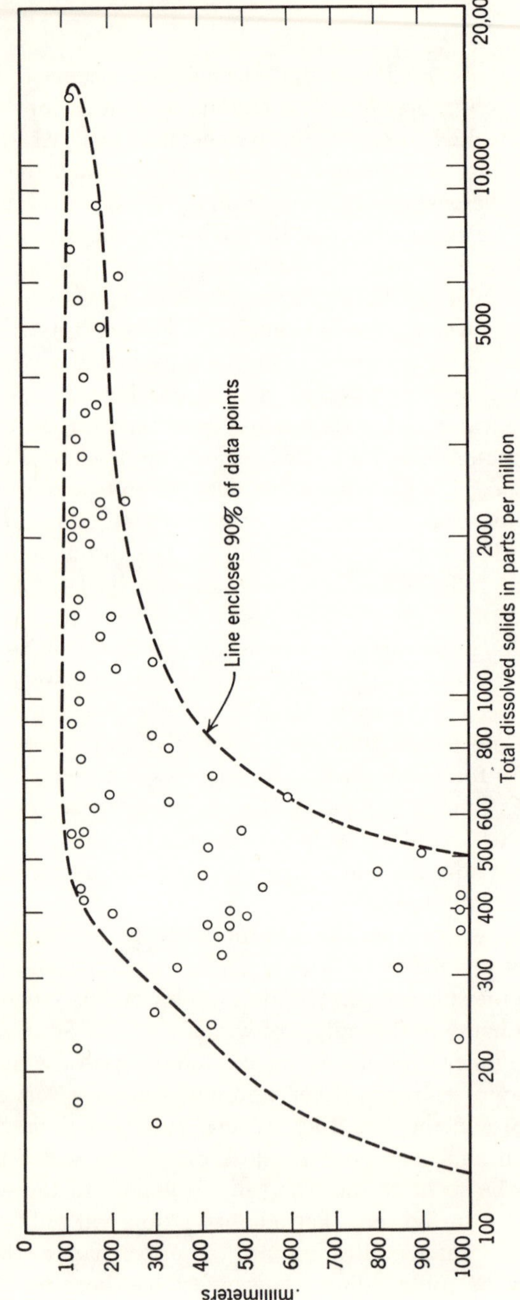


Figure 12.7 Relationship between mean annual precipitation and total dissolved solids for 70 samples of ground water from Syria. The precipitation is for the area in which the sample was collected and differs in many samples from the precipitation over the infiltration-recharge area. (From Burdon and Mazloum [7].)

*Table 12.1 Chemical Analyses of Water from Arid and Semiarid Regions, analyses in parts per million*

| | Water Leached from Saline Soil of the U.S.S.R. | Fresh Well Water from Alluvium, Northern Chile | Brackish Well Water from Alluvium, Northern Chile | Brackish Spring Water from Rocks Rich in Gypsum and Calcite, New Mexico | Fresh Spring Water from Rhyolite Aquifer near Socorro, New Mexico |
|---|---|---|---|---|---|
| $SiO_2$ | ... | 36 | 4 | 21 | 27 |
| Ca | 1540 | 7 | 295 | 456 | 18 |
| Mg | 360 | 0.5 | 58 | 113 | 4 |
| Na | 2010 | 160 | 474 | 47 | 54 |
| K | ... | 8 | 61 | ... | 2.8 |
| $HCO_3$ | 61 | 248 | 63 | 200 | 154 |
| $SO_4$ | 2430 | 96 | 415 | 1460 | 28 |
| Cl | 5040 | 53 | 1130 | 15 | 15 |
| $NO_3$ | ... | 4 | 0.3 | 0.2 | 1.2 |
| B | ... | ... | 18 | ... | ... |
| Total dissolved solids | ... | 487 | 2469 | ... | 224 |
| References | [26] | [8] | [8] | [15] | [15] |

[1] as well as countless other smaller projects throughout the world [12].

## EXPLORATION FOR GROUND WATER

Desert areas generally provide the maximum opportunity for the application of scientific methods to the study of water. Rock outcrops are fully exposed without the troublesome mantle of vegetation encountered in other regions, human habitations are sparse so that geophysical operations are not hampered by restrictions on the use of explosives or by metal pipes and the like that make resistivity surveys difficult, and points of natural discharge of potable water are clearly marked by vegetation. Conventional hydrogeologic methods of exploration are used in desert regions. Some modifications in operating procedures, however, are commonly needed because of greater depths to aquifers and because existing wells and bore holes are rarely sufficient in number to give the needed hydrologic and geologic control. More emphasis is given to seismic equipment coupled with extensive test drilling.

Many wells have been drilled into pediments in unsuccessful efforts to find ground water. Pediments are relatively smooth, sloping surfaces typically developed along the lower parts of desert mountains. Pediments are mantled by thin and sometimes discontinuous patches of alluvium. The strong resemblance between pediments and alluvial fans has misled many people into drilling for water only to find impermeable bedrock a short distance below the surface.

Phreatophytes are one of the most useful surface indicators of ground-water conditions [21]. The total area of vegetation gives some rough indication of the total amount of water being discharged at the surface. By identifying the species and measuring the density of vegetation, a rather close estimate can be made. Even without this work, a simple measurement of area from aerial photographs can be multiplied by a reasonable value for transpiration to obtain a first estimate of quantity [20]. For example, if 250 acres are covered with dense vegetation, the transpiration most probably lies between the values of 2.0 feet and 7.0 feet of water per year. If a reasonable value of 3.5 feet per year is assumed then the estimated discharge would be 875 acre-feet per year, or a constant discharge of about 540 gpm, enough to supply a town of more than one thousand inhabitants.

Phreatophytes give some indication of depth to water. Grasses thrive where the watertable is generally less than 10 feet below the surface, shrubs where the watertable is less than 30 feet below the surface, and trees where the watertable is less than 90 feet below the surface. These are rough figures at best. Actual depths depend on the width of the capillary fringe, amount of surface rain, actual plant species, and other factors. Phreatophytes also give some suggestion of water quality. Willows and cottonwood usually grow where the water is potable and pickleweed (*Allenrolfea occidentalis*) usually grows in places where the soil is saturated with saline water. Many other phreatophytes such as palms, mesquite, and salt grass may grow with roots in either potable or brackish water.

## **12.4** *Ground Water in Regions of Perennially Frozen Ground*

### INTRODUCTION

Perennially frozen ground, or permafrost, is present almost everywhere within the Arctic Circle in the Northern Hemisphere with the exception of large parts of northern Scandinavia (Figure 12.8). The area underlain by permafrost extends to latitude 48° in both central Asia and central North America. Permafrost is generally continuous north of the Arctic Circle; as one moves to the south, the permafrost becomes first discontinuous then sporadic along the southern limits. The southern Hemisphere is free from permafrost except on the high mountains of the Andes, isolated islands, and on Antarctica.

In general, the presence of permafrost, even though it may not be continuous, profoundly affects the circulation of ground water, and in extreme cases makes the recovery of ground water virtually impossible. Other sources of fresh water, nevertheless, exist in arctic regions (Figure 12.9). The large amount of energy needed to melt ice makes it almost imperative that sources of unfrozen water be utilized for population centers of more than a few score individuals. Large springs and lakes deeper than about 10 feet afford the only common year-long sources of water, other than well water, that persist throughout the winter. Suitable springs are not common in the Arctic, and lakes are subject to contamination. Wells, if they can be developed, are, therefore, the best general sources of potable water.

### FACTORS CONTROLLING THE DISTRIBUTION OF PERMAFROST

The broad cause of permafrost is obvious; it is simply an extremely cold climate that causes the ground temperature to be maintained below freezing. The details of the various physical factors that control ground temperature, however, are complex, and, in most cases, impossible to evaluate fully. Heat flow is generally outward from the interior of the earth. The amount of heat flowing to the surface is very small so that temperatures within a few hundred feet of the surface will be controlled primarily by the amount of solar heat absorbed at the surface and the

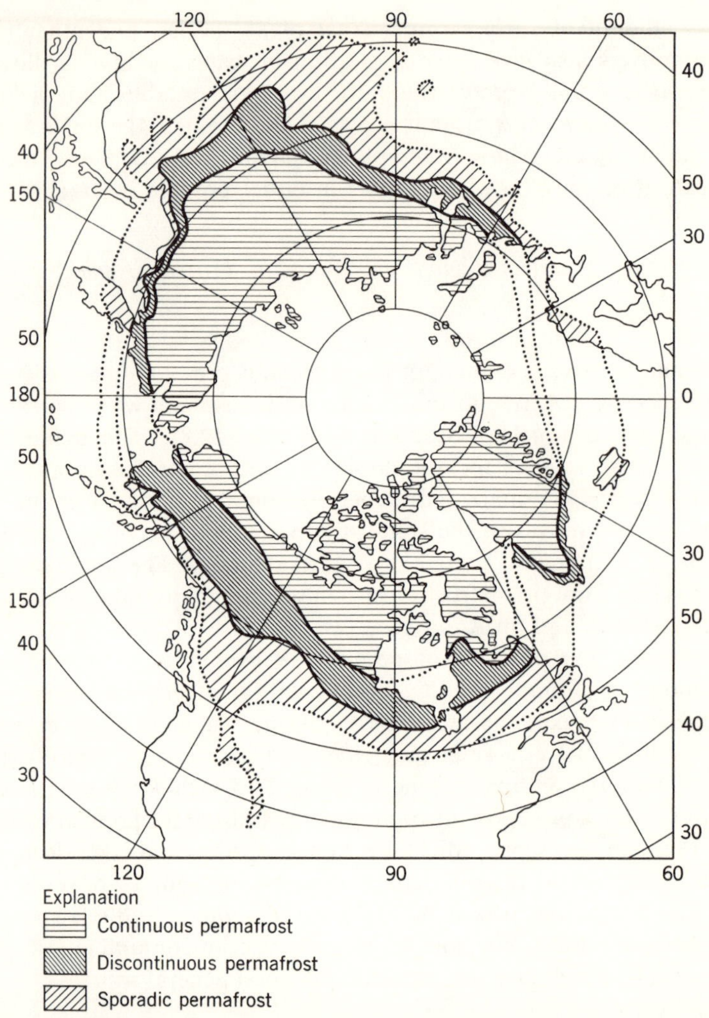


Figure 12.8 Distribution of Permafrost in the northern hemisphere (3,16,18,27).

thermal conductivity of the natural materials (Figure 12.10). The direction of heat flow, and hence the orientation of the isothermal surfaces, is determined by the spatial arrangement of rocks and sediments of varying thermal conductivity, by the configuration of the ground surface, by the transport of heat by circulating water, and by the temperature of the surface.

Thermal effects of rapidly infiltrating or rapidly discharging ground water are most important in regions of sporadic discontinuous permafrost. Rapid infiltration of water will take place on talus, sand dunes,

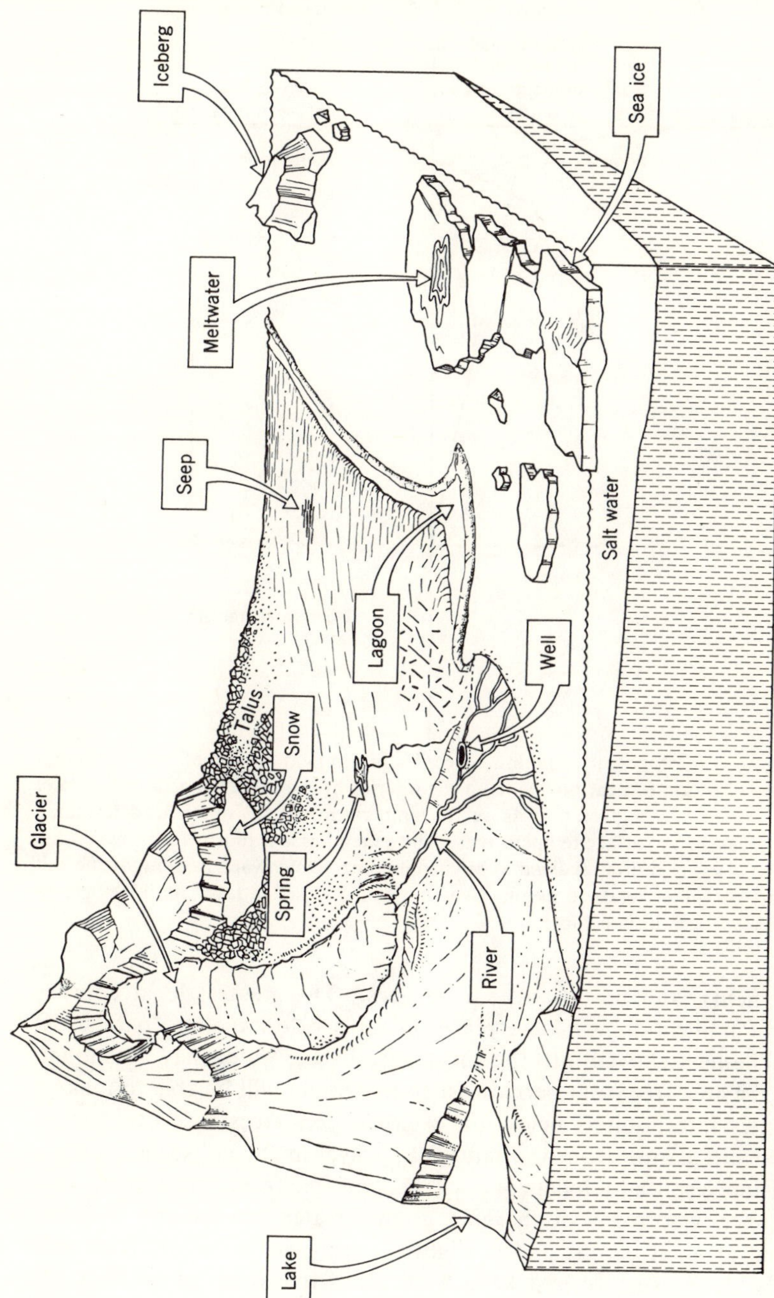


Figure 12.9 Sources of water in the polar regions. In addition to sea water, lagoonal water and sea ice may be saline. Melt water on sea ice may be saline to brackish if melting is well advanced. Snow cover on sea ice commonly provides fresh water on the upper parts of thick pack ice.

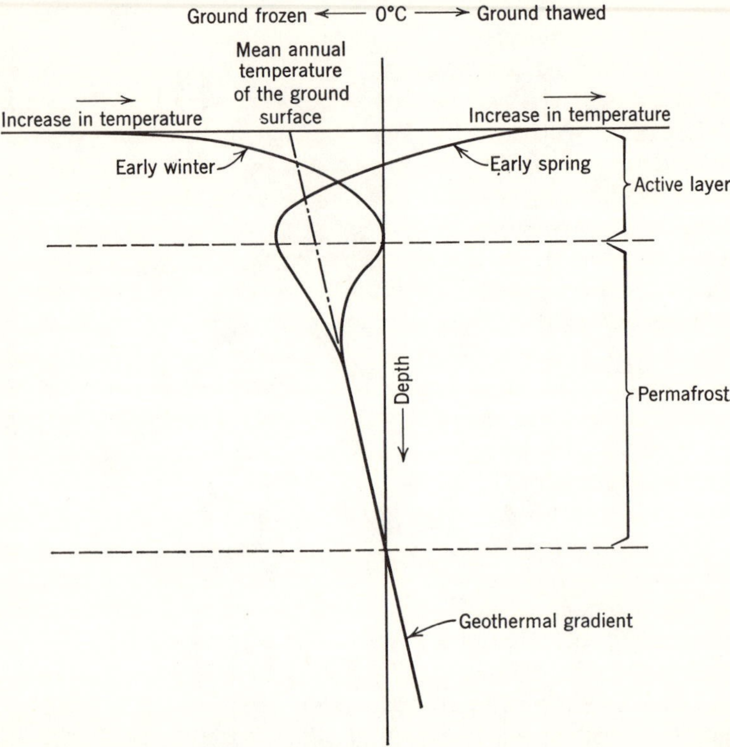


Figure 12.10 Hypothetical temperature profiles in a region of permafrost. Although the average direction of heat flow is from the interior of the earth towards the surface, the actual gradients at a given time as well as temperatures at any depth are determined in a large measure by surface temperatures. The general case is illustrated in which the active layer is never entirely thawed at any single time. Downward freezing of the soil at the start of winter tends to confine water within the active layer until it migrates laterally or is frozen in late winter.

boulder fields, and gravel-covered terraces. The greatest local concentration of infiltration will take place on talus below melting patches of snow. The heat carried into the subsurface by the water may be enough to keep the area free from permafrost so that water will be able to make a direct entry into aquifers below the regional permafrost. In other areas the water circulating from the talus may travel in the subsurface only a short distance so that a local zone will be kept free from permafrost. In a like manner, vigorously discharging ground water, particularly if it has circulated at great depths and is slightly warm, will transport heat continuously to the surface and keep local areas free from permafrost. It should be noted that many seemingly important springs evident in the summer are merely the result of shallow circulation of snow melt and may freeze during the winter.

The loss of heat by conduction through the near-surface part of unconsolidated material is in a large measure controlled by the moisture content of the material. Water has a thermal conductivity of roughly fifteen times the conductivity of dry air. Ice has a thermal conductivity of about four times the conductivity of liquid water. This means that a porous organic soil may have a thermal conductivity in a wet state of more than ten times the dry conductivity. If the interstitial water is frozen, the increase in conductivity is even greater. In practical terms it can be seen that if a soil dries in late winter and remains dry throughout the summer only a minor amount of heat will enter the ground and permafrost will be retained near the surface. On the other hand, soils that remain dry throughout the winter and that become saturated during the summer will tend to produce higher subsurface temperatures which may be high enough to prevent the formation of permafrost.

Factors that control the surface temperature are most complex. Vegetative cover, soil-evaporation rates, temperature and velocity of winds, orientation of sloping surfaces, winter snow cover, and soil color are all important. Low soil temperatures will be favored by shaded slopes during the summer, light colored soils that reflect solar energy, and barren winter surfaces exposed to cold winds of high velocity. High soil temperatures are favored by thick winter snow covers that insulate the underlying ground, by sunny exposures during the summer, and by dark surfaces that tend to absorb solar energy.

The unfrozen surfaces of lakes and rivers will absorb most of the sun's energy and will generally become much warmer than the surrounding ground. Heat will, therefore, flow from the water into the ground and tend to keep it above freezing. The only near-surface zones that remain above freezing in the northernmost parts of Asia and North America are probably the normal active zone and deeper zones near major bodies of water.

### INDICATORS OF PERMAFROST

The search for ground water in arctic regions is essentially a search for permeable and unfrozen zones that are saturated with potable water. Patches of sporadic permafrost can be avoided in many places by noting typical surface indications of permafrost. Vegetation patterns and types, microrelief patterns, hydrolaccoliths, features resulting from thawing, and various hydrologic phenomena can all be useful keys to the presence of permafrost [16,22]. Many of the features, particularly the microtopography, can be best studied using aerial photographs. Other features such as ice exposed in newly cut steam banks can best be investigated from the ground.

Vegetation can be an important aid to the recognition of permafrost but it should not be relied on exclusively. In Alaska it has been found that trees such as larch and black spruce can tolerate permafrost near the surface because of their shallow root systems. Tall willows on flood plains and mature aspen have deeper root systems and are more likely to be found where permafrost is absent. North of the forested regions, tall shrubs commonly mark local thawed areas.

Polygonal patterns are common indicators of permafrost. These patterns are formed on relatively flat areas by the development of large polygonal cracks that fill with ice wedges. The diameter of the polygons varies from about 20 feet to as much as 400 feet. Inasmuch as the ice in the ice-wedge polygons is rarely exposed at the surface, the topographic form can be confused with similar forms that are not necessarily related to permafrost. Most of the other forms are smaller in diameter and some of them may develop in areas of some relief. Frost-crack polygons can form by the contraction of seasonally frozen ground during an exceptionally cold period. The frost-crack polygon may or may not be underlain by permafrost. It is commonly less well defined and slightly larger than ice-wedge polygons in the same area. Stone polygons formed by frost stirring are smaller and are an indication of intense frost action, but do not necessarily indicate permafrost. Desiccation cracks in arid regions may form cracks several feet across. These are generally in playas subject to infrequent flooding. Although they are sometimes large enough to distinguish on aerial photographs, there are few regions in the arctic that could develop large desiccation cracks because of an annual surplus of moisture in the soil.

Hydrolaccoliths, or pingos, are hills that form by the accumulation of ice lenses which force the overlying soil upward and outward. The hills may be as much as 300 feet high and 800 feet in diameter, although smaller forms are more typical. The summit areas of many hydrolaccoliths are cracked open to form crater-like depressions which give the hills a vague resemblance to small volcanic cones. Hydrolaccoliths are reliable indicators of permafrost.

Extensive evidence of thawing generally indicates underlying permafrost. Annual freezing rarely results in soil filled with interstitial ice or ice wedges that are so thick as to produce on melting thaw lakes or ponds, beaded drainage, or estensive ground collapse near roads, buildings, and other cultural features.

Several hydrologic phenomena suggest the local absence of permafrost. Abrupt decrease in stream discharge as it crosses permeable areas, large streams that end in small lakes, or ponds that lack outlets, and unusually large fluctuations of lake levels would all suggest active infiltration into the subsurface and hence the possibility that the areas are not frozen even at some depth [16]. In a similar manner, springs, patches of luxuriant vegetation, and other indications of ground-water discharge suggest unfrozen aquifers.

### AREAS FAVORABLE FOR GROUND WATER

Some of the more favorable locations for the recovery of ground water in the Arctic are shown in Figures 12.11 and 12.12. Thawed zones near large bodies of water and aquifers under the regional body of permafrost are the most reliable sources of large amounts of water [10,11,22]. Where permafrost is very thick, the best potential aquifers may be frozen and the well must be completed in bedrock, or the wells may extend into deep aquifers saturated with connate water of such poor quality as to be of little use. The deepest well through permafrost recorded in Alaska yielded 45 gpm of potable water from beneath 603 feet of material that was almost all frozen [28]. With continued development, areas having many such wells could experience depletion of supplies because recharge below thick and continuous permafrost must be very small.

The easiest ground water to develop is in the active zone during late summer and early winter. Such a source is rarely a dependable year-long supply, although Cederstom [9] describes a shallow well under a heated building in Kotzebue, Alaska, that derives water from the thawed zone under the building as well as a surrounding thawed zone that persists between the winter frost zone and the permafrost. Shallow interceptor drains have also been used in thawed zones near streams [13]. The biological purity of such shallow water should always be open to question, particularly around small centers of habitation that lack adequate facilities for the disposal of waste material.

Thawed zones along feeder channels for hydrolaccoliths may be favorable for ground-water recovery. Muller [23], however, concludes that most hydrolaccoliths are nourished by relatively small flows of ground water. Larger flows will reach the surface as perennial springs and will not accumulate as ice in the subsurface. Such springs will cause large sheets of ice to form on top of the ground during the coldest time of the year.

Terrane and weather conditions make the exploration and development of ground water in the Arctic the most difficult of all geographic regions

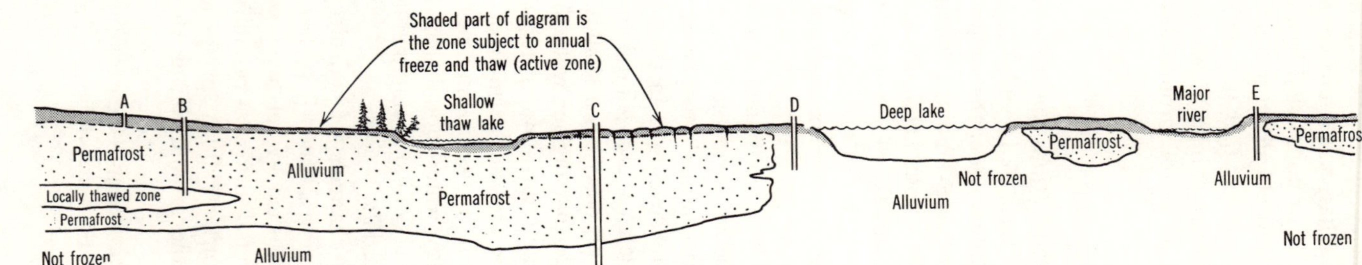


Figure 12.11 Recovery of ground water from alluvial aquifers: (A) above the permafrost within the active layer, (B) within a local thawed zone surrounded by permafrost, (C) below the permafrost, (D) near a lake, and (E) near a major river.

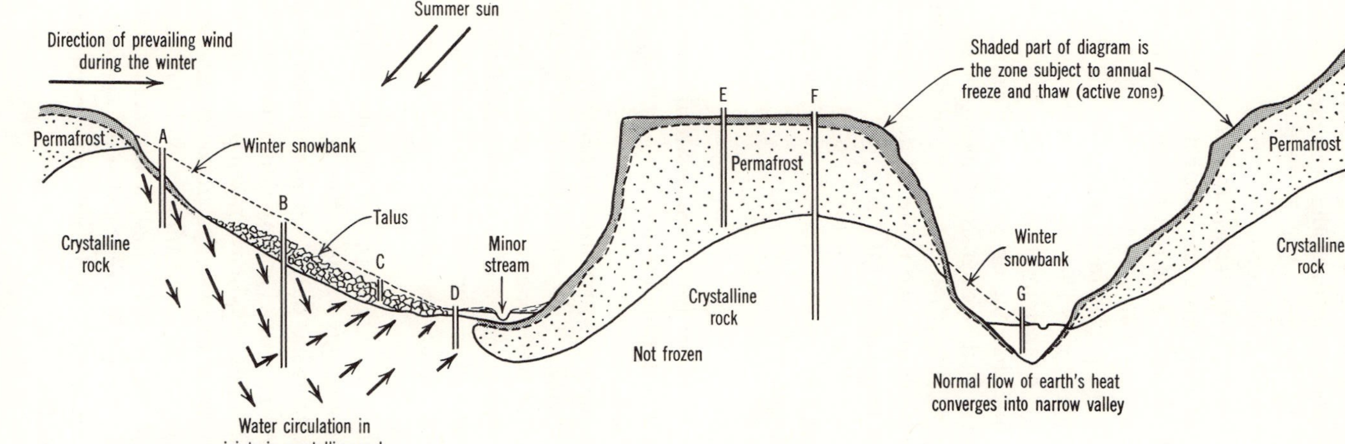


Figure 12.12 Wells in a mountainous area. (A) Well having small seasonal yield. This well is not deep enough for a dependable year-long supply. (B) Deep bedrock well with modest yield during the entire year. (C) Shallow well with high yield during periods of active recharge. Late summer yield may be small. (D) Shallow bedrock well having small but dependable yield. (E) Deep bedrock well with no yield because it is still in permafrost. (F) Deep bedrock well far below permafrost. The yield is probably very small because of a lack of permeability at great depth. (G) Good yield from a shallow well near a large stream.

[11]. Remoteness and freezing winds are only part of the difficulty. If drilling encounters permafrost many other problems arise. Cable-tool drilling may be stopped by the casing being frozen to the sides of the hole. Continuous drilling without long periods between individual advances of the casing helps. Warm water introduced into the hole also helps. Excessive thaw near the hole, however, may cause troublesome collapse of the ground around the well. This is a greater potential problem with rotary-drilled holes. Drilling with air in the winter or with brine or oil-based muds will enable the permafrost portion of the hole to be drilled with fluids at temperatures below freezing. After the well is constructed, continuous care needs to be taken in order to prevent freezing in the well, pump, storage tanks, and distribution systems. Heating coils or steam pipes can be used provided that extensive thawing of the permafrost is prevented. A less expensive method commonly used is simply to pump water from the well continuously and waste whatever water is not needed. Inasmuch as the water temperature may be a few degrees above freezing and the specific heat as well as heat of fusion of water is high, the total amount of water needed to prevent freezing in a moderately insulated water system is quite small.

### WATER QUALITY

The chemical quality of water from surface streams and lakes is generally excellent in the Arctic. Ground water that is actively recharged from these sources is also good.

Iron and manganese have been reported as being troublesome in some ground waters in the Arctic. Favorable conditions for the entrance of these constituents into the water are probably created by the presence of excessive amounts of organic material in alluvial aquifers. The preservation of the organic material is favored by the cold environment. The organic material in turn will create reducing conditions in the subsurface when it undergoes slow decay.

The organic material in the aquifers can also cause unsightly coloration of the water and, less commonly, disagreeable tastes.

Permafrost itself probably has only a secondary effect on water quality. Partial freezing leaves a residue of more saline water so that some concentration of dissolved solids could have taken place over the long period of time that it has taken for the water to freeze in permafrost areas. In most aquifers, however, water circulation is probably rapid enough to remove the more saline water faster than it can accumulate. Effects of seasonal freezing cycles have been reported from one region where the winter concentration of some constituents was much larger than the summer concentrations [14].

## REFERENCES


1. Ahmad, N., 1961, Soil salinity in West Pakistan and means to deal with it, in *Salinity problems in the arid zones:* Paris, United Nations Educational, Scientific, and Cultural Organization, pp. 117–125.
2. Bergstrom, R. E., and R. E. Aten, 1964, Natural recharge and localization of fresh groundwater in Kuwait (abstract): *Am. Geophys. Union Trans.*, v. 45, pp. 52.
3. Black, R. F., 1954, Permafrost—A review: *Geol. Soc. Am. Bull.*, v. 65, pp. 839–856.
4. Blasquez, L., 1953, Estudio hydrogeológico de la región desértica y subdesértica de Sonora, Mexico: Internat. Geol. Congress, 19th Session, Section 8, pp. 15–23.
5. Bogomolov, G. V., 1961, Conditions of formation of fresh waters under pressure in certain desert zones of North Africa, the U.S.S.R., and South-West Asia, in *Salinity problems in the arid zones:* Paris, United Nations Educational, Scientific, and Cultural Organization, pp. 37-41.
6. Bull, W. B., 1961, Causes and mechanics of near-surface subsidence in western Fresno County, California: *U.S. Geol. Survey Prof. Paper* 424-B, pp. 187–189.
7. Burdon, D. J., and S. Mazloum, 1961, *Some chemical types of ground-water from Syria:* Paris, United Nations Educational, Scientific, and Cultural Organization, pp. 73–90.
8. Castillo, O., 1960, El agua subterránea en el norte de la Pampa del Tamarugal (Chile): *Instituto de Investigaciones Geológicas Boletín No.* 5, 89 pp.
9. Cederstrom, D. J., 1952, *Summary of ground-water development in Alaska*, 1950: *U.S. Geol. Survey Circ.* 169, 37 pp.
10. ——, 1963, Ground-water resources of the Fairbanks area, Alaska: *U.S. Geol. Survey Water-Supply Paper* 1590, 84 pp.
11. Cederstrom, D. J., and others, 1953, Occurrence and development of ground water in permafrost regions: *U.S. Geol. Survey Circ.* 275, 30 pp.
12. Dixey, F., and S. H. Shaw, 1961, Introduction: hydrology with reference to salinity in *Salinity problems in the arid zones:* Paris, United Nations Educational, Scientific, and Cultural Organization, pp. 15–23.
13. Feulner, A. J., 1964, Galleries and their use for development of shallow ground-water supplies, with special reference to Alaska: *U.S. Geol. Survey Water-Supply Paper* 1809-E, 16 pp.
14. Feulner, A. J., and R. G. Schupp, 1963, Seasonal changes in the chemical quality of shallow ground water in northwestern Alaska: *U.S. Geol. Survey Prof. Paper* 475 B, pp. 189–191.
15. Hall, F. R., 1963, Springs in the vicinity of Socorro, New Mexico: *New Mexico Geol. Soc., Guidebook for 14th Field Conference*, pp. 160–179.
16. Hopkins, D. M., and others, 1955, Permafrost and ground water in Alaska: *U.S. Geol. Survey Prof. Paper* 264-F, pp. 113–146.
17. Keppel, R. V., and K. G. Renard, 1962, Transmission losses in ephermeral stream beds: *Am. Soc. Civil Eng. Proc., Hydraulics Div. No.* 3, v. 88, pp. 59–68.
18. Legget, R. F., and others, 1961, Permafrost investigations in Canada: *Natl. Research Council of Canada Div. of Building Research Tech. Paper* 122, pp. 956–969.
19. Löwy, H., 1953, Senkwasser in Sandwüsten: Neues Jahrb. *Geologie u. Paläontologie Monatshefte*, No. 6, pp. 241–243.
20. Mann, J. F., Jr., 1958, *Estimating quantity and quality of ground water in dry regions using airphotographs:* Internat. Assoc. Sci. Hydrology, General Assembly of Toronto, v. 2, pp. 125–134.

21. Meinzer, O. E., 1927, Plants as indicators of ground water: *U.S. Geol. Survey Water-Supply Paper* 577, 95 pp.
22. Muller, S. W., 1947, *Permafrost or permanently frozen ground and related engineering problems:* Ann Arbor, Michigan, J. W., Edwards, Inc., 231 pp.
23. Müller, F., 1959, Observations on pingos: *Meddelelser om Grønland*, v. 153, No. 3, 127 pp.
24. Robaux, A., 1953, Physical and chemical properties of ground water in the arid countries, in *Ankara symposium on arid zone hydrology:* United Nations Educational, Scientific, and Cultural Organization, pp. 17-28.
25. Schoeller, H., 1959, *Arid zone hydrology, recent developments:* Paris, United Nations Educational, Scientific, and Cultural Organization, 125 pp.
26. Siline-Bektchourine, A. I., 1961, Conditions of formation of saline waters in arid zones: in *Salinity problems in the arid zones:* Paris, United Nations Educational, Scientific, and Cultural Organization, pp. 43–47.
27. Temkin, L. E. (editor), 1963, Technical considerations in designing foundations in permafrost: *Nat. Research Council of Canada Tech. Translations* 1033, 87 pp.
28. Tisdel, F. W., 1964, Water supply from ground water sources in permafrost areas of Alaska; in *Science in Alaska* 1963, Fourteenth Alaskan Science Conference: Am. Assoc. Advancement of Sci., Alaska Division, pp. 113–124.
29. Walker, G. E., and T. E. Eaking, 1963, Geology and ground water of Amargosa Desert, Nevada-California: Nevada Dept. of Conservation and Natural Resources, *Ground-Water Resources-Reconnaissance Series Report* 14, 57 pp.
30. White, N. D., 1963, Annual report on ground water in Arizona, Spring 1962 to Spring 1963: Arizona State Land Dept., *Water Resources Rept. No.* 15, 136 pp.
31. Winograd, Isaac J., 1962, Interbasin movement of ground water at the Nevada Test Site, Nevada: *U.S. Geol. Survey Prof. Paper* 450-C, pp. 108–111.

# *appendix*

# USEFUL CONVERSION FACTORS

*Prefixes to designate magnitude*

| Multiple | Prefix |
|---|---|
| $10^6$ | mega |
| $10^3$ | kilo |
| $10^2$ | hecto |
| 10 | deka |
| $10^{-1}$ | deci |
| $10^{-2}$ | centi |
| $10^{-3}$ | milli |
| $10^{-6}$ | micro |
| $10^{-9}$ | nano |
| $10^{-12}$ | pico |

*Length*

1 meter—1.0936 yards
—3.2808 feet
—39.370 inches

1 foot—0.3048 meter

1 mile—1.6094 kilometers
—5280 feet

1 kilometer—0.62137 mile

*Area*

1 $cm^2$—0.1550 $in^2$
1 $in^2$—6.452 $cm^2$
1 $m^2$—10.764 $ft^2$
1 $ft^2$—929.0 $cm^2$

1 acre—43,560 $ft^2$
—4047 $m^2$

1 hectare—10,000 $m^2$
—2.471 acres

1 $mi^2$—2.590 $km^2$
—640 acres

*Volume*

1 $m^3$—1000 liters
—35.314 $ft^3$
—264 gal (U.S.)

1 $ft^3$—28.320 liters
—7.481 gal (U.S.)

1 gal—3.785 liters

1 acre foot—43,560 $ft^3$
—$3.259 \times 10^5$ gal
—1234 $m^3$

*Discharge*

1 $ft^3$/min—0.472 liters/sec
1 acre foot/day—$3.259 \times 10^5$ gal/day
1 $ft^3$/sec—448.8 gal/min
—724 acre feet/year

*Density*

Water 1.000 g/$cm^3$ at 4°C
0.998 g/$cm^3$ at 20°C
Sea water 1.025 g/$cm^3$ at 15°C
Mercury 13.55 g/$cm^3$ at 20°C
Air $1.29 \times 10^{-3}$ g/$cm^3$ at 20°C and atmospheric pressure

*Specific weight water in air*

8.335 lb/gal at 0°C
8.328 lb/gal at 60°F
8.322 lb/gal at 20°C
62.18 lb/$ft^3$ at 60°F

*Pressure*

1 bar—0.9869 atmosphere
—$10^6$ dynes/cm$^2$
—14.50 lb/in$^2$

pressure developed from static liquid
1 cm mercury—0.01316 atmosphere
1 ft water—0.02950 atmosphere
33.90 ft water—1.00 atmosphere

*Gravity field at sea level*

| Latitude | cm/sec$^2$ | ft/sec$^2$ |
|---|---|---|
| 0° | 978.04 | 32.09 |
| 20° | 978.64 | 32.11 |
| 40° | 980.17 | 32.16 |
| 60° | 981.92 | 32.22 |
| 80° | 983.06 | 32.25 |

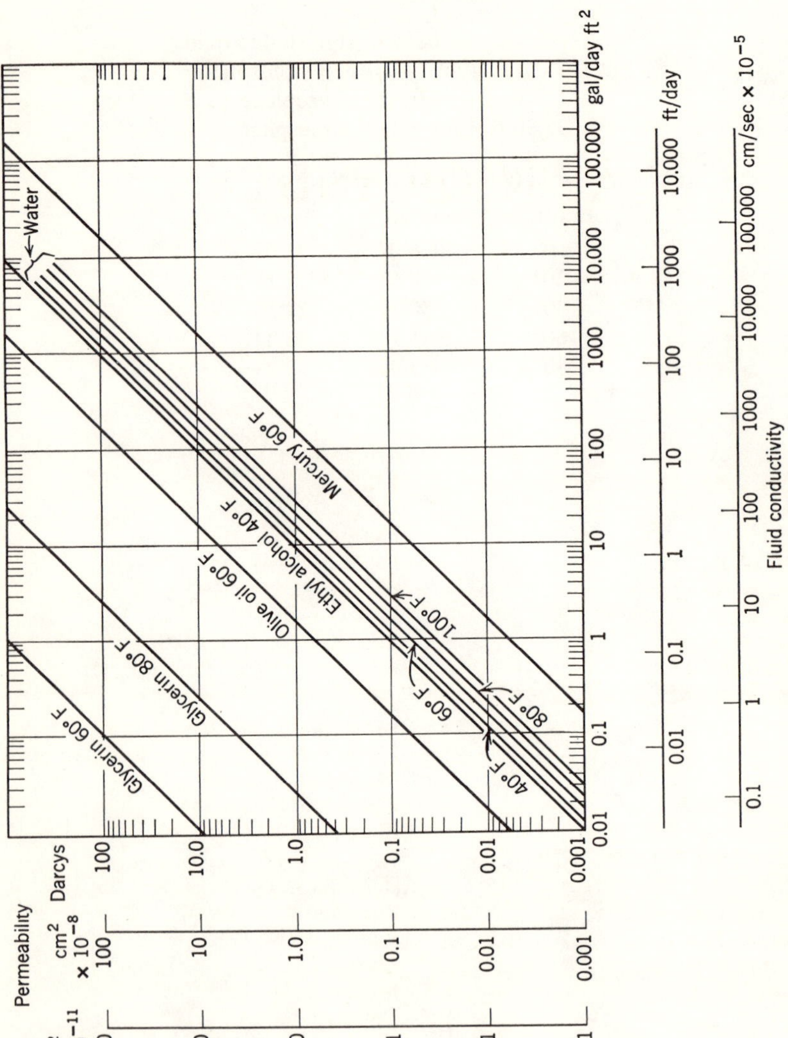


Figure 1 General numerical relation between permeability and conductivity for fluids having various densities and viscosities.

# AUTHOR INDEX

# SUBJECT INDEX